Design of Structural Elements

W.M.C. McKenzie BSc, PhD, CPhys, MInstP, CEng.
Teaching Fellow, Napier University, Edinburgh

palgrave
macmillan

First published 2004 by
PALGRAVE MACMILLAN
Houndmills, Basingstoke, Hampshire RG21 6XS and
175 Fifth Avenue, New York, N.Y. 10010
Companies and representatives throughout the world

PALGRAVE MACMILLAN is the global academic imprint of the Palgrave Macmillan division
of St. Martin's Press LLC and of Palgrave Macmillan Ltd. Macmillan© is a registered trademark
in the United States, United Kingdom and other countries. Palgrave is a registered trademark in
the European Union and other countries.

ISBN-10: 1–4039–1224–6 paperback
ISBN-13: 978–1–4039–1224–4 paperback

This book is printed on paper suitable for recycling and made from fully managed and sustained
forest sources

A catalogue record for this book is available from the British Library.

Library of Congress Cataloging-in-Publication Data

    p. cm.
    Includes bibliographical references and index.
    ISBN 0–333–00000–0
    1.

10   9   8   7   6   5

12   11   10   09   08   07

Printed in China

2021001603

Contents

xii

Contents

Preface

It has been suggested that structural engineering is:

'The science and art of designing and making with economy and elegance buildings, bridges, frameworks and other similar structures so that they can safely resist the forces to which they may be subjected.' (ref. 49).

The development of new building materials/structural systems and improved understanding of their behaviour has increased at a faster rate during the latter half of the 20th century than at any time in history. The most commonly used materials are reinforced concrete, structural steelwork, timber and masonry. In all developed countries of the world structural design codes/codes of practice have been formulated or adopted to enable engineers to design structures using these materials which should be safe and suitable for the purpose for which they are intended. Despite this, the fact remains that structural failures do, and always will, occur.

In order to minimize the occurrence of failure it is necessary to incorporate education, training, experience and quality control in all aspects of a design project. To provide merely familiarity with design codes as part of an educational program is clearly inadequate; understanding and competence will only develop with further training and experience.

This text is intended to introduce potential engineers to the design requirements of the codes for the four materials mentioned above and illustrates the concepts and calculations necessary for the design of the most frequently encountered basic structural elements.

Whilst the emphasis of the text is on reinforced concrete and structural steelwork, there are sufficient explanation and worked examples relating to timber and masonry design for most undergraduate courses. Comprehensive and detailed information, including numerous, worked examples in both timber and masonry design is available, in the following texts by the author; *Design of Structural Timber* ISBN 0-333-79236-X and *Design of Structural Masonry* ISBN 0-333-79237-8, published by Palgrave Macmillan

The most commonly used hand-analysis techniques, design philosophies and structural loadings are summarized and illustrated in Section One, whilst specific design requirements for reinforced concrete, structural steelwork, timber and masonry are given in Sections Two, Three, Four and Five respectively.

Structural stability, which is fundamental to the success of any building, is discussed in Section One. A number of 'Data Sheets' providing general information for analysis and design are given in the Appendices.

This book is intended for use by architectural, building and civil engineering students studying B.Sc./B.Eng./HND/HNC level courses in structural design. In addition it will be suitable for recently qualified, practising engineers or those who require a refresher course in structural design.

Extensive reference is made to the appropriate Clauses in the relevant design codes, most of which are given in 'Extracts from British Standards for students of structural design' (ref. 39).

W.M.C. McKenzie

To Roy Corp (former Head of Department of Civil Engineering at Napier University, Edinburgh) for his inspiration and John McNeill (Napier University) for his support and advice.

Acknowledgements

I wish to thank Caroline, Karen and Gordon for their endless support and encouragement.

1. Structural Analysis Techniques

Objective: *to provide a resumé of the elastic methods of structural analysis, most commonly used when undertaking structural design.*

1.1 *Resumé of Analysis Techniques*

The following resumé gives a brief summary of the most common manual techniques adopted to determine the forces induced in the members of statically determinate structures. There are numerous structural analysis books available which give comprehensive detailed explanations of these and other more advanced techniques.

The laws of structural mechanics are well established in recognised 'elastic theory' using the following assumptions:

♦ The material is *homogeneous,* which implies, its constituent parts have the same physical properties throughout its entire volume.
♦ The material is *isotropic,* which implies that the elastic properties are the same in all directions.
♦ The material obeys *Hooke's Law*, i.e. when subjected to an external force system the deformations induced will be directly proportional to the magnitude of the applied force.
♦ The material is *elastic*, which implies that it will recover completely from any deformation after the removal of load.
♦ The *modulus of elasticity* is the same in tension and compression.
♦ *Plane sections remain plane* during deformation. During bending this assumption is violated and is reflected in a non-linear bending stress diagram throughout cross-sections subject to a moment; this is normally neglected.

1.2 *Method of Sections for Pin-Jointed Frames*

The *method of sections* involves the application of the three equations of static equilibrium to two-dimensional plane frames. The sign convention adopted to indicate ties (i.e. tension members) and struts (i.e. compression members) in frames is as shown in Figure 1.1.

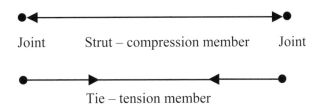

Joint   Strut – compression member   Joint

Tie – tension member

Figure 1.1

The method involves considering an imaginary section line which cuts the frame under consideration into two parts A and B as shown in Figure 1.4.

Since only three independent equations of equilibrium are available any section taken through a frame must not include more than three members for which the internal force is unknown.

Consideration of the equilibrium of the resulting force system enables the magnitude and sense (i.e. compression or tension) of the forces in the cut members to be determined.

1.2.1 Example 1.1: Pin-Jointed Truss

A pin-jointed truss simply supported by a pinned support at A and a roller support at E carries three loads at nodes G, H and I as shown in Figure 1.4. Determine the magnitude and sense of the forces induced in members X, Y and Z as indicated.

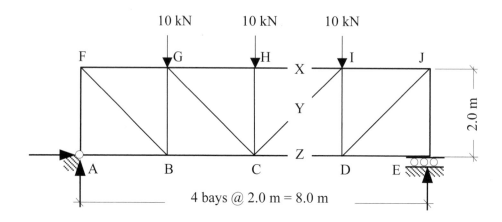

Figure 1.2

Step 1: Evaluate the support reactions. It is not necessary to know any information regarding the frame members at this stage other than dimensions as shown in Figure 1.3, since only **externally** applied loads and reactions are involved.

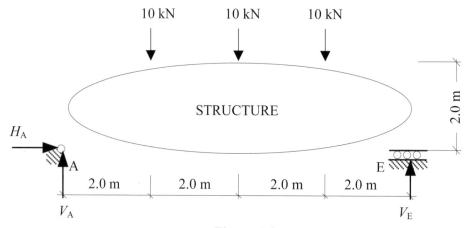

Figure 1.3

Apply the three equations of static equilibrium to the force system:

$$+ve \uparrow \quad \Sigma F_y = 0 \quad V_A - (10 + 10 + 10) + V_E = 0 \qquad V_A + V_E = 30 \text{ kN}$$

$$+ve \rightarrow \quad \Sigma F_x = 0 \qquad\qquad\qquad\qquad\qquad\qquad\qquad H_A = 0$$

$$+ve \curvearrowright \quad \Sigma M_A = 0 \quad (10 \times 2.0) + (10 \times 4.0) + (10 \times 6.0) - (V_E \times 8.0) = 0$$

$$V_E = 15 \text{ kN}$$

$$\text{hence} \quad V_A = 15 \text{ kN}$$

Step 2: Select a section through which the frame can be considered to be cut and using the same three equations of equilibrium determine the magnitude and sense of the unknown forces (i.e. the internal forces in the cut members).

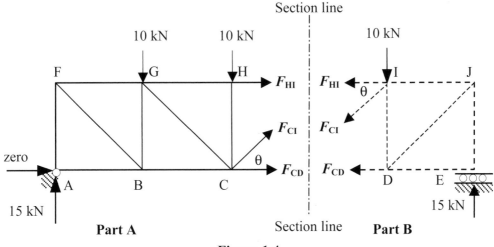

Figure 1.4

It is convenient to **assume** all unknown forces to be tensile and hence at the cut section their direction and lines of action are considered to be pointing away from the joint (refer to Figure 1.4). If the answer results in a negative force this means that the assumption of a tie was incorrect and the member is actually in compression, i.e. a strut.

The application of the equations of equilibrium to either part of the cut frame will enable the forces X, Y and Z to be evaluated.

Note: The section considered must not cut through more than three members with unknown internal forces since only three equations of equilibrium are applicable.

Consider Part A:

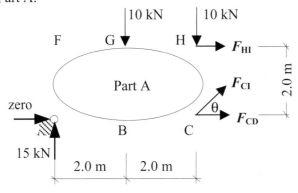

Figure 1.5

Design of Structural Elements

Note: $\sin\theta = \dfrac{1}{\sqrt{2}} = 0.707,$ $\cos\theta = \dfrac{1}{\sqrt{2}} = 0.707,$

+ve ↑ $\Sigma F_y = 0$ $15.0 - (10.0 + 10.0) + F_{CI}\sin\theta = 0$

$$F_{CI} = \frac{5.0}{\sin\theta} = \textbf{+7.07 kN}$$

Member CI is a tie

+ve → $\Sigma F_x = 0$ $F_{HI} + F_{CD} + F_{CI}\cos\theta = 0$

+ve ⟳ $\Sigma M_C = 0$ $(15.0 \times 4.0) - (10.0 \times 2.0) + (F_{HI} \times 2.0) = 0$

$$F_{HI} = \textbf{- 20.0 kN}$$

Member HI is a strut

hence $F_{CD} = -F_{HI} - F_{CI}\cos\theta = -(-20.0) - (7.07 \times \cos\theta)$ $= \textbf{+15.0 kN}$

Member CD is a tie

These answers can be confirmed by considering Part B of the structure and applying the equations as above.

1.3   *Method of Joint Resolution for Pin-Jointed Frames*

Considering the same frame using *joint resolution* highlights the advantage of the method of sections when only a few member forces are required.

In this technique (which can be considered as a special case of the method of sections), sections are taken which isolate each individual joint in turn in the frame, e.g.

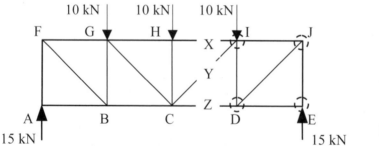

Figure 1.6

In Figure 1.6 four sections are shown, each of which isolates a joint in the structure as indicated in Figure 1.7.

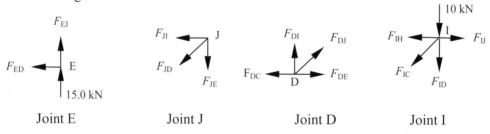

Joint E   Joint J   Joint D   Joint I

Figure 1.7

Since in each case the forces are coincident, the moment equation is of no value, hence only two independent equations are available. It is necessary when considering the equilibrium of each joint to do so in a sequence which ensures that there are no more than two unknown member forces in the joint under consideration. This can be carried out until all member forces in the structure have been determined.

Consider Joint E:

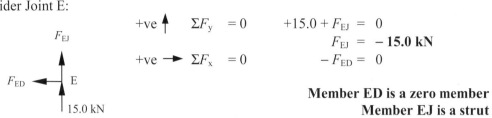

$$+ve \uparrow \quad \Sigma F_y = 0 \qquad +15.0 + F_{EJ} = 0$$
$$F_{EJ} = -\textbf{15.0 kN}$$
$$+ve \rightarrow \Sigma F_x = 0 \qquad -F_{ED} = 0$$

Member ED is a zero member
Member EJ is a strut

Consider Joint J: substitute for calculated values, i.e. F_{JE} (direction of force is into the joint)

$$+ve \uparrow \quad \Sigma F_y = 0 \quad +15.0 - F_{JD} \cos\theta = 0$$
$$F_{JD} = +15.0 / 0.707$$
$$F_{JD} = +\textbf{21.21 kN}$$
$$+ve \rightarrow \Sigma F_x = 0 \quad -F_{JI} - F_{JD} \sin\theta = 0$$
$$F_{JI} = -21.21 \times 0.707$$
$$F_{JI} = -\textbf{15.0 kN}$$

Member JD is a tie
Member JI is a strut

Consider Joint D: substitute for calculated values, i.e. F_{DJ} and F_{DE}

$$+ve \uparrow \quad \Sigma F_y = 0 \quad + F_{DI} + 21.21\sin\theta = 0$$
$$F_{DI} = -21.21 \times 0.707$$
$$F_{DI} = -\textbf{15.0 kN}$$
$$+ve \rightarrow \Sigma F_x = 0 \quad -F_{DC} + 21.21 \cos\theta = 0$$
$$F_{DC} = +21.21 \times 0.707$$
$$F_{DC} = +\textbf{15.0 kN}$$

Member DI is a strut
Member DC is a tie

Consider Joint I: substitute for calculated values, i.e. F_{ID} and F_{IJ}

$$+ve \uparrow \quad \Sigma F_y = 0 \quad +15.0 - 10.0 - F_{IC} \cos\theta = 0$$
$$F_{IC} = +5.0 / 0.707$$
$$F_{IC} = +\textbf{7.07 kN}$$
$$+ve \rightarrow \Sigma F_x = 0 \quad -F_{IH} -15.0 - F_{IC} \sin\theta = 0$$
$$F_{IH} = -\textbf{20.0 kN}$$

Member IC is a tie
Member IH is a strut

1.4   *Unit Load Method to Determine the Deflection of Pin-Jointed Frames*

The *unit load method* is an energy method which can be used to determine the deflection at any node in a pin-jointed frame as follows:

$$\delta = \sum \frac{PuL}{AE}$$

where:

δ is the displacement of the point of application of any load, along the line of action of that load,

P is the force in a member due to the externally applied loading system,

u is the force in a member due to a **unit load** acting at the position of and in the direction of the desired displacement,

L/A is the ratio of the length to the cross-sectional area of the members,

E is the modulus of elasticity of the material for each member (i.e. Young's modulus).

1.4.1 *Example 1.2: Deflection of a Pin-Jointed Truss*

A pin-jointed truss ABCD is shown in Figure 1.8 in which both a vertical and a horizontal load are applied at joint B as indicated. Determine the magnitude and direction of the resultant deflection at joint B and the vertical deflection at joint D.

Assume the cross-sectional area of all members is equal to A and all members are made from the same material, i.e. have the same modulus of elasticity E

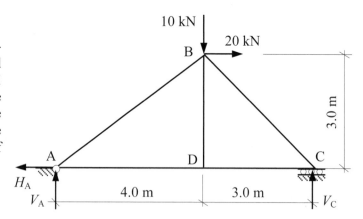

Figure 1.8

Step 1: Evaluate the support reactions. It is not necessary to know any information regarding the frame members at this stage other than dimensions as shown in Figure 1.8, since only **externally** applied loads and reactions are involved.

The reader should follow the procedure given in Example 1.1 to determine the following results:

Horizontal component of reaction at support A H_A = + 20 kN
Vertical component of reaction at Support A V_A = − 4.29 kN
Vertical component of reaction at Support C V_C = + 14.28 kN

Step 2:   Use the *method of sections* or *joint resolution* as indicated in Sections 1.2 and 1.3 respectively to determine the magnitude and sense of the unknown member forces (i.e. the *P* forces).

The reader should complete this calculation to determine the member forces as indicated in Figure 1.9.

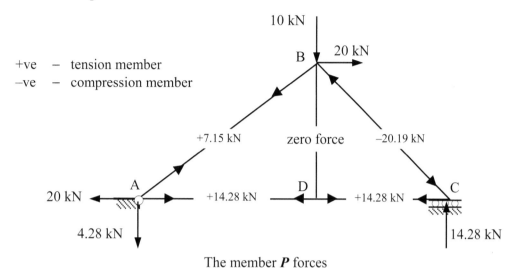

The member *P* forces

Figure 1.9

Step 3:   To determine the vertical deflection at joint B remove the externally applied load system and apply a **unit load only** in a vertical direction at joint B as shown in Figure 1.10. Use the *method of sections* or *joint resolution* as before to determine the magnitude and sense of the unknown member forces (i.e. the *u* forces).

The reader should complete this calculation to determine the member forces as indicated in Figure 1.10.

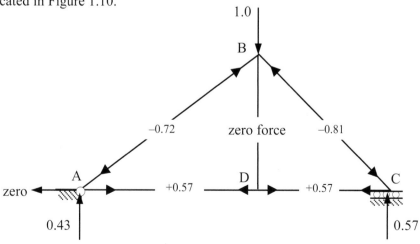

The member *u* forces for **vertical** deflection at joint B

Figure 1.10

The vertical deflection $\quad \delta_{vB} \quad = \quad \sum \dfrac{PuL}{AE}$

This is better calculated in tabular form as shown in Table 1.1.

Member	Length (L)	X-Section (A)	Modulus (E)	P forces (kN)	u forces	PuL (kNm)
AB (tie)	5.0 m	A	E	+ 7.15	− 0.72	− 25.75
BC (strut)	4.24 m	A	E	− 20.19	− 0.81	+ 69.32
AD (tie)	4.0 m	A	E	+ 14.28	+ 0.57	+ 32.56
CD (tie)	3.0 m	A	E	+ 14.28	+ 0.57	+ 24.42
BD (zero)	3.0 m	A	E	0.0	0.0	0.0
					Σ	**+ 100.6**

Table 1.1

The +ve sign indicates that the deflection is in the same direction as the applied unit load.

Hence the vertical deflection $\quad \delta_{vB} \quad = \quad \sum \dfrac{PuL}{AE} = +(100.6/AE) \quad \downarrow$

Note: Where the members have different cross-sectional areas and/or modulii of elasticity each entry in the last column of the table should be based on (PuL/AE) and not only (PuL).

Step 4: A similar calculation can be carried out to determine the horizontal deflection at joint B.

The reader should complete this calculation to determine the member forces as indicated in Figure 1.11.

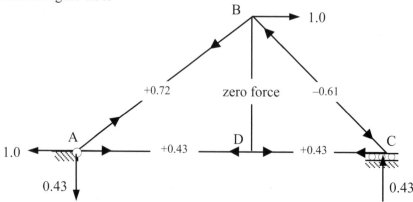

The member *u* forces for **horizontal** deflection at joint B

Figure 1.11

The horizontal deflection $\quad \delta_{hB} \quad = \quad \sum \dfrac{PuL}{AE}$

Member	Length (L)	X-Section (A)	Modulus (E)	P forces (kN)	u forces	PuL (kNm)
AB (tie)	5.0 m	A	E	+ 7.15	+ 0.72	+ 25.75
BC (strut)	4.24 m	A	E	− 20.19	− 0.61	+ 52.23
AD (tie)	4.0 m	A	E	+ 14.28	+ 0.43	+ 24.56
CD (tie)	3.0 m	A	E	+ 14.28	+ 0.43	+ 18.42
BD (zero)	3.0 m	A	E	0.0	0.0	0.0
					Σ	**+ 120.9**

Table 1.2

Hence the horizontal deflection $\delta_{hB} \;=\; \sum \dfrac{PuL}{AE} \;=\; + (120.9/AE) \longrightarrow$

The resultant deflection at joint B can be determined from the horizontal and vertical components evaluated above, i.e.

$$R \;=\; \sqrt{\left(100.6^2 + 120.9^2\right)} \,/AE \;=\; 157.3/AE$$

$(120.9/AE)$ $(100.6/AE)$ Resultant '*R*'

A similar calculation can be carried out to determine the vertical deflection at joint D.

The reader should complete this calculation to determine the member forces as indicated in Figure 1.12.

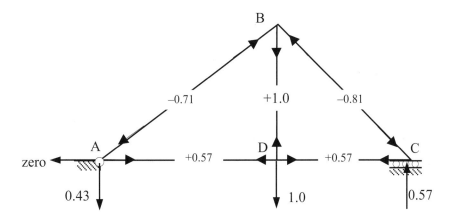

The member *u* forces for **vertical** deflection at joint D

Figure 1.12

The vertical deflection $\delta_{vD} \;=\; \sum \dfrac{PuL}{AE}$

Member	Length (L)	X-Section (A)	Modulus (E)	P forces (kN)	u forces	PuL (kNm)
AB (tie)	5.0 m	A	E	+ 7.15	− 0.71	− 25.35
BC (strut)	4.24 m	A	E	− 20.19	− 0.81	+ 69.32
AD (tie)	4.0 m	A	E	+ 14.28	+ 0.57	+ 32.56
CD (tie)	3.0 m	A	E	+ 14.28	+ 0.57	+ 24.42
BD (zero)	3.0 m	A	E	0.0	+1.0	0.0
					Σ	**+ 100.9**

Table 1.3

Hence the vertical deflection $\quad \delta_{vD} = \sum \dfrac{PuL}{AE} = +(100.9/AE)$ ↓

1.5   *Shear Force and Bending Moment*

Two parameters which are fundamentally important to the design of beams are ***shear force*** and ***bending moment***. These quantities are the result of internal forces acting on the material of a beam in response to an externally applied load system.

Consider a simply supported beam as shown in Figure 1.13 carrying a series of secondary beams each imposing a point load of 4 kN.

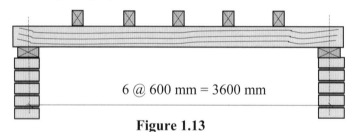

6 @ 600 mm = 3600 mm

Figure 1.13

This structure can be represented as a line diagram as shown in Figure 1.14:

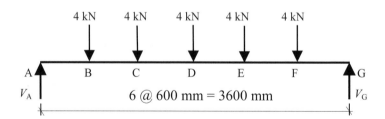

Figure 1.14

Since the externally applied force system is in equilibrium, the three equations of static equilibrium must be satisfied, i.e.

+ve ↑ $\Sigma F_y = 0$ The sum of the vertical forces must equal zero.

+ve $\curvearrowright$ $\Sigma M = 0$ The sum of the moments of all forces about *any* point on the plane of the forces must equal zero.

+ve $\rightarrow$ $\Sigma F_x = 0$ The sum of the horizontal forces must equal zero.

The assumed positive direction is as indicated. In this particular problem there are no externally applied horizontal forces and consequently the third equation is not required.

(**Note:** It is still necessary to provide horizontal restraint to a structure since it can be subject to a variety of load cases, some of which may have a horizontal component.)

Consider the vertical equilibrium of the beam:

+ve $\uparrow$ $\Sigma F_y = 0$

$+ V_A - (5 \times 4.0) + V_G = 0$ $\therefore$ $V_A + V_G = 20$ kN   (i)

Consider the rotational equilibrium of the beam:

+ve $\curvearrowright$ $\Sigma M_A = 0$

Note: The sum of the moments is taken about one end of the beam (end A) for convenience. Since one of the forces (V_A) passes through this point it does not produce a moment about A and hence does not appear in the equation. It should be recognised that the sum of the moments could have been considered about *any* known point in the same plane.

$+ (4.0 \times 0.6) + (4.0 \times 1.2) + (4.0 \times 1.8) + (4.0 \times 2.4) + (4.0 \times 3.0) - (V_G \times 3.6) = 0$

$\therefore$ $V_G = 10$ kN   (ii)

Substituting into equation (i) gives $\therefore$ $V_A = 10$ kN

This calculation was carried out considering only the externally applied forces, i.e.

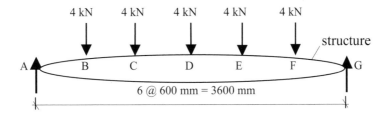

Figure 1.15

The structure itself was ignored, however the applied loads are transferred to the end supports through the material fibres of the beam. Consider the beam to be cut at section X–X producing two sections each of which is in equilibrium as shown in Figure 1.16.

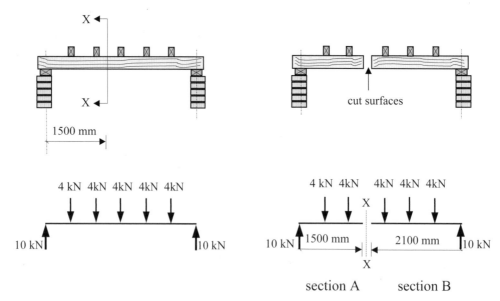

Figure 1.16

Clearly if the two sections are in equilibrium there must be internal forces acting on the cut - surfaces to maintain this; these forces are known as the ***shear force*** and the ***bending moment***, and are illustrated in Figure 1.17.

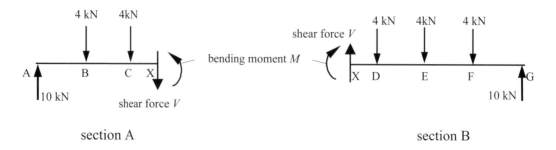

Figure 1.17

The force V and moment M are equal and opposite on each surface. The magnitude and direction of V and M can be determined by considering two equations of static equilibrium for either of the cut sections; both will give the same answer.

Consider the left-hand section with the 'assumed' directions of the internal forces V and M as shown in Figure 1.18.

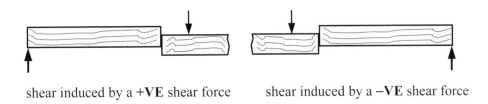

$+ve \uparrow \Sigma F_y = 0$

$+ 10 - 4.0 - 4.0 - F = 0 \qquad \therefore \; V = 2 \text{ kN}$

$+ve \curvearrowright \Sigma M_A = 0$

$+ (4.0 \times 0.6) + (4.0 \times 1.2) - (V \times 1.5) - M = 0$
$\therefore \; M = 10.2 \text{ kNm}$

Figure 1.18

1.5.1 Shear Force Diagrams

In a statically determinate beam, the numerical value of the shear force can be obtained by evaluating the algebraic sum of the vertical forces to one side of the section being considered. The convention normally adopted to indicate positive and negative shear forces is shown in Figure 1.19.

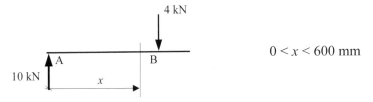

shear induced by a +**VE** shear force   shear induced by a −**VE** shear force

Figure 1.19

The calculation carried out to determine the shear force can be repeated at various locations along a beam and the values obtained plotted as a graph; this graph is known as the ***shear force diagram***. The shear force diagram indicates the variation of the shear force along a structural member.

Consider any section of the beam between A and B:

$0 < x < 600$ mm

Note: The value immediately under the point load at the cut section is not being considered.

The shear force at any position x = Σ vertical forces to one side
            = +10.0 kN

This value is a constant for all values of x between zero and 600 mm, the graph will therefore be a horizontal line equal to 10.0 kN. This force produces a +ve shear effect, i.e.

+ve shear effect

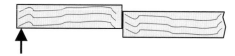

Consider any section of the beam between B and C:

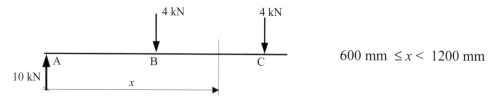

600 mm $\leq x <$ 1200 mm

The shear force at any position x =   Σ vertical force to one side
                                    =   +10.0 − 4.0 = 6.0 kN

This value is a constant for all values of x between 600 mm and 1200 mm, the graph will therefore be a horizontal line equal to 6.0 kN. This force produces a +ve effect shear effect.

Similarly for any section between C and D:

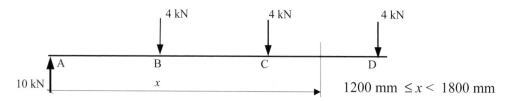

1200 mm $\leq x <$ 1800 mm

The shear force at any position x =   Σ vertical forces to one side
                                    =   +10.0 − 4.0 − 4.0 = 2.0 kN

Consider any section of the beam between D and E:

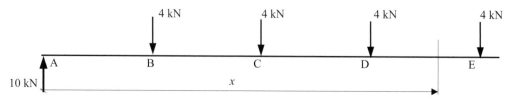

1800 mm $\leq x <$ 2400 mm

The shear force at any position x =   Σ vertical forces to one side
                                    =   +10.0 − 4.0 − 4.0 − 4.0 = − 2.0 kN

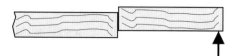

In this case the shear force is negative:

Similarly between E and F   2400 mm $< x <$ 3000 mm
The shear force at any position x = Σ vertical forces to one side
$$= +10.0 - 4.0 - 4.0 - 4.0 - 4.0 = -6.0 \text{ kN}$$

and

between F and G   3000 mm $< x <$ 3600 mm
The shear force at any position x = Σ vertical forces to one side
$$= +10.0 - 4.0 - 4.0 - 4.0 - 4.0 - 4.0 = -10.0 \text{ kN}$$

In each of the cases above the value has not been considered at the point of application of the load.

Consider the location of the applied load at B shown in Figure 1.20.

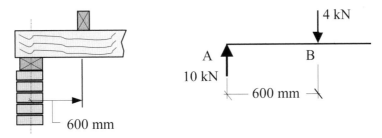

Figure 1.20

The 4.0 kN is not instantly transferred through the beam fibres at B but instead over the width of the actual secondary beam. The change in value of the shear force between $x <$ 600 mm and $x >$ 600 mm occurs over this width, as shown in Figure 1.21.

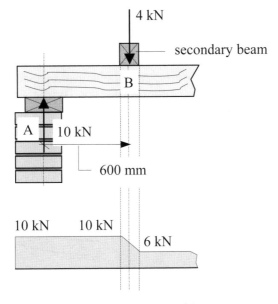

Figure 1.21

The width of the secondary beam is insignificant when compared with the overall span, and the shear force is assumed to change instantly at this point, producing a vertical line on the shear force diagram as shown in Figure 1.22.

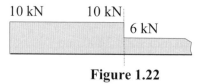

Figure 1.22

The full shear force diagram can therefore be drawn as shown in Figure 1.23.

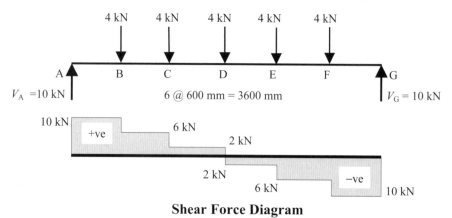

Shear Force Diagram

Figure 1.23

The same result can be obtained by considering sections from the right-hand side of the beam.

1.5.2 *Bending Moment Diagrams*

In a statically determinate beam the numerical value of the bending moment (i.e. moments caused by forces which tend to bend the beam) can be obtained by evaluating the algebraic sum of the moments of the forces to one side of a section. In the same manner as with shear forces either the left-hand or the right-hand side of the beam can be considered. The convention normally adopted to indicate positive and negative bending moments is shown in Figures 1.24(a) and (b).

Bending inducing **tension on the underside** of a beam is considered **positive**.

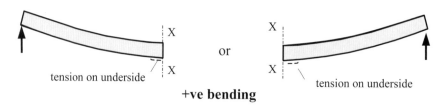

+ve bending

Figure 1.24 (a)

Bending inducing **tension on the top** of a beam is considered **negative**.

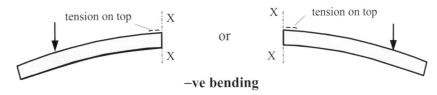

−ve bending

Figure 1.24 (b)

*Note: Clockwise/anti-clockwise moments do **not** define +ve or −ve **bending** moments. The sign of the bending moment is governed by the location of the tension surface at the point being considered.*

As with shear forces the calculation for bending moments can be carried out at various locations along a beam and the values plotted on a graph; this graph is known as the '*bending moment diagram*'. The bending moment diagram indicates the variation in the bending moment along a structural member.

Consider sections between A and B of the beam as before:

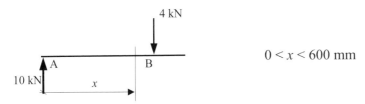

$0 < x < 600$ mm

In this case when $x = 600$ mm the 4.0 kN load passes through the section being considered and does not produce a bending moment, and can therefore be ignored.

Bending moment $= \Sigma$ algebraic sum of the moments of the forces to one side of a section

$\quad = \Sigma$ (Force $\times$ lever arm)

$M = 10.0 \times x = 10.0\,x$ kNm

Unlike the shear force, this expression is not a constant and depends on the value of 'x' which varies between the limits given. This is a linear expression which should be reflected in the calculated values of the bending moment.

$x = 0$	$M = 10.0 \times 0$	$= 0$ kNm
$x = 100$ mm	$M = 10.0 \times 0.1$	$= 1.0$ kNm
$x = 200$ mm	$M = 10.0 \times 0.2$	$= 2.0$ kNm
$x = 300$ mm	$M = 10.0 \times 0.3$	$= 3.0$ kNm
$x = 400$ mm	$M = 10.0 \times 0.4$	$= 4.0$ kNm
$x = 500$ mm	$M = 10.0 \times 0.5$	$= 5.0$ kNm
$x = 600$ mm	$M = 10.0 \times 0.6$	$= 6.0$ kNm

Clearly the bending moment increases linearly from zero at the simply supported end to a value of 6.0 kNm at point B.

Consider sections between B and C of the beam:

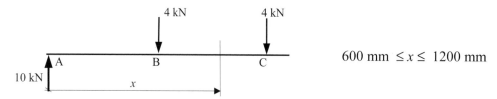

600 mm $\leq x \leq$ 1200 mm

Bending moment = Σ algebraic sum of the moments of the forces to 'one' side of a section

$$M = +(10.0 \times x) - (4.0 \times [x - 0.6])$$

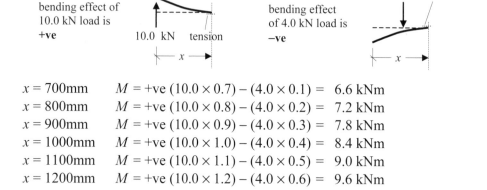

$x = 700$mm	$M = +$ve $(10.0 \times 0.7) - (4.0 \times 0.1) = $ 6.6 kNm
$x = 800$mm	$M = +$ve $(10.0 \times 0.8) - (4.0 \times 0.2) = $ 7.2 kNm
$x = 900$mm	$M = +$ve $(10.0 \times 0.9) - (4.0 \times 0.3) = $ 7.8 kNm
$x = 1000$mm	$M = +$ve $(10.0 \times 1.0) - (4.0 \times 0.4) = $ 8.4 kNm
$x = 1100$mm	$M = +$ve $(10.0 \times 1.1) - (4.0 \times 0.5) = $ 9.0 kNm
$x = 1200$mm	$M = +$ve $(10.0 \times 1.2) - (4.0 \times 0.6) = $ 9.6 kNm

As before the bending moment increases linearly, i.e. from 6.6 kNm at $x = 700$ mm to a value of 9.6 kNm at point C.

Since the variation is linear it is only necessary to evaluate the magnitude and sign of the bending moment at locations where the slope of the line changes, i.e. each of the point load locations.

Consider point D:

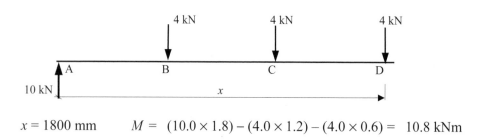

$x = 1800$ mm $M = (10.0 \times 1.8) - (4.0 \times 1.2) - (4.0 \times 0.6) = $ 10.8 kNm

Consider point E:

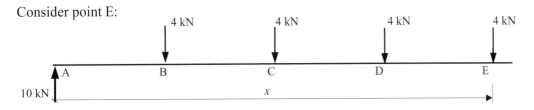

$x = 2400$ mm $\quad M = (10.0 \times 2.4) - (4.0 \times 1.8) - (4.0 \times 1.2) - (4.0 \times 0.6) \quad = 9.6$ kNm

Similarly at point F:
$x = 3000$ mm $\quad M = (10.0 \times 3.0) - (4.0 \times 2.4) - (4.0 \times 1.8) - (4.0 \times 1.2) - (4.0 \times 0.6)$
$\qquad\qquad\qquad = 6.0$ kNm

The full bending moment diagram can therefore be drawn as shown in Figure 1.25.

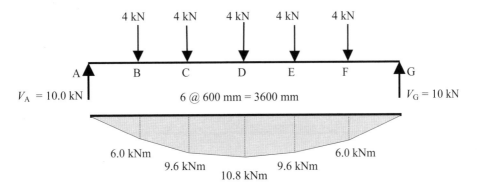

Bending Moment Diagram

Figure 1.25

The same result can be obtained by considering sections from the right-hand side of the beam. The value of the bending moment at any location can also be determined by evaluating the area under the shear force diagram.

Consider point B:

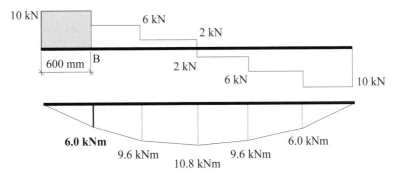

Bending moment at B = shaded area on the shear force diagram

$$M_B = (10.0 \times 0.6) = 6.0 \text{ kNm as before}$$

Consider a section at a distance of $x = 900$ mm along the beam between D and E:

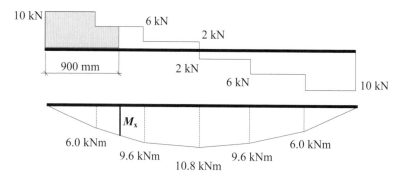

Bending moment at x = shaded area on the shear force diagram
$$M_x = (10.0 \times 0.6) + (6.0 \times 0.3) = 7.8 \text{ kNm as before}$$

Consider a section at a distance of $x = 2100$ mm along the beam between D and E:

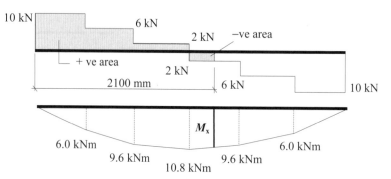

Bending moment at x = shaded area on the shear force diagram
$$M_x = (10.0 \times 0.6) + (6.0 \times 0.6) + (2.0 \times 0.6) - (2.0 \times 0.3)$$
$$= 10.2 \text{ kNm}$$

Note: A maximum bending moment occurs at the same position as a zero shear force.

1.5.3 Example 1.3: Beam with Uniformly Distributed Load (UDL)

Consider a simply - supported beam carrying a uniformly distributed load of 5 kN/m, as shown in Figure 1.26.

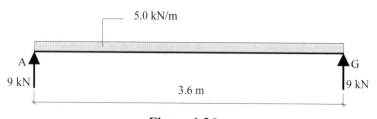

Figure 1.26

The shear force at any section a distance x from the support at A is given by:
$V_x =$ algebraic sum of the vertical forces

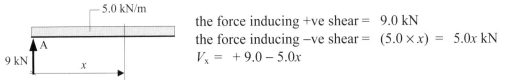

the force inducing +ve shear = 9.0 kN
the force inducing −ve shear = $(5.0 \times x) = 5.0x$ kN
$V_x = +9.0 - 5.0x$

This is a linear equation in which V_x decreases as x increases. The points of interest are at the supports where the maximum shear forces occur, and at the locations where the maximum bending moment occurs, i.e. the point of zero shear.

$$V_x = 0 \quad \text{when} \quad +9.0 - 5.0x = 0 \quad \therefore \quad x = 1.8 \text{ m}$$

Any intermediate value can be found by substituting the appropriate value of 'x' in the equation for the shear force; e.g.

$x = 600$ mm $\quad V_x = +9.0 - (5.0 \times 0.6) = +6.0$ kN
$x = 2100$ mm $\quad V_x = +9.0 - (5.0 \times 2.1) = -1.5$ kN

The shear force can be drawn as before as shown in Figure 1.27.

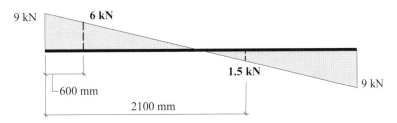

Shear Force Diagram

Figure 1.27

The bending moment can be determined as before, **either** using an equation or evaluating the area under the shear force diagram.
Using an equation:

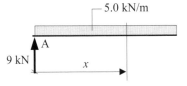

Bending moment at $x = \quad M_x = +(9.0 \times x) - [(5.0 \times x) \times (x/2)] = (9.0x - 2.5x^2)$

In this case the equation is *not* linear, and the bending moment diagram will therefore be *curved*.

Consider several values:

$x = 0$ $M_x = 0$

$x = 600$ mm $M_x = +(9.0 \times 0.6) - (2.5 \times 0.6^2) = 4.5$ kNm

$x = 1800$ mm $M_x = +(9.0 \times 1.8) - (2.5 \times 1.8^2) = 8.1$ kNm

$x = 2100$ mm $M_x = +(9.0 \times 2.1) - (2.5 \times 2.1^2) = 7.88$ kNm

Using the shear force diagram:

$x = 600$ mm

$M_x = $ shaded area $= +[0.5 \times (9.0 + 6.0) \times 0.6] = 4.5$ kNm

$x = 1800$ mm

$M_x = $ shaded area $= +[0.5 \times 9.0 \times 1.8] = 8.1$ kNm

$x = 2100$ mm

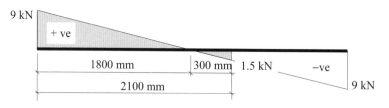

$M_x = $ shaded area $= +[8.1 - (0.5 \times 0.3 \times 1.5)] = 7.88$ kNm

The bending moment diagram is shown in Figure 1.28.

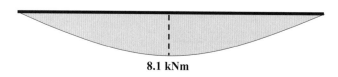

8.1 kNm

Bending Moment Diagram

Figure 1.28

The UDL loading is a 'standard' load case which occurs in numerous beam designs and can be expressed in general terms using L for the span and w for the applied load/metre or W_{total} $(= wL)$ for the total applied load, as shown in Figure 1.29.

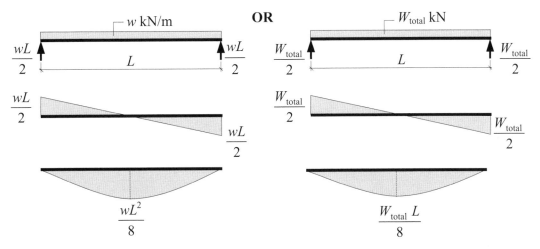

Figure 1.29

Clearly both give the same magnitude of support reactions, shear forces and bending moments.

1.5.4 Example 1.4: Beam with Combined Point Loads and UDLs

A simply supported beam ABCD carries a uniformly distributed load of 3.0 kN/m between A and B, point loads of 4 kN and 6 kN at B and C respectively, and a uniformly distributed load of 5.0 kN/m between B and D, as shown in Figure 1.30. Determine the support reactions, sketch the shear force diagram, and determine the position and magnitude of the maximum bending moment.

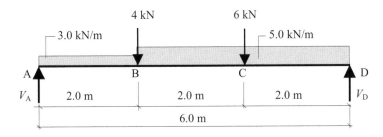

Figure 1.30

Consider the vertical equilibrium of the beam:

+ve $\uparrow \Sigma F_y = 0$

$V_A - (3.0 \times 2.0) - 4.0 - 6.0 - (5.0 \times 4.0) + V_D = 0 \qquad \therefore V_A + V_D = 36$ kN   (i)

Consider the rotational equilibrium of the beam:

$+ve \curvearrowright \Sigma M_A = 0$

$(3.0 \times 2.0 \times 1.0) + (4.0 \times 2.0) + (6.0 \times 4.0) + (5.0 \times 4.0 \times 4.0) - (V_D \times 6.0) = 0$ (ii)

$\therefore V_D = 19.67 \text{ kN}$

Substituting into equation (i) gives $\therefore V_A = 16.33 \text{ kN}$

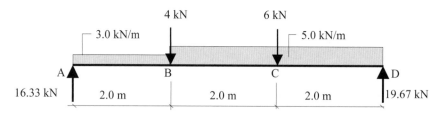

Shear force $=$ algebraic sum of the vertical forces
Consider the shear force at a section 'x' from the left-hand end of the beam:

$x = 0$ $V_x = 0$

At position x to the left of B before the 4.0 kN load
$\qquad V_x = + 16.33 - (3.0 \times 2.0) \quad = + 10.33 \text{ kN}$

At position x to the right of B after the 4.0 kN load
$\qquad V_x = + 10.33 - 4.0 \qquad\qquad = + 6.33 \text{ kN}$

At position x to the left of C before the 6.0 kN load
$\qquad V_x = + 6.33 - (5.0 \times 2.0) \quad = - 3.67 \text{ kN}$

At position x to the right of C after the 6.0 kN load
$\qquad V_x = - 3.67 - 6.0 \qquad\qquad = - 9.67 \text{ kN}$

$x = 6.0 \text{ m} \quad V_x = - 9.67 - (5.0 \times 2.0) \quad = - 19.67 \text{ kN}$

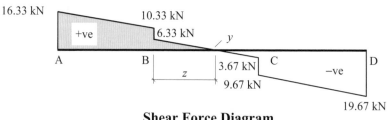

Shear Force Diagram

The maximum bending moment occurs at the position of zero shear, i.e. point y on the shear force diagram. The value of z can be determined from the shear force and applied loads:

$$z = \frac{6.33}{5.0} = 1.266 \text{ m} \qquad \text{(i.e. shear force/the value of the load per m length)}$$

Note: The slope of the shear force diagram between B and C is equal to the UDL of 5 kN/m.

Maximum bending moment M_y = shaded area of the shear force diagram
$$= [0.5 \times (16.33 + 10.33) \times 2.0] + [0.5 \times 1.266 \times 6.33]$$
$$= 30.67 \text{ kNm}$$

Alternatively, consider the beam cut at this section:

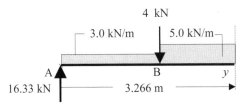

M_y = $+(16.33 \times 3.266) - (3.0 \times 2.0 \times 2.266) - (4.0 \times 1.266) - [(5.0 \times 1.266) \times (0.633)]$
= $+30.67 \text{ kNm}$

1.6 *McCaulay's Method for the Deflection of Beams*

In elastic analysis the deflected shape of a simply supported beam is normally assumed to be a circular arc of radius R (R is known as the radius of curvature), as shown in Figure 1.31.

Consider the beam AB to be subject to a variable bending moment along its length. The beam is assumed to deflect as indicated.

R is the radius of curvature,
L is the span,
I is the second moment of area about the axis of bending,
E is the modulus of elasticity,
ds is an elemental length of beam measured a distance of x from the left-hand end
M is the value of the bending moment at position x.

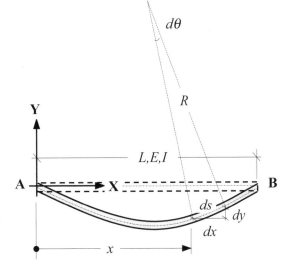

Figure 1.31

The slope of the beam at position x is given by:
$$\text{slope} = \frac{dy}{dx} = \int \frac{M}{EI} dx$$

Differentiating the slope with respect to x gives:

$$\frac{d^2 y}{dx^2} = \frac{M}{EI} \quad \text{and hence:}$$

$$EI\frac{d^2 y}{dx^2} = M \qquad \qquad \text{Equation (1) – \textbf{bending moment}}$$

Integrating Equation (1) with respect to x gives

$$EI\frac{dy}{dx} = \int M dx \qquad \qquad \text{Equation (2) – \textbf{slope}}$$

Integrating Equation (2) with respect to x gives

$$EI y = \iint \left(\frac{M}{EI} dx \right) dx \qquad \text{Equation (3) – \textbf{deflection}}$$

Equations (1) and (2) result in two constants of integration C1 and C2; these are determined by considering boundary conditions such as known values of slope and/or deflection at positions on the beam.

1.6.1 Example 1.5: Beam with Point Loads

Consider a beam supporting three point loads as shown in Figure 1.32.

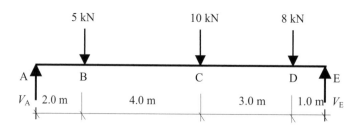

Figure 1.32

Step 1: Formulate an equation which represents the value of the bending moment at a position measured x from the left-hand end of the beam. This expression must include all of the loads and x should therefore be considered between points D and E.

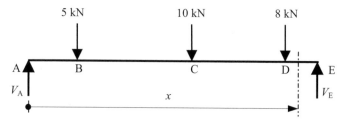

Figure 1.33

Consider the vertical equilibrium of the beam:

+ve $\uparrow$ $\Sigma F_y = 0$

$V_A - 5.0 - 10.0 - 8.0 + V_E = 0$ $\therefore$ $V_A + V_E = 23$ kN   (i)

Consider the rotational equilibrium of the beam:

+ve $\curvearrowright$ $\Sigma M_A = 0$

$(5.0 \times 2.0) + (10.0 \times 6.0) + (8.0 \times 9.0) - (V_E \times 10.0) = 0$ (ii)

$\therefore$ $V_E = 14.2$ kN

Substituting into equation (i) gives $\therefore$ $V_A = 8.8$ kN

The equation for the **bending moment** at x is:

$$EI\frac{d^2 y}{dx^2} = M_x = +8.8x - 5.0[x-2] - 10.0[x-6] - 8.0[x-9]$$

Equation (1)

The equation for the **slope** at x is:

$$EI\frac{dy}{dx} = \int M dx = +\frac{8.8}{2}x^2 - \frac{5.0}{2}[x-2]^2 - \frac{10.0}{2}[x-6]^2 - \frac{8.0}{2}[x-9]^2 + A$$

Equation (2)

The equation for the **deflection** at x is:

$$EI\,y = \iint\left(\frac{M}{EI}dx\right)dx = +\frac{8.8}{6}x^3 - \frac{5.0}{6}[x-2]^3 - \frac{10.0}{6}[x-6]^3 - \frac{8.0}{6}[x-9]^3 + Ax + B$$

Equation (3)

where A and B are constants of integration related to the boundary conditions.

Note: It is common practice to use square brackets, i.e. [ ], to enclose the lever arms for the forces as shown. These brackets are integrated as a unit and during the calculation for deflection they are ignored if the contents are –ve, i.e. the position x being considered is to the left of the load associated with the bracket.

Boundary Conditions
The boundary conditions are known values associated with the slope and/or deflection. In this problem, assuming no settlement occurs at the supports then the **deflection** is equal to zero at these positions, i.e.

$$\begin{matrix} x = 0 \\ y = 0 \end{matrix} \qquad +\frac{8.8}{6}x^3 - \frac{5.0}{6}[x-2]^3 - \frac{10.0}{6}[x-6]^3 - \frac{8.0}{6}[x-9]^3 + Ax + B = 0$$

Substituting for x and y in equation (3) $\therefore$ $B = 0$

$$x = 10.0 \quad + \frac{8.8}{6}10^3 - \frac{5.0}{6}[10-2]^3 - \frac{10.0}{6}[10-6]^3 - \frac{8.0}{6}[10-9]^3 + (A \times 10) \quad = \quad 0$$
$$y = 0$$
$$+ (1.466 \times 10^3) - (0.426 \times 10^3) - (0.106 \times 10^3) - 1.33 + 10A = 0$$
$$\therefore A = -93.265$$

The general equations for the slope and deflection at any point along the length of the beam are given by:

The equation for the **slope** at x is:

$$EI\frac{dy}{dx} = +\frac{8.8}{2}x^2 - \frac{5.0}{2}[x-2]^2 - \frac{10.0}{2}[x-6]^2 - \frac{8.0}{2}[x-9]^2 + A$$

Equation (4)

The equation for the **deflection** at x is:

$$EIy = +\frac{8.8}{6}x^3 - \frac{5.0}{6}[x-2]^3 - \frac{10.0}{6}[x-6]^3 - \frac{8.0}{6}[x-9]^3 + Ax$$

Equation (5)

e.g. the deflection at the mid-span point can be determined from equation (5) by substituting the value of $x = 5.0$ and ignoring the [ ] when their contents are − ve, i.e.

$$EIy = +\frac{8.8}{6}5^3 - \frac{5.0}{6}[5-2]^3 - \frac{10.0}{6}[5-6]^3 - \frac{8.0}{6}[5-9]^3 - (93.265 \times 5)$$

ignore   *ignore*

$$EIy = +183.33 - 22.5 - 466.325 \qquad \therefore y = -\frac{305.5}{EI}\ \mathrm{m} = -\left\{\frac{305.5 \times 10^3}{EI}\right\}\mathrm{mm}$$

The maximum deflection can be determined by calculating the value of x when the slope, i.e. equation (4) is equal to zero and substituting the calculated value of x into equation (5) as above.

In most simply supported spans the maximum deflection occurs near the mid-span point this can be used to estimate the value of x in equation (4) and hence eliminate some of the [ ] brackets, e.g. if the maximum deflection is assumed to occur at a position less than 6.0 m from the left-hand end the last two terms in the [ ] brackets need not be used to determine the position of zero slope. This assumption can be checked and if incorrect a subsequent calculation carried out including an additional bracket until the correct answer is found.

Assume $y_{maximum}$ occurs between 5.0 m and 6.0 m from the left-hand end of the beam, then:

The equation for the **slope** at x is:

$$EI\frac{dy}{dx} = +\frac{8.8}{2}x^2 - \frac{5.0}{2}[x-2]^2 - \frac{10.0}{2}[x-6]^2 - \frac{8.0}{2}[x-9]^2 - 93.265 = 0 \quad \text{for } y_{maximum}$$

ignore   *ignore*

This equation reduces to:

$$1.9x^2 + 10x - 103.265 = 0 \qquad \text{and hence} \qquad x = 5.2 \text{ m}$$

since x was assumed to lie between 5.0 m and 6.0 m ignoring the two [ ] terms was correct.

The maximum deflection can be found by substituting the value of $x = 5.2$ m in equation (5) and ignoring the [ ] when their contents are –ve, i.e.

$$EI\, y_{\text{maximum}} = +\frac{8.8}{6}5.2^3 - \frac{5.0}{6}[5.2-2]^3 - \frac{10.0}{6}[5.2-6]^3 - \frac{8.0}{6}[5.2-9]^3 - (93.265 \times 5.2)$$

$$\qquad\qquad\qquad\qquad\qquad\qquad\qquad\qquad\qquad\qquad ignore \qquad\qquad ignore$$

$$EI\, y_{\text{maximum}} = +206.23 - 27.31 - 484.98 \qquad \therefore\; y_{\text{maximum}} = -\frac{306}{EI}\text{ m}$$

Note: There is no significant difference from the value calculated at mid-span.

1.6.2 *Example 1.6: Beam with Combined Point Loads and UDLs*

A simply supported beam ABCD carries a uniformly distributed load of 3.0 kN/m between A and B, point loads of 4 kN and 6 kN at B and C respectively, and a uniformly distributed load of 5.0 kN/m between B and D as shown in Figure 1.34. Determine the position and magnitude of the maximum deflection.

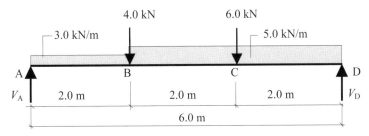

Figure 1.34

Consider the vertical equilibrium of the beam:

$$+ve \uparrow \Sigma F_y = 0$$

$$V_A - (3.0 \times 2.0) - 4.0 - 6.0 - (5.0 \times 4.0) + V_D = 0 \qquad \therefore\; V_A + V_D = 36 \text{ kN} \qquad (i)$$

Consider the rotational equilibrium of the beam:

$$+ve \curvearrowright \Sigma M_A = 0$$

$$(3.0 \times 2.0 \times 1.0) + (4.0 \times 2.0) + (6.0 \times 4.0) + (5.0 \times 4.0 \times 4.0) - (V_D \times 6.0) = 0 \qquad (ii)$$

$$\therefore\; V_D = 19.67 \text{ kN}$$

Substituting into equation (i) gives $\qquad\qquad\qquad\qquad \therefore\; V_A = 16.33 \text{ kN}$

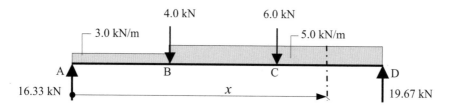

Figure 1.35

In the case of a **UDL** when a term is written in the moment equation in square brackets, [ ], this effectively applies the load for the full length of the beam. For example, in Figure 1.35 the 3.0 kN/m load is assumed to apply from A to D and consequently only an additional 2.0 kN/m need be applied from position B onwards as shown in Figure 1.36.

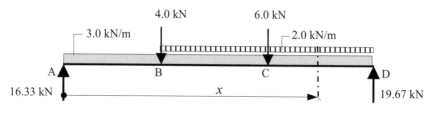

Figure 1.36

The equation for the **bending moment** at x is:

$$EI\frac{d^2y}{dx^2} = +16.33x - 3.0\frac{x^2}{2} - 4.0[x-2.0] - 2.0\frac{[x-2]^2}{2} - 6.0[x-4]$$

Equation (1)

The equation for the **slope** at x is:

$$EI\frac{dy}{dx} = +16.33\frac{x^2}{2} - 3.0\frac{x^3}{6} - 4.0\frac{[x-2]^2}{2} - 2.0\frac{[x-2]^3}{6} - 6.0\frac{[x-4]^2}{2} + A$$

Equation (2)

The equation for the **deflection** at x is:

$$EI\,y = +16.33\frac{x^3}{6} - 3.0\frac{x^4}{24} - 4.0\frac{[x-2]^3}{6} - 2.0\frac{[x-2]^4}{24} - 6.0\frac{[x-4]^3}{6} + Ax + B$$

Equation (3)

where A and B are constants of integration related to the boundary conditions.

Boundary Conditions
In this problem, assuming no settlement occurs at the supports then the **deflection** is equal to zero at these positions, i.e.

$$\begin{array}{l} x = 0 \\ y = 0 \end{array} \quad +16.33\frac{x^3}{6} - 3.0\frac{x^4}{24} - 4.0\frac{[x-2]^3}{6} - 2.0\frac{[x-2]^4}{24} - 6.0\frac{[x-4]^3}{6} + Ax + B$$

Substituting for x and y in equation (3)   ∴ $B = 0$

$$x = 6.0 \quad +16.33\frac{x^3}{6} - 3.0\frac{x^4}{24} - 4.0\frac{[x-2]^3}{6} - 2.0\frac{[x-2]^4}{24} - 6.0\frac{[x-4]^3}{6} + Ax = 0$$
$$y = 0$$

$$+16.33\frac{6.0^3}{6} - 3.0\frac{6.0^4}{24} - 4.0\frac{4.0^3}{6} - 2.0\frac{4.0^4}{24} - 6.0\frac{2.0^3}{6} + 6.0A = 0$$

$$\therefore A = -58.98$$

The general equations for the slope and bending moment at any point along the length of the beam are given by:

The equation for the **slope** at x is:

$$EI\frac{dy}{dx} = +16.33\frac{x^2}{2} - 3.0\frac{x^3}{6} - 4.0\frac{[x-2]^2}{2} - 2.0\frac{[x-2]^3}{6} - 6.0\frac{[x-4]^2}{2} + A$$

Equation (4)

The equation for the **deflection** at x is:

$$EI\,y = +16.33\frac{x^3}{6} - 3.0\frac{x^4}{24} - 4.0\frac{[x-2]^3}{6} - 2.0\frac{[x-2]^4}{24} - 6.0\frac{[x-4]^3}{6} + A\,x$$

Equation (5)

Assume y_{maximum} occurs between 2.0 m and 4.0 m from the left-hand end of the beam, then:
The equation for the **slope** at 'x' is:

$$EI\frac{dy}{dx} = +16.33\frac{x^2}{2} - 3.0\frac{x^3}{6} - 4.0\frac{[x-2]^2}{2} - 2.0\frac{[x-2]^3}{6} - 6.0\frac{[x-4]^2}{2} - 58.98 = 0$$

ignore

This cubic can be solved by iteration.
Guess a value for x, e.g. 3.1 m
$$(16.33 \times 3.1^2)/2 - (3.0 \times 3.1^3)/6 - (4.0 \times 1.1^2)/2 - (2.0 \times 1.1^3)/6 - 58.98 = 1.77 > 0$$

The assumed value of 3.1 is slightly high, try $x = 3.05$ m
$$(16.33 \times 3.05^2)/2 - (3.0 \times 3.05^3)/6 - (4.0 \times 1.05^2)/2 - (2.0 \times 1.05^3)/6 - 58.98 = 0.21$$
This value is close enough. $x = 3.05$ m and since x was assumed to lie between 2.0 m and 4.0 m, ignoring the $[x - 4]$ term was correct.

The maximum deflection can be found by substituting the value of $x = 3.05$ m in equation (5) and ignoring the [ ] when their contents are −ve, i.e.

$$EI\,y_{\text{maximum}} = +16.33\frac{x^3}{6} - 3.0\frac{x^4}{24} - 4.0\frac{[x-2]^3}{6} - 2.0\frac{[x-2]^4}{24} - 6.0\frac{[x-4]^3}{6} - 58.98\,x$$

ignore

$$EI\,y_{\text{maximum}} = +77.22 - 10.82 - 0.77 - 0.1 - 179.89 \qquad \therefore y_{\text{maximum}} = -\frac{114.4}{EI}\text{ m}$$

1.7   *Equivalent UDL Technique for the Deflection of Beams*

In a simply supported beam, the maximum deflection induced by the applied loading always approximates the mid-span value if it is not equal to it. A number of standard frequently used load cases for which the elastic deformation is required are given in Appendix 3 in this text.

In many cases beams support complex load arrangements which do not lend themselves either to an individual load case or to a combination of the load cases given in Appendix 3. Provided that deflection is not the governing design criterion, a calculation which gives an approximate answer is usually adequate. The equivalent UDL method is a useful tool for estimating the deflection in a simply supported beam with a complex loading.

Consider a single-span, simply supported beam carrying a *non-uniform* loading which induces a maximum bending moment of *M* as shown in Figure 1.37.

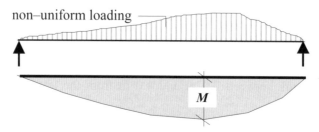

non–uniform loading

M

Bending Moment Diagram

Figure 1.37

The equivalent UDL (w_e) which would induce the same *magnitude* of maximum bending moment (**note:** the position may be different) on a simply supported span carrying a *uniform* loading can be determined from:

$$\text{Maximum bending moment} \qquad M \;=\; \frac{w_e L^2}{8}$$

$$\therefore \qquad w_e \;=\; \frac{8\,M}{L^2}$$

where w_e is the equivalent uniform distributed load.

The maximum deflection of the beam carrying the uniform loading will occur at the mid-

span and will be equal to $\qquad\qquad\qquad \delta \;=\; \dfrac{5 w_e L^4}{384 EI} \qquad\qquad$ (see Appendix 3)

Using this expression, the maximum deflection of the beam carrying the non-uniform loading can be estimated by substituting for the w_e term, i.e.

$$\delta \approx \frac{5w_eL^4}{384E\,I} = \frac{5\times\left(\frac{8M}{L^2}\right)L^4}{384\,EI} = \frac{0.104\,M\,L^2}{EI}$$

The maximum bending moments in Examples 1.5 and 1.6 are 32.8 kNm and 30.67 kNm respectively (the reader should check this as shown in Section 1.5.2).

Using the equivalent UDL method to estimate the maximum deflection in each case gives:

Example 1.5 $\quad \delta_{\text{maximum}} \approx \dfrac{0.104\,M\,L^2}{EI} = -\dfrac{341.1}{EI}\,\text{m}$ (actual value $= \dfrac{305.5}{EI}\,\text{m}$)

Example 1.6 $\quad \delta_{\text{maximum}} \approx \dfrac{0.104\,M\,L^2}{EI} = -\dfrac{114.9}{EI}\,\text{m}$ (actual value $= \dfrac{114.4}{EI}\,\text{m}$)

Note: The estimated deflection is more accurate for beams which are predominantly loaded with distributed loads.

1.8 *Elastic Shear Stress Distribution*

The shear forces induced in a beam by an applied load system generate shear stresses in both the horizontal and vertical directions. At any point in an elastic body, the shear stresses in two mutually perpendicular directions are equal to each other in magnitude.

Consider an element of material subject to shear stresses along the edges, as shown in Figure 1.38.

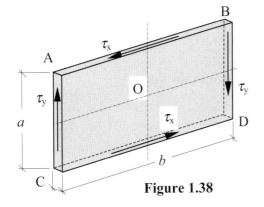

Force on surface AB $= F_{AB} = \tau \times bt$
Force on surface CD $= F_{CD} = \tau \times bt$
Force on surface AC $= F_{AC} = \tau \times at$
Force on surface BD $= F_{BD} = \tau \times at$

+ve $\uparrow$ $\Sigma F_y = 0$

$$F_{AC} = F_{BD}$$

+ve $\rightarrow$ $\Sigma F_x = 0$

$$F_{AB} = F_{CD}$$

Figure 1.38

+ve $\searensuremath$ $\Sigma M_O = 0$

$$-\left(F_{AB}\times \frac{a}{2}\right) - \left(F_{CD}\times\frac{a}{2}\right) + \left(F_{AC}\times\frac{b}{2}\right) + \left(F_{BD}\times\frac{b}{2}\right) = 0$$

Substitute for F_{CD} and F_{BD}:

$$-(F_{AB}\times a) + (F_{AC}\times b) = 0 \qquad \therefore F_{AB}\,a = F_{AC}\,b$$

$$(\tau_x bt)\,a = (\tau_y at)\,b$$

$$\tau_x = \tau_y$$

The two shear stresses are equal and complementary. The magnitude of the shear stress at any vertical cross-section on a beam can be determined using:

$$\tau = \frac{VA\bar{y}}{Ib}$$ Equation (1)

where:
V the vertical shear force at the section being considered,
A the area of the cross-section above (or below) the 'horizontal' plane being considered (note that the shear stress varies throughout the depth of a cross-section for any given value of shear force),
$\bar{y}$ the distance from the elastic neutral axis to the centroid of the area A,
b the breadth of the beam at the level of the horizontal plane being considered,
I the second moment of area of the full cross-section about the elastic neutral axis.

The intensity of shear stress throughout the depth of a section is not uniform and is a maximum at the level of the neutral axis.

1.8.1 Example 1.7: Shear Stress Distribution in a Rectangular Beam

The rectangular beam shown in Figure 1.39 is subject to a vertical shear force of 3.0 kN. Determine the shear stress distribution throughout the depth of the section.

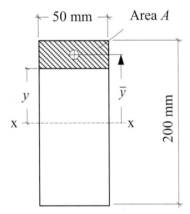

The shear stress at any horizontal level a distance y from the neutral axis is given by:

$$\tau = \frac{VA\bar{y}}{Ib}$$

$V = $ shear force $= 3.0$ kN

$$I = \frac{bd^3}{12} = \frac{50 \times 200^3}{12} = 33.33 \times 10^6 \text{ mm}^4$$

$b = 50$ mm   (for all values of y)

Figure 1.39

Consider the shear stress at a number of values of y,

$y = 100$ mm $A\bar{y} = 0$ (since $A = 0$) $\tau_{100} = 0$

$y = 75$ mm

$A = 50 \times 25 = 1250 \text{ mm}^2$
$\bar{y} = 75 + 12.5 = 87.5$ mm
$A\bar{y} = 109.375 \times 10^3 \text{ mm}^3$

$$\tau_{75} = \frac{3 \times 10^3 \times 109.375 \times 10^3}{33.3 \times 10^6 \times 50} = 0.197 \text{ N/mm}^2$$

y = 50 mm

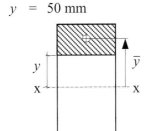

A = 50×50 =   2500 mm^2
$\bar{y}$ = $50 + 25$ =   75 mm
$A\bar{y}$ = 187.5×10^3 mm^3

$$\tau_{50} = \frac{3 \times 10^3 \times 187.5 \times 10^3}{33.3 \times 10^6 \times 50} = 0.338 \text{ N/mm}^2$$

y = 25 mm

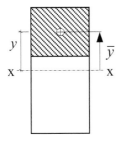

A = 50×75 =   3750 mm^2
$\bar{y}$ = $25 + 37.5$=   62.5 mm
$A\bar{y}$ = 234.375×10^3 mm^3

$$\tau_{25} = \frac{3 \times 10^3 \times 234.375 \times 10^3}{33.3 \times 10^6 \times 50} = 0.422 \text{ N/mm}^2$$

y = 0 mm

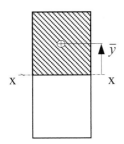

A = 50×100 =   5000 mm^2
$\bar{y}$ = 50 mm
$A\bar{y}$ = 250×10^3 mm^3

$$\tau_{0} = \frac{3 \times 10^3 \times 250 \times 10^3}{33.3 \times 10^6 \times 50} = 0.45 \text{ N/mm}^2$$

y = − 25 mm

A = 50×125 =   6250 mm^2
$\bar{y}$ = 37.5 mm
$A\bar{y}$ = 234.375×10^3 mm^3

$$\tau_{75} = \frac{3 \times 10^3 \times 23375 \times 10^3}{33.3 \times 10^6 \times 50} = 0.422 \text{ N/mm}^2$$

This is the same value as for y = + 25 mm

The cross-section (and hence the stress diagram) is symmetrical about the elastic neutral axis, as shown in Figure 1.40.

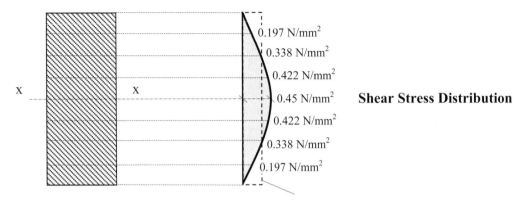

Figure 1.40

The maximum value occurs at the same level as the elastic neutral axis. The 'average' shear stress for a cross-section is equal to the applied force distributed uniformly over the entire cross-section, i.e.

$$\tau_{average} = \frac{\text{Force}}{\text{Area}} = \frac{V}{A} = \frac{3.0 \times 10^3}{50 \times 200} = 0.3 \text{ N/mm}^2$$

For a *rectangular* section:

$$\tau_{maximum} = 1.5 \times \tau_{average} = \frac{1.5V}{A} = 1.5 \times 0.3 = 0.45 \text{ N/mm}^2$$

1.9 *Elastic Bending Stress Distribution*

The bending moments induced in a beam by an applied load system generate bending stresses in the material fibres which vary from a maximum in the extreme fibres to a minimum at the level of the neutral axis as shown in Figure 1.41.

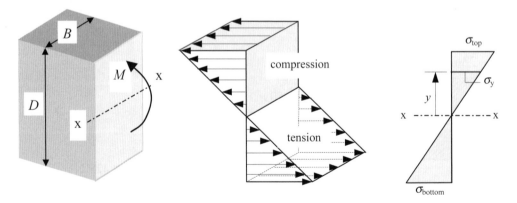

Bending Stress Distribution   **Stress Diagram**

Figure 1.41

The magnitude of the bending stresses at any vertical cross-section can be determined using the simple theory of bending from which the following equation is derived:

$$\frac{M}{I} = \frac{E}{R} = \frac{\sigma}{y} \qquad \therefore \ \sigma = \frac{My}{I} \qquad\qquad \text{Equation} \quad (2)$$

where:
M the applied bending moment at the section being considered,
E the value of Young's modulus of elasticity,
R the radius of curvature of the beam,
σ the bending stress,
y the distance measured from the elastic neutral axis to the level on the cross-section at which the stress is being evaluated,
I the second moment of area of the full cross-section about the elastic neutral axis.

It is evident from Equation (2) that for any specified cross-section in a beam subject to a known value of bending moment (i.e. M and I constant), the bending stress is directly proportional to the distance from the neutral axis; i.e.

$$\sigma = \text{constant} \times y \qquad \therefore \qquad \sigma \ \alpha \ y$$

This is shown in Figure 1.41, in which the maximum bending stress occurs at the extreme fibres, i.e. $y_{maximum} = D/2$.

In design it is usually the extreme fibre stresses relating to the $y_{maximum}$ values at the top and bottom which are critical. These can be determined using:

$$\sigma_{top} = \frac{M}{Z_{top}} \qquad \text{and} \qquad \sigma_{bottom} = \frac{M}{Z_{bottom}}$$

where:
σ and M are as before,

Z_{top} is the elastic section modulus relating to the top fibres and defined as $\dfrac{I_{xx}}{y_{top}}$

Z_{bottom} is the elastic section modulus relating to the top fibres and defined as $\dfrac{I_{xx}}{y_{bottom}}$

If a cross-section is symmetrical about the x–x axis then $Z_{top} = Z_{bottom}$. In asymmetric sections the maximum stress occurs in the fibres corresponding to the smallest Z value. For a rectangular cross-section of breadth B and depth D subject to a bending moment M about the major x–x axis, the appropriate values of I, y and Z are:

$$I = \frac{BD^3}{12} \qquad y_{maximum} = \frac{D}{2} \qquad Z_{minimum} = \frac{BD^2}{6}$$

In the case of bending about the minor y–y axis:

$$I = \frac{DB^3}{12} \qquad y_{maximum} = \frac{B}{2} \qquad Z_{minimum} = \frac{DB^2}{6}$$

The maximum stress induced in a cross-section subject to bi-axial bending is given by:

$$\sigma_{maximum} = \frac{M_x}{Z_{x\ minimum}} + \frac{M_y}{Z_{y\ minimum}}$$

where M_x and M_y are the applied bending moments about the x and y axes respectively.

1.9.1 Example 1.8: Bending Stress Distribution in a Rectangular Beam

The rectangular beam shown in Figure 1.42 is subject to a bending moment of 2.0 kNm. Determine the bending stress distribution throughout the depth of the section.

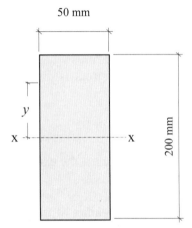

50 mm

200 mm

y

x — — — x

Figure 1.42

The bending stress at any horizontal level a distance y from the neutral axis is given by:

$$\sigma = \frac{My}{I}$$

M = bending moment = 2.0 kNm

$$I = \frac{bd^3}{12} = \frac{50 \times 200^3}{12} = 33.33 \times 10^6 \ mm^4$$

Consider the bending stress at a number of values of y,

$y = 100$ mm $\sigma_{100} = \dfrac{2.0 \times 10^6 \times 100}{33.33 \times 10^6}$ = 6.0 N/mm^2

$y = 75$ mm $\sigma_{75} = \dfrac{2.0 \times 10^6 \times 75}{33.33 \times 10^6}$ = 4.5 N/mm^2

$y = 50$ mm $\sigma_{50} = \dfrac{2.0 \times 10^6 \times 50}{33.33 \times 10^6}$ = 3.0 N/mm^2

$y = 25$ mm $\sigma_{25} = \dfrac{2.0 \times 10^6 \times 25}{33.33 \times 10^6}$ = 1.5 N/mm^2

$y = 0$ $\sigma_0 = 0$

$y = -25$ mm $\sigma_{-25} = -\dfrac{2.0 \times 10^6 \times 25}{33.33 \times 10^6}$ = -1.5 N/mm^2

The cross-section (and hence the stress diagram) is symmetrical about the elastic neutral axis, as shown in Figure 1.43.

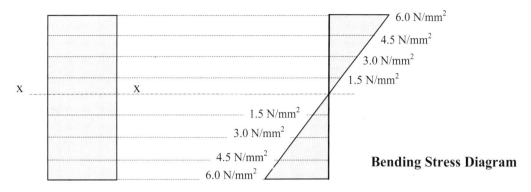

Bending Stress Diagram

Figure 1.43

1.10 *Transformed Sections*

Beams such as ply-web and box-beams in timber are generally fabricated from different materials, e.g. plywood webs and softwood flanges fastened together. During bending, the stresses induced in such sections are shared among all the component parts. The extent to which sharing occurs is dependent on the method of connection at the interfaces. This connection is normally designed such that *no slip* occurs between the different materials during bending. The resulting structural element is a composite section which is non-homogeneous. (**Note:** *This invalidates the simple theory of bending in which homogeneity is assumed.*)

A useful technique often used when analysing such composite sections is the *transformed section* method. When using this method, an equivalent homogeneous section is considered in which all components are assumed to be the same material. The simple theory of bending is then used to determine the stresses in the transformed sections, which are subsequently modified to determine the stresses in the *actual* materials.

Consider the composite section shown in Figure 1.44(a) in which a steel plate has been securely fastened to the underside face. There are two possible transformed sections which can be considered:

(i)   an equivalent section in terms of timber; Figure 1.44(b)   or
(ii) an equivalent section in terms of steel; Figure 1.44(c).

To obtain an equivalent section made from timber, the same material as the existing timber must replace the steel plate. The dimension of the replacement timber must be modified to reflect the different material properties. The equivalent *transformed* section properties are shown in Figure 1.44(b).

The overall depth of both sections is the same $(D + t)$. The *strain* in element δA_s in the original section is equal to the strain in element δA_t of the transformed section:

$$\zeta_{steel} = \zeta_{timber}$$

but $\text{strain} = \dfrac{stress}{modulus\ (E)} = \dfrac{\sigma}{E}$ $\therefore \zeta = \dfrac{\sigma}{E}$

$$\frac{\sigma_s}{E_s} = \frac{\sigma_t}{E_t} \qquad \therefore \quad \frac{\sigma_s}{\sigma_t} = \frac{E_s}{E_t}$$

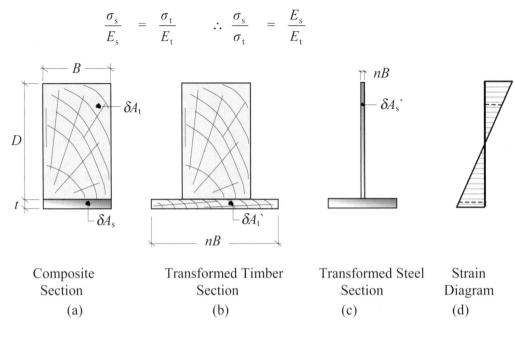

Composite Section	Transformed Timber Section	Transformed Steel Section	Strain Diagram
(a)	(b)	(c)	(d)

Figure 1.44

The force in each element must also be equal:

$$\text{Force} = (\text{stress} \times \text{area}) \qquad P_s = P_t$$
$$\sigma_s \delta A_s = \sigma_t \delta A_t`$$
$$\delta A_t` = \frac{\sigma_s}{\sigma_t}\delta A_s = \frac{E_s}{E_t}\delta A_s$$

This indicates that in the transformed section:

Equivalent area of transformed timber $= (n \times \text{original area of steel})$

where: n is the '*modular ratio*' of the materials and is equal to $\dfrac{E_s}{E_t}$

The equivalent area of timber must be subject to the same value of strain as the original material it is replacing, i.e. it is positioned at the same distance from the elastic neutral axis. The simple equation of bending (see Equation (2) in section 1.9) can be used with the transformed section properties to determine the bending stresses. The *actual* stresses in the steel will be equal to ($n \times$ equivalent timber stresses.) The use of this method is illustrated in Example 1.9. A similar, alternative analysis can be carried out using a transformed steel section, as shown in Figure 1.46.

1.10.1 Example 1.9: Composite Timber/Steel Section

A timber beam is enhanced by the addition of two steel plates as shown in Figure 1.45. Determine the maximum timber and steel stresses induced in the cross-section when the beam is subjected to a bending moment of 70 kNm.

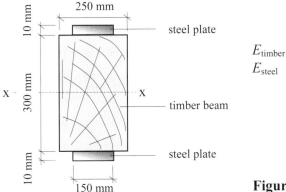

$E_{\text{timber}} = 8200 \text{ N/mm}^2$

$E_{\text{steel}} = 205 \times 10^3 \text{ N/mm}^2$

Figure 1.45

(a)   Transformed section based on *timber*

Equivalent width of timber to replace the steel plate $= (n \times 150)$ mm

where:

$$n = \frac{E_{\text{steel}}}{E_{\text{timber}}} = \frac{205 \times 10^3}{8200} = 25 \qquad nB = (25 \times 150) = 3750 \text{ mm}$$

The maximum stresses occur in the timber when $y = 150$ mm, and in the steel (or equivalent replacement timber) when $y = 160$ mm.

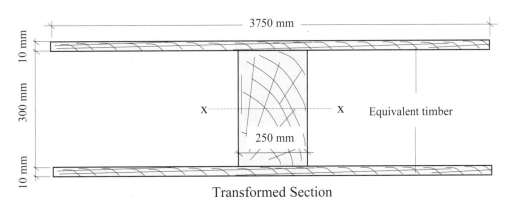

Transformed Section

$$I_{\text{xx transformed}} = \left\{ \frac{3750 \times 320^3}{12} - \frac{3500 \times 300^3}{12} \right\} = 2.365 \times 10^9 \text{ mm}^4$$

Maximum bending stress in the *timber* is given by:

$$\sigma_{\text{timber}} = \frac{My_{150}}{I} = \frac{70 \times 10^6 \times 150}{2.365 \times 10^9} = 4.43 \text{ N/mm}^2$$

Maximum bending stress in the *equivalent timber* is given by:

$$\sigma = \frac{My_{160}}{I} = \frac{70 \times 10^6 \times 160}{2.365 \times 10^9} = 4.74 \text{ N/mm}^2$$

This value of stress represents a maximum value of stress in the steel plates given by:

$$\sigma_{steel} \;=\; n \times \sigma \;\;=\; (25 \times 4.74) \;\;\;\; = \; 118.5 \; \text{N/mm}^2$$

(b)   Transformed section based on *steel*
Equivalent width of steel to replace the timber beam $= (n \times 150)$ mm
where:

$$n \;=\; \frac{E_{timber}}{E_{steel}} \;=\; \frac{1}{25} \qquad\qquad nB \;=\; \frac{1 \times 250}{25} \;=\; 10 \text{ mm}$$

The maximum stresses occur in the timber (or equivalent replacement steel) when $y = 150$ mm, and in the steel when $y = 160$ mm.

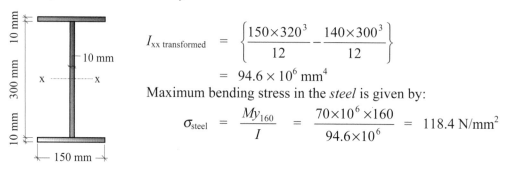

$$I_{xx \text{ transformed}} \;=\; \left\{ \frac{150 \times 320^3}{12} - \frac{140 \times 300^3}{12} \right\}$$

$$= \; 94.6 \times 10^6 \text{ mm}^4$$

Maximum bending stress in the *steel* is given by:

$$\sigma_{steel} \;=\; \frac{My_{160}}{I} \;=\; \frac{70 \times 10^6 \times 160}{94.6 \times 10^6} \;=\; 118.4 \text{ N/mm}^2$$

Figure 1.46

Maximum bending stress in the *equivalent steel* is given by:

$$\sigma \;=\; \frac{My_{150}}{I} \;=\; \frac{70 \times 10^6 \times 150}{94.6 \times 10^6} \;=\; 111 \text{ N/mm}^2$$

This value of stress represents a maximum value of stress in the timber given by:

$$\sigma_{timber} = n \times \sigma \;\;\; = \; \left(\frac{1}{25} \times 111 \right) \;\;\; = \; 4.44 \text{ N/mm}^2$$

Normally when using this method in the design of ply-web beams the plywood webs are replaced by softwood timber of equivalent thickness to give the required transformed section; as indicated in Example 1.9, both transformed sections produce the same result.

1.11   *Moment Distribution*

The methods of analysis illustrated in the previous sections of this chapter are suitable for **determinate** structures such as single-span beams in which the support reactions and the member forces, e.g. shear forces and bending moments, can be evaluated using stability considerations and the three equations of static equilibrium, i.e. $\Sigma F_y = 0$, $\Sigma F_x = 0$ and $\Sigma M = 0$. In many instances multi-span beams and rigid-jointed frames are used in design, and consequently it is necessary to consider the effects of continuity on the support reactions and member forces. Such structures are **indeterminate**, i.e. there are more unknown variables than can be solved using only the three equations of equilibrium. A few examples of such structures are shown in Figure 1.47.

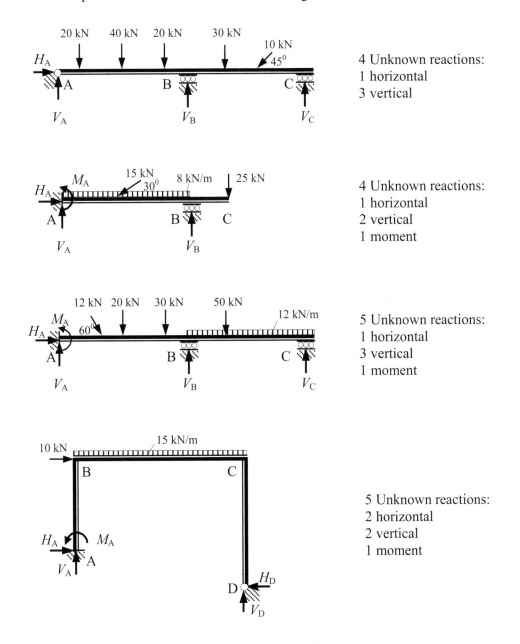

Figure 1.47

This section deals with continuous beams and propped cantilevers. An American engineer, Professor Hardy Cross, developed a very simple, elegant and practical method of analysis for such structures called *Moment Distribution*. This technique is one of developing successive approximations and is based on several basic concepts of structural behaviour which are illustrated in sections 1.11.1 to 1.11.10.

1.11.1 Bending (Rotational) Stiffness

A fundamental relationship which exists in the elastic behaviour of structures and structural elements is that between an applied force system and the displacements which are induced by that system, i.e.

$$\text{Force} = \text{Stiffness} \times \text{Displacement}$$
$$P = k\delta$$

where:
P is the applied force,
k is the stiffness,
δ is the displacement.

A definition of stiffness can be derived from this equation by rearranging it such that:

$$k = P / \delta$$

when $\delta = 1.0$ (i.e. unit displacement) the stiffness is: '***the force necessary to maintain a UNIT displacement, all other displacements being equal to zero.***'

The displacement can be a shear displacement, an axial displacement, a bending (rotational) displacement or a torsional displacement, each in turn producing the shear, axial, bending or torsional stiffness.

When considering beam elements in continuous structures using the moment distribution method of analysis, the bending stiffness is the principal characteristic which influences behaviour.

Consider the beam element AB shown in Figure 1.48 which is subject to a **UNIT** rotation at end A and is fixed at end B as indicated.

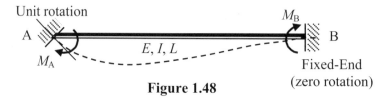

Unit rotation

A   E, I, L M_B B

M_A

Fixed-End
(zero rotation)

Figure 1.48

The force (M_A) necessary to maintain this displacement can be shown (e.g. using McCaulay's Method) to be equal to $(4EI)/L$. From the definition of stiffness given previously, the bending stiffness of the beam is equal to (Force/1.0), therefore $k = (4EI)/L$. This is known as the ***absolute*** bending stiffness of the element. Since most elements in continuous structures are made from the same material, the value of Young's Modulus (E) is constant throughout and $4E$ in the stiffness term is also a constant. This constant is normally ignored, to give $k = I/L$ which is known as the ***relative*** bending stiffness of the element. It is this value of stiffness which is normally used in the method of Moment Distribution.

It is evident from Figure 1.48 that when the beam element deforms due to the applied rotation at end A, an additional moment (M_B) is also transferred by the element to the remote end if it has zero slope (i.e. is fixed) The moment M_B is known as the ***carry-over*** moment.

1.11.2 Carry-Over Moment

Using the same analysis as that to determine M_A, it can be shown that $M_B = (2EI)/L$, i.e. ($\frac{1}{2} \times M_A$). It can therefore be stated that '***if a moment is applied to one end of a beam then a moment of the same sense and equal to half of its value will be transferred to the remote end provided that it is fixed.***'

   If the remote end is '**pinned**', then the beam is less stiff and there is no carry-over moment.

1.11.3 Pinned End

Consider the beam shown in Figure 1.49 in which a unit rotation is imposed at end A as before but the remote end B is pinned.

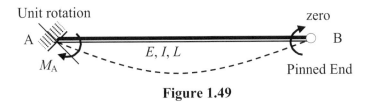

Figure 1.49

The force (M_A) necessary to maintain this displacement can be shown (e.g. using McCaulay's Method) to be equal to $(3EI)/L$, which represents the reduced absolute stiffness of a pin-ended beam. It can therefore be stated that '***the stiffness of a pin-ended beam is equal to ¾ × the stiffness of a fixed-end beam.***' In addition it can be shown that there is no carry-over moment to the remote end. These two cases are summarised in Figure 1.50.

Remote End Fixed:

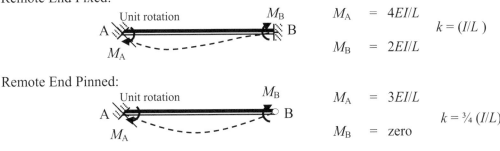

$$M_A = 4EI/L$$
$$k = (I/L)$$
$$M_B = 2EI/L$$

Remote End Pinned:

$$M_A = 3EI/L$$
$$k = ¾\,(I/L)$$
$$M_B = \text{zero}$$

Figure 1.50

1.11.4 Free and Fixed Bending Moments

When a beam is free to rotate at both ends as shown in Figures 1.51(a) and (b) such that no bending moment can develop at the supports, then the bending moment diagram resulting from the applied loads on the beam is known as the ***Free Bending Moment Diagram***.

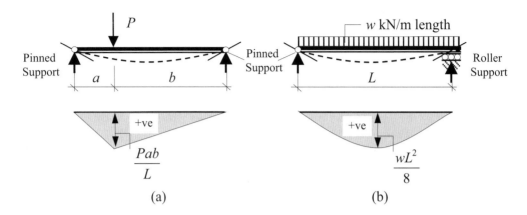

Figure 1.51 – Free Bending Moment Diagrams

When a beam is fixed at the ends (encastre) such that it cannot rotate, i.e. zero slope at the supports, as shown in Figure 1.52, then bending moments are induced at the supports and are called *Fixed-End Moments*. The bending moment diagram associated **only** with the fixed-end moments is called the ***Fixed Bending Moment Diagram***.

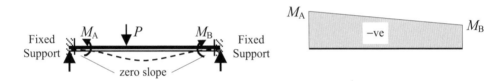

Figure 1.52 – Fixed Bending Moment Diagram

Using the principle of superposition, this beam can be considered in two parts in order to evaluate the support reactions and the **Final** bending moment diagram:

(i)   *The fixed-reactions (moments and forces) at the supports*

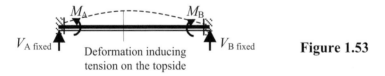

Figure 1.53

(ii)   *The free reactions at the supports and the bending moments throughout the length due to the applied load, assuming the supports to be pinned*

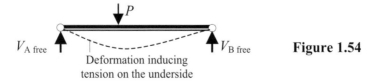

Figure 1.54

Combining (i) + (ii) gives the final bending moment diagram as shown in Figure 1.55:

$$V_A = (V_{A\ fixed} + V_{A\ free}); \qquad V_B = (V_{B\ fixed} + V_{B\ free})$$

$$M_A = (\ M_A + 0\ ); \qquad M_B = (\ M_B + 0\ )$$

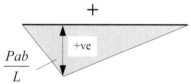

Fixed Bending Moment Diagram

Note: $M = [M_B + (M_A - M_B)b/L\ ]$

$+$

$=$

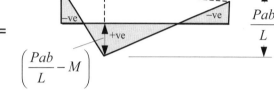

Free Bending Moment Diagram

Final Bending Moment Diagram

Figure 1.55

The values of M_A and M_B for the most commonly applied load cases are given in Appendix 3. These are standard *Fixed-End Moments* relating to single-span encastre beams and are used extensively in structural analysis.

1.11.5 Example 1.10: Single-span Encastre Beam

Determine the support reactions and draw the bending moment diagram for the encastre beam loaded as shown in Figure 1.56.

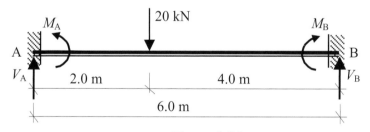

Figure 1.56

Solution:
Consider the beam in two parts.

(i) *Fixed Support Reactions*
The values of the fixed-end moments are given in Appendix 3.

$$M_A = -\frac{Pab^2}{L^2} = -\frac{20 \times 2 \times 4^2}{6^2} = -17.78 \text{ kNm}$$

$$M_B = +\frac{Pa^2b}{L^2} = +\frac{20 \times 2^2 \times 4}{6^2} = +8.89 \text{ kNm}$$

These moments induce tension in the top of the beam

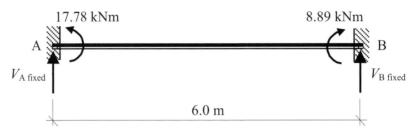

Consider the rotational equilibrium of the beam:

$+ve \curvearrowright \Sigma M_A = 0$

$-(17.78) + (8.89) - (6.0 \times V_{B \text{ fixed}}) = 0$ Equation (1)

$\therefore V_{B \text{ fixed}} = -1.48 \text{ kN} \downarrow$

Consider the vertical equilibrium of the beam:

$+ve \uparrow \Sigma F_y = 0$

$+ V_{A \text{ fixed}} + V_{B \text{ fixed}} = 0 \quad \therefore V_{A \text{ fixed}} = -(-1.48 \text{ kN}) = +1.48 \text{ kN} \uparrow$ Equation (2)

(ii) *Free Support Reactions*

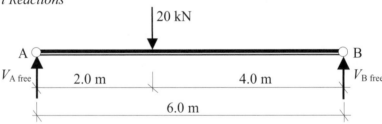

Consider the rotational equilibrium of the beam:

$+ve \curvearrowright \Sigma M_A = 0$

$+ (20 \times 2.0) - (6.0 \times V_{B \text{ free}}) = 0 \quad \therefore V_{B \text{ free}} = +6.67 \text{ kN} \uparrow$ Equation (1)

Consider the vertical equilibrium of the beam:

$+ve \uparrow \Sigma F_y = 0$

$+ V_{A \text{ free}} + V_{B \text{ free}} - 20 = 0 \qquad \therefore V_{A \text{ free}} = +13.33 \text{ kN} \uparrow$ Equation (2)

Bending Moment under the point load $= (+13.33 \times 2.0) = +26.67 \text{ kNm}$

This induces tension in the bottom of the beam

The final vertical support reactions are given by (i) + (ii):

$V_A = V_{A \text{ fixed}} + V_{A \text{ free}} = (+1.48 + 13.33) = +14.81 \text{ kN}$

$V_B = V_{B \text{ fixed}} + V_{B \text{ free}} = (-1.48 + 6.67) = +5.19 \text{ kN}$

Check the vertical equilibrium: Total vertical force $= +14.81 + 5.19 = +20 \text{ kN}$

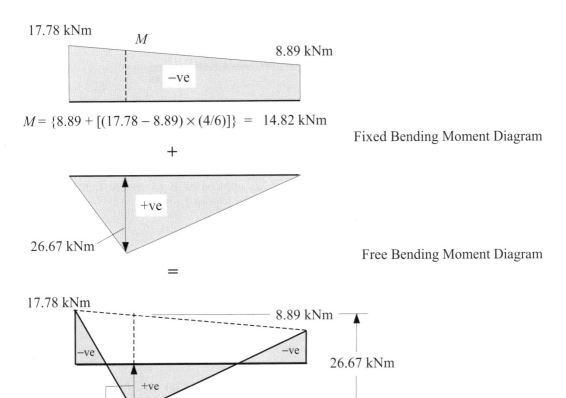

17.78 kNm

M

8.89 kNm

−ve

$M = \{8.89 + [(17.78 - 8.89) \times (4/6)]\} = 14.82$ kNm

Fixed Bending Moment Diagram

+

+ve

26.67 kNm

Free Bending Moment Diagram

=

17.78 kNm

8.89 kNm

−ve

−ve

26.67 kNm

+ve

$(26.67 - 14.82) = 11.87$ kNm

Final Bending Moment Diagram

Figure 1.57

Note the similarity between the shape of the bending moment diagram and the final deflected shape as shown in Figure 1.58.

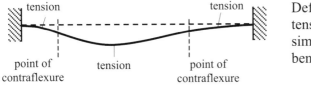

tension

tension

point of contraflexure

tension

point of contraflexure

Deflected shape indicating tension zones and the similarity to the shape of the bending moment diagram

Figure 1.58

1.11.6 Propped Cantilevers

The fixed-end moment for propped cantilevers (i.e. one end fixed and the other end simply supported) can be derived from the standard values given for encastre beams as follows. Consider the propped cantilever shown in Figure 1.59, which supports a uniformly distributed load as indicated.

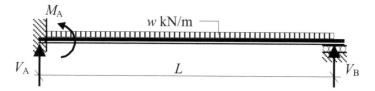

Figure 1.59

The structure can be considered to be the superposition of an encastre beam with the addition of an equal and opposite moment to M_B applied at B to ensure that the final moment at this support is equal to zero, as indicated in Figure 1.60.

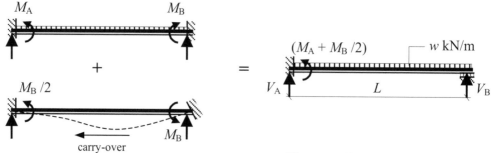

Figure 1.60

1.11.7 Example 1.11: Propped Cantilever

Determine the support reactions and draw the bending moment diagram for the propped cantilever shown in Figure 1.61.

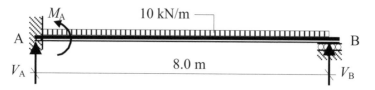

Figure 1.61

Solution

Fixed-End Moment for Propped Cantilever:
Consider the beam fixed at both supports.

The values of the fixed-end moments for encastre beams are given in Appendix 3.

$$M_A = - \frac{wL^2}{12} = - \frac{10 \times 8^2}{12} = -53.33 \text{ kNm}$$

$$M_B = + \frac{wL^2}{12} = + \frac{10 \times 8^2}{12} = +53.33 \text{ kNm}$$

The moment M_B must be cancelled out by applying an equal and opposite moment at B which in turn produces a carry-over moment equal to ($\frac{1}{2} \times M_B$) at support A.

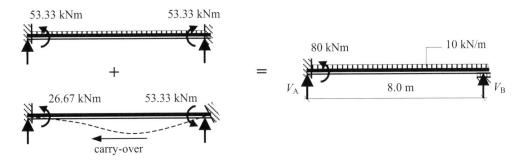

(i) *Fixed Support Reactions*

Consider the rotational equilibrium of the beam:

$+ve \curvearrowright \Sigma M_A = 0$

$-(80) - (8.0 \times V_{B \text{ fixed}}) \quad = \quad 0$ Equation (1)

$\therefore V_{B \text{ fixed}} \quad = \quad -10.0 \text{ kN} \downarrow$

Consider the vertical equilibrium of the beam:

$+ve \uparrow \Sigma F_y = 0$

$+ V_{A \text{ fixed}} + V_{B \text{ fixed}} = 0 \quad \therefore V_{A \text{ fixed}} = -(-10.0 \text{ kN}) \quad = \quad +10.0 \text{ kN} \uparrow$ Equation (2)

(ii) *Free Support Reactions*

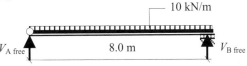

Consider the rotational equilibrium of the beam:

$+ve \curvearrowright \Sigma M_A = 0$

$+ (10 \times 8.0 \times 4.0) - (8.0 \times V_{B \text{ free}}) = 0 \quad \therefore V_{B \text{ free}} = +40.0 \text{ kN} \uparrow$ Equation (1)

Consider the vertical equilibrium of the beam:

$+ve \uparrow \Sigma F_y = 0$

$+ V_{A \text{ free}} + V_{B \text{ free}} - (10 \times 8.0) \quad = 0 \quad \therefore V_{A \text{ free}} = +40.0 \text{ kN} \uparrow$ Equation (2)

The final vertical support reactions are given by (i) + (ii):
$V_A = V_{A \text{ fixed}} + V_{A \text{ free}} = (+10.0 + 40.0) = +50.0 \text{ kN}$
$V_B = V_{B \text{ fixed}} + V_{B \text{ free}} = (-10.0 + 40.0) = +30.0 \text{ kN}$
Check the vertical equilibrium: Total vertical force $= +50.0 + 30.0 = +80 \text{ kN}$

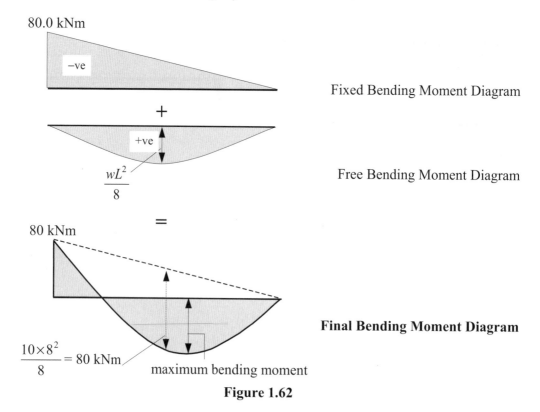

80.0 kNm

−ve

Fixed Bending Moment Diagram

+

+ve

$\dfrac{wL^2}{8}$

Free Bending Moment Diagram

=

80 kNm

Final Bending Moment Diagram

$\dfrac{10\times8^2}{8} = 80$ kNm

maximum bending moment

Figure 1.62

Note the similarity between the shape of the bending moment diagram and the final deflected shape as shown in Figure 1.63

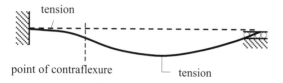

tension

point of contraflexure   tension

Deflected shape indicating tension zones and the similarity to the shape of the bending moment diagram

Figure 1.63

The position of the maximum bending moment can be determined by finding the point of zero shear force.

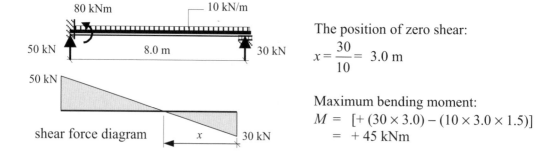

80 kNm   10 kN/m

50 kN   8.0 m   30 kN

50 kN

shear force diagram   x   30 kN

The position of zero shear:

$x = \dfrac{30}{10} = 3.0$ m

Maximum bending moment:

$M = [+(30\times3.0) - (10\times3.0\times1.5)]$

$= +45$ kNm

1.11.8 Distribution Factors

Consider a uniform two-span continuous beam, as shown in Figure 1.64.

Figure 1.64

If an external moment M is applied to this structure at support B it will produce a rotation of the beam at the support; part of this moment is absorbed by each of the two spans BA and BC, as indicated in Figure 1.65.

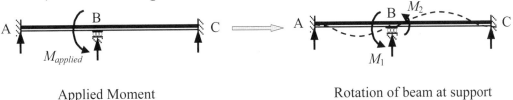

Applied Moment   Rotation of beam at support
$$(M_{applied} = M_1 + M_2)$$

Figure 1.65

The proportion of each moment induced in each span is directly proportional to the relative stiffnesses, e.g.

$$\text{Stiffness of span BA} = k_{BA} = (I_1/L_1)$$
$$\text{Stiffness of span BC} = k_{BC} = (I_2/L_2)$$

$$\text{Total stiffness of the beam at the support} = k_{total} = (k_{BA} + k_{BC}) = [(I_1/L_1) + (I_2/L_2)]$$

The moment absorbed by beam BA $\quad M_1 = M_{applied} \times \left(\dfrac{k_{BA}}{k_{total}} \right)$

The moment absorbed by beam BC $\quad M_2 = M_{applied} \times \left(\dfrac{k_{BC}}{k_{total}} \right)$

The ratio $\left(\dfrac{k}{k_{total}} \right)$ is known as the **Distribution Factor** for the member at the joint where the moment is applied.

As indicated in Section 1.11.2, when a moment (M) is applied to one end of a beam in which the other end is fixed, a carry-over moment equal to 50% of M is induced at the remote fixed-end and consequently moments equal to $\frac{1}{2} M_1$ and $\frac{1}{2} M_2$ will develop at supports A and C respectively, as shown in Figure 1.66.

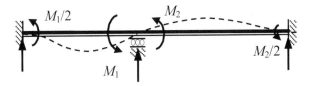

Figure 1.66

1.11.9 Application of the Method

All of the concepts outlined in Sections 1.11.1 to 1.11.8 are used when analysing indeterminate structures using the method of moment distribution. Consider the two *separate* beam spans indicated in Figure 1.67.

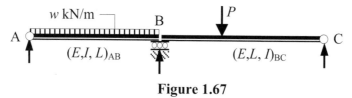

Figure 1.67

Since the beams are **not** connected at the support B they behave independently as simply supported beams with separate reactions and bending moment diagrams, as shown in Figure 1.68.

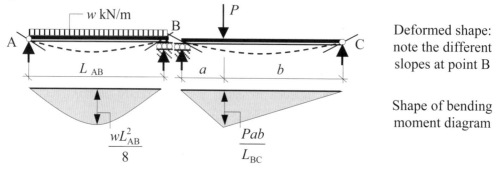

Deformed shape: note the different slopes at point B

Shape of bending moment diagram

Figure 1.68

When the beams are continuous over support B as shown in Figure 1.69(a), a continuity moment develops for the continuous structure as shown in Figures 1.69(b) and (c). Note the similarity of the bending moment diagram for member AB to the propped cantilever in Figure 1.62. Both members AB and BC are similar to propped cantilevers in this structure.

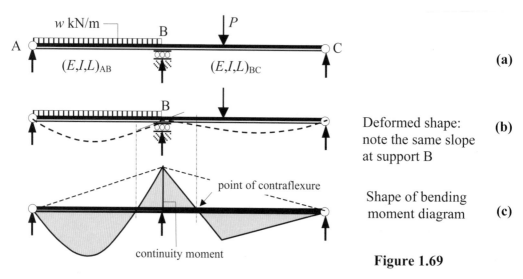

(a)

Deformed shape: note the same slope at support B   (b)

Shape of bending moment diagram   (c)

Figure 1.69

Moment distribution enables the evaluation of the continuity moments. The method is ideally suited to tabular representation and is illustrated in Example 1.12.

1.11.10 Example 1.12: Three-span Continuous Beam

A non-uniform, three span beam ABCD is fixed at support A and pinned at support D, as illustrated in Figure 1.70. Determine the support reactions and sketch the bending moment diagram for the applied loading indicated.

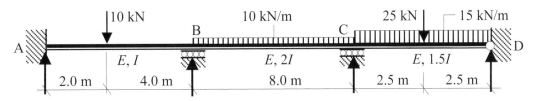

Figure 1.70

Solution:

Step 1

The first step is to assume that all supports are fixed against rotation and evaluate the 'fixed-end moments'.

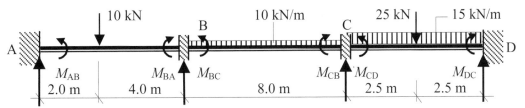

The values of the fixed-end moments for encastre beams are given in Appendix 3.

Span AB

$$M_{AB} = - \frac{Pab^2}{L^2} = - \frac{10 \times 2 \times 4^2}{6.0^2} = -8.89 \text{ kNm}$$

$$M_{BC} = + \frac{Pa^2 b}{L^2} = + \frac{10 \times 2^2 \times 4}{6.0^2} = +4.44 \text{ kNm}$$

Span BC

$$M_{BC} = - \frac{wL^2}{12} = - \frac{10 \times 8^2}{12} = -53.33 \text{ kNm}$$

$$M_{CB} = + \frac{wL^2}{12} = + \frac{10 \times 8^2}{12} = +53.33 \text{ kNm}$$

Span CD*

$$M_{CD} = - \frac{wL^2}{12} - \frac{PL}{8} = - \frac{15 \times 5^2}{12} - \frac{25 \times 5}{8} = -46.89 \text{ kNm}$$

$$M_{DC} = + \frac{wL^2}{12} + \frac{PL}{8} = + \frac{15 \times 5^2}{12} + \frac{25 \times 5}{8} = +46.89 \text{ kNm}$$

* Since support D is pinned, the fixed-end moments are $(M_{CD} + M_{DC}/2)$ at C and zero at D (see Figure 1.60): $(M_{CD} + M_{DC}/2) =$ $[46.89 + (0.5 \times 46.89)]$ = 70.34 kNm.

Step 2
The second step is to evaluate the member and total *stiffness* at each internal joint/support and determine the *distribution factors* at each support. Note that the applied force system is **not** required to do this.

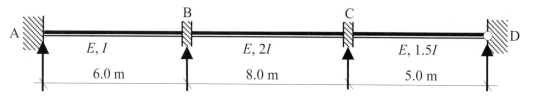

Support B

Stiffness of BA = k_{BA} = $(I/6.0)$ = $0.167I$
Stiffness of BC = k_{BC} = $(2I/8.0)$ = $0.25I$ k_{total} = $(0.167 + 0.25)I$ = $0.417I$

Distribution factor (DF) for BA = $\dfrac{k_{BA}}{k_{total}}$ = $\dfrac{0.167I}{0.417I}$ = 0.4

Distribution factor (DF) for BC = $\dfrac{k_{BC}}{k_{total}}$ = $\dfrac{0.25I}{0.417I}$ = 0.6 ΣDF's = 1.0

Support C

Stiffness of CB = k_{CB} = k_{BC} = $0.25I$ **Note:** the remote end D is
Stiffness of CD = k_{CD} = $\tfrac{3}{4} \times (1.5I/5.0)$= $0.225I$ pinned and $k = \tfrac{3}{4}(I/L)$

k_{total} = $(0.25 + 0.225)I$ = $0.475I$

Distribution factor (DF) for CB = $\dfrac{k_{CB}}{k_{total}}$ = $\dfrac{0.25I}{0.475I}$ = 0.53

Distribution factor (DF) for CD = $\dfrac{k_{BC}}{k_{total}}$ = $\dfrac{0.141I}{0.475I}$ = 0.47 ΣDF's = 1.0

The structure and the distribution factors can be represented in tabular form, as shown in Figure 1.71.

Joints/Support	A		B			C		D
Member	AB		BA	BC		CB	CD	DC
Distribution Factors	0		0.4	0.6		0.53	0.47	1.0

Figure 1.71

The distribution factor for fixed supports is equal to zero since any moment is resisted by an equal and opposite moment within the support and no balancing is required. In the case of pinned supports the distribution factor is equal to 1.0 since 100% of any applied moment, e.g. by a cantilever overhang, must be balanced and a carry-over of ½ × the balancing moment transferred to the remote end at the internal support.

Step 3
The fixed-end moments are now entered into the table at the appropriate locations, taking care to ensure that the signs are correct.

Joints/Support	A		B			C			D
Member	AB		BA	BC		CB	CD		DC
Distribution Factors	0		0.4	0.6		0.53	0.47		1.0
Fixed-End Moments	− 8.89		+ 4.44	− 53.33		+ 53.33	− 70.34		zero

Step 4
When the structure is restrained against rotation there is normally a resultant moment at a typical internal support. For example, consider the moments B:
$M_{BA} = +4.44$ kNm ↺ and $M_{BC} = -53.33$ kNm ↻

The 'out-of-balance' moment is equal to the algebraic difference between the two:
The out-of-balance moment $= (+4.44 - 53.33) = -48.89$ kNm ↻

If the imposed fixity at **one** support (all others remaining fixed), e.g. support B, is released, the beam will rotate sufficiently to induce a balancing moment such that equilibrium is achieved and the moments M_{BA} and M_{BC} are equal and opposite. The application of the balancing moment is distributed between BA and BC in proportion to the *distribution factors* as calculated previously.

Moment applied to BA $= +(48.89 \times 0.4) = +19.56$ kNm
Moment applied to BC $= +(48.89 \times 0.6) = +29.33$ kNm

Joints/Support	A		B			C			D
Member	AB		BA	BC		CB	CD		DC
Distribution Factors	0		0.4	0.6		0.53	0.47		1.0
Fixed-End Moments	− 8.89		+ 4.44	− 53.33		+ 53.33	− 70.34		zero
Balance Moment			+ 19.56	+ 29.33					

As indicated in Section 1.11.2, when a moment is applied to one end of a beam whilst the remote end is fixed, a carry-over moment equal to ($\frac{1}{2} \times$ applied moment) and of the same sign is induced at the remote end. This is entered into the table as shown.

Joints/Support	AB	BA	BC	CB	CD	DC
	A	B		C		D
Member	AB	BA	BC	CB	CD	DC
Distribution Factors	0	0.4	0.6	0.53	0.47	1.0
Fixed-End Moments	− 8.89	+ 4.44	− 53.33	+ 53.33	− 70.34	zero
Balance Moment		+ 19.56	+ 29.33			
Carry-over to Remote Ends	+ 9.78			+ 14.67		

Step 5

The procedure outline above is then carried out for each restrained support in turn. The reader should confirm the values given in the table for support C.

Joints/Support	AB	BA	BC	CB	CD	DC
	A	B		C		D
Member	AB	BA	BC	CB	CD	DC
Distribution Factors	0	0.4	0.6	0.53	0.47	1.0
Fixed-End Moments	− 8.89	+ 4.44	− 53.33	+ 53.33	− 70.34	zero
Balance Moment		+ 19.56	+ 29.33			
Carry-over to Remote Ends	+ 9.78			+ 14.67		
Balance Moment				+ 1.27	+ 1.12	Note: No carry-over to the pinned end
Carry-over to Remote Ends			+ 0.64			

If the total moments at each internal support are now calculated they are:

$M_{BA} = (+ 4.44 + 19.56) = + 24.0$ kNm } The difference = 0.64 kNm   i.e.
$M_{BC} = (− 53.33 + 29.33 + 0.64) = − 23.36$ kNm } the value of the carry-over moment
$M_{CB} = (+ 53.33 + 14.67 + 1.27) = + 69.27$ kNm }
$M_{CD} = (− 70.34 + 1.12) = − 69.27$ kNm }   The difference = 0

It is evident that after one iteration of each support moment the true values are nearer to 23.8 kNm and 69.0 kNm for B and C respectively. The existing out-of-balance moments which still exist, 0.64 kNm, can be distributed in the same manner as during the first iteration. This process is carried out until the desired level of accuracy has been achieved, normally after three or four iterations.

A slight modification to carrying out the distribution process which still results in the same answers is to carry out the balancing operation for **all** supports simultaneously and the carry-over operation likewise. This is quicker and requires less work. The reader should complete a further three/four iterations to the solution given above and compare the results with those shown in Figure 1.72.

Joints/Support	A		B			C			D
Member	AB		BA	BC		CB	CD		DC
Distribution Factors	**0**		**0.4**	**0.6**		**0.53**	**0.47**		**1.0**
Fixed-End Moments	– 8.89		+ 4.44	– 53.33		+ 53.33	– 70.34		zero
Balance Moment			+ 19.56	+ 29.33		+ 9.01	+ 7.99		
Carry-over to Remote Ends	+ 9.78			+ 4.50		+ 14.67			
Balance Moment			– 1.80	– 2.70		– 7.78	– 6.89		
Carry-over to Remote Ends	–0.91			– 3.89		– 1.35			
Balance Moment	+ 0.78	carry-over*	+ 1.56	+ 2.33		+ 0.72	+ 0.63		
Total	**+ 0.76**		**+ 23.76**	**– 23.76**		**+ 68.60**	**– 68.61**		**zero**

* The final carry-over, to the fixed support only, means that this value is one iteration more accurate than the internal joints.

Figure 1.72

The continuity moments are shown in Figure 1.73.

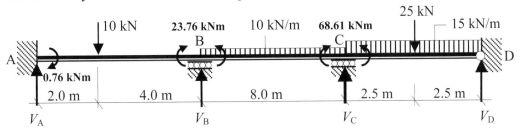

Figure 1.73

The support reactions and the bending moment diagrams for each span can be calculated using superposition as before by considering each span separately.

(i) *Fixed Support Reactions*

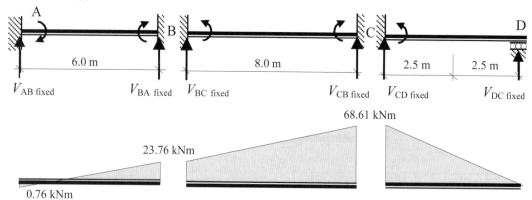

Consider span AB:

+ve $\circlearrowright$ $\Sigma M_A = 0$

$+ 0.76 + 23.76 - (6.0 \times V_{BA \text{ fixed}}) = 0$ Equation (1)

$\therefore V_{BA \text{ fixed}} = + 4.09 \text{ kN} \uparrow$

Consider the vertical equilibrium of the beam:

+ve $\uparrow$ $\Sigma F_y = 0$

$+ V_{AB \text{ fixed}} + V_{BA \text{ fixed}} = 0 \therefore V_{AB \text{ fixed}} = - 4.09 \text{ kN} \downarrow$ Equation (2)

Consider span BC:

+ve $\circlearrowright$ $\Sigma M_B = 0$

$-23.76 + 68.61 - (8.0 \times V_{CB \text{ fixed}}) = 0$ Equation (1)

$\therefore V_{CB \text{ fixed}} = + 5.61 \text{ kN} \uparrow$

Consider the vertical equilibrium of the beam:

+ve $\uparrow$ $\Sigma F_y = 0$

$+ V_{BC \text{ fixed}} + V_{CB \text{ fixed}} = 0 \therefore V_{BC \text{ fixed}} = - 5.61 \text{ kN} \downarrow$ Equation (2)

Consider span CD:

+ve $\circlearrowright$ $\Sigma M_C = 0$

$- 68.61 - (5.0 \times V_{DC \text{ fixed}}) = 0$ Equation (1)

$\therefore V_{DC \text{ fixed}} = - 13.72 \text{ kN} \downarrow$

Consider the vertical equilibrium of the beam:

+ve $\uparrow$ $\Sigma F_y = 0$

$+ V_{CD \text{ fixed}} + V_{DC \text{ fixed}} = 0 \therefore V_{CD \text{ fixed}} = + 13.72 \text{ kN} \uparrow$ Equation (2)

Fixed vertical reactions

The total vertical reaction at each support due to the continuity moments is equal to the algebraic sum of the contributions from each beam at the support.

$$
\begin{aligned}
V_{\text{A fixed}} &= V_{\text{AB fixed}} && = -4.09 \text{ kN} \\
V_{\text{B fixed}} &= V_{\text{BA fixed}} + V_{\text{BC fixed}} &&= (+4.09 - 5.61) &&= -1.52 \text{ kN} \\
V_{\text{C fixed}} &= V_{\text{CB fixed}} + V_{\text{CD fixed}} &&= (+5.61 + 13.72) &&= +19.33 \text{ kN} \\
V_{\text{D fixed}} &= V_{\text{DC fixed}} && = -13.72 \text{ kN}
\end{aligned}
$$

(ii) *Free Support Reactions*

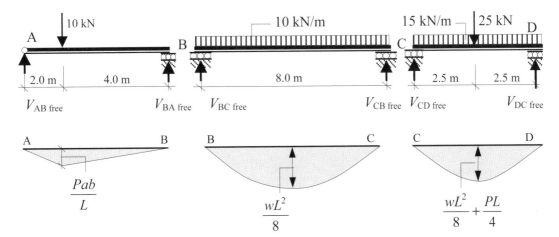

Free bending moments

Span AB $\dfrac{Pab}{L} = \dfrac{10 \times 2 \times 4}{6} = 13.3 \text{ kNm}$

Span BC $\dfrac{wL^2}{8} = \dfrac{10 \times 8^2}{8} = 80.0 \text{ kNm}$

Span CD $\left[\dfrac{wL^2}{8} + \dfrac{PL}{4}\right] = \left[\dfrac{15 \times 5^2}{8} + \dfrac{25 \times 5}{4}\right] = 78.13 \text{ kNm}$

Consider span AB:

$+ve \circlearrowright \Sigma M_A = 0$

$+(10 \times 2.0) - (6.0 \times V_{\text{BA free}}) = 0 \qquad \therefore V_{\text{BA free}} = +3.33 \text{ kN} \uparrow \quad$ Equation (1)

Consider the vertical equilibrium of the beam:

$+ve \uparrow \Sigma F_y = 0$

$+ V_{\text{BA free}} + V_{\text{BA free}} - 10.0 = 0 \qquad \therefore V_{\text{AB free}} = +6.67 \text{ kN} \uparrow \quad$ Equation (2)

Consider span BC:

$$+ve \, \curvearrowright \, \Sigma M_B = 0$$

$$+ (10 \times 8.0 \times 4.0) - (8.0 \times V_{CB \, free}) = 0 \qquad \therefore V_{CB \, free} \; = \; +40.0 \; kN \; \uparrow \quad Equation \; (1)$$

Consider the vertical equilibrium of the beam:

$$+ve \, \uparrow \, \Sigma F_y = 0$$

$$+ V_{BC \, free} + V_{CB \, free} - (10 \times 8.0) \quad = 0 \qquad \therefore V_{BC \, free} \; = \; +40.0 \; kN \; \uparrow \quad Equation \; (2)$$

Consider span CD:

$$+ve \, \curvearrowright \, \Sigma M_C = 0$$

$$+ (25 \times 2.5) + (15 \times 5.0 \times 2.5) - (5.0 \times V_{DC \, free}) = 0 \qquad \qquad \uparrow$$
$$\therefore V_{DC \, free} \; = \; +50.0 \; kN \qquad Equation \; (1)$$

Consider the vertical equilibrium of the beam:

$$+ve \, \uparrow \, \Sigma F_y = 0$$

$$+ V_{CD \, free} + V_{DC \, free} - 25.0 - (15 \times 5.0) = 0 \quad \therefore V_{CD \, free} \; = \; +50.0 \; kN \; \uparrow \quad Equation \; (2)$$

Free vertical reactions:

$$V_{A \, free} = \; V_{AB \, free} \qquad\qquad\qquad\qquad = \; +6.67 \; kN$$
$$V_{B \, free} = \; V_{BA \, free} + V_{BC \, free} \; = \; (+3.33 + 40.0) \quad = \; +43.33 \; kN$$
$$V_{C \, free} = \; V_{CB \, free} + V_{CD \, free} \; = \; (+40.0 + 50.0) \quad = \; +90.0 \; kN$$
$$V_{D \, free} = \; V_{DC \, free} \qquad\qquad\qquad\qquad = \; +50.0 \; kN$$

The final vertical support reactions are given by (i) + (ii):

$$V_A = \; V_{A \, fixed} + V_{A \, free} \; = \; (-4.09 + 6.67) \quad = \; +2.58 \; kN$$
$$V_B = \; V_{B \, fixed} + V_{B \, free} \; = \; (-1.58 + 43.33) \quad = \; +41.81 \; kN$$
$$V_C = \; V_{C \, fixed} + V_{C \, free} \; = \; (+19.33 + 90.0) \quad = \; +109.33 \; kN$$
$$V_D = \; V_{D \, fixed} + V_{D \, free} \; = \; (-13.72 + 50.0) \quad = \; +36.28 \; kN$$

Check the vertical equilibrium: Total vertical force $\quad = \; +2.58 + 41.81 + 109.33 + 36.28$
$$= \; +190 \; kN \; (= total \; applied \; load)$$

The final bending moment diagram is shown in Figure 1.74.

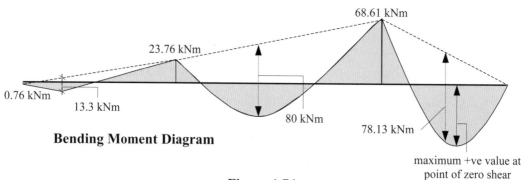

Bending Moment Diagram

Figure 1.74

2. Design Philosophies

Objective: *to provide a resumé of the design philosophies most commonly used when undertaking structural design.*

2.1 Introduction

The successful completion of any structural design project is dependent on many variables, however, there are a number of fundamental objectives which must be incorporated in any design philosophy to provide a structure which throughout its intended lifespan:

  (i)   will possess an acceptable margin of safety against collapse whilst in use,
  (ii)   is serviceable and perform its intended purpose whilst in use,
  (iii)   is sufficiently robust such that damage to an extent disproportionate to the original cause will not occur,
  (iv)   is economic to construct,   and
  (v)   is economic to maintain.

Historically, structural design was carried out on the basis of intuition, trial and error, and experience which enabled empirical *design rules*, generally relating to structure/member proportions, to be established. These rules were used to minimise structural failures and consequently introduced a *margin-of-safety* against collapse. In the latter half of the 19th century the introduction of modern materials and the development of mathematical modelling techniques led to the introduction of a design philosophy which incorporated the concept of *factor-of-safety* based on known material strength, e.g. ultimate tensile stress; this is known as ***permissible stress design***. During the 20th century two further design philosophies were developed and are referred to as ***load-factor*** design and ***limit-state*** design; each of the three philosophies is discussed separately in Sections 2.2 to 2.4.

2.2 Permissible Stress Design

When using permissible stress design, the margin of safety is introduced by considering structural behaviour under working/service load conditions and comparing the stresses under these conditions with permissible values. The permissible values are obtained by dividing the failure stresses by an appropriate factor of safety. The applied stresses are determined using elastic analysis techniques, i.e.

$$stress\ induced\ by\ working\ loads \ \leq \ \frac{failure\ stress}{factor\ of\ safety}$$

2.3 Load Factor Design

When using load factor design, the margin of safety is introduced by considering structural behaviour at collapse load conditions. The ultimate capacities of sections based on yield strength (e.g. axial, bending moment and shear force capacities) are compared with the design effects induced by the ultimate loads. The ultimate loads are determined by multiplying the working/service loads by a factor of safety. Plastic methods of analysis are used to determine section capacities and design load effects. Despite being acceptable, this method has never been widely used.

$$\begin{matrix} \textit{Ultimate design load effects due to} \\ \textit{(working loads} \times \textit{factor of safety)} \end{matrix} \quad \leq \quad \begin{matrix} \textit{Ultimate capacity based on the} \\ \textit{failure stress of the material} \end{matrix}$$

2.4   Limit State Design

The limit state design philosophy, which was formulated for reinforced concrete design in Russia during the 1930s, achieves the objectives set out in Section 2.1 by considering two 'types' of limit state under which a structure may become unfit for its intended purpose. They are:

1.   the ***Serviceability Limit State*** in which a condition, e.g. deflection, vibration or cracking, occurs to an extent, which is unacceptable to the owner, occupier, client etc. and
2.   the ***Ultimate Limit State*** in which the structure, or some part of it, is unsafe for its intended purpose, e.g. compressive, tensile, shear or flexural failure or instability leading to partial or total collapse.

The basis of the approach is statistical and lies in assessing the probability of reaching a given limit state and deciding upon an acceptable level of that probability for design purposes. The method in most codes is based on the use of ***characteristic values*** and ***partial safety factors***.

Partial Safety Factors: The use of partial safety factors, which are applied separately to individual parameters, enables the degree of risk for each one to be varied, reflecting the differing degrees of control which are possible in the manufacturing process of building structural materials/units (e.g. steel, concrete, timber, mortar and individual bricks) and construction processes such as steel fabrication, in-situ/pre-cast concrete, or building in masonry.

Characteristic Values: The use of characteristic values enables the statistical variability of various parameters such as material strength, different load types etc. to be incorporated in an assessment of the acceptable probability that the value of the parameter will be exceeded during the life of a structure. The term 'characteristic' in current design codes normally refers to a value of such magnitude that statistically only a 5% probability exists of its being exceeded.

In the design process the characteristic loads are multiplied by the partial safety factors to obtain the design values of design effects such as axial or flexural stress, and the design strengths are obtained by dividing the characteristic strengths by appropriate partial safety factors for materials. To ensure an adequate margin of safety the following must be satisfied:

Design strength $\geq$ **Design load effects**

e.g. $\quad \dfrac{f_k}{\gamma_m} \geq [(\text{stress due to } G_k \times \gamma_{f\,\text{dead}}) + (\text{stress due to } Q_k \times \gamma_{f\,\text{imposed}})]$

where:

f_k	is the characteristic compressive strength,
γ_m	is the partial safety factor for materials,
G_k	is the characteristic dead load,
Q_k	is the characteristic imposed load,
$\gamma_{f\,\text{dead}}$	is the partial safety factor for dead loads,
$\gamma_{f\,\text{imposed}}$	is the partial safety factor for imposed loads.

The limit state philosophy can be expressed with reference to frequency distribution curves for design strengths and design effects, as shown in Figure 2.1.

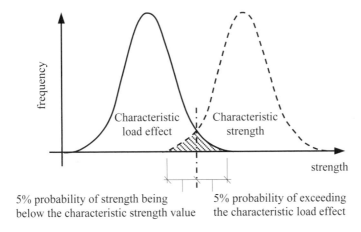

Figure 2.1

The shaded area represents the probability of failure, i.e. the level of design load effect which can be expected to be exceeded by 5% and the level of design strength which 5% of samples can be expected to fall below. The point of intersection of these two distribution curves represents the ultimate limit state, i.e. the design strength equals the design load effects.

The partial safety factors represent the uncertainty in the characteristic values. The lack of detailed statistical data on all of the parameters considered in design and the complexity of the statistical analysis has resulted in the use of a more subjective assessment of the values of partial safety factors than is mathematically consistent with the philosophy.

2.5 *Design Codes*

The design philosophies outlined above are reflected in structural design standards and codes of practice which are used by designers in producing safe and economic structures. The design 'rules' and guidance given within them are specific to individual materials, e.g.

- ◆ **BS 8110** – **Structural use of concrete,**
- ◆ **BS 5950** – **Structural use of steelwork,**
- ◆ **BS 5268** – **Structural use of timber,**
- ◆ **BS 5628** – **Structural use of masonry,**

and are based on material characteristics such as the stress–strain relationship, the modulus of elasticity, Poisson's ratio and the inherent variability both within the manufacture of the materials and the processes adopted during construction.

Currently the structural **timber** design code is a **permissible stress** design code and those for concrete, steelwork and masonry are based on the 'limit state' design philosophy.

The following values for the partial safety factors (γ_f) applied to loads are given in most limit state design codes used in the UK.

(a) *Dead and imposed load*
design dead load $= 0.9\,G_k$ or $1.4\,G_k$
design imposed load $= 1.6Q_k$
design earth and
water load $= 1.4\,E_n$

(b) *Dead and wind load*
design dead load $= 0.9\,G_k$ or $1.4\,G_k$
design wind load $= 1.4\,W_k$ or $0.015G_k$
                                 whichever is the larger
design earth and
water load $= 1.4\,E_n$

(c) *Dead, imposed and wind load*
design dead load $= 1.2\,G_k$
design imposed load $= 1.2\,Q_k$
design wind load $= 1.2\,W_k$ or $0.015G_k$ whichever is the larger
design earth and water load $= 1.2\,E_n$

where:
G_k is the characteristic dead load,
Q_k is the characteristic imposed load,
W_k is the characteristic wind load,
E_n is the earth load as described in 'Earth retaining structures', Civil Engineering Code of Practice No.2.

Note: Upper case letters (e.g. G_k) are normally used for concentrated loads and lower case letters (e.g. g_k) for distributed loads.

The partial safety factors relating to material properties (γ_m) are also given in the relevant codes and considered in Chapters 5, 6, 7 and 8 for reinforced concrete, steelwork, timber and masonry respectively.

2.6 Eurocodes

The European Standards Organisation, CEN, is the umbrella organisation under which a set of common structural design standards (EC1, EC2, EC3, etc.) are being developed. The Structural Eurocodes are the result of attempts to eliminate barriers to trade throughout the European Union. Separate codes exist for each structural material, as indicated in Table 2.1. The basis of design and loading considerations are included in EC1.

Each country publishes its own European Standards (EN), e.g. in the UK the British Standards Institution (BSI) issues documents (which are based on the Eurocodes developed under CEN), with the designation BS EN.

Structural Eurocodes are currently issued as Pre-standards (ENV) which can be used as an alternative to existing national rules (with the exception of EN 1990 and EN 1991-1-1). In the UK the BSI has used the designation DD ENV; the pre-standards are equivalent to the traditional 'Draft for development' Documents.

CEN Number	Title of Eurocode
EN 1990	*Eurocode: Basis of structural design*
ENV 1991	*Eurocode 1: Actions on structures*
ENV 1992	*Eurocode 2: Design of concrete structures*
ENV 1993	*Eurocode 3: Design of steel structures*
ENV 1994	*Eurocode 4: Design of composite steel and concrete structures*
ENV 1995	*Eurocode 5: Design of timber structures*
ENV 1996	*Eurocode 6: Design of masonry structures*
ENV 1997	*Eurocode 7: Geotechnical design*
ENV 1998	*Eurocode 8: Design of structures for earthquake resistance*
ENV 1999	*Eurocode 9: Design of aluminium structures*

Table 2.1

After approval of an EN by CEN the Standard reaches the **Date of Availability** (DAV). During the two years following this date, National Calibration is expected to be carried out in each country to determine the NDPs **Nationally Determined Parameters** (previously referred to as 'boxed values'). The NDPs represent the safety factors to be adopted when using the EN, and setting their values remains the prerogative for each individual country.

Following this will be **Co-Existence Period** for a maximum of three years during which each country will withdraw all of the existing National Codes which have a similar scope to the EC being introduced.

2.6.1 National Annex

Each country which issues a European Standard also issues a **National Annex** currently published in the UK as a 'NAD' – National Application Document, for use with the EN. There is no legal requirement for a country to produce a National Annex; however, it is recommended to do so. The purpose of the National Annex is to provide information to designers relating to product standards for materials, partial safety factors and any additional rules and/or supplementary information specific to design within that country. (**Note:** The contents of the core document of an EC cannot be changed to include the NDPs being adopted: these should be provided if required in a separate document.)

2.6.2 Normative and Informative

There are two types of information contained within the core documents: **normative** and **informative**. Normative Annexes have the same status as the main body of the text whilst Informative Annexes provide additional information. The Annexes generally contain more detailed material or material which is used less frequently.

2.6.3 Terminology, Symbols and Conventions

The terminology, symbols and conventions used in EC6: Part 1.1 differ from those used by current British Standards. The code indicates **Principles** which are general statements and definitions which must be satisfied, and **Rules** which are recommended design procedures which comply with the Principles. The Rules can be substituted by alternative procedures provided that these can be shown to be in accordance with the Principles.

2.6.3.1 Decimal Point

Standard ISO practice has been adopted in representing a decimal point by a comma, i.e. $5,3 \equiv 5.3$.

2.6.3.2 Symbols and Subscripts

As in British Standards there are numerous symbols[*] and subscripts used in the codes. They are too numerous to include here, but some of the most frequently used ones are given here for illustration purposes:

F: A*ction,* A force (load) applied to a structure or an imposed deformation (indirect action), such as temperature effects or settlement

$\quad$ **G**: permanent action such as dead loads due to self-weight, e.g.
$\quad$ Characteristic value of a permanent action $\quad = G_k$
$\quad$ Design value of a permanent action $\quad\quad = G_d$
$\quad$ Lower design value of a permanent action $\quad = G_{d,inf}$
$\quad$ Upper design value of a permanent action $\quad = G_{d,sup}$

$\quad$ **Q**: variable actions such as imposed, wind or snow loads,
$\quad$ Characteristic value of a variable action $\quad = Q_k$
$\quad$ Design value of a variable action $\quad\quad = Q_d$

$\quad$ **A**: accidental actions such as explosions, fire or vehicle impact.

$\quad$ **E**: *effect of actions* on static equilibrium or of gross displacements etc., e.g.
$\quad$ Design effect of a destabilising action $\quad = E_{d,dst}$
$\quad$ Design effect of a stabilising action $\quad = E_{d,stb}$

$\quad$ **Note:** $\quad E_{d,dst} \leq E_{d,stb}$

---

[*] In most cases the Eurocode does not use italics for variables.

R: *Design resistance* of structural elements, e.g.

Design vertical load resistance of a wall $\quad = N_{Rd} \quad = \dfrac{\Phi_{i,m} t f_k}{\gamma_M}$

Design shear resistance of a wall $\quad\quad = V_{Rd} \quad = \dfrac{f_{vk} t l_c}{\gamma_M}$

Design moment of resistance of a section $= M_{Rd} \quad = \dfrac{A_s f_{yk} z}{\gamma_s}$

S: *Design value of actions* Factored values of externally applied loads or load effects such as axial load, shear force, bending moment etc., e.g.

Design vertical load $\qquad\qquad N_{Sd} \quad \leq \quad N_{Rd}$
Design shear force $\qquad\qquad\quad V_{Sd} \quad \leq \quad V_{Rd}$
Design bending moment $\qquad\quad M_{Sd} \quad \leq \quad M_{Rd}$

X: *Material property* Physical properties such as tension, compression, shear and bending strength, modulus of elasticity etc., e.g.

Characteristic compressive strength of masonry $= X_k \quad = f_k = K f_b^{0,65}$ N/mm^2

Design compressive strength of masonry $\quad = X_d = \dfrac{X_k}{\gamma_M} \; = f_d = \dfrac{K f_b^{0,65}}{\gamma_M}$ N/mm^2

2.6.4 Limit State Design

The limit states are states beyond which a structure can no longer satisfy the design performance requirements (see Section 2.4). The two classes of limit state are:

♦ *Ultimate limit states:* These include failures such as full or partial collapse due to e.g. rupture of materials, excessive deformations, loss of equilibrium or development of mechanisms. Limit states of this type present a direct risk to the safety of individuals.

♦ *Serviceability limit states:* Whilst not resulting in a direct risk to the safety of people, serviceability limit states still render the structure unsuitable for its intended purpose. They include failures such as excessive deformation resulting in unacceptable appearance or non-structural damage, loss of durability or excessive vibration causing discomfort to the occupants.

The limit states are quantified in terms of design values for actions, material properties and geometric characteristics in any particular design. Essentially the following conditions must be satisfied:

Ultimate limit state:

Rupture $\qquad\qquad S_d \; \leq \; R_d$

where:
S_d is the design value of the effects of the actions imposed on the structure/structural elements,
R_d is the design resistance of the structure/structural elements to the imposed actions.

Stability $S_{d,dst} \leq R_{d,stb}$

where:
$S_{d,dst}$ is the design value of the destabilising effects of the actions imposed on the structure (including self-weight where appropriate).
$S_{d,stb}$ is the design value of the stabilising effects of the actions imposed on the structure (including self-weight where appropriate).

Serviceability limit state:

Serviceability $S_d \leq C_d$

where:
S_d is the design value of the effects of the actions imposed on the structure/structural elements,
C_d is a prescribed value, e.g. a limit of deflection.

2.6.5 Design Values

The term *design* is used for factored loading and member resistance

Design loading (F_d) = partial safety factor $(\gamma_F) \times$ characteristic value (F_k)

e.g. $G_d = \gamma_G G_k$

where:
γ_G is the partial safety factor for permanent actions,
G_k is the characteristic value of the permanent actions.

Note: $G_{d,sup}$ $(= \gamma_{G,sup}G_{k,sup}$ or $\gamma_{G,sup}G_k)$ represents the 'upper' design value of a permanent action,
$G_{d,inf}$ $(= \gamma_{G,inf}G_{k,inf}$ or $\gamma_{G,inf}G_k)$ represents the 'lower' design value of a permanent action.

Design resistance (R_d) $= \dfrac{\text{material characteristic strength } (X_k)}{\text{material partial safety factor } (\gamma_m)}$

e.g.   ***Design flexural strength*** $= f_{x,d} = \dfrac{f_{x,k}}{\gamma_m}$

The design values of the actions vary depending upon the limit state being considered. All of the possible load cases should be considered in different combinations as given in the codes, e.g. for persistent and transient design situations:

$$F_d = \Sigma\gamma_{G,j}\, G_{k,j} + \gamma_{Q,1}Q_{k,1} + \Sigma\gamma_{Q,i}\, \psi_{0,i}\, Q_{k,i} \qquad\qquad \text{Equation (1)}$$
$$i > 1$$

where:
$\gamma_{G,j}$ partial safety factor for permanent actions,   (Table 1 of the NAD)

$G_{k,j}$ characteristic values of permanent actions,

$\gamma_{Q,1}$ partial safety factor for *'one'* of the variable actions, (Table 1 of the NAD)

$Q_{k,1}$ characteristic value of *'one'* of the variable actions,

$\gamma_{Q,i}$ partial safety factor for the other variable actions, (Table 1 of the NAD)

$\psi_{0,i}$ combination factor which is applied to the characteristic value Q_k of an action not being considered as $Q_{k,1}$, (Eurocode 1 and NAD – Table 2)

$Q_{k,i}$ characteristic value of *'other'* variable actions.

Extract from NAD Table 1:

Table 1. Partial safety factors (γ factors)						
Reference in ENV 1996-1-1	**Definition**	**Symbol**	**Condition**	**Value**		
					Boxed ENV 1996-1-1	**UK**
2.3.3.1	Partial factors for variable actions	γ_A	Accidental		1,0	1,0
		$\gamma_{F,inf}$	Favourable		0,0	0,0
		γ_Q	Unfavourable		1,5	1,5
		γ_Q	Reduced Favourable		0,0	0,0
		γ_Q	Reduced Unfavourable		1,35	1,35
2.3.3.1	Partial factors for permanent actions	γ_{GA}	Accidental		1,0	1,0
		γ_G	Favourable		1,0	1,0
		γ_G	Unfavourable		1,35	1,35
		$\gamma_{G,inf}$	Favourable		0,9	0,9
		$\gamma_{G,sup}$	Unfavourable		1,1	1,1
		γ_p	Favourable		0,9	0,9
		γ_p	Unfavourable		1,2	1,2

Figure 2.2

Extract from NAD Table 2:

Table 2. Combination factors (ψ factors)				
Variable action	**Building type**	ψ_0	ψ_1	ψ_2
Imposed floor loads	Dwellings	0,5	0,4	0,2
	Other occupancy classes[1]	0,7	0,6	0,3
	Parking	0,7	0,7	0,6
Imposed roof loads	All occupancy classes[1]	0,7	0,2	0,0
Wind loads	All occupancy classes[1]	0,7	0,2	0,0
[1] As listed and defined in Table 1 of BS 6399 : Part 1 : 1984				

Figure 2.3

2.6.5.1 *Partial Safety Factors*

The Eurocode provides indicative values for various safety factors: these are shown in the text as 'boxed values' e.g. $\boxed{1,35}$. Each country defines NDPs – Nationally Determined Parameters within the National Annex document to reflect the levels of safety required by the appropriate authority of the national government; in the UK this is the British Standards Institution.

The boxed values of partial safety factors for actions in building structures for persistent and transient design situations are given in Table 2.2 of EC6: Part 1.1 and in Figure 2.4 of this text (these values are also given in Table 1 of the NAD).

Extract from EC6: Table 2.2

Table 2.2 Partial safety factors for actions in building structures for persistent and transient design situations.

	Permanent actions (γ_G) (see note)	Variable actions (γ_Q)		Prestressing (γ_p)
		One with its characteristic value	Others with their combination value	
Favourable effect	$\boxed{1,0}$	$\boxed{0}$	$\boxed{0}$	$\boxed{0,9}$
Unfavourable effect	$\boxed{1,35}$	$\boxed{1,5}$	$\boxed{1,35}$	$\boxed{1,2}$

Note: See also paragraph 2.3.3.1 (3)

Figure 2.4

Consider a design situation in which there are two characteristic dead loads, G_1 and G_2, in addition to three characteristic imposed loads, Q_1, Q_2 and Q_3. Assume the partial safety factors and combination factor are $\gamma_{G,j} = 1,35,$ $\gamma_{Q,1} = 1,5$ $\gamma_{Q,i} = 1,5$ $\psi_{0,i} = 0,7.$

Combination 1: $F_d = (1,35G_1 + 1,35G_2) + 1,5\,Q_1 + (1,5 \times 0,7 \times Q_2) + (1,5 \times 0,7 \times Q_3)$
　　　　　　　　$\mathbf{F_d = 1,35(G_1 + G_2) + 1,5Q_1 + 1,05(Q_2 + Q_3)}$
Combination 2: $\mathbf{F_d = 1,35(G_1 + G_2) + 1,5Q_2 + 1,05(Q_1 + Q_3)}$
Combination 3: $\mathbf{F_d = 1,35(G_1 + G_2) + 1,5Q_3 + 1,05(Q_1 + Q_2)}$

When developing these combinations permanent effects are represented by their upper design values, i.e. $G_{d,sup} = \gamma_{G,sup}\,G_{k,sup}$ or $\gamma_{G,sup}\,G_k$

Those which decrease the effect of the variable actions (i.e. favourable effects) are replaced by their lower design values, i.e. $G_{d,inf} = \gamma_{G,inf}\,G_{k,inf}$ or $\gamma_{G,inf}\,G_k$

In most situations either the upper or lower design values are applied throughout the structure; specifically in the case of continuous beams, the same design value of self-weight is applied on all spans. A similar approach is used when dealing with accidental actions.

Two simplified expressions using the Table 2.2 values to replace Equation (1) are:

(i)   considering the most unfavourable variable action

$$F_d = \Sigma \gamma_{G,j}\, G_{k,j} + 1{,}5\, Q_{k,1} \qquad\qquad \text{Equation (2)}$$

considering all unfavourable variable actions

$$F_d = \Sigma \gamma_{G,j}\, G_{k,j} + 1{,}35\, \Sigma Q_{k,i} \qquad\qquad \text{Equation (3)}$$
$$\phantom{F_d = \Sigma \gamma_{G,j}\, G_{k,j} + 1{,}35\, \Sigma Q}{}_{i>1}$$

whichever gives the larger value.

2.6.6 Conventions

The difference in conventions most likely to cause confusion with UK engineers is the change in the symbols used to designate the major and minor axes of a cross-section. Traditionally in the UK the **y-y axis** has represented the minor axis; in EC6 this represents the **MAJOR axis** and the minor axis is represented by the **z-z** axis. The **x-x axis** defines the **LONGITUDINAL axis**. All three axes are shown in Figure 2.5.

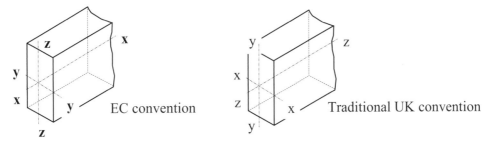

Figure 2.5

A summary of the abbreviations used in relation to Eurocodes is given in Table 1.2.

Abbreviation	Meaning
CEN	European Standards Organisation
EC	Eurocode produced by CEN
EN	European Standard based on Eurocode and issued by member countries
ENV	Pre-standard of Eurocode issued by member countries
DD ENV	UK version of Pre-standard (BSI)
National Annex (NAD)	National Application Document issued by member countries (BSI)
NSB	National Standards Bodies
DAV	Date of Availability
NDP	Nationally Determined Parameter (contained with National Annexes)

Table 2.2

3. Structural Loading

> **Objective:** to introduce the principal forms of structural loading and their distribution in one-way and two-way spanning floor slabs.

3.1 Introduction

All structures are subjected to loading from various sources. The main categories of loading are: dead, imposed and wind loads. In some circumstances there may be other loading types which should be considered, such as settlement, fatigue, temperature effects, dynamic loading, or impact effects (e.g. when designing bridge decks, crane-gantry girders or maritime structures). In the majority of cases design considering combinations of dead, imposed and wind loads is the most appropriate.

The definition of dead loading is given in BS 648:1964, that of imposed floor loading in BS 6399-1:1996, that of wind loading in BS 6399-2:1997, and that for imposed roof loads in BS 6399-3:1988.

3.1.1 Dead Loads: BS 648:1964

Dead loads are loads which are due to the effects of gravity, i.e. the self-weight of all permanent construction such as beams, columns, floors, walls, roofs and finishes.

If the position of permanent partition walls is known, their weight can be assessed and included in the dead load. In speculative developments, internal partitions are regarded as imposed loading.

3.1.2 Imposed Loads: BS 6399-1:1996 (Clauses 5.0 and 6.0)

Imposed loads are loads which are due to variable effects such as the movement of people, furniture, equipment and traffic. The values adopted are based on observation and measurement and are inherently less accurate than the assessment of dead loads.

In the code, Clause 5.0 and Table 1 define the magnitude of uniformly distributed and concentrated point loads which are recommended for the design of floors, ceilings and their supporting elements. Loadings are considered in the following categories:

A Domestic and residential activities
B Offices and work areas not covered elsewhere
C Areas where people may congregate
D Shopping areas
E Areas susceptible to the accumulation of goods (e.g. warehouses)
F Vehicle and traffic areas (Light)
G Vehicle and traffic areas (Heavy)

Most floor systems are capable of lateral distribution of loading and the recommended concentrated load need not be considered. In situations where lateral distribution is not possible, the effects of the concentrated loads should be considered with the load applied at locations which will induce the most adverse effect, e.g. maximum bending moment, shear and deflection. In addition, local effects such as crushing and punching should be considered where appropriate.

In multi-storey structures it is very unlikely that all floors will be required to carry the full imposed load at the same time. Statistically it is acceptable to reduce the total floor loads carried by a supporting member by varying amounts depending on the number of floors or floor area carried. This is reflected in Clause 6.2 and Tables 2 and 3 of BS 6399: Part 1 in which a percentage reduction in the total distributed imposed floor loads is recommended when designing columns, piers, walls, beams and foundations. Parapet, barrier and balustrade loads are given in Table 4 of the code.

3.1.3 Imposed Roof Loads: BS 6399-3:1988

Imposed loading caused by snow is included in the values given in this part of the code, which relates to imposed roof loads. Flat roofs, sloping roofs and curved roofs are also considered.

3.1.4 Wind Loads: BS 6399-2:1997

Environmental loading such as wind loading is clearly variable and its source is outwith human control. In most structures the dynamic effects of wind loading are small, and static methods of analysis are adopted. The nature of such loading dictates that a statistical approach is the most appropriate in order to quantify the magnitudes and directions of the related design loads. The main features which influence the wind loading imposed on a structure are:

- geographical location   – London, Edinburgh, Inverness, Chester, ...
- physical location   – city centre, small town, open country, ...
- topography   – exposed hill top, escarpment, valley floor, ...
- altitude   – height above mean sea level
- building shape   – square, rectangular, cruciform, irregular, ...
- roof pitch   – shallow, steep, mono-pitch, duo-pitch, multi-bay…
- building dimensions
- wind speed and direction
- wind gust peak factor.

3.2 Floor Load Distribution

Tabulated procedures enable these features to be evaluated and hence to produce a system of equivalent static forces which can be used in the analysis and design of a structure. The application of the load types discussed in Sections 3.1.1 to 3.1.4 to structural beams and frames results in axial loads, shear forces, bending moments and deformations being induced in the floor/roof slabs, beams, columns and other structural elements which comprise a structure. The primary objective of structural analysis is to determine the distribution of internal moments and forces throughout a structure such that they are in equilibrium with the applied design loads.

As indicated in Chapter 1, there are a number of manual mathematical models (computer-based models are also available) which can be used to idealise structural behaviour. These methods include: two- and three-dimensional elastic behaviour, elastic behaviour considering a redistribution of moments, plastic behaviour and non-linear behaviour. Detailed explanations of these techniques can be found in the numerous structural analysis text books which are available.

In braced structures (see Chapter 4, Section 4.5) where floor slabs and beams are considered to be simply supported, vertical loads give rise to three basic types of beam loading condition:

(i)   uniformly distributed line loads,
(ii) triangular and trapezoidal loads,
(iii) concentrated point loads.

These load types are illustrated in Examples 3.1 to 3.6 (self-weights have been ignored).

3.3   *Example 3.1:   Load Distribution – One-way Spanning Slabs*

Consider the floor plan shown in Figure 3.1(a) where two one-way spanning slabs are supported on three beams AB, CD and EF. Both slabs are assumed to be carrying a uniformly distributed design load of 5 kN/m^2.

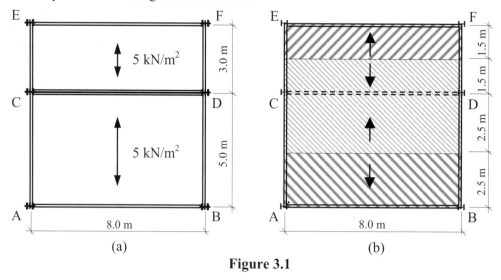

Figure 3.1

Both slabs have continuous contact with the top flanges of their supporting beams and span in the directions indicated. The floor area supported by each beam is indicated in Figure 3.1(b).

Beam AB: Total load = (floor area supported × magnitude of distributed load/m^2)
= $(2.5 \times 8.0) \times (5.0)$ =   100 kN

Beam CD: Total load = $(4.0 \times 8.0) \times (5.0)$ =   160 kN

Beam EF: Total load = $(1.5 \times 8.0) \times (5.0)$ =   60 kN
Check: Total load on both slabs = $(8.0 \times 8.0 \times 5.0)$ = 320 kN

3.4 *Example 3.2: Load Distribution – Two-way Spanning Slabs*

Consider the same floor plan as in Example 3.1 but now with the floor slabs two-way spanning, as shown in Figure 3.2(a).

Since both slabs are two-way spanning, their loads are distributed to supporting beams on all four sides assuming a 45° dispersion as indicated in Figure 3.2(b).

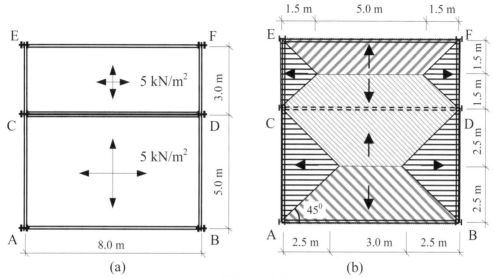

Figure 3.2

Beam AB: Load due to slab ACDB $= \left(\dfrac{8.0+3.0}{2} \times 2.5\right) \times (5.0) \quad = 68.75 \text{ kN}$

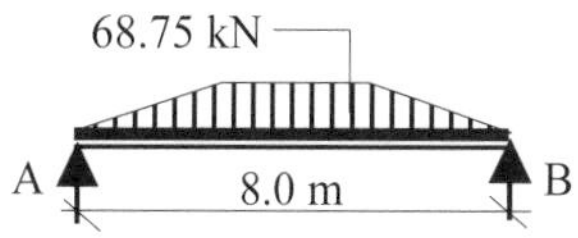

Figure 3.2.1

Beam EF: Load due to slab CEFD $= \left(\dfrac{8.0+5.0}{2} \times 1.5\right) \times (5.0) \quad = 48.75 \text{ kN}$

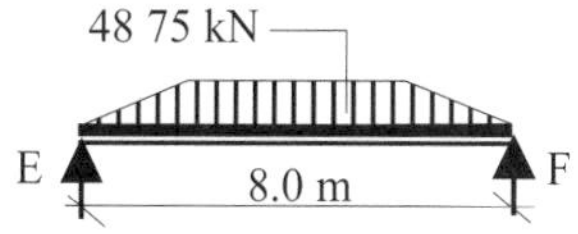

Figure 3.2.2

Beams AC and BD: Load due to slab ACDB $= \left(\dfrac{5.0}{2} \times 2.5\right) \times (5.0) = 31.25 \text{ kN}$

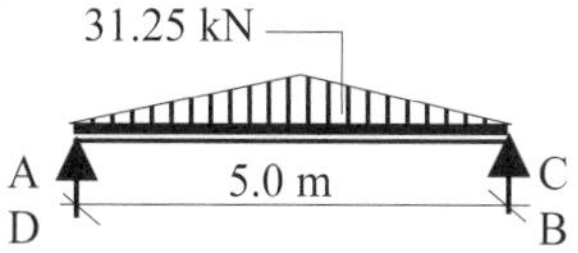

Figure 3.2.3

Beams CE and DF: Load due to slab CEFD $= \left(\dfrac{3.0}{2} \times 1.5\right) \times (5.0) = 11.25 \text{ kN}$

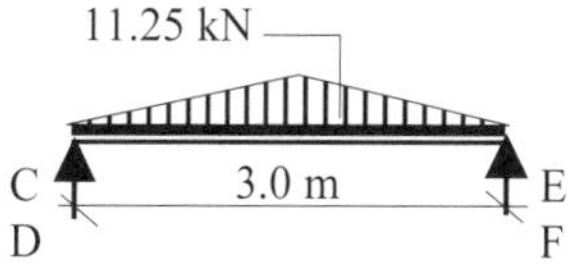

Figure 3.2.4

The loading on beam CD can be considered to be the addition of two separate loads, i.e.
Load due to slab ACDB   =   68.75 kN   (as for beam AB)
Load due to slab CEFD   =   48.75 kN   (as for beam EF)
Note: Both loads are trapezoidal, but they are different.

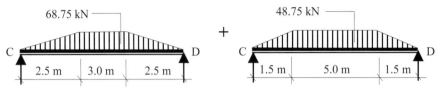

Figure 3.2.5

Check: Total load on all beams =   2(68.75 + 48.75 + 31.25 + 11.25)   =   320 kN

3.5   *Example 3.3: Load Distribution – Secondary Beams*

Consider the same floor plan as in Example 3.2 with the addition of a secondary beam spanning between beams AB and CD as shown in Figure 3.3(a). The load carried by this new beam imposes a concentrated load at the mid-span position of beams CD and AB at the mid-span points G and H respectively.

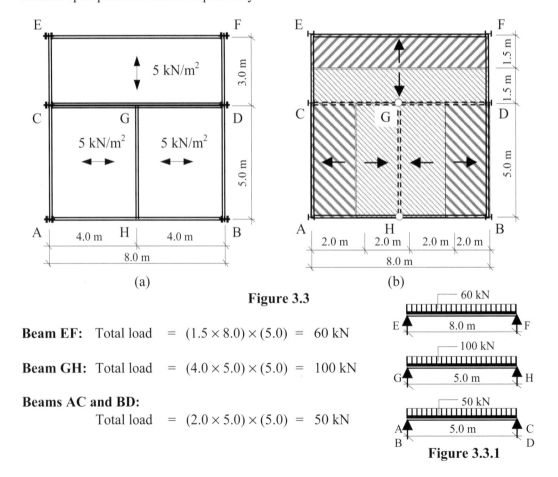

(a)   (b)

Figure 3.3

Beam EF: Total load $= (1.5 \times 8.0) \times (5.0)$ $=$ 60 kN

Beam GH: Total load $= (4.0 \times 5.0) \times (5.0)$ $=$ 100 kN

Beams AC and BD:
          Total load $= (2.0 \times 5.0) \times (5.0)$ $=$ 50 kN

Figure 3.3.1

Beam AB: Total load = End reaction from beam GH = 50 kN

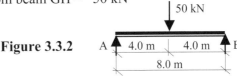

Figure 3.3.2

Beam CD:

The loading on beam CD can be considered to be the addition of two separate loads, i.e.
Load due to slab CEFD = 60 kN   (as for beam EF)
Load due to beam GH = 50 kN   (as for beam AB)

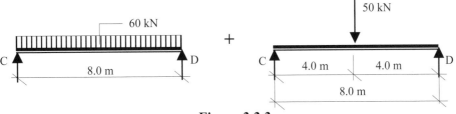

Figure 3.3.3

3.6   Example 3.4: *Combined One-way Slabs, Two-way Slabs and Beams – 1*

Considering the floor plan shown in Figure 3.4(a), with the one-way and two-way spanning slabs indicated, determine the type and magnitude of the loading on each of the supporting beams.

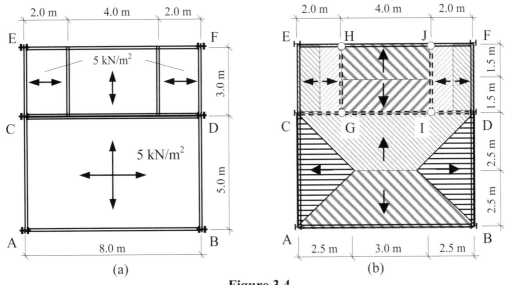

(a)   (b)

Figure 3.4

The loads on beams AB, AC and BD are the same as in Example 3.2.

Beams CE, DF, GH and IJ:
Total load = $(3.0 \times 1.0) \times (5.0)$ = 15 kN

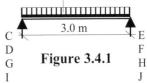

Figure 3.4.1

Beam EF: The loads on EF are due to the end reactions from beams GH and IJ and a distributed load from GHJI.

End reaction from beam GH = 7.5 kN
End reaction from beam JI = 7.5 kN

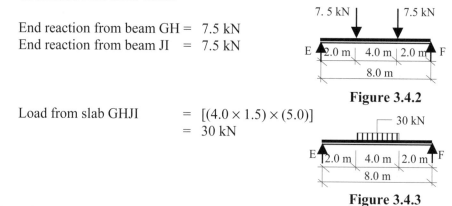

Figure 3.4.2

Load from slab GHJI   = $[(4.0 \times 1.5) \times (5.0)]$
                       = 30 kN

Figure 3.4.3

Total loads on beam EF due to beams GH, JI and slab GHJI:

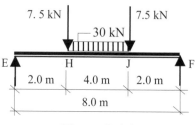

Figure 3.4.4

Beam CD: The loads on CD are due to the end reactions from beams GH and IJ, a distributed load from GHIJ and a trapezoidal load from slab ABCD as in member AB of Example 3.2.

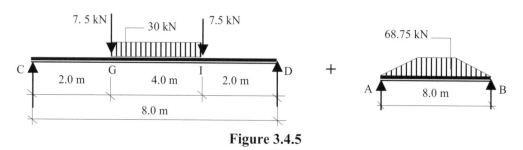

Figure 3.4.5

3.7   *Example 3.5: Combined One-way Slabs, Two-way Slabs and Beams – 2*

The floor plan of an industrial building is shown in Figure 3.5. Using the characteristic dead and imposed loads given, determine:

   (i)   the design loads carried by beams B1 and B2,
   (ii) the maximum shear force and maximum bending moment for beam B1.

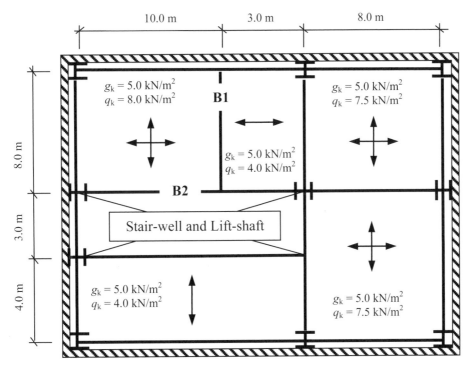

Figure 3.5

Solution
Beam B1

The load on beam B1 is equal to a triangular load from the two-way spanning slab combined with a uniformly distributed load from the one-way spanning slab.

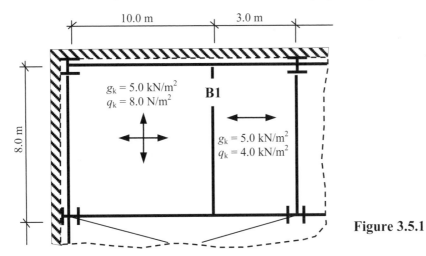

Figure 3.5.1

Design load $= [(1.4 \times g_k) + (1.6 \times q_k)]$

Triangular area $= (4.0 \times 4.0) = 16.0 \text{ m}^2$

Design load $= \{[(1.4 \times 5.0) + (1.6 \times 8.0)] \times 16.0\} = 316.8 \text{ kN}$

Rectangular area   = (1.5×8.0) = 12.0 m^2
Design load   = $\{[(1.4 \times 5.0) + (1.6 \times 4.0)] \times 2.0\}$ = 160.8 kN

Total design load = $(316.8 + 160.8)$ = 477.6 kN

Beam B2
The load on beam B2 is equal to a trapezoidal load from the two-way spanning slab combined with a point load from beam B1.

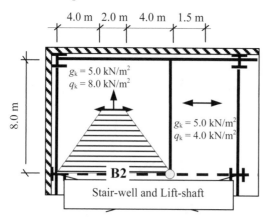

Figure 3.5.2

Trapezoidal area   = $[0.5 \times (2.0 + 10.0) \times 4.0]$ = 24.0 m^2
Design load   = $\{[(1.4 \times 5.0) + (1.6 \times 8.0)] \times 24.0\}$ = 475.2 kN
Beam B1 end reaction = 238.8 kN

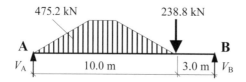

Figure 3.5.3

Beam B1: Shear Force and Bending Moment
Total design load   =   477.6 kN
Since the beam is symmetrical the maximum shear force is equal to the end reaction.

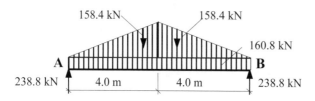

Figure 3.5.4

Maximum shear force   = $(477.6 / 2.0)$ = 238.8 kN
The maximum bending moment occurs at the mid-span.
Design bending moment = $[(238.8 \times 4.0) - (80.4 \times 2.0) - (158.4 \times 4.0/3.0)]$
                       = 583.2 kNm

3.8 Example 3.6: Combined One-way Slabs, Two-way Slabs and Beams – 3

A multi-storey framed building with an octagonal floor plan as indicated is shown in Figure 3.6. Determine the loads carried by, the maximum shear force and maximum bending moment in beam **BE** in a typical floor plan.

Design data:

Characteristic dead loads	$g_k = 5.0 \text{ kN/m}^2$
Characteristic imposed loads	$q_k = 7.5 \text{ kN/m}^2$

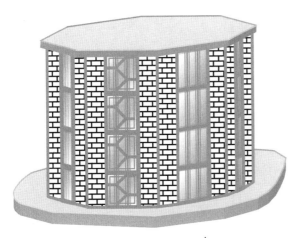

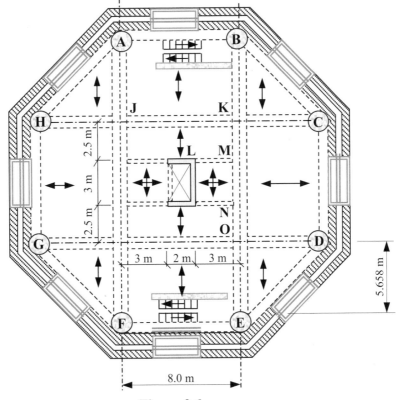

Figure 3.6

Solution

The contribution from each beam and floor slab supported by beam **BE** must be determined separately as indicated in Figure 3.6.1, i.e. W_1 to W_6. Since the structure is symmetrical, only one half need be considered to evaluate the loads and beam end reactions.

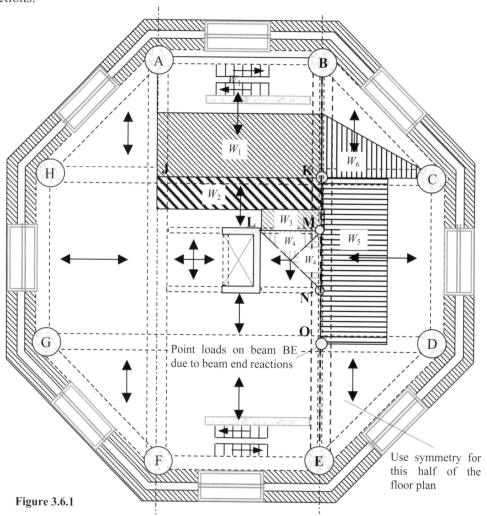

Figure 3.6.1

Point loads on beam BE due to beam end reactions

Use symmetry for this half of the floor plan

Design Floor Load = (Characteristic Load × Partial Safety Factor)

= $[(g_k \times \gamma_f) + (q_k \times \gamma_f)]$

= $[(1.4 \times 5.0) + (1.6 \times 7.5)] = 19.0 \text{ kN/m}^2$

W_1

Area supported = (8.0×2.83) = 22.64 m^2

Floor loading = (19.0×22.64) = 430.2 kN

430.2 kN

J   8.0 m   K   **Figure 3.6.2**

W_2

Area supported = (8.0×1.25) = 10.0 m^2

Floor loading = (19.0×10.0) = 190.0 kN

190 kN

J   8.0 m   K   **Figure 3.6.3**

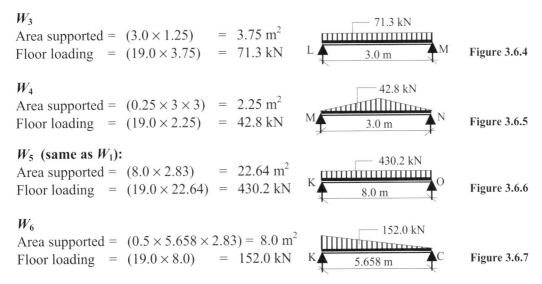

W_3
Area supported = (3.0×1.25) = 3.75 m^2
Floor loading = (19.0×3.75) = 71.3 kN

Figure 3.6.4

W_4
Area supported = $(0.25 \times 3 \times 3)$ = 2.25 m^2
Floor loading = (19.0×2.25) = 42.8 kN

Figure 3.6.5

W_5 (same as W_1):
Area supported = (8.0×2.83) = 22.64 m^2
Floor loading = (19.0×22.64) = 430.2 kN

Figure 3.6.6

W_6
Area supported = $(0.5 \times 5.658 \times 2.83) = 8.0$ m^2
Floor loading = (19.0×8.0) = 152.0 kN

Figure 3.6.7

Beams JK, LM, MN and KO are symmetrical, so their end reactions are the same and equal to ($\frac{1}{2} \times$ the total load). In the case of beam KC the end reaction at K imposes a point load on BE equal to ($^2/_3 \times$ the total load). The total loading on beam BE is as shown in Figure 3.7.

Point load at K = $\left(\dfrac{W_1}{2} + \dfrac{W_2}{2} + \dfrac{2W_6}{3} \right)$ = $[(430.2 + 190)/2 + (2 \times 152)/3]$ = 411.4 kN

Similarly for the point load at position O.

Point load at M = $\left(\dfrac{W_3}{2} + \dfrac{W_4}{2} \right)$ = $(71.3 + 42.8)$ = 57.1 kN

Similarly for the point load at position N.

The uniformly distributed load between points K and O is equal to W_5 = 430.2 kN

The triangular load between points M and N is equal to W_4 = 42.8 kN

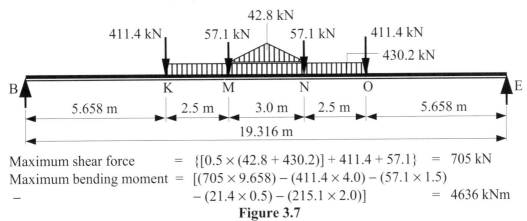

Maximum shear force   = $\{[0.5 \times (42.8 + 430.2)] + 411.4 + 57.1\}$ = 705 kN
Maximum bending moment = $[(705 \times 9.658) - (411.4 \times 4.0) - (57.1 \times 1.5)$
– $- (21.4 \times 0.5) - (215.1 \times 2.0)]$ = 4636 kNm

Figure 3.7

4. Structural Instability, Overall Stability and Robustness

Objective: *to introduce the principles of stability and elastic buckling in relation to overall buckling, local buckling and lateral torsional buckling.*

4.1   Introduction

Structural elements which are subjected to tensile forces are inherently stable and will generally fail when the stress in the cross-section exceeds the ultimate strength of the material. In the case of elements subjected to compressive forces, secondary bending effects caused by, for example, imperfections within materials and/or fabrication processes, inaccurate positioning of loads or asymmetry of the cross-section, can induce premature failure either in a part of the cross-section, such as the outstand flange of an **I** section, or of the element as a whole. In such cases the failure mode is normally buckling (i.e. lateral movement), of which there are three main types:

- ◆ overall buckling,
- ◆ local buckling, and
- ◆ lateral torsional buckling.

Each of these types of buckling is considered in this chapter.[*]

4.2   Overall Buckling

The design of most compressive members is governed by their overall buckling capacity, i.e. the maximum compressive load which can be carried before failure occurs by excessive deflection in the plane of greatest slenderness.

Typically this occurs in columns in building frames and columns in trussed frameworks as shown in Figure 4.1.

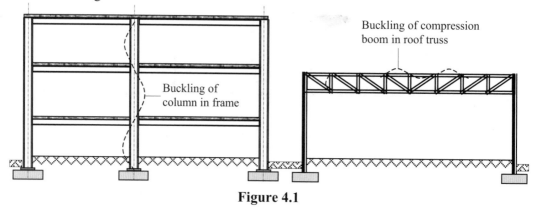

Buckling of compression boom in roof truss

Buckling of column in frame

Figure 4.1

[*] **Note:** torsional buckling (i.e. twisting about the longitudinal axis) is also possible, but is not a common mode of failure.

Compression elements can be considered to be sub-divided into three groups: short elements, slender elements and intermediate elements. Each group is described separately, in Sections 4.2.1, 4.2.2 and 4.2.3 respectively.

4.2.1 Short Elements

Provided that the **slenderness** of an element is **low,** e.g. the length is not greater than ($10 \times$ the least horizontal length), the element will fail by crushing of the material induced by predominantly axial compressive stresses as indicated in Figure 4.2(a). Failure occurs when the stress over the cross-section reaches a yield or crushing value for the material. The failure of such a column can be represented on a stress/slenderness curve as shown in Figure 4.2(b).

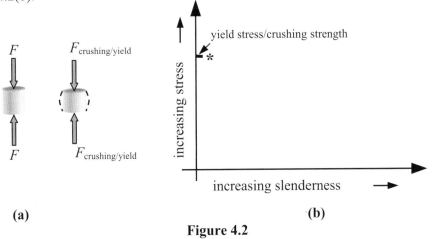

(a)   **(b)**

Figure 4.2

4.2.2 Slender Elements

When the **slenderness** of an element is **high,** the element fails by excessive lateral deflection (i.e. buckling) at a value of stress considerably less than the yield or crushing values as shown in Figures 4.3(a) and (b).

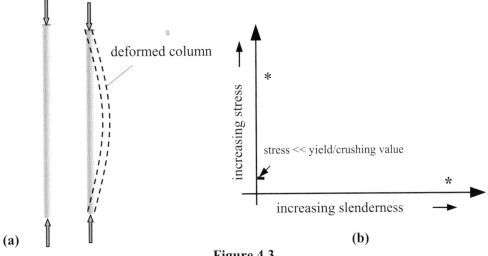

(a)   **(b)**

Figure 4.3

4.2.3 Intermediate Elements

The failure of an element which is neither short nor slender occurs by a combination of buckling and yielding/crushing as shown in Figures 4.4(a) and (b).

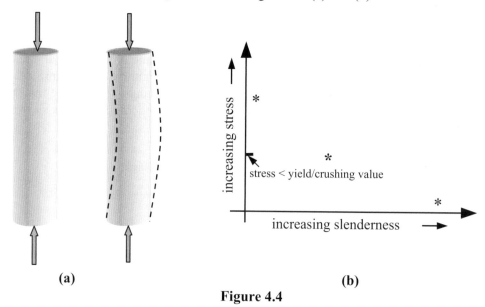

(a)   **(b)**

Figure 4.4

4.2.4 Secondary Stresses

As mentioned in Section 4.1, buckling is due to small imperfections within materials, application of load and the like, which induce secondary bending stresses which may or may not be significant depending on the type of compression element. Consider a typical column as shown in Figure 4.5 in which there is an actual centre-line, reflecting the variations within the element, and an *assumed* centre-line along which acts an applied compressive load, *assumed* to be concentric.

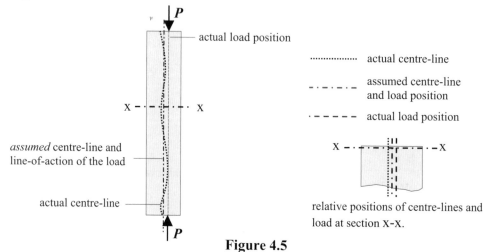

Figure 4.5

At any given cross-section the point of application of the load *P* will be eccentric to the actual centre-line of the cross-section at that point, as shown in Figure 4.6.

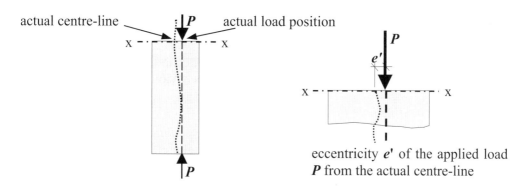

eccentricity e' of the applied load
P from the actual centre-line

Figure 4.6

The resultant eccentric load produces a secondary bending moment in the cross-section. The cross-section is therefore subject to a combination of an axial stress due to P and a bending stress due to (Pe) where e is the eccentricity from the *assumed* centre-line as indicated in Figure 4.7.

Figure 4.7

The combined axial and bending stress is given by: $\sigma = \left(\dfrac{P}{A} \pm \dfrac{Pe}{Z}\right)$

where:
σ is the combined stress,
P is the applied load,
e is the eccentricity from the assumed centre-line,
A is the cross-sectional area of the section, and
Z is the elastic section modulus about the axis of bending.

This equation, which includes the effect of secondary bending, can be considered in terms of each of the types of element.

4.2.4.1 *Effect on Short Elements*

In short elements the value of *bending* stress in the equation is insignificant when compared to the axial stress i.e. $\left(\dfrac{P}{A}\right) >> \left(\dfrac{Pe}{Z}\right)$ and consequently the lateral movement and buckling effects can be ignored.

4.2.4.2 *Effect on Slender Elements*

In slender elements the value of *axial* stress in the equation is insignificant when compared to the bending stress i.e. $\left(\dfrac{P}{A}\right) << \left(\dfrac{Pe}{Z}\right)$ particularly since the eccentricity during buckling is increased considerably due to the lateral deflection; consequently the lateral movement and buckling effects determine the structural behaviour.

4.2.4.3 *Effect on Intermediate Elements*

Most practical columns are considered to be in the *intermediate* group and consequently *both* the axial and bending effects are significant in the column behaviour, i.e. both terms in the equation $\sigma = \left(\dfrac{P}{A} \pm \dfrac{Pe}{Z}\right)$ are important.

4.2.5 Critical Stress ($\sigma_{critical}$)

In each case described in Sections 4.2.4.1 to 4.2.4.3 the critical load P_c (i.e. critical stress × cross-sectional area) must be estimated for design purposes. Since the critical stress depends on the slenderness it is convenient to quantify slenderness in mathematical terms as:

$$\text{slenderness } \lambda = \frac{L_e}{r}$$

where:
L_e is the buckling length,

r is the radius of gyration $= \sqrt{\dfrac{I}{A}}$ and

I and A are the second moment of area about the axis of bending and the cross-sectional area of the section as before.

4.2.5.1 *Critical Stress for Short Columns*

Short columns fail by yielding/crushing of the material and $\sigma_{critical} = p_y$, the yield stress of the material. If, as stated before, columns can be assumed short when the length is not greater than (10 × the least horizontal length) then for a typical rectangular column of cross-section ($b \times d$) and length $L \approx 10b$, a limit of slenderness can be determined as follows:

radius of gyration $\qquad r = \sqrt{\dfrac{I}{A}} = \sqrt{\dfrac{db^3}{12\times(b\times d)}} = \dfrac{b}{2\sqrt{3}}$

slenderness $\qquad\qquad \lambda = \dfrac{L}{r} \approx \dfrac{10b}{b/2\sqrt{3}} = 30-35$

From this we can consider that short columns correspond with a value of slenderness less than or equal to approximately 30 to 35.

4.2.5.2 *Critical Stress for Slender Columns*

Slender columns fail by buckling and the applied compressive stress $\sigma_{critical} << p_y$. The critical load in this case is governed by the bending effects induced by the lateral deformation.

Euler Equation

In 1757 the Swiss engineer/mathematician Leonhard Euler developed a theoretical analysis of premature failure due to buckling. The theory is based on the differential equation of the elastic bending of a pin-ended column which relates the applied bending moment to the curvature along the length of the column, i.e.

$$\text{Bending Moment} \; = \; EI\left(\frac{d^2 y}{dx^2}\right)$$

where $\left(\dfrac{d^2 y}{dx^2}\right)$ approximates to the curvature of the deformed column.

Since this expression for bending moments only applies to linearly elastic materials, it is only valid for stress levels equal to and below the elastic limit of proportionality. This therefore defines an *upper* limit of stress for which the Euler analysis is applicable. Consider the deformed shape of the assumed centre-line of a column in equilibrium under the action of its critical load P_c as shown in Figure 4.8.

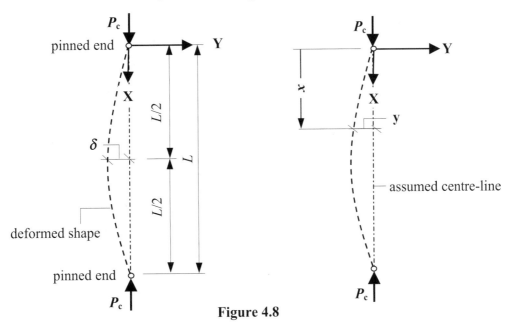

Figure 4.8

The bending moment at position x along the column is equal to $[P_c \times (-y)] = - P_c y$

and hence Bending Moment $= \; EI\left(\dfrac{d^2 y}{dx^2}\right) = - P_c y \qquad \therefore \quad EI\left(\dfrac{d^2 y}{dx^2}\right) + P_c y \; = \; 0$

This is a 2nd Order Differential Equation of the form: $\quad a\dfrac{d^2 y}{dx^2} + by = 0$

The solution of this equation can be shown to be: $P_c = n^2 \dfrac{EI\pi^2}{L^2}$

where:
n is 0,1,2,3 … etc.
EI and L are as before.

This expression for P_c defines the Euler Critical Load (P_E) for a pin-ended column. The value of $n = 0$ is meaningless since it corresponds to a value of $P_c = 0$. All other values of n correspond to the 1st, 2nd, 3rd …etc. harmonics (i.e. buckling mode shapes) for the sinusoidal curve represented by the differential equation. The first three harmonics are indicated in Figure 4.9.

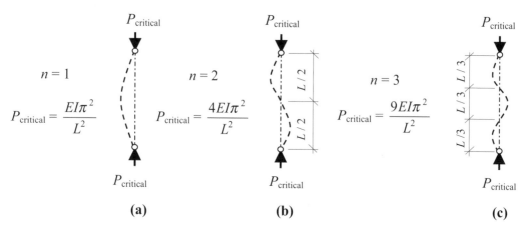

Figure 4.9 – Buckling mode-shapes for pin-ended columns

The higher level harmonics are only possible if columns are restrained at the appropriate levels, e.g. mid-height point in the case of the 2nd harmonic and the third-height points in the case of the 3rd harmonic.
The *fundamental* critical load (i.e. $n = 1$) for a *pin-ended* column is therefore given by:

$$\textbf{Euler Critical Load}\quad P_E = \frac{EI\pi^2}{L^2}$$

This fundamental case can be modified to determine the critical load for a column with different end-support conditions by defining an **effective buckling length** equivalent to that of a pin-ended column.

4.2.5.3 *Effective Buckling Length (L)*

The Euler Critical Load for the fundamental buckling mode is dependent on the *buckling length* between pins and/or points of contra-flexure as indicated in Figure 4.9. In the case of columns which are **not** pin-ended, a modification to the boundary conditions when solving the differential equation of bending given previously yields different mode shapes and critical loads as shown in Figure 4.10.

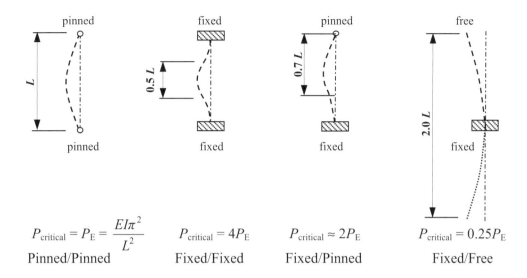

$$P_{\text{critical}} = P_E = \frac{EI\pi^2}{L^2}$$

Pinned/Pinned

$$P_{\text{critical}} = 4P_E$$

Fixed/Fixed

$$P_{\text{critical}} \approx 2P_E$$

Fixed/Pinned

$$P_{\text{critical}} = 0.25P_E$$

Fixed/Free

Figure 4.10 – Effective Buckling Lengths for Different End Conditions

The Euler *stress* corresponding to the Euler Buckling Load for a pin-ended column is given by:

$$\sigma_{\text{Euler}} = \frac{P_E}{\text{Area}(A)} = \frac{\pi^2 EI}{L^2 A}$$

Since the second moment of area $I = Ar^2$

$$\sigma_{\text{Euler}} = \frac{\pi^2 E}{(L/r)^2}$$

where (L/r) is the slenderness as before.

A lower limit to the slenderness for which the Euler Equation is applicable can be found by substituting the stress at the proportional limit σ_e for σ_{Euler} as shown in the following example with a steel column.
Assume that $\sigma_e = 200$ N/mm^2 and that $E = 205$ kN/mm^2

$$\therefore \quad 200 = \frac{\pi^2 \times 205 \times 10^3}{(L/r)^2} \quad \therefore (L/r) = \sqrt{\frac{\pi^2 \times 205 \times 10^3}{200}} \approx 100$$

In this case the Euler load is only applicable for values of slenderness $\geq \approx 100$ and can be represented on a stress/slenderness curve in addition to that determined in Section 4.2.5.1 for short columns as shown in Figure 4.11.
The Euler Buckling Load has very limited direct application in terms of practical design because of the following assumptions and limiting conditions:

- ♦ the column is subjected to a perfectly concentric axial load only,
- ♦ the column is pin-jointed at each end and restrained against lateral loading,

- ♦ the material is perfectly elastic,
- ♦ the maximum stress does not exceed the elastic limit of the material,
- ♦ there is no initial curvature and the column is of uniform cross-section along its length,
- ♦ lateral deflections of the column are small when compared to the overall length,
- ♦ there are no residual stresses in the column,
- ♦ there is no strain hardening of the material in the case of steel columns,
- ♦ the material is assumed to be homogeneous.

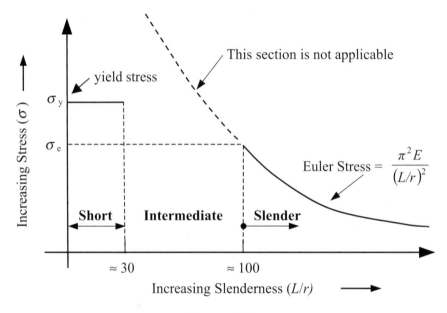

Figure 4.11

Practical columns do not satisfy these criteria, and in addition in most cases are considered to be intermediate in terms of slenderness.

4.2.5.4 Critical Stress for Intermediate Columns

Since the Euler Curve is unsuitable for values of stress greater than the elastic limit it is necessary to develop an analysis which overcomes the limitations outlined above and which can be applied between the previously established slenderness limits (see Figure 4.11) as shown in Figure 4.12.

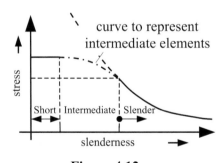

Figure 4.12

4.2.5.4.1. Tangent Modulus Theorem

Early attempts to develop a relationship for intermediate columns included the Tangent Modulus Theorem. Using this method a modified version of the Euler Equation is adopted to determine the stress/slenderness relationship in which the value of the modulus of elasticity at any given level of stress is obtained from the stress/strain curve for the material and used to evaluate the corresponding slenderness. Consider a column manufactured from a material which has a stress/strain curve as shown in Figure 4.13(a).

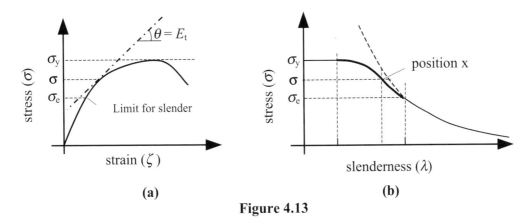

Figure 4.13

The slope of the tangent to the stress/strain curve at a value of stress equal to σ is equal to the value of the tangent modulus of elasticity E_t (**Note:** this is different from the value of E at the elastic limit). The value of E_t can be used in the Euler Equation to obtain a modified slenderness corresponding to the value of stress σ as shown at position x in Figure 4.13 (b):

$$\sigma = \frac{\pi^2 E_t}{(L/r)^2} \qquad \therefore \text{ Slenderness } \lambda \text{ at position x } = (L/r) = \sqrt{\frac{\pi^2 E_t}{\sigma}}$$

If successive values of λ for values of stress between σ_e and σ_y are calculated and plotted as shown, then a curve representing the intermediate elements can be developed. This solution still has many of the deficiencies of the original Euler equation.

4.2.5.4.2. *Perry-Robertson Formula*

The Perry-Robertson Formula was developed to take into account the deficiencies of the Euler equation and other techniques such as the Tangent Modulus Method. This formula evolved from the assumption that all practical imperfections could be represented by a hypothetical initial curvature of the column.

As with the Euler analysis a 2nd Order Differential Equation is established and solved using known boundary conditions, and the extreme fibre stress in the cross-section at mid-height (the assumed critical location) is evaluated. The extreme fibre stress, which includes both axial and bending effects, is then equated to the yield value. Clearly the final result is dependent on the initial hypothetical curvature.

Consider the deformed shape of the assumed centre-line of a column in equilibrium under the action of its critical load P_c and an assumed initial curvature as shown in Figure 4.14.

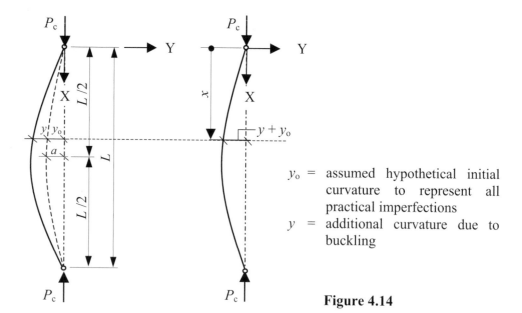

y_o = assumed hypothetical initial curvature to represent all practical imperfections

y = additional curvature due to buckling

Figure 4.14

The bending moment at position x along the column is equal to $= -P_c(y + y_o)$

and hence the bending moment $= EI\left(\dfrac{d^2 y}{dx^2}\right) = -P_c(y + y_o)$

$$\therefore \left(\frac{d^2 y}{dx^2}\right) + \left(\frac{P_c}{EI}\right)y = -\left(\frac{P_c}{EI}\right)y_o$$

If the initial curvature is assumed to be sinusoidal, then $y_o = a \operatorname{Sin}\left(\dfrac{\pi x}{L}\right)$ where a is the amplitude of the initial displacement and the equation becomes:

$$\therefore \left(\frac{d^2 y}{dx^2}\right) + \left(\frac{P_c}{EI}\right)y = -\left(\frac{P_c}{EI}\right)a \operatorname{Sin}\frac{\pi x}{L}$$

The solution to this differential equation is:

$$y = A \operatorname{Cos}\left(\frac{P_c}{EI}x\right) + B \operatorname{Sin}\left(\frac{P_c}{EI}x\right) + \frac{\dfrac{P_c}{EI}a}{\left(\dfrac{\pi^2}{L^2} - \dfrac{P_c}{EI}\right)} \operatorname{Sin}\left(\frac{\pi x}{L}\right)$$

The constants A and B are determined by considering the boundary values at the pinned ends, i.e. when $x = 0$ $y = 0$ and when $x = L$ $y = 0$.

Substitution of the boundary conditions in the equation gives:

$x = 0$ $y = 0$ $\quad \therefore$ $A = 0$

$x = L$ $y = 0$ $\quad \therefore$ $B \operatorname{Sin}\left(\dfrac{P_c}{EI}L\right) = 0$ $\quad$ For $\left(\dfrac{P_c}{EI}\right)$ not equal to zero, then $B = 0$

$$y = \frac{\dfrac{P_c}{EI}a}{\left(\dfrac{\pi^2}{L^2} - \dfrac{P_c}{EI}\right)} \operatorname{Sin}\left(\dfrac{\pi x}{L}\right) \qquad \text{If the equation is divided throughout by } \left(\dfrac{P_c}{EI}\right) \text{ then}$$

$$y = \frac{a\operatorname{Sin}\left(\dfrac{\pi x}{L}\right)}{\left(\dfrac{\pi^2 EI}{P_c L^2} - 1.0\right)} \qquad \text{The Euler load } P_E = \dfrac{\pi^2 EI}{L^2} \qquad \therefore y = \frac{a\operatorname{Sin}\left(\dfrac{\pi x}{L}\right)}{\left(\dfrac{P_E}{P_c} - 1.0\right)}$$

The value of the stress at mid-height is the critical value since the maximum eccentricity of the load (and hence maximum bending moment) occurs at this position;

when $\quad x = L/2, \quad \operatorname{Sin}\left(\dfrac{\pi x}{L}\right) = 1.0 \quad$ and $\quad y_{\text{mid-height}} = \dfrac{a\operatorname{Sin}\left(\dfrac{\pi x}{L}\right)}{\left(\dfrac{P_E}{P_c} - 1.0\right)} = \dfrac{a}{\left(\dfrac{P_E}{P_c} - 1.0\right)}$

(**Note:** y_o at mid-height is equal to the amplitude a of the assumed initial curvature).

The maximum bending moment $\quad M = P_c\,(a + y_{\text{mid-height}}) = P_c\,a\left[1 + \dfrac{1}{\left(\dfrac{P_E}{P_c} - 1.0\right)}\right]$

The maximum combined stress at this point is given by:

$$\sigma_{\text{maximum}} = \left(\frac{axial\ load}{A} + \frac{bending\ moment \times c}{I}\right) = \left(\frac{P_c}{A} + \frac{M \times c}{Ar^2}\right)$$

where c is the distance from the neutral axis of the cross-section to the extreme fibres.

The *maximum stress* is equal to the yield value, i.e. $\sigma_{\text{maximum}} = \sigma_y$

$$\therefore \sigma_y = \left(\frac{P_c}{A} + \frac{M \times c}{Ar^2}\right) = \frac{P_c}{A} + P_c\,a\left[1 + \frac{1}{\left(\dfrac{P_E}{P_c} - 1.0\right)}\right] \times \frac{c}{Ar^2}$$

The *average stress* over the cross-section is the load divided by the area, i.e. (P_c / A)

$$\therefore \quad \sigma_y = \sigma_{\text{average}} + \sigma_{\text{average}}\left(\frac{P_E}{P_E - P_C}\right) \times \frac{ac}{r^2} = \sigma_{\text{average}}\left[1 + \left(\frac{P_E}{P_E - P_c}\right) \times \frac{ac}{r^2}\right]$$

$$\sigma_{\text{average}} = (P_c / A) \quad \text{and} \quad \sigma_{\text{Euler}} = (P_{\text{Euler}} / A)$$

$$\therefore \quad \sigma_y = \sigma_{\text{average}}\left[1 + \left(\frac{\sigma_E}{\sigma_E - \sigma_{\text{average}}}\right) \times \frac{ac}{r^2}\right]$$

The (ac/r^2) term is dependent upon the assumed initial curvature and is normally given the symbol η.

$$\therefore \quad \sigma_y \quad = \quad \sigma_{average} \left[1 + \left(\frac{\eta \sigma_E}{\sigma_E - \sigma_{average}} \right) \right]$$

This equation can be rewritten as a quadratic equation in terms of the average stress:

$$\sigma_y \, (\sigma_E - \sigma_{average}) \quad = \quad \sigma_{average} \, [(1 + \eta)\sigma_E - \sigma_{average}]$$

$$\sigma^2_{average} - \sigma_{average} \, [\sigma_y + (1 + \eta)\sigma_E] + \sigma_y \sigma_E = 0$$

The solution of this equation in terms of $\sigma_{average}$ is:

$$\sigma_{average} \quad = \quad \frac{\left[\sigma_y + (1 + \eta)\sigma_E\right] - \sqrt{\left[\sigma_y + (1 + \eta)\sigma_E\right]^2 - 4\sigma_y \sigma_E}}{2.0}$$

This equation represents the *average* value of stress in the cross-section which will induce the yield stress at mid-height of the column for any given value of η. Experimental evidence obtained by Perry and Robertson indicated that the hypothetical initial curvature of the column could be represented by;

$$\eta \quad = \quad 0.3(L_{effective} / 100r^2)$$

which was combined with a load factor of 1.7 and used for many years in design codes to determine the critical value of average compressive stress below which overall buckling would not occur. The curve of stress/slenderness for this curve is indicated in Figure 4.15 for comparison with the Euler and Tangent Modulus solutions.

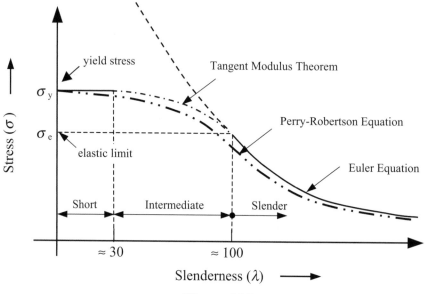

Figure 4.15

4.2.5.4.3.   BS 5950 – European Column Curves

Whilst the Perry-Robertson formula does take into account many of the deficiencies of the Euler and Tangent Modulus approaches it does not consider all of the factors which influence the failure of columns subjected to compressive stress, for example in the case of steel columns the effects of residual stresses induced during fabrication, the type of section being considered (i.e. the cross-section shape), the material thickness, the axis of buckling, the method of fabrication (i.e. rolled or welded), and so on.

A more realistic formula of the critical load capacity of columns has been established following extensive full-scale testing both in the UK and in other European countries. The Perry-Robertson formula has in effect been modified and is referred to in the current UK steel design code BS 5950 as the ***Perry strut formula*** and is given in the following form:

$$(p_E - p_c)(p_y - p_c) \ = \ \eta\, p_E\, p_c \quad \text{from which the value of } p_c \text{ may be obtained using:}$$

$$p_c \ = \ \frac{p_E\, p_y}{\phi + \left(\phi^2 - p_E\, p_y\right)^{0.5}} \quad \text{in which } \phi \ = \ \frac{p_y + (\eta + 1)\, p_E}{2} \quad \text{and} \quad p_E \ = \ (\pi^2 E / \lambda^2)$$

where:
p_y is the design strength
λ is the slenderness

The Perry factor η for flexural buckling under axial force should be taken as:

$$\eta \ = \ a(\lambda - \lambda_0)/1000 \geq 0 \quad \text{where} \quad \lambda_0 \ = \ 0.2\,(\pi^2 E / p_y)^{0.5}$$

The Robertson constant a should be taken as 2.0, 3.5, 5.5 or 8.0 as indicated in the code depending on the cross-section, thickness of material, axis of buckling and method of fabrication.

The European Column curves are represented in tabular form in the design code and indicated in graphical form in Figure 4.16.

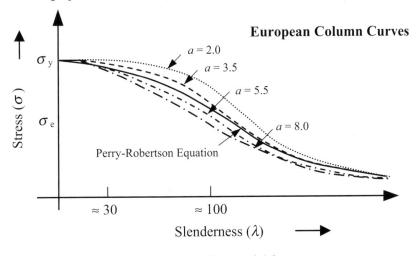

Figure 4.16

4.3   *Local Buckling*

Local instability can occur in a cross-section when one or more individual elements in a cross-section (e.g. the flange or web of an **I** section), as shown in Figure 4.17, buckles without any overall deflection.

An element within a cross-section which has a high width to thickness ratio (i.e. slender) is susceptible to local buckling, the effect of which is to reduce the load-carrying capacity of the section.

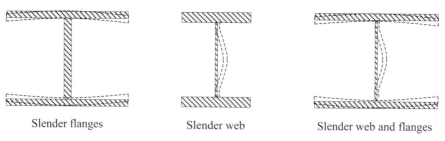

Slender flanges   Slender web   Slender web and flanges

Figure 4.17

Failure in a flange occurs due to excessive compression and in a web due to excessive shear or combined shear and bending. In addition, a web may buckle as a result of vertical compression due to the application of a concentrated load.

In the case of hot-rolled steel sections the flange and web proportions (i.e. flange outstand : thickness and web thickness : depth ratios) are normally selected to minimize the possibility of local buckling, although web stiffening is sometimes required at points of concentrated load such as reactions and column positions on beams. In the case of welded plate girders, additional web stiffening is usually necessary to prevent shear buckling of the web. The classification of steel sections to determine their susceptibility to local buckling is dealt with in detail in Chapter 6.

In design there are two approaches generally considered appropriate to allow for the possibility of local buckling. They are:

1. adopting a reduced design strength when calculating the member capacity,
2. adopting '*effective*' section properties in which an '*actual*' plate width is replaced by a narrower '*effective*' plate width which is then used to calculate modified section properties with which to determine the section capacity.

Both of these methods are illustrated in Chapter 6 in relation to structural steelwork.

4.4   *Lateral Torsional Buckling*

A beam subject to bending is partly in tension and partly in compression as shown in Figure 4.18. The tendency of an unrestrained compression flange in these circumstances is to deform sideways and to twist about the longitudinal axis as shown in Figure 4.19.

Deformed shape of beam

compression zone

tension zone

Figure 4.18

This type of failure is called ***lateral torsional buckling*** and will normally occur at a value of applied moment less than the moment capacity (M_c) of the section, based on the yield strength of the material. The reduced moment at failure is known as the ***buckling resistance moment*** and is discussed in detail in Chapter 6.

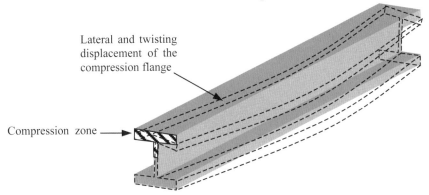

Lateral and twisting displacement of the compression flange

Compression zone

Figure 4.19 – Lateral Torsional Buckling of Beams

The tendency for the compression flange to deform is influenced by a number of factors such as lateral restraint, torsional restraint, flange thickness and effective buckling length.

4.4.1 *Lateral Restraint*

The lateral restraint to the compression flange of a beam prevents a sideways movement of the flange relative to the tension flange.

4.4.1.1 *Full Lateral Restraint*

It is always desirable where possible to provide full lateral restraint to the compression flange of a beam. The existence of either a cast-in-situ or precast concrete slab which is supported directly on the top flange, as indicated in Figure 4.20, or cast around it is normally considered to provide adequate restraint. A steel plate floor tack-welded or bolted to the flange also provides adequate restraint; steel floors which are fixed in a manner such that removal for access is required are not normally considered adequate for restraint. Timber floors and beams are frequently supported by steel beams. Generally unless they are fixed to the beam by cleats, bolts or some similar method and are securely held at their remote end or along their length they are **not** considered to provide adequate restraint.

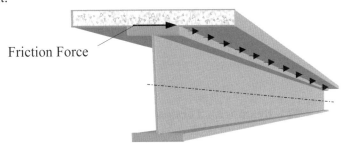

Friction Force

Figure 4.20 – Lateral Friction Force between Underside of Slab and Top Flange

4.4.1.2 Intermittent Lateral Restraint

Most beams in buildings which do not have full lateral restraint are provided with intermittent restraint in the form of secondary beams, ties or bracing members as shown in Figure 4.21.

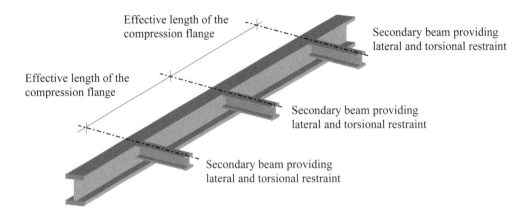

Figure 4.21

It is important to ensure that the elements providing restraint are an integral part of a braced structural system and are capable of transmitting the lateral force to the supporting structure.

4.4.1.3 Torsional Restraint

A beam is assumed to have torsional restraint about its longitudinal axis at any location where both flanges are held in their relative positions by external members during bending, as illustrated in Figure 4.22.

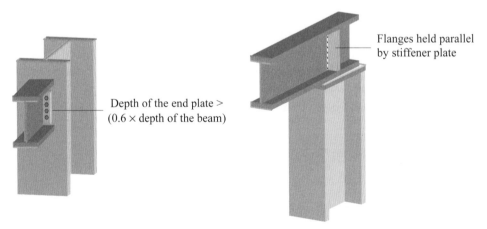

Figure 4.22 – Beam with Torsional Restraint

This type of restraint may be provided by load bearing stiffeners or by the provision of adequate end connection details.

4.4.1.4 Beams without Torsional Restraint

In situations where a beam is supported by a wall as in Figure 4.23, no torsional restraint is provided to the flanges and buckling is more likely to occur.

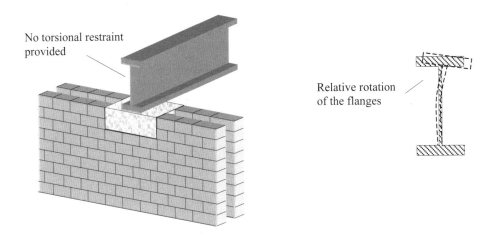

No torsional restraint provided

Relative rotation of the flanges

Figure 4.23

4.4.2 Effective Length L_E

The provision of lateral and torsional restraints to a beam introduces the concept of **effective length,** similar to that in columns. The effective length of a compression flange is the equivalent length between restraints over which a pin-ended beam would fail by lateral torsional buckling. The values adopted for effective length depend on three factors relating to the degree of lateral and torsional restraint at the position of the intermittent restraints, i.e. the existence of torsional restraints, the degree of lateral restraint of the compression flange and the type of loading. These factors are discussed Chapter 6.

4.5   Overall Structural Stability

In the subsequent chapters the requirements of strength, stiffness and stability of individual structural components have been considered in detail. It is also **essential** in any structural design to consider the requirements of **overall** structural stability.

The term **stability** has been defined in *Stability of Buildings* published by the Institution of Structural Engineers (ref. 49) in the following manner:

'Provided that displacements induced by normal loads are acceptable, then a building may be said to be stable if:
- *a minor change in its form, condition, normal loading or equipment would not cause partial or complete collapse and*
- *it is not unduly sensitive to change resulting from accidental or other actions.*
Normal loads include the permanent and variable actions for which the building has been designed.

> The phrase "is not unduly sensitive to change" should be broadly interpreted to mean that the building should be so designed that it will not be damaged by accidental or other actions to an extent disproportionate to the magnitudes of the original causes of damage.'

This publication, and the inclusion of stability, robustness and accidental damage clauses in current design codes, is largely a consequence of the overall collapse or significant partial collapse of structures, e.g. the collapse of precast concrete buildings under erection at Aldershot in 1963 (ref. 51) and notably the Ronan Point Collapse due to a gas explosion in 1968 (ref. 48).

The Ronan Point failure occurred in May 1968 in a 23-storey precast building. A natural gas explosion in a kitchen triggered the progressive collapse of all of the units in one corner above and below the kitchen. The spectacular nature of the collapse had a major impact on the philosophy of structural design resulting in important revisions of design codes world-wide.

This case stands as one of the few landmark failures which have had a sustained impact on structural thinking.

The inclusion of such clauses in codes and building regulations is not new. The following is an extract from the 'CODE OF LAWS OF HAMMURABI (2200 BC), KING OF BABYLONIA' (the earliest building code yet discovered):

A. *If a builder builds a house for a man and do not make its construction firm and the house which he has built collapse and cause the death of the owner of the house – that builder shall be put to death.*

B. *If it cause the death of the son of the owner of the house – they shall put to death a son of that builder.*

C. *If it cause the death of a slave of the owner of the house – he shall give to the owner of the house a slave of equal value.*

D. *If it destroy property, he shall restore whatever it destroyed, and because he did not make the house which he built firm and it collapsed, he shall rebuild the house which collapsed at his own expense.*

E. *If a builder build a house for a man and do not make its construction meet the requirements and a wall fall in, that builder shall strengthen the wall at his own expense.*

Whilst this code is undoubtedly harsh it probably did concentrate the designer's mind on the importance of structural stability!

An American structural engineer, Dr Jacob Feld, spent many years investigating structural failure and suggested ten basic rules to consider when designing and/or constructing any structure (ref. 43):

1. *Gravity always works, so if you don't provide permanent support, something will fail.*

2. *A chain reaction will make a small fault into a large failure, unless you can afford a fail-safe design, where residual support is available when one component fails. In the competitive private construction industry, such design procedure is beyond consideration.*

3. *It only requires a small error or oversight – in design, in detail, in material strength, in assembly, or in protective measures – to cause a large failure.*

4. *Eternal vigilance is necessary to avoid small errors. If there are no capable crew or group leaders on the job and in the design office, then supervision must take over the chore of local control. Inspection service and construction management cannot be relied on as a secure substitute.*

5. *Just as a ship cannot be run by two captains, a construction job cannot be run by a committee. It must be run by one individual, with full authority to plan, direct, hire and fire, and full responsibility for production and safety.*

6. *Craftsmanship is needed on the part of the designer, the vendor, and the construction teams.*

7. *An unbuildable design is not buildable, and some recent attempts at producing striking architecture are approaching the limit of safe buildability, even with our most sophisticated equipment and techniques.*

8. *There is no foolproof design, there is no foolproof construction method, without guidance and proper and careful control.*

9. *The best way to generate a failure on your job is to disregard the lessons to be learnt from someone else's failures.*

10. *A little loving care can cure many ills. A little careful control of a job can avoid many accidents and failures.*

An appraisal of the overall stability of a complete structure during both the design and construction stages should be carried out by, and be the responsibility of, one individual. In many instances a number of engineers will be involved in designing various elements or sections of a structure but never the whole entity. It is **essential**, therefore, that one identified engineer carries out this vital appraisal function, including consideration of any temporary measures which may be required during the construction stage. In Clause 20.1 of the code it is clearly stated:

'*The designer responsible for the overall stability of the structure should ensure the compatibility of the design and details of parts and components. There should be no doubt of this responsibility for overall stability when some or all of the design and details are not made by the same designer.*'

4.5.1 Structural Form

Generally, instability problems arise due to an inadequate provision to resist lateral loading (e.g. wind loading) on a structure. There are a number of well-established structural forms which, when used correctly, will ensure adequate stiffness, strength and stability. It is important to recognise that stiffness, strength and stability are three different characteristics of a structure. In simple terms:

- the *stiffness* determines the deflections which will be induced by the applied load system,
- the *strength* determines the maximum loads which can be applied before acceptable material stresses are exceeded and,
- the *stability* is an inherent property of the structural form which ensures that the building will remain stable.

The most common forms of structural arrangements which are used to transfer loads safely and maintain stability are:

- braced frames,
- unbraced frames,
- shear cores/walls,
- cross-wall construction,
- cellular construction,
- diaphragm action.

In many structures, a combination of one or more of the above arrangements is employed to ensure adequate load paths, stability and resistance to lateral loading. All buildings behave as complex three-dimensional structures with components frequently interacting compositely to resist the applied force system. Analysis and design processes are a simplification of this behaviour in which it is usual to analyse and design in two dimensions with wind loading considered separately in two mutually perpendicular directions.

4.5.2 Braced Frames

In braced frames lateral stability is provided in a structure by utilising systems of diagonal bracing in at least two vertical planes, preferably at right angles to each other. The bracing systems normally comprise a triangulated framework of members which are either in tension or compression. The horizontal floor or roof plane can be similarly braced at an appropriate level, as shown in Figure 4.24, or the floor/roof construction may be designed as a deep horizontal beam to transfer loads to the vertical, braced planes, as shown in Figure 4.25. There are a number of configurations of bracing which can be adopted to accommodate openings, services etc. and are suitable for providing the required load transfer and stability.

   In such systems the entire wind load on the building is transferred to the braced vertical planes and hence to the foundations at these locations.

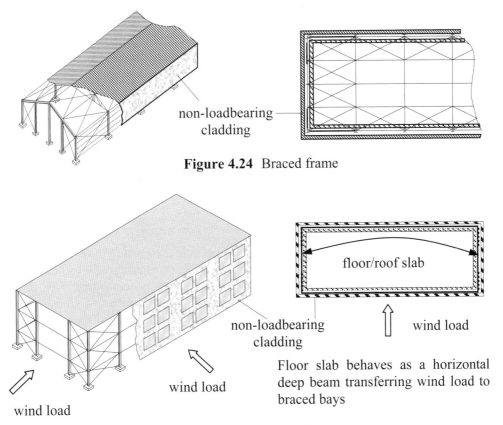

Figure 4.24 Braced frame

non-loadbearing cladding

floor/roof slab

non-loadbearing cladding

wind load

wind load

wind load

Floor slab behaves as a horizontal deep beam transferring wind load to braced bays

Figure 4.25 Braced frames

4.5.3 Unbraced Frames

Unbraced frames comprise structures in which the lateral stiffness and stability are achieved by providing an adequate number of rigid (moment-resisting) connections at appropriate locations. Unlike braced frames in which 'simple connections' only are required, the connections must be capable of transferring moments and shear forces. This is illustrated in the structure in Figure 4.26, in which stability is achieved in two mutually perpendicular directions using rigid connections. In wind direction A each typical transverse frame transfers its own share of the wind load to its own foundations through the moment connections and bending moments/shear forces/axial forces in the members. In wind direction B the wind load on either gable is transferred through the members and floors to stiffened bays (i.e. in the longitudinal section), and hence to the foundation at these locations. It is not necessary for *every* connection to be moment-resisting.

It is common for the portal frame action in a stiffened bay in wind direction B to be replaced by diagonal bracing whilst still maintaining the moment-resisting frame action to transfer the wind loads in direction A.

As with braced frames, in most cases the masonry cladding and partition walls are non-loadbearing.

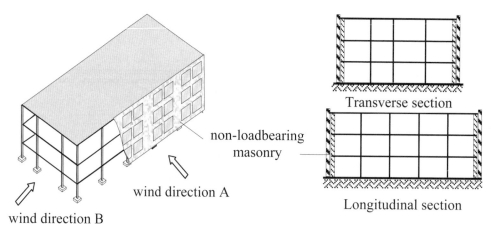

Transverse section

Longitudinal section

non-loadbearing
masonry

wind direction A

wind direction B

Figure 4.26 Unbraced frame

4.5.4 Shear Cores/Walls

The stability of modern high-rise buildings can be achieved using either braced or unbraced systems as described in Sections 4.5.2 and 4.5.3, or alternatively by the use of shear-cores and/or shear-walls. Such structures are generally considered as three-dimensional systems comprising horizontal floor plates and a number of strong-points provided by cores/walls enclosing stairs or lift shafts. A typical layout for such a building is shown in Figure 4.27.

In most cases the vertical loads are generally transferred to the foundations by a conventional skeleton of beams and columns whilst the wind loads are divided between several shear-core/wall elements according to their relative stiffness.

Where possible the plan arrangement of shear-cores and walls should be such that the centre-line of their combined stiffness is coincidental with the resultant of the applied wind load, as shown in Figure 4.28.

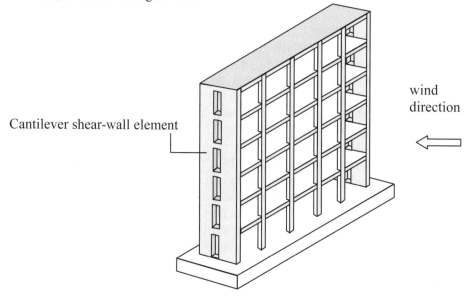

Cantilever shear-wall element

wind
direction

Figure 4.27 Typical shear-wall

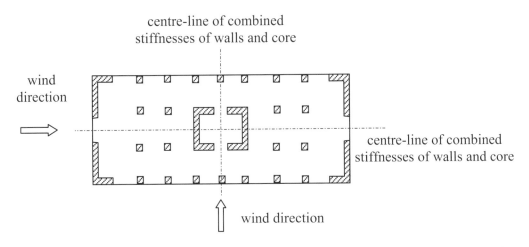

Figure 4.28   Efficient layout of shear-core/walls

If this is not possible and the building is much stiffer at one end than the other, as in Figure 4.29, then torsion may be induced in the structure and must be considered. It is better at the planning stage to avoid this situation arising by selecting a judicious floor-plan layout. The floor construction must be designed to transfer the vertical loads (which are perpendicular to their plane) to the columns/wall elements in addition to the horizontal wind forces (in their own plane) to the shear-core/walls. In the horizontal plane they are designed as deep beams spanning between the strong-points.

There are many possible variations, including the use of concrete, steel, masonry and composite construction, which can be used to provide the necessary lateral stiffness, strength and stability.

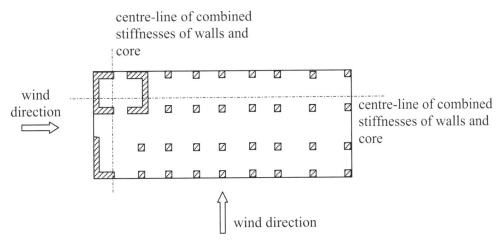

Figure 4.29   Inefficient layout of shear-core/walls

4.5.5 Cross-Wall Construction

In long rectangular buildings which have repetitive, compartmental floor plans such as hotel bedroom units and classroom blocks, as shown in Figure 4.30, masonry cross-wall construction is often used.

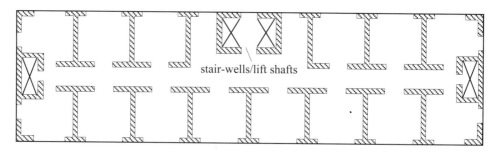

Figure 4.30 Cross-wall construction

Lateral stability parallel to the cross-walls is very high, with the walls acting as separate vertical cantilevers sharing the wind load in proportion to their stiffnesses. Longitudinal stability, i.e. perpendicular to the plane of the walls, must be provided by the other elements such as the box sections surrounding the stair-wells/lift-shafts, corridor and external walls.

4.5.6 Cellular Construction

It is common in masonry structures for the plan layout of walls to be irregular with a variety of exterior and interior walls, as shown in Figure 4.31.

The resulting structural form is known as 'cellular construction', and includes an inherent high degree of interaction between the intersecting walls. The provision of stair-wells and lift-shafts can also be integrated to contribute to the overall bracing of the structure.

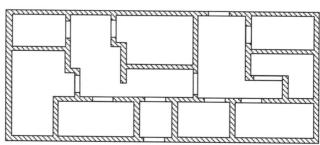

Figure 4.31 Cellular construction

It is important in both cross-wall and cellular masonry construction to ensure the inclusion of features such as:

- bonding or tying together of all intersecting walls,
- provision of returns where practicable at ends of load-bearing walls,
- provision of bracing walls to external walls,

- ◆ provision of internal bracing walls,
- ◆ provision of strapping of the floors and roof at their bearings to the load-bearing walls,

as indicated in *Stability of Buildings* (ref. 49).

4.5.7 *Diaphragm Action*

Floors, roofs, and in some cases cladding, behave as horizontal diaphragms which distribute lateral forces to the vertical wall elements. This form of structural action is shown in Figure 4.32.

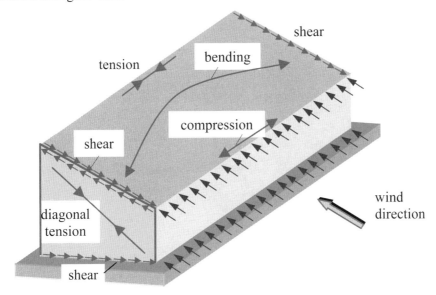

Figure 4.32 Diaphragm action

It is essential when utilising diaphragm action to ensure that each element and the connections between the various elements are capable of transferring the appropriate forces and providing adequate load-paths to the supports.

4.5.8 *Accidental Damage and Robustness*

It is inevitable that accidental loading such as vehicle impact or gas explosions will result in structural damage. A structure should be sufficiently robust to ensure that damage to small areas or failure of individual elements does not lead to progressive collapse or significant partial collapse. There are a number of strategies which can be adopted to achieve this, e.g.

- ◆ enhancement of continuity which includes increasing the resistance of connections between members and hence load-transfer capability,
- ◆ enhancement of overall structural strength including connections and members,
- ◆ provision of multiple load-paths to enable the load carried by any individual

member to be transferred through adjacent elements in the event of local failure,

♦   the inclusion of load-shedding devices such as venting systems to allow the escape of gas following an explosion, or specifically designed weak elements/details to prevent transmission of load.

The robustness required in a building may be achieved by 'tying' the elements of a structure together using peripheral and internal ties at each floor and roof level, as indicated in Figure 4.33.

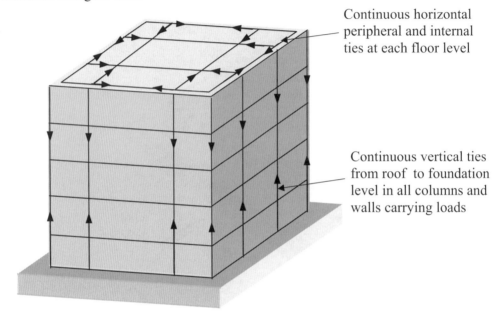

Continuous horizontal peripheral and internal ties at each floor level

Continuous vertical ties from roof to foundation level in all columns and walls carrying loads

Figure 4.33

An alternative to the 'fully tied' solution is one in which the consequences of the removal of each load-bearing member are considered in turn. If the removal of a member results in an unacceptable level of damage then this member must be strengthened to become a **protected member** (i.e. one which will remain intact after an accidental event), or the structural form must be improved to limit the extent of the predicted collapse. This process is carried out until all non-protected horizontal and vertical members have been removed one at a time.

5.   Design of Reinforced Concrete Elements (BS 8110)

Objective: to illustrate the process of design for reinforced concrete elements.

5.1   Introduction

Concrete is a widely used structural material with applications ranging from simple elements such as fence posts and railway sleepers to major structures such as bridges, offshore oil production platforms and high-rise buildings. In essence the material is a conglomerate of chemically inert aggregates (i.e. natural sands, crushed rock etc.) bonded together by a matrix of mineral cement. The aggregates and cement are mixed together with water to create an amorphous, plastic mass, i.e. concrete. A chemical reaction between the cement and the water (known as **the *hydration* process**) causes the cement to harden and the conglomerate to gain strength over a period of time. The process of hardening is known as **curing** of the concrete and is important in developing the final strength of the material. Prior to hardening, the concrete, which has been mixed into a plastic mass, can be moulded to virtually any desired shape and dimension enabling an almost limitless variation in architectural expression.

The constituents of concrete can be found throughout the world and its use is suited equally well to primitive, low-technology, labour-intensive applications frequently encountered in the developing world and to highly sophisticated, capital-intensive applications in the industrialised nations.

The success of concrete as a material is due to its versatility, particularly when combined with steel to act compositely as *reinforced* or *pre-stressed* concrete; only reinforced concrete is considered in this text. Whilst hardened concrete has a high compressive strength its tensile strength is very low (i.e. in the region of 10% of the compressive strength e.g. 2 N/mm^2; this is normally assumed to be equal to zero in reinforced concrete design). This minimal tensile strength restricted the use of concrete to circumstances in which the stress was almost entirely compressive until the late 19th century when methods were developed for reinforcing concrete to overcome its weakness in tension.

Consider an *unreinforced* concrete beam of rectangular cross-section which is simply supported at the ends and carries a distributed load, as shown in Figure 5.1.

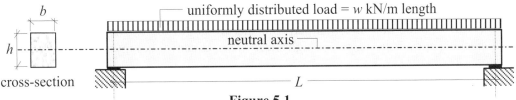

Figure 5.1

The beam will deflect due to the bending moments and shear forces induced by the applied loading, resulting in a curved shape as indicated in Figure 5.2.

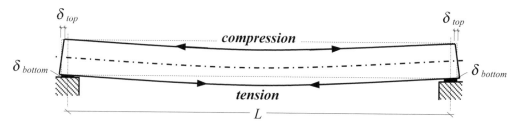

Original length of the beam before deformation   = L
Final length of the top edge after deformation   = $(L - 2\delta_{top})$ i.e. shortening
Final length of the bottom edge after deformation = $(L + 2\delta_{bottom})$ i.e. lengthening

Figure 5.2

Clearly if the ends of the beam are assumed to remain perpendicular to the longitudinal axis, then the material above this axis must be in compression, whilst that below it must be in tension. Since the strain in the material is directly proportional to the distance from the neutral axis (see Chapter 1, Figure 1.41), flexural tensile cracking will begin at the extreme bottom fibres and extend towards the neutral axis, as shown in Figure 5.3.

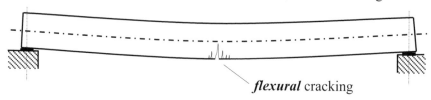

flexural cracking

Figure 5.3

The inverted 'V' shape is characteristic of flexural cracking in concrete.
    In addition to the tensile stresses caused by flexure, diagonal tensile stresses are induced by the shear forces, as shown in Figure 5.4.

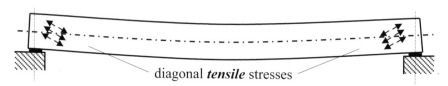

diagonal *tensile* stresses

Figure 5.4

Since cracks develop in a direction perpendicular to that of the tensile stresses, diagonal shear cracking appears in regions of high shear stress, as shown in Figure 5.5.

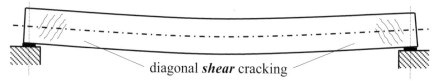

diagonal *shear* cracking

Figure 5.5

In addition to cracking caused by directly applied loads cracking can also occur due to factors such as settlement of the supports, temperature variations and/or shrinkage strains. The classification of cracks ranges from surface hairline cracks (approximately 0.13 mm wide), which are generally regarded as negligible, to severe penetrating cracks which can be as much as 15.0 mm to 25.0 mm wide. In the latter case extensive damage and possibly structural instability will be evident.

The cracking caused by flexure in the unreinforced beam in Figure 5.1 produces tensile failure at a very low value of *w*. Only 10% of the compressive strength capacity of the beam is being utilized.

The introduction of steel reinforcement bars in the tension zone of the beam (steel has a very high tensile strength compared to concrete, typically 460 N/mm^2) enables the applied load *w* to be increased considerably until the beam fails by yielding of the steel in the bottom in tension and crushing of the concrete in the top fibres in compression, as shown in Figure 5.6. Note that the neutral axis at failure has moved to a position nearer the top of the beam.[1]

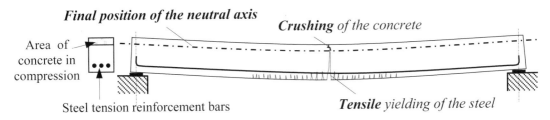

Figure 5.6

The amount of reinforcement steel required at any given section is dependent on the value of the bending moment at that point.

The cracking caused by shear in an unreinforced beam is prevented by providing *shear links* (also known as *stirrups*), as shown in Figure 5.7, which ensure that steel reinforcement is present to resist the diagonal tension indicated in Figure 5.4.

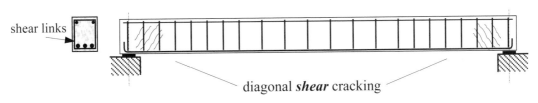

Figure 5.7

The spacing of the links varies depending on the magnitude of the shear force and the depth of the beam at any given section. At locations of high shear, e.g. the support points, the links are closer together than is required at regions where the shear is low, e.g. at mid-span in the beam indicated in Figure 5.1.

The design of reinforced concrete is governed by the requirements of *BS 8110 –1:1997 'Structural use of concrete – Part 1: Code of practice for design and construction'*.

---

[1] The failure of sections and provision of reinforcement are discussed in more detail in Section 5.3.

In addition to flexural and shear strength requirements it is necessary to ensure that elements have sufficient stiffness to avoid excessive cracking or deflection and possess other properties such as adequate durability and fire resistance. Each of these requirements is considered in detail in further sections of this chapter.

5.2 Material Properties

5.2.1 Concrete Compressive Strength: f_{cu} (Clause 2.4.2.1)

In structural terms the most important material property of concrete is its inherent compressive strength. In *BS 8110 – 1:1997, Part 1*, the *characteristic strength* (see Section 2.3) of concrete is defined in Clause 2.4.2.1 as the value of the **cube strength** of concrete f_{cu}. The cube strength is defined on the basis of test results carried out on 10 cm or 15 cm (4 inch or 6 inch) cubes cast and cured under rigid, specified conditions and loaded to failure in a standardized compression testing machine as indicated in *BS 1881:Testing Concrete*. The characteristic value is the value below which not more than 5% of all possible results fall, and is given by:

$$f_k = f_m - 1.64\,s$$

where:
f_k is the characteristic value,
f_m is the mean value – normally determined from cubes which are tested 28 days after casting,
s is the standard deviation of the test results.

In the design code (BS 8110) concrete is graded according to the **characteristic compressive strength** and designated as: C30, C35, C40, C45 and C50, where the numbers 30, 40, 45 and 50 represent compressive strengths in N/mm^2. Other grades of concrete are also used for specific purposes, e.g. low-strength concretes are often used to provide a base on which construction work can begin whilst high-strength concretes are often used in circumstances where high stresses are developed, such as in pre-stressed concrete.

It is important to realize that the characteristic cube strength represents the *potential* strength of the concrete. The material in a structural element is likely to be less than this value since it will have been created under less stringent manufacturing control and curing conditions than the sample cubes used for testing. The difference between the *potential* and *actual* strengths is reflected in the material partial safety factor γ_m.

5.2.2 Concrete Tensile Strength: f_t (Clauses 2.2.3.4.1 and 2.4.6.2)

In reinforced concrete design the tensile strength of the concrete is normally assumed to be zero. In serviceability calculations, e.g. for the determination of cracking strengths of pre-stressed concrete members, reference is made to Section 4 of the code; this is outwith the scope of this text. Calculations to determine crack widths of reinforced concrete members under serviceability limit state conditions are carried out assuming material properties as described in Clause 3.2.4 of *BS 8110:Part 2:1985*.

5.2.3 Concrete Stress-Strain Relationship (Clause 2.4.2.3)

The flexural strength of a reinforced concrete section is determined by consideration of the stress-strain relationship of both the concrete and the reinforcing steel. These characteristics are defined in Clause 2.4.2.3 and Figures 2.1 and 2.2 of the code.

A typical stress-strain curve for concrete is shown in Figure 5.8. This is a non-linear curve in which the peak stress is developed at a compressive strain of approximately 0.002 (depending upon the f_{cu} value) with an ultimate strain of approximately 0.0035. There is no clearly defined elastic range over which the stress varies linearly with the strain. Such stress/strain curves are typical of brittle materials.

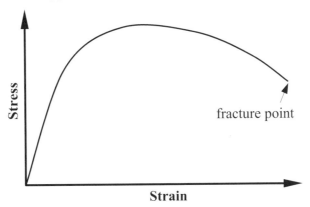

Figure 5.8 – Typical stress/strain curve for concrete

This curve is replaced in the design code by a simplified representation of the short-term design stress/strain curve for normal concrete shown in Figure 5.9 (see Figure 2.1 of the code).

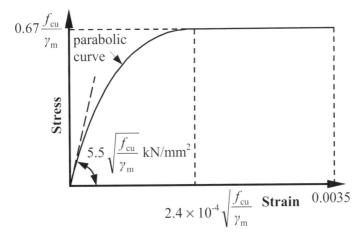

where:
f_{cu} is the cube strength in N/mm^2,
γ_m is the partial safety factor for concrete (taken as 1.5 in the code),
0.67 is a coefficient to allow for the difference in compressive strength as determined using a cube in axial compression and the compressive strength developed in a section due to flexure.

Figure 5.9 – BS 8110 Short-term design stress/strain curve for reinforced concrete

5.2.4 Concrete Modulus of Elasticity: E_c (Clause 2.5.4)

Generally, it is satisfactory to determine shear forces, bending moments and axial loads in structural members by using standard methods of linear elastic analysis. If carried out by computer this requires a value of modulus of elasticity as input data; typical mean values for the '*static modulus of elasticity at 28 days for normal-weight concrete* $(E_{c,28})$' are given in Table 7.2 of *BS 8110:Part 2:1985* which is reproduced in Figure 5.10.

Table 7.2 Typical range for the static modulus of elasticity at 28 days of normal-weight concrete		
$f_{cu,28}$	$E_{c,28}$	
	Mean value	Typical range
N/mm^2	kN/mm^2	kN/mm^2
20	24	18 to 30
25	25	19 to 31
30	26	20 to 32
40	28	22 to 34
50	30	24 to 36
60	32	26 to 38

Figure 5.10 Extract from *BS 8110:Part 2:1985*

5.2.5 Concrete Poisson's Ratio: v_c (Clause 2.4.2.4)

The value of Poisson's ratio (v_c) of concrete for use in linear elastic analysis is given in Clause 2.4.2.4 as 0.2.

5.2.6 Steel Reinforcement Strength: f_y (Clause 3.1.7.4)

The characteristic strengths of reinforcement are given in Table 3.1 of the code as:

f_y = 250 N/mm^2 for hot rolled mild steel (MS) and

   = 460 N/mm^2 for hot rolled or cold worked high yield steel (HYS).

5.2.7 Steel Reinforcement Stress-Strain Relationship (Clause 2.4.2.3)

A typical stress-strain curve for hot-rolled mild steel is shown in Figure 5.11. When a test specimen of mild steel reinforcing bar is subjected to an axial tension in a testing machine, the stress/strain relationship is linearly elastic until the value of stress reaches a yield value, e.g. 250 N/mm^2.

At this point an appreciable increase in the stretching of the sample occurs at constant load: this is known as **yielding**. During the process of yielding a molecular change takes place in the material which has the effect of hardening the steel. After approximately 5% strain has occurred sufficient *strain-hardening* will have developed to enable the steel to carry a further increase in load until a maximum load is reached.

The stress-strain curve falls after this point due to a local reduction in the diameter of the sample (known as necking, see Figure 5.11) with a consequent smaller cross-sectional area and load carrying capacity. Eventually the sample fractures at approximately 35% strain.

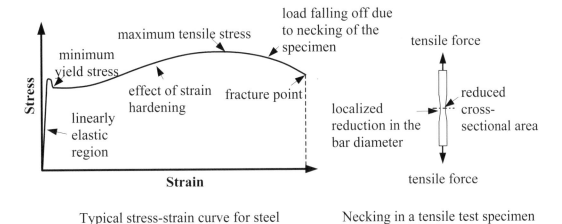

Typical stress-strain curve for steel   Necking in a tensile test specimen

Figure 5.11

This curve is replaced in the design code by a simplified, bi-linear representation of the short-term design stress-strain curve for reinforcement as shown in Figure 5.12 (see Figure 2.2 of the code) in which the 'characteristic' yield stress divided by an appropriate partial safety factor is used to define the limit of the range within which the steel stress is permitted. The same behaviour is assumed in both tension and compression, as indicated.

It should be noted from Figure 5.11 that a reinforcing bar still has a considerable margin of safety within its maximum load-carrying capacity and a significant amount of stretching capability beyond the yield point. In design this is very useful, since tensile cracking of the concrete will develop and give a warning that overloading and possibly failure is about to occur.

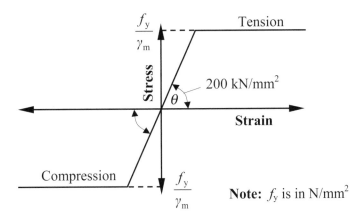

Figure 5.12 – BS 8110 Short-term design stress-strain curve for reinforcement

5.2.8 Steel Reinforcement Modulus of Elasticity: E_s (Clause 2.4.2.3)

The modulus of elasticity of reinforcement is equal to $\tan\theta$, where θ is the angle of the linear section of the stress-strain curve given in Figure 2.2 of the code. This should be taken as 200 kN/mm^2 as indicated in Figure 5.12.

5.2.9 Material Partial Safety Factors: γ_m (Clause 2.4.4.1)

As indicated in Section 2.4 when using limit state design:

$$Design\ strength = \frac{Characteristic\ strength}{\gamma_m}$$

The appropriate values of γ_m for concrete and reinforcement are given in Table 2.2 of Clause 2.4.4.1 as:

γ_m = 1.5   for concrete in flexure or axial load,
    = 1.25   for concrete shear strength *without* reinforcement,
    = 1.4   for bond strength between the reinforcement and the concrete,
    = 1.05   for reinforcement,
    ≥ 1.5   for other conditions e.g. bearing stresses.

5.2.10 Durability (Clauses 2.2.4 and 3.1.5)

The integrity of reinforced concrete depends on its ability to prevent corrosion of the reinforcement when exposed to a wide range of environmental conditions, e.g. ranging from '*mild exposure*' such as concrete surfaces protected against weather or aggressive conditions to '*most severe*' or '*abrasive*' conditions in which concrete surfaces may be frequently exposed to sea water spray, de-icing salts or the abrasive actions of machinery.

The classification of **exposure conditions** in terms of '*moderate*', '*severe*', '*very severe*', '*most severe*' and '*abrasive*' is defined in Table 3.2 of the code.

In addition to **protection against corrosion** of the steel (Table 3.3), **fire resistance** requirements (Table 3.4) are also necessary to allow sufficient time to evacuate a building and prevent premature failure, spalling of the concrete must be avoided and adequate bond forces must develop between the reinforcement and the concrete.

The essential elements of design which ensure adequate durability are the structural form/detailing and the amount of concrete cover provided to protect the steel, as illustrated in Figure 5.13.

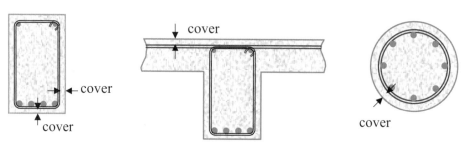

Figure 5.13

Concrete cover is defined as the thickness of concrete between the outer surface of the steel reinforcement and the nearest concrete surface. The *actual* concrete cover provided varies due to a number of factors such as:

♦ construction tolerances inherent in building the formwork (i.e. the mould into which the concrete is cast),

♦ variations in dimensions of the reinforcement resulting from the cutting and bending of the steel, and

♦ errors occurring during the fixing of the steel in the formwork.

The limiting values of cover given in Tables 3.3 and 3.4 of the code which ensure adequate provision to satisfy durability and fire protection are specified in terms of *'nominal'* cover to all reinforcement including the links. As indicated in Clause 3.3.1.1, the nominal cover is: '… *the dimension used in design and indicated on the drawings. The actual cover to all reinforcement should never be less than the nominal cover minus 5 mm …*'

There a number of criteria to be considered when determining the nominal cover. They are:

♦ **Bar size** *(Clause 3.3.1.2)*

Single bars: nominal cover $\geq$ main bar diameter d_1

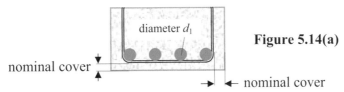

Figure 5.14(a)

Paired bars: nominal cover $\geq$ $\sqrt{2}\,d_1$ where d_1 is the main bar diameter

Figure 5.14(b)

Bundled bars: nominal cover $\geq$ $2\sqrt{\dfrac{A_{\text{equivalent}}}{\pi}}$

where $A_{\text{equivalent}}$ is the cross-sectional area equal to the sum of the cross-sectional areas of the bars in the bundle

$$\left(\text{e.g. In Figure 5.14(c)} \quad A_{\text{equivalent}} = 4 \times \left[\frac{\pi d_1^2}{4}\right]\right)$$

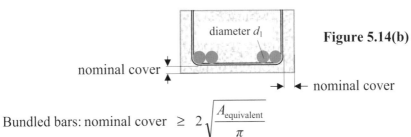

Figure 5.14(c)

◆ **Nominal maximum aggregate size** *(Clause 3.3.1.3)*
   Nominal cover   ≥   nominal maximum size of aggregate

$$\text{i.e. normally} \leq \frac{\text{minimum thickness of concrete section}}{4}$$

   In most cases, 20 mm aggregate is suitable.

◆ **Uneven surfaces** *(Clause 3.3.1.4)*
   When concrete is cast on uneven surfaces (e.g. earth or blinding, which is finely
   crushed aggregate rolled on the top of compacted fill such as hardcore) additional
   cover to that indicated in Table 3.3 should be provided as shown in Figure 5.15.

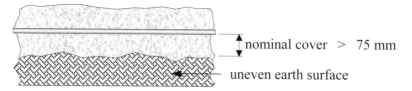

   nominal cover > 75 mm
   uneven earth surface

Concrete cast directly on the earth – nominal cover from average soil level

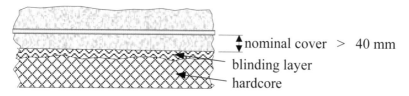

   nominal cover > 40 mm
   blinding layer
   hardcore

Concrete cast on an adequate blinding layer (e.g. 50 mm thick)

Figure 5.15

◆ **Ends of straight bars** *(Clause 3.3.2)*
   Normally 40 mm cover is provided at the ends of straight bars, as shown in
   Figure 5.16, however as indicated in this clause where the end of a floor or roof
   unit is not exposed to the weather or to condensation, cover is not mandatory.

nominal cover

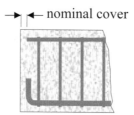

Figure 5.16

In Clause 7.3 of the code recommendations are given to ensure that the reinforcement is
properly placed and the required cover obtained. This is achieved during construction by
inserting *spacer blocks* and *chairs* in the formwork, on the reinforcement as indicated in

Figures 5.17 and 5.18. The spacers must be designed such that they are durable and will not lead to corrosion of the reinforcement or to spalling of the concrete. The use of spacer blocks constructed on site from concrete is ***not*** permitted.

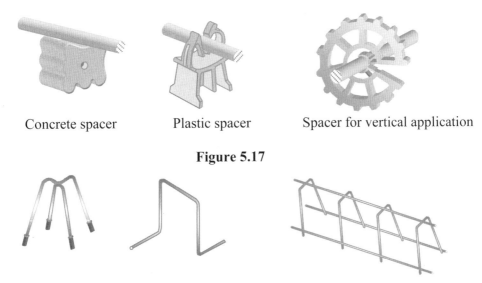

    Concrete spacer   Plastic spacer   Spacer for vertical application

Figure 5.17

Single and continuous high wire chairs for top steel – particularly in cantilevers

Figure 5.18

5.2.10.1 Minimum Dimensions (Clause 3.3.6 and Figure 3.2)

In addition to nominal cover requirements, the code also specifies ***minimum dimensions*** (i.e. beam widths, rib widths, floor and wall thicknesses, and column widths) for some structural elements to provide adequate fire resistance. The dimensions are given to ensure minimum periods of fire resistance ranging from 0.5 hours to 4 hours and relate specifically to the covers given in Table 3.4.

5.2.11 *Example 5.1: Nominal Cover 1*

A rectangular reinforced concrete beam inside a building is simply supported and is required to support precast concrete units as shown in Figure 5.19. Using the data given, determine:

    (i)   the nominal cover required to the underside of the beam, and
    (ii)   the minimum width of beam required.

Data:

Exposure condition	mild
Characteristic strength of concrete (f_{cu})	40 N/mm^2
Nominal maximum aggregate size (h_{agg})	20 mm
Diameter of main tension steel	25 mm
Diameter of shear links	8 mm
Minimum required fire resistance	1.5 hours

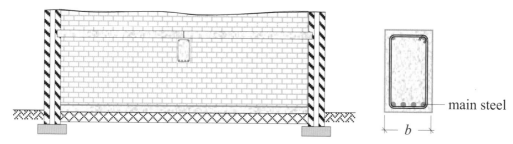

main steel

Figure 5.19

Solution:
(i)   Clause 3.3.1.2   Nominal cover $\geq$ (main bar diameter – link diameter)
                                                    $\geq$ (25 – 8) = 17 mm
         Clause 3.3.1.3   Nominal cover $\geq$ nominal maximum aggregate size > 20 mm
         Clause 3.3.3   Exposure condition is mild
                                        Grade of concrete is C40
         Table 3.3   Nominal cover $\geq$ 20 mm*
         Clause 3.3.6   Minimum fire resistance = 1.5 hr
                                        The beam is simply supported
         Table 3.4   Nominal cover $\geq$ 20 mm*
The required nominal cover = 20 mm
Note: Under these conditions this value can be reduced to 15 mm when the maximum
            aggregate size does not exceed 15 mm.

(ii)   Clause 3.3.6   The minimum beam width b to satisfy the required 1.5 hours.
         Figure 3.2   fire resistance = 200 mm

5.2.12 Example 5.2: Nominal Cover 2

A continuous, ribbed floor slab covering a car parking area is exposed on the underside
and protected on the topside as shown in Figure 5.20. Using the data given determine:

        (i)   the nominal cover required to the underside of the rib,
        (ii)   the nominal cover required to the topside of the floor, and
        (iii)   the minimum floor thickness and width of rib required.

Data:
Characteristic strength of concrete   (f_{cu})   35 N/mm^2
Nominal maximum aggregate size   (h_{agg})   20 mm
Diameter of main tension steel   32 mm
Diameter of shear links   10 mm
Minimum required fire resistance   2.0 hours

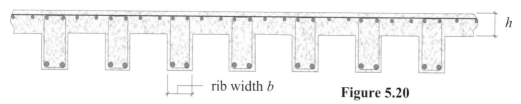

h

rib width b **Figure 5.20**

Solution:

(i) Clause 3.3.1.2 Nominal cover $\geq$ (main bar diameter − link diameter)

$$\geq (32 - 10) = 22 \text{ mm}$$

Clause 3.3.1.3 Nominal cover $\geq$ nominal maximum aggregate size

$$\geq 20 \text{ mm}$$

Table 3.2 Exposure condition is moderate

Grade of concrete is C35

Table 3.3 Nominal cover $\geq$ 35 mm

Clause 3.3.6 Minimum fire resistance = 2.0 h

The rib is continuous

Table 3.4 Nominal cover $\geq$ 35 mm*

The required nominal cover to the underside of the ribs = 35 mm

(ii) The topside of the slab is protected and hence the exposure condition is mild:

Table 3.3 Nominal cover $\geq$ 20 mm

Clause 3.3.6 Minimum fire resistance = 2.0 h

The floor is continuous

Table 3.4 Nominal cover $\geq$ 25 mm

The required nominal cover to the topside of the floor = 25 mm

(iii) Clause 3.3.6 The minimum rib width b to satisfy the required 2.0 hours

Figure 3.2 fire resistance = **125 mm**

This can be achieved if the bars are placed vertically, as shown. The minimum width b required to accommodate the bars and the cover = (35 + 10 + 32 + 10 + 35) = 122 mm < 125 mm

If the bars are placed horizontally, assuming a gap of 25 mm (i.e. [h_{agg} + 5 mm], see Clause 3.12.11.1), the minimum width required = [2 × (35 + 10 + 32) + 25] ≈ 180 mm > 125 mm

Clause 3.3.6 The minimum floor thickness h to satisfy the required

Figure 3.2 2.0 hours fire resistance = **125 mm**

*****Note:** In the case of the cover exceeding 40 mm for dense or 50 mm for lightweight aggregate concrete there is a danger of spalling. If the ribbed slab were simply supported then Note 2 in Table 3.4 indicates that additional measures would be necessary to reduce the risks of spalling, as indicated in section 4 of *BS 8110:Part 2:1985*.

Possible measures include an applied finish by hand or spray of plaster, the provision of a false ceiling as a fire barrier, the use of lightweight aggregates, the use of sacrificial tensile steel, or the provision of supplementary reinforcement in the form of welded steel fabric placed within the cover at 20 mm from the concrete face.

5.2.13   Example 5.3: Nominal Cover 3

A ground floor slab in a warehouse building is constructed on a blinded, compacted layer of hardcore as shown in Figure 5.21. Using the data given, determine the nominal cover required to the underside of the slab.

Data:

Characteristic strength of concrete (f_{cu})	35 N/mm^2
Nominal maximum aggregate size (h_{agg})	20 mm
Diameter of main tension steel	20 mm

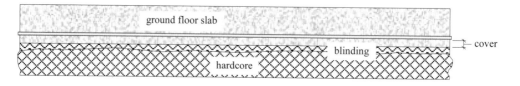

ground floor slab

blinding — cover

hardcore

Figure 5.21

Solution:

(i)   Clause 3.3.1.2   Nominal cover $\geq$ main bar diameter
                                                  $\geq$ 20 mm
        Clause 3.3.1.3   Nominal cover $\geq$ nominal maximum aggregate size
                                                  $\geq$ 20 mm
        Clause 3.3.3   Exposure condition is mild
                                      Grade of concrete is C35
        Table 3.3   Nominal cover $\geq$ 20 mm
        Clause 3.3.1.4   Since this slab is cast against an adequate blinding
        Table 3.4   Nominal cover $\geq$ 40 mm
        The required nominal cover = 40 mm

5.3   *Flexural Strength of Sections*

The flexural strength (i.e. the ultimate moment of resistance of a cross-section) is determined assuming the following conditions as given in Clause 3.4.4.1, *BS8110:Part 1*.

♦   Plane sections remain plane, i.e.

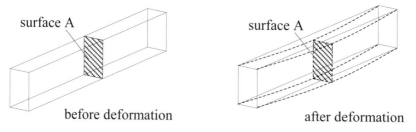

surface A   surface A

before deformation   after deformation

The surface of any cross-section does **not** distort out-of-plane during deformation.

Figure 5.22

♦ The compressive stresses in the concrete may be derived from the stress-strain curve in Figure 2.1 (Figure 5.9 in this text) with $\gamma_m = 1.5$, i.e. resulting in a rectangular-parabolic stress block.

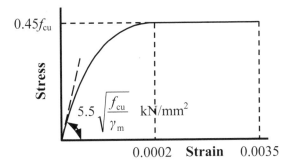

The total compressive force in the concrete is equal to the sum of F_{c1} and F_{c2} each considered acting through their respective centroids and on the areas as indicated in Figure 5.23.

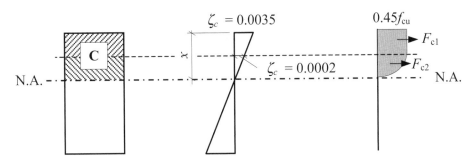

Figure 5.23

or alternatively

Using the simplified rectangular stress block as indicated in Figure 3.3 of the code as shown in Figure 5.24.

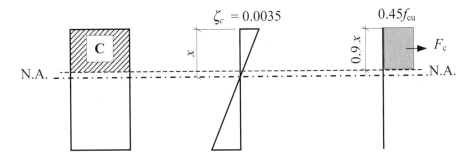

Figure 5.24

The total compressive force in the concrete, F_c, is considered as acting through the centroid of a reduced depth of stress block which is equal to 90% of the depth of the neutral axis from the compression face.

The alternative simplified rectangular stress block is normally used in design since it is more convenient when evaluating the magnitude of the total compressive force in the concrete and the position of its centroid. This simplification produces results which are very similar to those given by the rectangular-parabolic stress block. (It should be noted that Part 3 of BS 8110 gives design charts which are based on the rectangular-parabolic stress block.)

♦   The tensile strength of the concrete is ignored, i.e. all concrete below the level of the neutral axis is considered ineffective.

♦   The stresses in the reinforcement are derived from the stress-strain curve in Figure 2.2 (Figure 5.12 in this text) with $\gamma_m = 1.05$.

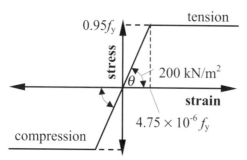

The characteristic yield strength for high yield steel reinforcement is given in Table 3.1 of the code as 460 N/mm² resulting in an design ultimate yield strength equal to 437 N/mm² and the strain (ξ_s) corresponding to the limit of elasticity equal to 0.00219.

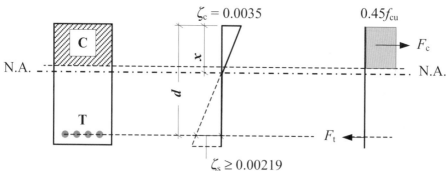

Figure 5.25

The dimension from the extreme compression face to the centroid of the tensile force is known as the *effective depth* of the cross-section and given the symbol *d*. From the strain diagram in Figure 5.25:

$$\frac{\zeta_{steel}}{(d-x)} = \frac{\zeta_{concrete}}{x} \qquad \therefore \; \zeta_s = \left(\frac{d-x}{x}\right)\zeta_{concrete}$$

The strain and consequently the stress in the steel are dependent on the depth of the neutral axis *x* from the compression face.

When $x \leq d/2$ then $\zeta_s \geq \zeta_c$ (i.e. ≥ 0.0035) and the steel has yielded, the steel stress is given by:

f_{steel} $=$ $0.95f_y$ and the cross-section will fail by yielding of the steel.

The design ultimate moment of resistance will be governed by the capacity of the steel in the section.

When $x \geq d/2$ then $\zeta_s \leq \zeta_c$

In this case it is possible for f_{steel} to be less than $0.95f_y$ and the cross-section can fail by crushing of the concrete. The critical value of x at which the steel stress becomes less than $0.95f_y$ can be found as follows:

$$\frac{\zeta_{steel}}{(d-x)} = \frac{\zeta_{concrete}}{x} \qquad \therefore \quad \frac{0.00219}{(d-x)} = \frac{0.0035}{x}$$

$0.00219x = 0.0035d - 0.0035x$ $\therefore$ $x = 0.615d$

Since this type of failure occurs without warning and must be avoided, the value of x to the neutral axis is limited to $\leq \mathbf{0.5d}$ as indicated in Clause 3.4.4.4 of the code, and hence limits the permitted design ultimate moment of resistance when based on the concrete strength.

♦   The lever arm (i.e. the distance z between the centroids of the total compressive force and the tensile force) should be $\leq 0.95d$.

This effectively defines a lower limit on the depth of concrete which is considered to act in compression. It limits the maximum strain which can be induced in the reinforcement to a value of 0.0283 as shown in Figure 5.26, and in addition avoids the reliance on any poor quality concrete material which may be present at the top of the beam.

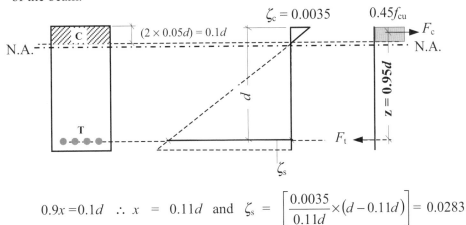

$$0.9x = 0.1d \quad \therefore \quad x = 0.11d \quad \text{and} \quad \zeta_s = \left[\frac{0.0035}{0.11d} \times (d - 0.11d)\right] = 0.0283$$

Figure 5.26

♦   In the analysis of a cross-section resisting a small axial load ≤ $(0.1f_{cu} \times$ area), the
    effects of the load may be ignored.
        The presence of a small co-existent axial thrust slightly increases the calculated
    ultimate moment of resistance of a section; however the complexity of the
    calculations to determine it do not justify the effort required.

The conditions of Clause 3.4.4.1 have been illustrated using cross-sections in which there
is sufficient concrete above the neutral axis to resist the required compressive force
induced by the applied design moment, i.e. ***singly-reinforced*** sections. Generally an
increase in the applied design bending moment can be resisted by increasing the depth of
the section (i.e. increasing the area available to resist compression).
    There are circumstances in which this cannot be done, e.g. to satisfy client/architectural
requirements or existing physical constraints, and the capacity to resist compressive forces
is increased by introducing reinforcing steel in the compression zone, as shown in
Figure 5.27. Such sections are known as ***doubly-reinforced*** sections.

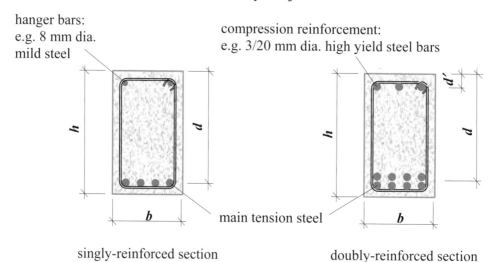

singly-reinforced section   doubly-reinforced section

where:
h is the overall depth of a section,
b is the breadth of a section,
d is the effective depth from the compression face of the concrete to the centroid of the
    tension reinforcement,
d' is the depth from the compression face of the concrete to the centroid of the
    compression reinforcement.

Figure 5.27

In singly-reinforced sections the steel in the compression zone is to enable the fabrication
of a reinforcing cage comprising the main tension steel and the shear links.
    In doubly-reinforced sections the steel in the compression zone is required to resist
additional compressive forces; note the higher area of tensile steel in this section:

force in tensile steel   =   (force in compressive steel + compressive force in concrete)

5.3.1 Singly-reinforced Sections

The ultimate moment of resistance of singly-reinforced rectangular beams can be determined in terms of:

    (a) the concrete capacity, and
    (b) the steel capacity,

the smaller of these two values being the critical case.

(a) Concrete capacity
The maximum compressive force which can be resisted by the concrete corresponds to the maximum depth permitted for the neutral axis, as shown in Figure 5.28 (i.e. $x = d/2$).

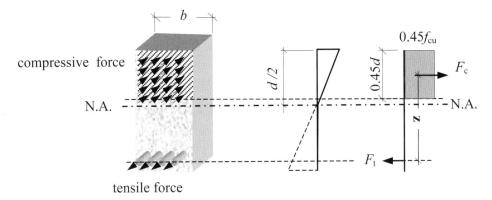

Figure 5.28

The moment of resistance of the section is developed by the compressive and tensile forces F_c and F_t separated by the lever arm z.

Consider the moment of the compressive force about the line of action of F_t :

$$M_{ult,concrete} = (F_c \times z)$$

where F_c = compressive force   = (stress × area)

$$= [0.45f_{cu} \times (b \times 0.9x)] \quad \text{(for the maximum concrete force } x = d/2)$$

$$= [0.45f_{cu} \times (b \times 0.45d)] = 0.2bdf_{cu}$$

and z = lever arm

$$= [d - (0.5 \times 0.45d)] \quad = 0.775d$$

{**Note**: In general $z = [d - (0.5 \times 0.9x)]$ }

$$M_{ult,concrete} = (0.2bdf_{cu}) \times (0.775d) = 0.156bd^2f_{cu}$$

This equation can be rewritten as $0.156 = \dfrac{M}{bd^2 f_{cu}}$

The term $\dfrac{M}{bd^2 f_{cu}}$ is given the symbol K in Clause 3.4.4.4 and represents the moment capacity of the section based on the concrete area in compression. The value of 0.156 is the limiting value of K and is given the symbol K'.

When $K \le K'$ for a singly-reinforced section the maximum moment permitted, based on the concrete strength, is equal to $0.156\,bd^2 f_{cu}$

When $K > K'$ a section requires compression reinforcement.

(b)   Steel capacity
As discussed previously, if the maximum neutral axis depth is limited to $0.5d$ (see Clause 3.4.4.4) the steel stress will reach its design strength of $0.95f_y$.

Consider the moment of the tensile force about the line of action of F_c:

$$
\begin{aligned}
M_{ult,steel} &= (F_s \times z) \\
\text{where}\quad F_s. &= \text{tensile force} \quad = (\text{stress} \times \text{area}) \\
&= (0.95f_y \times A_s) \\
\text{and}\quad z &= \text{lever arm} \\
M_{ult,steel} &= 0.95f_y A_s z
\end{aligned}
$$

In the code this equation is presented as $A_s = M/0.95f_y z$

Consider z, the lever arm:
$$ z = [d - (0.5 \times 0.9x)] \qquad \therefore\ 0.9x = 2(d - z) $$

The *general* equation for the moment of resistance in terms of the concrete is:
$$
\begin{aligned}
M_{ult,concrete} &= [0.45f_{cu} \times (b \times 0.9x)] \times z \qquad\qquad \text{(substitute for } 0.9x) \\
&= 0.9f_{cu}\, b\, (d - z)\, z
\end{aligned}
$$
also
$$
\begin{aligned}
M_{ult,concrete} &= K\,bd^2 f_{cu} \\
\therefore\ K\,bd^2 f_{cu} &= 0.9f_{cu}\, b\,(d - z)\, z \qquad \therefore\ Kd^2 = 0.9f_{cu}\,dz - 0.9z^2
\end{aligned}
$$

$z^2 - dz + \dfrac{K}{0.9}d^2 = 0$ The solution of this quadratic equation $(ax^2 + bx + c = 0)$ gives an expression which can be used to determine the lever arm, z.

$$
z = \frac{-b \pm \sqrt{b^2 - 4ac}}{2a} = \frac{d \pm \sqrt{d^2 - \dfrac{4Kd^2}{0.9}}}{2} = 0.5d + \frac{d}{2}\sqrt{0.25 - \frac{K}{0.9}}
$$

$$
z = d\left[0.5 + \sqrt{0.25 - \frac{K}{0.9}}\, \right] \qquad \text{This is the expression given in Clause 3.4.4.4 of BS 8110}
$$

with the condition that $z \le 0.95$.

The maximum ultimate moment of resistance of a singly-reinforced beam in which the dimensions b and d and the area of reinforcing steel A_s are known is given by the lesser of the following equations:

$$M_{\text{ult,concrete}} = \mathbf{0.156bd^2f_{cu}} \qquad \text{based on the concrete strength}$$

$$M_{\text{ult,steel}} = \mathbf{0.95f_yA_sz} \qquad \text{based on the steel strength where } z = 0.775d$$

5.3.2 Example 5.4: Singly-reinforced Rectangular Beam 1

A rectangular beam section is shown in Figure 5.29. Using the data given, determine the maximum ultimate moment which can be applied to the section assuming it to be singly-reinforced.

Data:
Characteristic strength of concrete (f_{cu}) $\qquad$ 40 N/mm^2
Characteristic strength of steel (f_y) $\qquad$ 460 N/mm^2

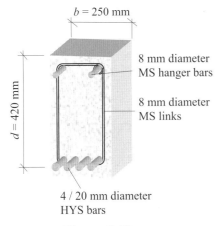

$b = 250$ mm

8 mm diameter MS hanger bars

8 mm diameter MS links

$d = 420$ mm

4 / 20 mm diameter HYS bars

Figure 5.29

Solution:
Strength based on concrete:
$$\begin{aligned} M_{\text{ult}} &= 0.156bd^2f_{cu} \\ &= (0.156 \times 250 \times 420^2 \times 40)/10^6 \\ &= 275.2 \text{ kNm} \end{aligned}$$

Strength based on steel:
$$\begin{aligned} A_s &= 1260 \text{ mm}^2 \quad \text{(see Appendix 6)} \\ M_{\text{ult}} &= 0.95f_yA_sz \\ &= (0.95 \times 460 \times 1260 \times 0.775 \times 420)/10^6 \\ &= 179.5 \text{ kNm} \end{aligned}$$

The maximum design moment which can be applied is: $\quad \boldsymbol{M_{\text{ult}} = \textbf{179.5 kNm}}$

5.3.3 Example 5.5: Singly-reinforced Rectangular Beam 2

The cross-section of a simply supported rectangular beam is shown in Figure 5.30. Using the data given, determine the maximum ultimate moment which can be applied to the section assuming it to be singly-reinforced.

Data:
Characteristic strength of concrete (f_{cu}) $\qquad$ 30 N/mm^2
Characteristic strength of steel (f_y) $\qquad$ 460 N/mm^2
Nominal maximum aggregate size (h_{agg}) $\qquad$ 20 mm
Diameter of main tension steel $\qquad$ 32 mm
Diameter of shear links $\qquad$ 8 mm
Exposure condition $\qquad$ mild
Minimum required fire resistance $\qquad$ 1.0 hour

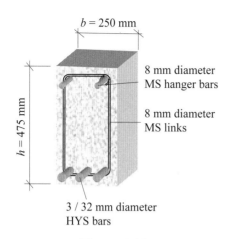

$b = 250$ mm

8 mm diameter
MS hanger bars

8 mm diameter
MS links

$h = 475$ mm

3 / 32 mm diameter
HYS bars

Figure 5.30

Solution:

Clause 3.3.1   Nominal cover to all steel
Clause 3.3.1.2   ≥ bar size
                      cover ≥ (32 – 8) = 24 mm
Clause 3.3.1.3   ≥ Nominal maximum
                      aggregate size
                      cover ≥ 20 mm
Clause 3.3.3   Exposure condition: mild
Table 3.3   and $f_{cu} = 30$ N/mm^2
                      cover ≥ 25 mm
Clause 3.3.6   Min. fire resistance: 1 hr
                      beam is simply supported
Table 3.4   cover ≥ 20 mm

The required nominal cover to all steel   = 25 mm

Figure 3.2   Minimum width b for 1 hour fire resistance   = 200 mm   ∴ adequate

Effective depth d = (h – cover – link diameter – bar diameter/2)
                    = (475 – 25 – 8 – 16)   = 426 mm

Strength based on concrete:
M_{ult} = $0.156bd^2f_{cu}$ = $(0.156 \times 250 \times 426^2 \times 30)/10^6$ = 212.3 kNm

Strength based on steel:
A_s = 2410 mm^2 (see Appendix 6)
M_{ult} = $0.95f_yA_sz$ = $(0.95 \times 460 \times 2410 \times 0.775 \times 426)/10^6$ = 349.9 kNm

The maximum design moment which can be applied is: $\boldsymbol{M_{ult}}$ = **212.3 kNm**

5.3.4 Example 5.6: Singly-reinforced Rectangular Beam 3

The cross-section of a simply supported rectangular beam is shown in Figure 5.31. Using the data given, and assuming the section to be singly-reinforced, determine the area of tension reinforcement required to resist an applied ultimate bending moment of 150 kNm.

Data:
Characteristic strength of concrete   (f_{cu})   40 N/mm^2
Characteristic strength of steel   (f_y)   460 N/mm^2
Nominal maximum aggregate size   (h_{agg})   20 mm
Diameter of main tension steel   Assume 25 mm
Diameter of shear links   10 mm
Exposure condition   moderate
Minimum required fire resistance   2.0 hours

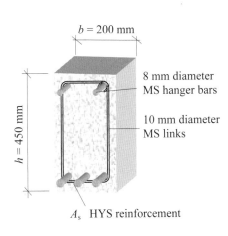

b = 200 mm

8 mm diameter
MS hanger bars

10 mm diameter
MS links

h = 450 mm

A_s HYS reinforcement

Figure 5.31

Solution:

Clause 3.3.1   Nominal cover to all steel
Clause 3.3.1.2   ≥ bar size
                  cover = (25 – 10) = 15 mm
Clause 3.3.1.3   ≥ Nominal maximum
                       aggregate size
                  cover = 20 mm
Clause 3.3.3   Exposure condition:
                       moderate
Table 3.3 $f_{cu} = 40$ N/mm^2
                  cover ≥ 30 mm
Clause 3.3.6   Min. fire resistance: 2.0 hr
                       beam is simply supported
Table 3.4   cover ≥   40 mm

The required nominal cover to the main steel   = 40 mm
Figure 3.2   Minimum width *b* for 2 hours, fire resistance   = 200 mm ∴ adequate

Effective depth *d* = (*h* – cover – link diameter – bar diameter/2)
                     = (450 – 40 – 10 – 13)   = 387 mm

Clause 3.4.4.4
Check that the section is singly-reinforced:

$$K = \frac{M}{bd^2 f_{cu}} = \left[\frac{150 \times 10^6}{200 \times 387^2 \times 40}\right] = 0.125 \le 0.156$$

Since $K < K'$ the section is singly-reinforced:

$$Z = d\left[0.5 + \sqrt{0.25 - \frac{K}{0.9}}\right] = d\left[0.5 + \sqrt{0.25 - \frac{0.125}{0.9}}\right] = 0.83d \le 0.95d$$

$A_s = M/0.95 f_y z = [(150 \times 10^6)/(0.95 \times 460 \times 0.83 \times 387)] = 1068$ mm^2

Appendix 6   **Adopt 4/ 20 mm diameter HYS bars providing 1260 mm^2.**

5.3.5 *Example 5.7: Singly-reinforced Rectangular Slab 1*

A rectangular floor slab is supported on two masonry walls as shown in Figure 5.32. Using the data given, determine the *main reinforcement* required:

  (i)   in span AB and
  (ii)   over support B.

(**Note:** slabs also have *secondary reinforcement* which is placed perpendicular to the main steel. This is discussed in Section 5.7.1 and has not been included in this example.)

Data:

Characteristic strength of concrete	(f_{cu})	40 N/mm^2
Characteristic strength of steel	(f_y)	460 N/mm^2
Nominal maximum aggregate size	(h_{agg})	20 mm
Diameter of main tension steel		Assume 20 mm
Exposure condition		mild
Minimum required fire resistance		1.0 hours
Characteristic dead load	(g_k)	1.5 kN/m^2
Characteristic imposed load	(q_k)	5.0 kN/m^2
Characteristic dead load	$(\gamma_{concrete})$	24.0 kN/m^3

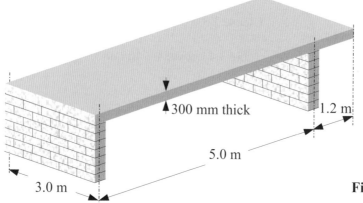

Figure 5.32

The design of slabs is normally carried out on the basis of a 1.0 metre wide strip and the area of steel (A_s/metre width) required is calculated and given as bars at specified centres.

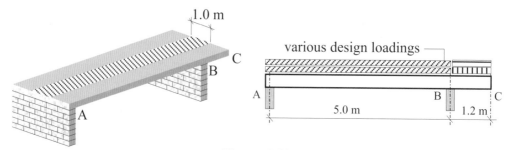

Figure 5.33

The shape of the bending moment diagram for this slab is indicated in Figure 5.34 indicating tension occurring in the bottom of the concrete between A and B and at the top over support B and along the cantilever length BC.

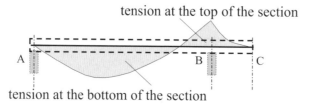

Figure 5.34

This results in the provision of main steel in the bottom of the slab between A and B and in the top of the slab over support B and the cantilever span as shown in Figure 5.34.

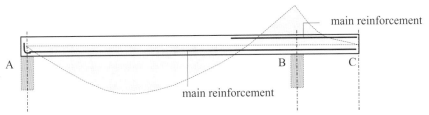

Figure 5.34

The ultimate design bending moments for which the main steel is determined between A and B and over support B are calculated considering the appropriate load cases, as follows.

Load case 1: Loads required for maximum bending moment in span AB

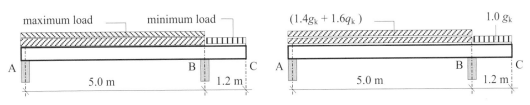

Figure 5.35(a)

Load case 2: Loads required for maximum bending moment over support B and in span BC:

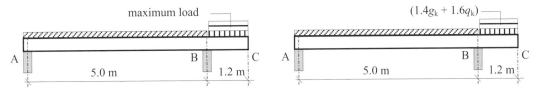

Note that in this load case the loading on span AB does not need to be considered to determine the maximum value of bending moment over support B.

Figure 5.36(b)

Design loads:

Self-weight of the slab	$= (24 \times 0.3)$	$= 7.2 \text{ kN/m}^2$	
Characteristic applied dead load		$= 1.5 \text{ kN/m}^2$	
Maximum ultimate design dead load		$= 1.4 \times (1.5 + 7.2)$	$= 12.2 \text{ kN/m}^2$
		$= 12.2 \text{ kN/m length for a 1.0 m wide strip}$	
Minimum ultimate design dead load		$= 1.0 \times (1.5 + 7.2)$	$= 8.7 \text{ kN/m}^2$
		$= 8.7 \text{ kN/m length for a 1.0 m wide strip}$	
Maximum ultimate design imposed load	$= (1.6 \times 5.0)$	$= 8.0 \text{ kN/m}^2$	
		$= 8.0 \text{ kN/m length for a 1.0 m wide strip}$	

Load case 1:

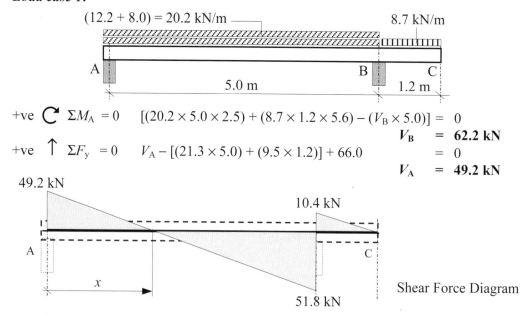

$+ve \; \circlearrowleft \; \Sigma M_A \; = 0 \quad [(20.2 \times 5.0 \times 2.5) + (8.7 \times 1.2 \times 5.6) - (V_B \times 5.0)] \; = \; 0$

$$V_B \; = \; \textbf{62.2 kN}$$

$+ve \; \uparrow \; \Sigma F_y \; = 0 \quad V_A - [(21.3 \times 5.0) + (9.5 \times 1.2)] + 66.0 \qquad = \; 0$

$$V_A \; = \; \textbf{49.2 kN}$$

The position of the maximum bending moment $\quad x \; = \; \dfrac{49.2}{20.2} \; = \; 2.44 \text{ m}$

The maximum bending moment is given by the area under the shear force diagram:

Ultimate applied bending moment $\; M_{\text{ult}} = \quad (0.5 \times 2.44 \times 49.2) = \; 60.02$ kNm

Clause 3.3.1	Nominal cover to all steel	
Clause 3.3.1.2 ≥	Bar size	cover ≥ 20 mm
Clause 3.3.1.3 ≥	Nominal maximum aggregate size	cover = 20 mm
Clause 3.3.3	Exposure condition: mild $\quad f_{cu} = 40$ N/mm^2	
Table 3.3		cover ≥ 20 mm
Clause 3.3.6	Minimum fire resistance: 1.0 hr slab is continuous	
Table 3.4		cover ≥ 20 mm

The required nominal cover to the main steel   = 20 mm

Figure 3.2   Minimum thickness h for 1 hour fire resistance = 95 mm   ∴ adequate

Effective depth $d \; = \;$ (h − cover − bar diameter/2)

$\qquad\qquad\qquad = \;$ (300 − 20 − 10)   = 270 mm

Clause 3.4.4.4

Check that the section is singly-reinforced:

$$K \; = \; \frac{M}{bd^2 f_{cu}} \; = \; \left[\frac{60.02 \times 10^6}{1000 \times 270^2 \times 40} \right] \; = \; 0.021 \; \leq \; 0.156$$

Since $K \; < \; K'$ the section is singly-reinforced.

$$Z = d\left[0.5 + \sqrt{0.25 - \frac{K}{0.9}}\right] = d\left[0.5 + \sqrt{0.25 - \frac{0.021}{0.9}}\right] = 0.98d \quad > \quad 0.95d$$

The lever arm is restricted to $0.95d$

$$A_s = M/0.95 f_y z = [(60.02 \times 10^6)/(0.95 \times 460 \times 0.95 \times 270)] = 536 \text{ mm}^2 / \text{metre width}$$

Appendix 6   **Adopt 10 mm diameter HYS bars @ 125 mm centres providing 628 mm^2/m width.**

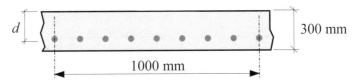

The assumed bar diameter used to determine the effective depth was 20 mm and the bars selected are 10 mm diameter. This makes a small difference to the calculation of the effective depth. Theoretically the calculation should be repeated with the modified value: however it is generally acceptable to neglect this correction (i.e. d is slightly increased).

Load case 2:

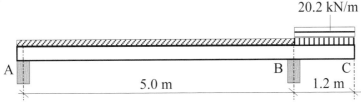

Ultimate applied bending moment $M_{ult} = (20.2 \times 1.2 \times 0.6) = 14.5 \text{ kNm}$

The required nominal cover to the main steel is the same as before $= 20 \text{ mm}$

Effective depth $d = (h - \text{cover} - \text{bar diameter}/2)$ Assume 10 mm diameter bars
$$= (300 - 20 - 5) = 275 \text{ mm}$$

Clause 3.4.4.4:
Check that the section is singly-reinforced:

$$K = \frac{M}{bd^2 f_{cu}} = \left[\frac{14.5 \times 10^6}{1000 \times 275^2 \times 40}\right] = 0.005 \ll 0.156$$

Since $K < K'$ the section is singly-reinforced:

$$Z = d\left[0.5 + \sqrt{0.25 - \frac{K}{0.9}}\right] = d\left[0.5 + \sqrt{0.25 - \frac{0.005}{0.9}}\right] = 0.99d \quad > \quad 0.95d$$

The lever arm is restricted to $0.95d$.

$A_s = M/0.95f_yz = [(14.5 \times 10^6)/(0.95 \times 460 \times 0.95 \times 270)] = 130 \text{ mm}^2/\text{metre width}$

This is a very small area of steel, and the code requires a minimum percentage of reinforcement in a section, as indicated in Section 5.7.1. In the case of solid rectangular slabs the minimum area when using high yield steel is 0.13% of the total concrete cross-sectional area.

The minimum area of steel $A_{s, \text{minimum}} = \left[\dfrac{0.13}{100} \times (1000 \times 300) \right] = 390 \text{ mm}^2/\text{m width}$

Appendix 6   **Adopt 10 mm diameter HYS bars @ 200 mm centres providing 393 mm²/m width.**

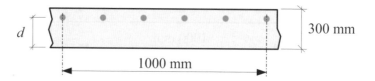

5.3.6   *Doubly-reinforced Sections*

When the applied design bending moment exceeds the concrete capacity (i.e. $0.156\,bd^2f_{cu}$) compression reinforcement is required. Consider the rectangular beam shown in Figure 5.36 in which the neutral axis depth is equal to $d/2$ and both tension (A_s) and compression (A_s') reinforcement are present:

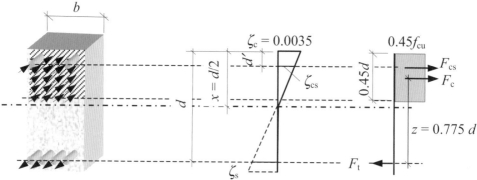

where:
A_s is the area of tension reinforcement,
A_s' is the area of compression reinforcement,
F_t is the force in the tensile reinforcement,
F_c is the compression force in the concrete,
F_{cs} is the force in the compression reinforcement,
ζ_{cs} is the strain in the compression reinforcement,
b, d, f_{cu} and ζ_s are as before.

Figure 5.36

The force in the compression reinforcement is dependent on the stress and consequently the strain ζ_{cs}. In order to ensure that this is greater than or equal to the yield stress, the ratio of (d'/x) is limited to 0.37 as shown in Figure 5.37.

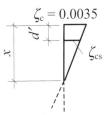

$$\frac{(x-d')}{\zeta_{cs}} = \frac{x}{\zeta_c} \qquad \therefore (x-d') = x\frac{\zeta_{cs}}{\zeta_c}$$

$$d' = x\left(1-\frac{\zeta_{cs}}{\zeta_c}\right) = x\left(1-\frac{0.002}{0.0035}\right) = 0.43x$$

$$d'/x = 0.43$$

Figure 5.37

This value is given as **0.37** in the code to accommodate redistribution of moments in excess of 10%.

The moment of resistance of the section is developed by the action of the combined compressive forces (F_c and F_{cs}) and the tensile force (F_t), separated by the lever arms z and $(d-d')$ respectively.

Consider the moment of the compressive forces about the line of action of F_t :

$$M = \{(F_c \times z) + [F_{cs} \times (d-d')]\}$$

where:

$$F_c = [0.45f_{cu} \times (b \times 0.45d)] = 0.2bdf_{cu}$$
$$z = [d-(0.5 \times 0.45d)] \quad = 0.775d$$
$$F_{cs} = 0.95f_y A_s'$$

$$M = \{[(0.2bdf_{cu}) \times (0.775d)] + [(0.95f_y A_s') \times (d-d')]\}$$
$$M = 0.156bd^2f_{cu} + [0.95f_y A_s'(d-d')]$$

This equation can be rewritten as:

$$A_s' = \frac{\left[M - \left(0.156\,bd^2 f_{cu}\right)\right]}{0.95 f_y (d-d')}$$

When considering singly-reinforced sections the symbol K is defined in terms of the applied moment as $K = \dfrac{M}{bd^2 f_{cu}}$ and hence $M = K\,bd^2 f_{cu}$. This value of M can be substituted in the equation for A_s' and rewritten as:

$$A_s' = [K\,bd^2 f_{cu} - (0.156bd^2 f_{cu})] / [0.95f_y (d-d')]$$

The limiting value of K when defining singly-reinforced sections is $K' = 0.156$, giving:

$$A_s' = [K\,bd^2 f_{cu} - K'bd^2 f_{cu}] / [0.95f_y (d-d')]$$

$$\therefore \; A_s' = (K-K')\,f_{cu}\,bd^2 / 0.95f_y (d-d') \qquad \text{as presented in Clause 3.4.4.4 of the code.}$$

This equation is given in the code to determine the required area of compression reinforcement when $d' / x \leq 0.37$ and the redistribution of moments is $\leq 10\%$ (see Section 5.11.2).

The required area of tension reinforcement can be determined by equating the compressive and the tensile forces acting on the cross-section.

$$
\begin{aligned}
\text{Tensile force} \quad &= \quad \text{Compressive force} \\
0.95 f_y A_s \quad &= \quad (0.2bdf_{cu}) \quad + \quad (0.95 f_y A_s') \\
A_s \quad &= \quad \frac{0.2 f_{cu} bd}{0.95 f_y} \quad + \quad A_s'
\end{aligned}
$$

In the code this equation is presented as:

$$
A_s \quad = \quad (K' f_{cu}\, bd^2 / 0.95 f_y\, z) + A_s'
$$

where $K' = 0.156$ and

$$
z = d\left[0.5 + \sqrt{0.25 - \frac{K}{0.9}}\right] \quad \text{In this case } K = K' \text{ and hence } z = 0.775d
$$

$$
\therefore \quad A_s \quad = \quad \frac{0.156 f_{cu} bd^2}{0.95 f_y \times 0.78d} + A_s' \quad = \quad \frac{0.2 f_{cu} bd}{0.95 f_y} + A_s' \text{ as before.}
$$

In most cases designers position the compression steel to ensure that $d'/ x \leq 0.37$. If this is not the case, then the value of $0.95 f_y$ should be replaced by the stress (f_{sc}) corresponding to the calculated strain (ζ_{sc}) in the compression steel, i.e.

$$
f_{sc} = (E \times \zeta_{sc}) = (200,000 \times \zeta_{sc}) \text{ N/mm}^2
$$

The modified equations then become:

$$
\therefore \quad A_s' = (K - K') f_{cu}\, bd^2 / f_{sc}\, (d - d') \text{ for the compression reinforcement} \quad \text{and}
$$

$$
A_s = (K' f_{cu}\, bd^2 / 0.95 f_y\, z) + A_s' \times \left(\frac{f_{sc}}{0.95 f_y}\right) \quad \text{for the tension reinforcement.}
$$

This is illustrated in Example 5.10.

5.3.7 Example 5.8: Doubly-reinforced Rectangular Beam 1

The rectangular beam shown in Figure 5.38 is required to resist an ultimate design bending moment of 340 kNm. Using the data given, determine the required areas of main reinforcing steel.

Data:

Characteristic strength of concrete (f_{cu})	40 N/mm^2
Characteristic strength of steel (f_y)	460 N/mm^2
Diameter of main tension steel	Assume 32 mm
Diameter of main compression steel	Assume 12 mm

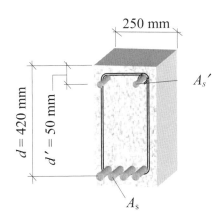

Figure 5.38

Solution:
Clause 3.4.4.4
$$K = M/bd^2f_{cu}$$
$$= (340 \times 10^6)/(250 \times 420^2 \times 40)$$
$$= 0.193$$
$K > K'$ (= 0.156), the section requires compression reinforcement.

$$z = d\left[0.5 + \sqrt{0.25 - \frac{K'}{0.9}}\right] = 0.775d$$

$$= (0.775 \times 420) = 327.6 \text{ mm}$$
$$x = (d-z)/0.45 = 0.5d$$
$$= (0.5 \times 420) = 210 \text{ mm}$$
$$d'/x = 50/210 = 0.24 \leq 0.37$$

The compression steel will yield and the stress is equal to $0.95f_y$.

$$A_s' = (K-K')f_{cu}\,bd^2/0.95f_y\,(d-d')$$
$$= [(0.193 - 0.156) \times 40 \times 250 \times 420^2]/[0.95 \times 460 \times (420 - 50)]$$
$$= 403 \text{ mm}^2$$

Appendix 6 **Adopt 4/12 mm diameter HYS bars providing 452 mm².**

$$\therefore A_s = (K'f_{cu}\,bd^2/0.95f_y\,z) + A_s'$$
$$= [(0.156 \times 40 \times 250 \times 420^2)/(0.95 \times 460 \times 327.6)] + 403^*$$
$$= 2325 \text{ mm}^2$$

Appendix 6 **Adopt 3/32 mm diameter HYS bars providing 2410 mm².**

*Note: The A_s' required is used here and *not* the A_s' provided.

5.3.8 Example 5.9: Doubly-reinforced Rectangular Beam 2

The cross-section of a two-span continuous beam is 300 mm wide with an overall depth of 500 mm. Using the data given determine suitable main reinforcement such that the beam can resist:

(i) a bending moment equal to + 600 kNm at mid-span and
(ii) a bending moment equal to − 475 kNm over the central support.

Data:

Characteristic strength of concrete (f_{cu})	45 N/mm²
Characteristic strength of steel (f_y)	460 N/mm²
Nominal maximum aggregate size (h_{agg})	20 mm
Diameter of main tension steel	Assume 32 mm
Diameter of main compression steel	Assume 20 mm
Diameter of shear links	10 mm
Exposure condition	severe
Minimum required fire resistance	1.0 hour

Solution:

Clause 3.3.1   Nominal cover to all steel

Clause 3.3.1.2 ≥ Bar size   = (32 – 10)   cover ≥ 22 mm

Clause 3.3.1.3 ≥ Nominal maximum aggregate size   cover = 20 mm

Clause 3.3.3   Exposure condition: severe f_{cu} = 45 N/mm²

Table 3.3   cover ≥ 30 mm

Clause 3.3.6   Minimum fire resistance: 1.0 hr since beam is continuous

Table 3.4   cover ≥ 20 mm

The required nominal cover to the main steel = 30 mm

Figure 3.2   Minimum width b for 1 hour fire resistance = 200 mm ∴ adequate

(i) *Consider the mid-span position:*

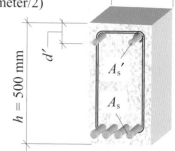

Effective depth d = (h – cover – link diameter – bar diameter/2)

          = (500 – 30 – 10 – 16) = 444 mm

Clause 3.4.4.4

Check if the section is singly-reinforced:

$$K = \frac{M}{bd^2 f_{cu}} = \left[\frac{600 \times 10^6}{300 \times 444^2 \times 45}\right] = 0.225 > 0.156$$

Since $K > K'$ the section is doubly-reinforced.

$$z = d\left[0.5 + \sqrt{0.25 - \frac{K'}{0.9}}\right] = 0.775d = (0.775 \times 444) = 346.3 \text{ mm}$$

x = $(d – z)/0.45$ = $0.5d$ = (0.5×444) = 222 mm

d' = (cover + link diameter + bar diameter/2) = (30 + 10 + 10) = 50 mm

d'/x = 50 / 222 = 0.23 ≤ 0.37

The compression steel will yield and the stress is equal to $0.95f_y$.

A_s' = $(K – K') f_{cu} bd^2 / 0.95f_y (d – d')$

    = $[(0.225 – 0.156) \times 45 \times 300 \times 444^2] / [0.95 \times 460 \times (444 – 50)]$

    = 1067 mm²

Appendix 6   **Adopt 4/20 mm diameter HYS bars providing 1260 mm².**

∴ A_s = $(K' f_{cu} bd^2 / 0.95f_y z) + A_s'$

    = $[(0.156 \times 45 \times 300 \times 444^2) / (0.95 \times 460 \times 346.3)] + 1067$

    = 3811 mm²

Appendix 6   **Adopt 8/25 mm diameter HYS bars providing 3930 mm².**

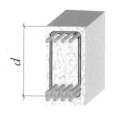

If the bars are placed horizontally assuming a gap of 25 mm (i.e. [h_{agg} + 5 mm], see Clause 3.12.11.1), the minimum width required = $[(2 \times 30) + (4 \times 25) + (3 \times 25) + (2 \times 10)]$

          = 255 mm.

          < 300 mm The width is therefore adequate.

The actual effective depth is slightly less since the reinforcing bars are placed in two rows. The reader should re-calculate the modified areas of steel resulting from this and compare them with A_s and A_s' provided above.

(ii)   *Consider over the central support position:*
Effective depth d = (h − cover − link diameter − bar diameter/2)
$\qquad\qquad\qquad$ = $(500 - 30 - 10 - 16)$ = 444 mm

Clause 3.4.4.4:
Check if the section is singly-reinforced:

$$K = \frac{M}{bd^2 f_{cu}} = \left[\frac{475\times10^6}{300\times444^2\times45}\right] = 0.178 > 0.156$$

Since $K > K'$ the section is doubly-reinforced.

$$z = d\left[0.5+\sqrt{0.25-\frac{K'}{0.9}}\right] = 0.775d = (0.775\times444) = 346.3 \text{ mm}$$

x = $(d-z)/0.45$ = $0.5d$ = (0.5×444) = 222 mm
d' = (cover + link diameter + bar diameter/2) = $(30+10+10)$ = 50 mm

d'/x = 50 / 222 = 0.23 ≤ 0.37

The compression steel will yield and the stress is equal to $0.95f_y$.

$\quad A_s'$ = $(K-K') f_{cu} bd^2 / 0.95f_y (d-d')$
$\qquad$ = $[(0.178 - 0.156)\times45\times300\times444^2] / [0.95\times460\times(444-50)]$
$\qquad$ = 340 mm^2
Appendix 6 $\qquad$ **Adopt 3/12 mm diameter HYS bars providing 339 mm^2.**

$\therefore A_s$ = $(K' f_{cu} bd^2 / 0.95f_y z) + A_s'$
$\qquad$ = $[(0.156\times45\times300\times444^2) / (0.95\times460\times346.3)] + 340$
$\qquad$ = 3083 mm^2
Appendix 6 $\qquad$ **Adopt 4/32 mm diameter HYS bars providing 3220 mm^2.**

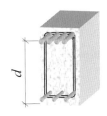

If the bars are placed horizontally assuming a gap of 25 mm (i.e. [h_{agg} + 5 mm], see Clause 3.12.11.1), the minimum width required = $[(2\times30)+(4\times32)+(3\times25)+(2\times10)]$ = 283 mm.
$\qquad$ < 300 mm The width is therefore adequate.

5.3.9 Example 5.10: Doubly-reinforced Rectangular Beam 3

The cross-section of a rectangular beam which is subjected to a bending moment of 160 kNm is shown in Figure 5.39. Using the data given, determine the areas of main steel which are required.

Data:

Characteristic strength of concrete (f_{cu})   50 N/mm^2
Characteristic strength of steel (f_y)   460 N/mm^2
Diameter of main tension steel   Assume 32 mm
Diameter of main compression steel   Assume 20 mm
Diameter of shear links   10 mm

Solution:

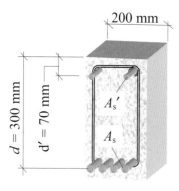

Figure 5.39

$$K = \frac{M}{bd^2 f_{cu}} = \left[\frac{160 \times 10^6}{200 \times 300^2 \times 50}\right] = 0.18 > 0.156$$

Since $K > K'$ the section is doubly-reinforced.

$$z = d\left[0.5 + \sqrt{0.25 - \frac{K'}{0.9}}\right] = 0.775d$$
$$= (0.775 \times 300) = 234 \text{ mm}$$

$$x = (d-z)/0.45 = 0.5d = (0.5 \times 300) = 150 \text{ mm}$$

$d'/x = 70/150 = 0.46 > 0.37$ ∴ the strain in the compression steel is less than the yield value. Since $\zeta_{cs} < \zeta_{yield}$ the compression steel will *not* yield and the stress f_{cs} must be calculated and used in the modified equations for the areas of reinforcement.

$$\zeta_{cs} = 0.0035 \times \frac{(150 - 70)}{150} = 0.0019$$

stress $= E \times$ strain

$$f_{sc} = (200,000 \times 0.0019) = 380 \text{ N/mm}^2$$

$A_s' = (K - K') f_{cu} bd^2 / f_{sc}(d - d')$ for the compression reinforcement
$= [(0.18 - 0.156) \times 50 \times 200 \times 300^2] / [380 \times (300 - 70)]$
$= 247 \text{ mm}^2$

Appendix 6   **Adopt 3/12 mm diameter HYS bars providing 339 mm^2.**

$\therefore A_s = (K' f_{cu} bd^2 / 0.95 f_y z) + A_s' \times \left(\dfrac{f_{sc}}{0.95 f_y}\right)$ for the tension reinforcement.

$$= [(0.156 \times 50 \times 200 \times 300^2) / (0.95 \times 460 \times 234)] + \left[247 \times \left(\frac{380}{460}\right)\right]$$

$$= 1577 \text{ mm}^2$$

Appendix 6   **Adopt 2/32 mm diameter HYS bars providing 1610 mm^2.**

5.4 *Shear Strength of Sections*

As indicated in Figure 5.7 reinforced concrete beams are subject to diagonal tension as a result of shear forces. Consider a simply supported beam in which a series of squares have been painted on the side as shown in Figure 5.40(a). Before deformation of the beam due to the applied load of $2V$ the typical diagonals AB and CD are of equal length as indicated.

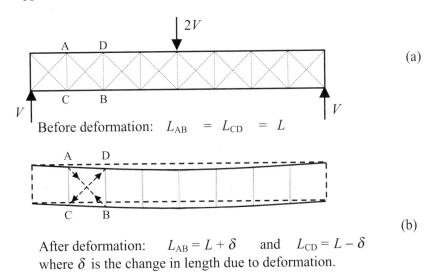

(a)

Before deformation: $L_{AB} = L_{CD} = L$

(b)

After deformation: $L_{AB} = L + \delta$ and $L_{CD} = L - \delta$
where δ is the change in length due to deformation.

Figure 5.40

After deformation due to the load, the squares become distorted such that AB *increases* in length and CD *decreases* in length as shown in Figure 5.40(b). The resulting tensile forces caused by the increase in length will induce diagonal cracking perpendicular to AB when the tensile stress exceeds the tensile strength of the concrete.

Experimental evidence has indicated that shear failure usually occurs on a diagonal plane occurring at 45° to the longitudinal axis of a beam as shown in Figure 5.41.

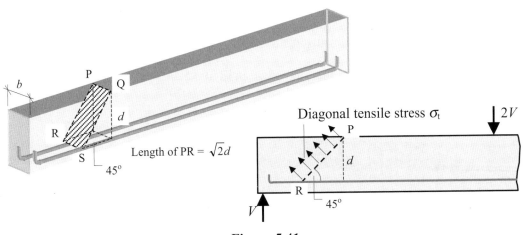

Figure 5.41

The surface area over which the diagonal tensile stress acts is the area of the plane PQRS:

$$\text{Surface area A} = \sqrt{2}d \times b = \sqrt{2}bd$$

The total tensile force on this diagonal plane $T = (\text{stress} \times \text{area}) = (\sigma_t \times \sqrt{2}bd)$

Consider the beam to be cut along this diagonal tension plane as shown in Figure 5.42:

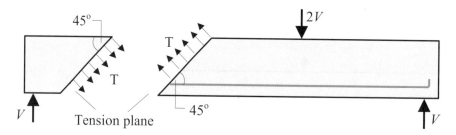

Figure 5.42

The vertical equilibrium equation of either the left-hand or the right-hand section is:

$$+ve \uparrow \Sigma F_y = 0 \qquad V - (T/\sqrt{2}) = 0 \quad \therefore V = (T/\sqrt{2})$$

$$V = [(\sigma_t \times \sqrt{2}bd)/\sqrt{2}] = \sigma_t\, bd \quad \therefore \sigma_t = \frac{V}{bd}$$

This equation is given in the code as:

$$v = \frac{V}{b_v d} \qquad\qquad \textbf{equation 3}$$

where:
v is the design shear stress,
V is the design shear force due to ultimate loads,
b_v is the breadth of the section,
d is the effective depth.

The actual mechanism by which a reinforced concrete beam transfers shear is very complex and a rigorous analytical treatment is not justified. The analysis of extensive experimental data has enabled the derivation of an empirical design procedure which has been adopted in the code to determine a suitable area of shear reinforcement. The use of '*equation 3*' is a mathematical convenience used in developing this simplified design method.

Consider the effect of vertical links as shown in Figure 5.43.

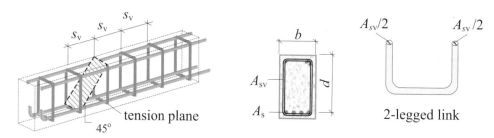

Figure 5.43

If the link spacing $s_v = d$, then it is just possible for a 45° diagonal crack to develop between adjacent links. In these circumstances the beam could fail in shear with the links being ineffective. To prevent this occurring the spacing of the links is limited to a **maximum value of 0.75d** as indicated in Clause 3.4.5.5 of the code.

When $s_v < d$ any tension plane at 45° will cut through more than one link, as shown in Figure 5.44. The load carried by each link will be proportionately smaller the closer the links are together.

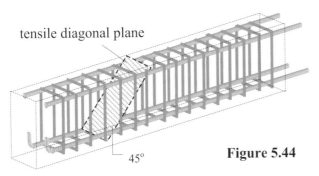

Figure 5.44

The design shear force V applied to a concrete beam is resisted by a combination of the design concrete shear stress v_c and the shear reinforcement. The *truss analogy* is often used as a basis for the design method. Consider the concrete beam shown in Figure 5.45:

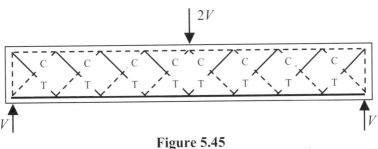

Figure 5.45

It is assumed that the tensile forces are resisted by the shear links and the longitudinal reinforcement whilst the compressive forces are resisted by the strength of the concrete. The concrete strength develops from three sources:

- ♦ the compressive strength of the concrete,
- ♦ the **aggregate interlock** which occurs when the opposite faces of the fracture plane slide past each other   and,
- ♦ the **dowel action** of the longitudinal reinforcement in the tension zone of the cross-section.

Experimental evidence has shown that the design concrete shear stress v_c is dependent on the percentage of longitudinal steel ($100A_s/b_vd$) and the effective depth of the section d. In the code values of v_c are given in Table 3.8 for a characteristic concrete strength of 25 N/mm². When using other concrete strengths greater than 25 N/mm², these values may be modified by multiplying by $\sqrt[3]{(f_{cu}/25)}$ with a maximum value of $f_{cu} = 40$ N/mm².

Design of Structural Elements

A maximum value of shear stress is defined in Clause 3.4.5.2 of the code to prevent crushing failure of the concrete. This is given as: $v \leq 0.8\sqrt{f_{cu}}$

$$\leq 5 \text{ N/mm}^2$$

Consider the vertical equilibrium of the section shown in Figure 5.46:

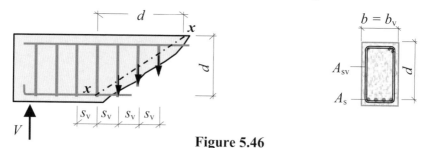

Figure 5.46

$+ve \uparrow \Sigma F_y = 0$ V = Resistance due to concrete + Resistance due to reinforcement

$$= \text{(concrete stress} \times \text{area)} + \text{(steel stress} \times \text{area of links)}$$

The force in each link = (stress × area) $= \left(\dfrac{f_{yv}}{\gamma_m} \times A_{sv} \right) = (0.95 f_{yv} \times A_{sv})$

The number of links cut by the tensile diagonal plane at 45° $= d/s_v$

Total force in the cut links $= \left(0.95 f_{yv} A_{sv}\right) \times \dfrac{d}{s_v}$

$$V = (v_c \times b_v d) + \left[\left(0.95 f_{yv} A_{sv}\right) \times \dfrac{d}{s_v} \right]$$

$$(0.95 f_{yv} A_{sv}) = [V - (v_c \times b_v d)] \times \dfrac{s_v}{d}$$

From equation 3: $V = v b_v d$

$$(0.95 f_{yv} A_{sv}) = [(v b_v d) - (v_c b_v d)] \times \dfrac{s_v}{d} = [b_v s_v (v - v_c)]$$

and hence $A_{sv} \geq \dfrac{b_v s_v (v - v_c)}{0.95 f_{yv}}$

This is the equation given in Table 3.7 of the code for the cross-sectional area of *designed links* when the design shear stress v at a cross-section is greater than $(v_c + 0.4)$. It is recognised that the truss analogy produces conservative results and the code specifies that:

designed links are required when:

$$(v_c + 0.4) < v < 0.8\sqrt{f_{cu}} \text{ or } 5 \text{ N/mm}^2$$

It is important to provide minimum areas of steel in concrete to minimize thermal and shrinkage cracking, etc. Minimum links are specified in the code which provide a shear resistance of 0.4N/mm^2 in addition to the design concrete shear stress v_c, and consequently designed links are required when $v > (v_c + 0.4)$ as indicated. In Table 3.7:

minimum links are required when $\mathbf{0.5v_c < v < (v_c + 0.4)}$

The cross-sectional area of the links required is given by:

$$A_{sv} \geq \frac{\mathbf{0.4b_v s_v}}{\mathbf{0.95 f_{yv}}}$$

In addition, in members of minor importance such as lintel beams or where $v < 0.5v_c$, links may be omitted. These requirements are shown graphically in Figure 5.47. The steel stress f_{yv} for links normally relates to mild steel since this is easier to bend into shape.

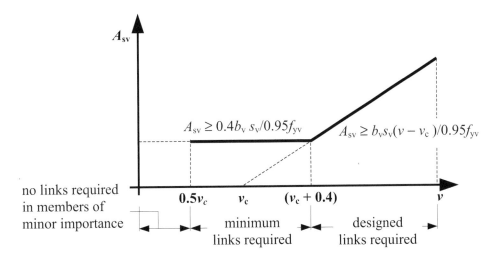

Figure 5.47

The area of longitudinal tension steel which is used in Table 3.8 is that in which the reinforcement continues for a distance at least equal to d beyond the section being considered, as indicated in Clause 3.4.5.4 of the code. At supports, this area of steel must be fully anchored as indicated in Clause 3.12.9 (see Section 5.7.5).

5.4.1   *Example 5.11: Shear Links Beam 1*

A concrete beam is simply supported over a 5.0 m span as shown in Figure 5.48. Using the data given, determine suitable shear reinforcement.

Data:
Characteristic strength of concrete (f_{cu})	40 N/mm^2
Characteristic strength of mild steel (f_{yv})	250 N/mm^2
Maximum shear force at the support	40 kN

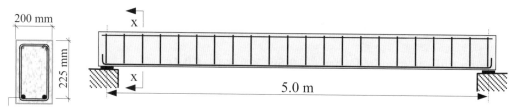

200 mm

225 mm

2 / 16 mm HYS bars

Section x-x

Figure 5.48

Solution:

Consider the end of the beam at the support where the shear force is a maximum = 40 kN

Clause 3.4.5.2 $v = \dfrac{V}{b_v d} = \dfrac{40\times10^3}{200\times225} = 0.89 \text{ N/mm}^2$

$$0.8\sqrt{f_{cu}} = \left(0.8\times\sqrt{40}\right) = 5.06 \text{ N/mm}^2$$

$$\therefore v \leq 0.8\sqrt{f_{cu}}$$

$$\leq 5.0 \text{ N/mm}^2$$

Table 3.7   To determine the required reinforcement, evaluate v_c and either adopt minimum links throughout or use designed links.

Table 3.8 A_s = area of 2/16 mm diameter bars = 402 mm^2

$\dfrac{100 A_s}{b_v d} = \dfrac{100\times402}{200\times225} = 0.893;$ $d = 225$ mm

$\dfrac{100 A_s}{b_v d}$	Effective depth *d*			
	-	200	225	250	-
	-	N/mm^2	N/mm^2	N/mm^2	-
-	-	-	-	-	-
0.5	-	0.60	0.58	0.56	-
0.75	-	0.68	0.66	0.65	-
1.00	-	0.75	0.73	0.71	-
-	-	-	-	-	-

Extract from Table 3.8 of *BS 8110-1*

$v_c = \{0.66 + [(0.73 - 0.66) \times (0.893 - 0.75)/0.25]\} = 0.7 \text{ N/mm}^2$

Since $f_{cu} > 25$ N/mm^2 v_c can be multiplied by $\sqrt[3]{(f_{cu}/25)}$ (see note in the bottom of Table 3.8 of the code).

$v_c = 0.7 \times \sqrt[3]{(40/25)} = 0.82 \text{ N/mm}^2$

$(v_c + 0.4) = (0.82 + 0.4) = 1.22 \text{ N/mm}^2$

Table 3.7   Since $0.5v_c < v < (v_c + 0.4)$, minimum links are required for the whole length of the beam.

The cross-sectional area of the links required is given by $A_{sv} \geq \dfrac{0.4b_v s_v}{0.95 f_{yv}}$

There are two unknowns: A_{sv} and s_v, the spacing of the links. It is necessary to assume one of these and to calculate the other.

Option 1:

Assume 2-legged / 6 mm diameter mild steel links. $\therefore$ A_{sv} = 56.6 mm^2

Table 3.7   The spacing required $s_v = \dfrac{0.95 f_{yv} A_{sv}}{0.4b_v} = \dfrac{0.95 \times 250 \times 56.6}{0.4 \times 200}$ = 168 mm

Clause 3.4.5.5

The maximum spacing of the links $\leq$ 0.75d
$$= (0.75 \times 225) = 169 \text{ mm}$$
Adopt 6 mm diameter mild steel links @ 150 mm centres throughout the length of the beam.

Option 2:

Clause 3.4.5.5

The maximum spacing of the links $\leq$ 0.75d
$$= (0.75 \times 225) = 169 \text{ mm}$$
Assume s_v = 150 mm.

Table 3.7   The cross-sectional area required $A_{sv} \geq \dfrac{0.4b_v s_v}{0.95 f_{yv}} = \dfrac{0.4 \times 200 \times 150}{0.95 \times 250}$
$$= 50.5 \text{ mm}^2$$
Adopt 6 mm diameter mild steel links (56.6 mm^2) @ 150 mm centres throughout the length of the beam as in option 1.

5.4.2   Example 5.12: Shear Links Beam 2

A concrete beam is simply supported over a 7.0 m span as shown in Figure 5.49. Using the data given, determine suitable shear reinforcement.

Data:

Characteristic strength of concrete	(f_{cu})	40 N/mm^2
Characteristic strength of mild steel	(f_{yv})	250 N/mm^2
Characteristic dead load	(g_k)	5.0 kN/m
Characteristic imposed load	(q_k)	30.0 kN/m

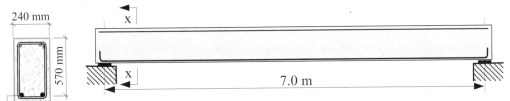

240 mm

570 mm

X

X

7.0 m

2 / 25 mm HYS bars

Section x-x

Figure 5.49

Solution:

Ultimate design load $= [(1.4 \times 5.0) + (1.6 \times 30.0)] = 55.0$ kN/m

Ultimate design shear force at the support $\quad V = (55.0 \times 3.5) = 192.5$ kN

Clause 3.4.5.2 $\qquad v = \dfrac{V}{b_v d} = \dfrac{192.5 \times 10^3}{240 \times 570} = 1.41$ N/mm^2

$$0.8 \sqrt{f_{cu}} = (0.8 \times \sqrt{40}) = 5.06 \text{ N/mm}^2$$

$$\therefore v \leq 0.8 \sqrt{f_{cu}}$$

$$\leq 5.0 \text{ N/mm}^2$$

Table 3.7 To determine the required reinforcement evaluate v_c and to determine the requirement for either minimum links or designed links.

Table 3.8 A_s = area of 2/25 mm diameter bars = 982 mm^2

$\dfrac{100 A_s}{b_v d} = \dfrac{100 \times 982}{240 \times 570} = 0.72;$ $\qquad d = 570$ mm ≥ 400 mm

$\dfrac{100 A_s}{b_v d}$	**Effective depth d**				
	-	-	-	300	≥ 400
	-	-	-	N/mm^2	N/mm^2
-	-	-	-	-	-
0.25	-	-	-	0.43	0.40
0.50	-	-	-	0.54	0.50
0.75	-	-	-	0.62	0.57
-	-	-	-	-	-

Extract from Table 3.8 of *BS 8110-1*

$v_c = \{0.50 + [(0.57 - 0.50) \times (0.72 - 0.5)/0.25]\} = 0.56$ N/mm^2

Since $f_{cu} > 25$ N/mm^2 v_c can be multiplied by $\sqrt[3]{(f_{cu}/25)}$

$v_c = 0.56 \times \sqrt[3]{(40/25)} = 0.66$ N/mm^2

$(v_c + 0.4) = (0.66 + 0.4) = 1.06$ N/mm^2

Table 3.7 Since $(v_c + 0.4) < v < 0.8 \sqrt{f_{cu}}$

$$< 5 \text{ N/mm}^2$$

designed links are required.

The *maximum shear force* which can be resisted by the provision of *minimum links* is given by:

$$V = (v_c + 0.4) b_v d = (1.06 \times 240 \times 570)/10^3 = 145 \text{ kN}$$

Consider the shear force diagram shown in Figure 5.50. Minimum links are adequate in the central portion where the shear force is less than 145 kN. The end sections in which the shear force is greater than 145 kN require designed links.

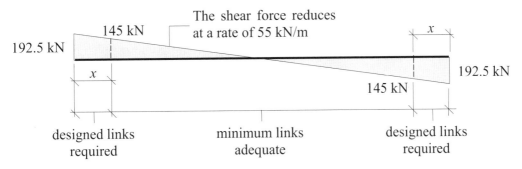

Figure 5.50

$$x = \frac{(192.5 - 145)}{55} = 0.86 \text{ m} \qquad \text{say } 1.0 \text{ m}$$

Provide designed links for the first metre from each end and minimum links elsewhere.

The cross-sectional area of the designed links required is given by $A_{sv} \geq \dfrac{b_v s_v (v - v_c)}{0.95 f_{yv}}$

Assuming 8 mm diameter mild steel links, $A_{sv} = 101 \text{ mm}^2$ and $f_{yv} = 250 \text{ N/mm}^2$.

$$s_v \leq \frac{A_{sv} 0.95 f_{yv}}{b(v - v_c)} = \frac{(101 \times 0.95 \times 250)}{240 \times (1.41 - 0.66)} = 133.3 \text{ mm}$$

Clause 3.4.5.5

The maximum spacing of the links $\leq 0.75d$

$$= (0.75 \times 570) = 428 \text{ mm}$$

Adopt 8 mm diameter mild steel links @ 125 mm centres for 1.0 m from each end of the beam.

The cross-sectional area of the minimum links required is given by $A_{sv} \geq \dfrac{0.4 b_v s_v}{0.95 f_{yv}}$

Table 3.7 The spacing required $s_v \leq \dfrac{0.95 f_{yv} A_{sv}}{0.4 b_v} = \dfrac{0.95 \times 250 \times 101}{0.4 \times 240} = 250 \text{ mm}$

Adopt 8 mm diameter mild steel links @ 250 mm centres throughout the central 5.0 m of the beam.

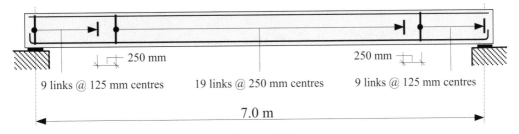

Figure 5.51

5.5   *Deflection of Beams*

Estimation of the deflection of a reinforced concrete beam, whilst being more complex than for example timber or steel, can still be carried out with reasonable accuracy using semi-empirical methods. The numerous variables affecting the result, – e.g. creep, shrinkage, varying E value, percentage of both tension and compression reinforcement, and steel stress – make any precise calculation both tedious and time consuming. In practical terms, the code provides tabulated values (Table 3.9) of basic (span/effective depth) ratios which can be modified to reflect the influence of tension (Table 3.10) and compression (Table 3.11) reinforcement.

   Limiting the (span/effective depth) ratio of a section also limits the deflection as a fraction of the span and hence the *curvature* of a beam. Excessive curvature can damage brittle finishes.

   Consider a simply-supported beam of rectangular cross-section supporting a distributed load as shown in Figure 5.52:

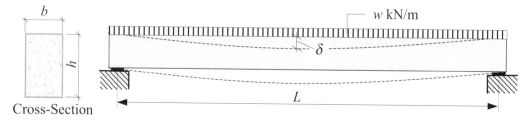

E is Young's modulus of elasticity,
f_e is the maximum elastic bending stress allowed,
z_e is the elastic section modulus of the cross-section,
I is the second moment of area of the cross-section,
δ is the mid-span deflection.

Figure 5.52

Under service conditions and assuming elastic behaviour:

$$\text{Maximum bending moment} \quad = \quad \frac{wL^2}{8}$$

Moment of resistance of the cross-section $= f_e z_{xx}$

The load can be expressed in terms of the section properties and length, i.e.

$$w = \frac{8f_e z_{xx}}{L^2} = \left(\frac{8f_e}{L^2} \times \frac{bh^2}{6}\right) = \frac{4f_e bh^2}{3L^2}$$

The mid-span deflection $\delta = \dfrac{5wL^4}{384EI} = \left(\dfrac{5wL^4}{384E} \times \dfrac{12}{bh^3}\right) = \dfrac{5wL^4}{32Ebh^3}$

Substituting for w gives:

$$\delta = \left[\frac{5L^4}{32Ebh^3} \times \frac{4f_e bh^2}{3L^2}\right] = \frac{5L^2 f_e}{24Eh}$$

The deflection can be expressed as a fraction of the span by dividing both sides by L.

$$\left(\frac{\delta}{L}\right) = \frac{5f_{\mathrm{e}}}{24E}\times\left(\frac{L}{h}\right) = k\times\left(\frac{L}{h}\right) \text{ where } k \text{ is the constant } \frac{5f_{\mathrm{e}}}{24E}$$

hence $\left(\dfrac{\delta}{L}\right) \propto$ (span/depth)   i.e. (span/depth) is related to the curvature.

The beam in Figure 5.52 is assumed to be elastic, however reinforced concrete is not: its behaviour is dependent on the percentage of reinforcing steel and the extent of cracking present. Both tension and compression steel influence the final deflection and hence curvature.

Allowances for these factors to reflect the actual behaviour are made in the code in terms of:

- ♦ using a basic 'span/*effective* depth' ratio (span/d) rather than a 'span/h ' ratio (Table 3.9),
- ♦ applying a modifying factor which is dependent on the steel strength (steel percentage and steel strength are directly related through the factor $K = M/bd^2$) and the steel stress (f_s) under service conditions (Table 3.10),
- ♦ applying a modifying factor which is dependent on the percentage of compression steel provided (Table 3.11).

The basic ratios given in Table 3.9 relate to three support conditions:

- ♦ cantilevers,
- ♦ simply supported beams,
- ♦ continuous beams.*

*Note: Continuous beams are considered to be any beam in which at least one end of the beam is continuous, i.e. this includes propped cantilevers at the end of a series of continuous beams.

The ratios are given for both rectangular and flanged sections and are based on limiting the total deflection to ≤ (span/250). This should ensure that any deflection occurring after construction of finishes and partitions ≤ (span/500) and ≤ (20 mm).

These ratios apply to spans up to and including 10.0 m in length. In the case of long span beams i.e. > 10.0 m the basic (span/d) ratios may permit excessive values of deflection which would be unacceptable for aesthetic and/or practical reasons such as excessive curvature. A fixed limit to d is included in the code to avoid this. In the case of a fixed value of δ, (δ/L) will decrease as the span increases and consequently (L/h) must also decrease to satisfy (δ/L) ∝ (L/h). This is accommodated in Clause 3.4.6.4 of the code for long span beams in which the Table 3.9 values must be multiplied by (10/span) for spans exceeding 10.0 m.

If designers require different limitations from those assumed in the code, e.g. total deflection ≤ (span/360) instead of (span/250), then the Table 3.9 values should be modified accordingly, e.g.

for the case of (span/360):   basic (span/d) = [Table 3.9 value × (250/360)]

5.5.1 Example 5.13: Deflection Beam 1

A rectangular concrete beam 250 mm wide × 475 mm overall depth is simply supported over a 7.0 m span. Using the data given, check the suitability of the beam with respect to deflection.

Data:

Characteristic strength of concrete (f_{cu})	40 N/mm²
Characteristic strength of main steel (f_y)	460 N/mm²
Design ultimate bending moment at mid-span (M)	150.0 kNm
Assume the distance to the centre of the main steel from the tension face is	50 mm

Solution:

Effective depth $d = (475 - 50) = 425$ mm

Check that section is singly-reinforced:

Clause 3.4.4.4 $K = \dfrac{M}{bd^2 f_{cu}} = \left[\dfrac{150 \times 10^6}{250 \times 425^2 \times 40} \right] = 0.083 \leq 0.156$

Since $K < K'$ the section is singly-reinforced.

$$Z = d\left[0.5 + \sqrt{0.25 - \frac{K}{0.9}} \right] = d\left[0.5 + \sqrt{0.25 - \frac{0.083}{0.9}} \right] = 0.9d$$

$$\leq 0.95d$$

$$A_s = M/0.95f_y z = [(150 \times 10^6)/(0.95 \times 460 \times 0.9 \times 425)]$$
$$= 897 \text{ mm}^2$$

Appendix 6   **Adopt 3/20 mm diameter HYS bars providing 943 mm²**

Clause 3.4.6   There is no redistribution of moments ∴ $\beta_b = 1.0$

Table 3.9   The beam is simply supported: ∴ Basic $\left(\dfrac{\text{span}}{d} \right) = 20$

Clause 3.4.6.5 $M/bd^2 = (K \times f_{cu}) = (0.082 \times 40) = 3.32$ N/mm²

Table 3.10 $f_s = \left(\dfrac{2 f_y A_{s\,req}}{3 A_{s\,prov}} \times \dfrac{1}{\beta_b} \right) = \left(\dfrac{2 \times 460 \times 897}{3 \times 943} \right) = 291.7$ N/mm²

Extract from Table 3.10

Table 3.10 Modification factor for tension reinforcement

Service Stress		M/bd^2			
	-	-	3.00	4.00	-
-	-	-	-	-	-
250	-	-	1.04	0.94	-
300	-	-	0.93	0.85	-
(f_y = 460) 307	-	-	0.91	0.84	-

The precise value of the modification factor can be obtained by interpolation from the four values identified in Table 3.10.

f_s		M/bd^2 3.32	
	-	3.0	4.0
	250	1.04	0.94
291.7	300	0.93	0.85

For $f_s = 291.7$ N/mm^2:

$$\text{Factors} = \left\{1.04 - \left[(1.04-0.93)\times\frac{(291.7-250)}{50}\right]\right\} = 0.95 \quad \text{and}$$

$$= \left\{0.94 - \left[(0.94-0.85)\times\frac{(291.7-250)}{50}\right]\right\} = 0.86$$

f_s	M/bd^2 3.32	
	3.0	4.0
291.7	0.95	0.86

For $M/bd^2 = 3.32$ N/mm^2 :

$$\text{Factors} = \left\{0.95 - \left[(0.95-0.86)\times\frac{(3.32-3.0)}{1.0}\right]\right\} = 0.92$$

$$\text{Basic} \left(\frac{\text{span}}{d}\right) \times \text{Table 3.10 value} = (20\times0.92) = 18.4$$

$$\text{Actual} \left(\frac{\text{span}}{d}\right) = \left(\frac{6000}{425}\right) = 14.12 < 18.4 \quad \therefore \textbf{The deflection is acceptable}$$

Note: In many cases it is sufficient to use the *lowest* value of the four identified in the table for interpolation, e.g. in this case 0.85, to determine an approximate value of the modification factor. If the deflection is acceptable when using this value then it must be acceptable using the precise value which is greater.
i.e.

$$\text{Basic} \left(\frac{\text{span}}{d}\right) \times \text{Table 3.10 value} \approx (20\times0.85) = 17.0$$

Clearly if the modified value is less than the actual (span/depth) ratio then a more precise calculation using interpolation is required.

5.5.2 Example 5.14: Deflection Beam 2

It is proposed to use the same beam given in Example 5.13 as a cantilever of 3.0 m span. If the section is subjected to an increased bending moment equal to 420 kNm, check the suitability of the beam with respect to deflection.

Solution:

Effective depth $d = (475 - 50) = 425$ mm

Clause 3.4.4.4 $K = \dfrac{M}{bd^2 f_{cu}} = \left[\dfrac{420 \times 10^6}{250 \times 425^2 \times 40}\right] = 0.23 > 0.156$

$K > K'$ the section is doubly-reinforced

$$z = d\left[0.5 + \sqrt{0.25 - \dfrac{K'}{0.9}}\right] = 0.775d = (0.775 \times 425) = 331.5 \text{ mm}$$

$x = (d - z)/0.45 = 0.5d = (0.5 \times 425) = 212.5$ mm
$d' = 50$ mm
$d'/x = 50/212.5 = 0.24 \leq 0.37$
The compression steel will yield and the stress is equal to $0.95f_y$.

$A_s' = (K - K') f_{cu} bd^2 / 0.95 f_y (d - d')$
$= [(0.23 - 0.156) \times 40 \times 250 \times 425^2] / [0.95 \times 460 \times (425 - 50)]$
$= 816$ mm^2

Appendix 6   **Adopt 3/20 mm diameter HYS bars providing 943mm^2.**

$A_s = (K' f_{cu} bd^2 / 0.95 f_y z) + A_s'$
$= [(0.156 \times 40 \times 250 \times 425^2)/(0.95 \times 460 \times 331.5)] + 816$
$= 2761$ mm^2

Appendix 6   **Adopt 4/32 mm diameter HYS bars providing 3220 mm^2.**

Clause 3.4.6   There is no redistribution of moments: $\therefore \beta_b = 1.0$

Table 3.9   The beam is a cantilever: $\therefore$ Basic $\left(\dfrac{\text{span}}{d}\right) = 7$

Clause 3.4.6.5 $M/bd^2 = (K \times f_{cu}) = (0.023 \times 40) = 9.2$ N/mm^2

Table 3.10 $f_s = \left(\dfrac{2f_y A_{s\,req}}{3A_{s\,prov}} \times \dfrac{1}{\beta_b}\right) = \left(\dfrac{2 \times 460 \times 2761}{3 \times 3220}\right) = 263.0$ N/mm^2

Extract from Table 3.10

Table 3.10 Modification factor for tension reinforcement									
Service Stress	M/bd^2								
	0.5	0.75	1.00	1.50	2.00	3.00	4.00	5.00	6.00
-	-	-	-	-	-	-	-	-	-
263.0 250	1.90	1.70	1.55	1.34	1.20	1.04	0.94	0.87	0.82
300	-	-	-	-	-	-	-	-	0.76
($f_y = 460$) 307	-	-	-	-	-	-	-	-	-

When M/bd^2 is greater than 6.0 N/mm^2 the values in the last column of Table 3.10 are used. A typical graph of M/bd^2 versus modification factor is shown in Figure 5.53. It is evident that the modification factor tends towards a constant with increasing M/bd^2 values.

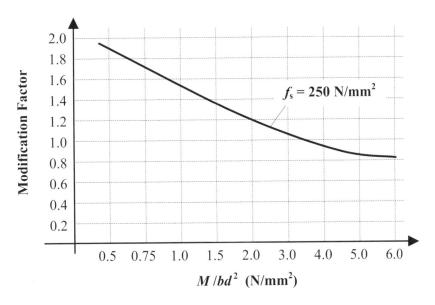

Figure 5.53

	f_s	M/bd^2
	-	6.0
	250	0.82
263.0	300	0.76

For f_s = 263.0 N/mm^2 :

Factors $= \left\{ 0.82 - \left[(0.82 - 0.76) \times \dfrac{(263.0 - 250)}{50} \right] \right\}$

$= 0.8$

Clause 3.4.6.6   The basic (span/d) ratio should also be modified to allow for the effects of compression reinforcement.

Table 3.11 $\dfrac{100 A'_{s\,prov}}{bd} = \left(\dfrac{100 \times 943}{250 \times 425} \right) = 0.89$

$\dfrac{100 A'_{s\,prov}}{bd}$	*Factor*
-	-
0.75	1.2
0.89 1.0	1.25
-	-

Modification factor $= 1.2 + (1.25 - 1.2) \times \dfrac{(0.89 - 0.75)}{0.25}$

$= 1.23$

Basic $\left(\dfrac{span}{d} \right) \times$ Table 3.10 value $\times$ Table 11 value

$= (7 \times 0.8 \times 1.23) = 6.9$

Actual $\left(\dfrac{span}{d} \right) = \left(\dfrac{3000}{425} \right) = 7.1 > 6.9$

The deflection is unacceptable

Since the actual (span/d) ratio exceeds the modified Table 3.9 value a deeper concrete section is required.

5.5.3 Example 5.15: Deflection Rectangular Slab

Check the suitability of the rectangular slab indicated in Example 5.7 with respect to deflection.

Solution:
The slab should be checked within the 5.0 m span and at the end of the cantilever. Using the data obtained from the previous calculations:

Within the span: $\quad A_{s,req'd}$ = 565 mm^2/m width; $A_{s,prov}$ = 628 mm^2/m width
$\qquad\qquad\qquad$ Effective depth d = 270 mm

Table 3.9 $\qquad$ The beam is considered to be simply supported: Basic $\left(\dfrac{span}{d}\right)$ = 20

Clause 3.4.6.5 $\quad M/bd^2$ = $(K \times f_{cu})$ = (0.0217×40) = 0.868 N/mm^2

Table 3.10 $\qquad f_s$ = $\left(\dfrac{2f_y A_{s\,req}}{3A_{s\,prov}} \times \dfrac{1}{\beta_b}\right)$ = $\left(\dfrac{2\times460\times565}{3\times628}\right)$ = 276 N/mm^2

Extract from Table 3.10

Assume the modification factor is approximately equal to the smallest value corresponding with the service stress and M/bd^2 values, i.e. 1.33

Table 3.10   Modification factor for tension reinforcement

Service Stress	M/bd^2	
	0.75	1.00
250	1.7	1.55
276 300	1.44	1.33

$\qquad$ Basic $\left(\dfrac{span}{d}\right)\times$ Table 3.10 value $\approx$ (20×1.33) = 26.6

$\qquad$ Actual $\left(\dfrac{span}{d}\right)$ = $\left(\dfrac{5000}{270}\right)$ = 18.5 < 26.6

$\qquad\qquad\qquad\qquad\qquad\qquad\qquad$ **The deflection is acceptable**

The cantilever span: $\quad A_{s,req'd}$ = 137 mm^2/m width; $A_{s,prov}$ = 390 mm^2/m width
$\qquad\qquad\qquad$ Effective depth d = 275 mm

Table 3.9 $\qquad$ For a cantilever $\quad$ Basic $\left(\dfrac{span}{d}\right)$ = 7

Clause 3.4.6.5 $\quad M/bd^2$ = $(K \times f_{cu})$ = (0.005×40)= 0.2 N/mm^2

Table 3.10 $\qquad f_s$ = $\left(\dfrac{2f_y A_{s\,req}}{3A_{s\,prov}} \times \dfrac{1}{\beta_b}\right)$ = $\left(\dfrac{2\times460\times137}{3\times390}\right)$ = 107.7 N/mm^2

Table 3.10 $\qquad$ The modification factor $\qquad$ = 2.0

$\qquad$ Basic $\left(\dfrac{span}{d}\right)\times$ Table 3.10 value = (7×2.0) = 14.0

$\qquad$ Actual $\left(\dfrac{span}{d}\right)$ = $\left(\dfrac{1200}{275}\right)$ = 4.4 << 14

$\qquad\qquad\qquad\qquad\qquad\qquad\qquad$ **The deflection is acceptable**

5.6 *Effective Span of Beams (Clause 3.4.1)*

The span of beams which is used for analysis purposes is known as the *effective span* and given the symbol *l*. Three situations are defined in Clause 3.4.1 of the code:

♦ simply supported beams (Clause 3.4.1.2)

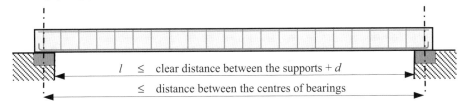

Figure 5.54

♦ continuous beams (Clause 3.4.1.3)

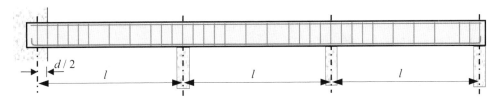

Figure 5.55

♦ cantilevers (Clause 3.4.1.4)

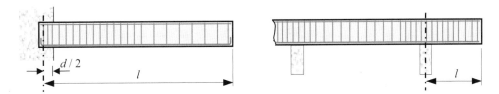

Figure 5.56

The definitions given in Clause 3.4.1 do not apply to deep beams (i.e. where the clear span is less than $2d$). The design of such beams requires reference to specialist literature.

5.7 *Detailing of Sections*

The success of any reinforced concrete element is dependent on efficient and practical techniques being adopted during casting of the concrete when detailing the type of steel, diameter of bar, shape of reinforcement and its location within the formwork.

Detailing considerations which are required by detailers in the design office to interpret the designer's instructions in the form of drawings and schedules for communication to the site are given in the publication *Standard method of detailing reinforced concrete* published by the Institution of Structural Engineers (55). This document is widely regarded within the UK and overseas as a standard reference work indicating the principles which should be followed both in general and in detail to produce consistent, clear, complete and unambiguous instructions to steel fixers on site. The reader is *strongly* recommended to make reference to this publication.

In addition, the standard of workmanship required during construction is important; advice is given relating to this in Section 6 of *BS 8110:Part 1:1997*.

In sections 5.7.1 to 5.7.7 of this text information is given relating to *design considerations*, e.g. minimum/maximum areas of steel, anchorage lengths and curtailment, as required by the code.

5.7.1 *Minimum Areas of Steel (Clause 3.12.5)*

Minimum areas of steel are required in structural elements to ensure that any unnecessary cracking due to thermal/shrinkage effects or tension induced by accidental loading can be minimized. The required minimum steel areas for different situations, i.e. tension or compression reinforcement, rectangular beams/slabs, flanged beams and columns/walls, are given in Table 3.25 of the code and illustrated in Example 5.16.

5.7.1.1 Example 5.16: Minimum Areas of Steel

Determine the minimum areas of steel required for the cross-sections shown in Figure 5.57; in all cases $f_y = 460$ N/mm^2.

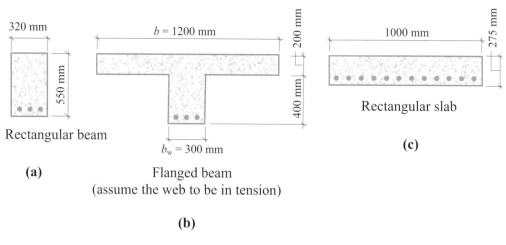

Figure 5.57

Solution:
  (a) *Rectangular beam*

Extract from Table 3.25

Table 3.25 Minimum percentage of reinforcement			
Situation	Definition of percentage	Minimum percentage	
		$f_y = 250$ N/mm^2 %	$f_y = 460$ N/mm^2 %
-	-	-	-
-	-	-	-
c) rectangular section (in solid slabs this minimum should be provided in both directions)	$100A_s/A_c$	0.24	0.13
-	-	-	-

The required area of steel is given by:

$$100A_s / A_c = 0.13$$

where:
A_s is the minimum recommended area of reinforcement,
A_c is the total area of concrete.

$$A_c = (550 \times 320) = 176 \times 10^3 \text{ mm}^2$$

$$A_s = \frac{(0.13 \times 176 \times 10^3)}{100} = 229 \text{ mm}^2$$

(b) *Flanged beam*

Extract from Table 3.25

Table 3.25 Minimum percentage of reinforcement

Situation	Definition of percentage	Minimum percentage	
		$f_y = 250$ N/mm^2 %	$f_y = 460$ N/mm^2 %
-	-	-	-
Sections subject to flexure:	-	-	-
a) flanged beams, web in tension:			
1) $b_w / b < 0.4$	$100A_s/b_wh$	0.32	0.18
2) $b_w / b \geq 0.4$	$100A_s/b_wh$	0.24	0.13
b) flanged beams, flange in tension:			
1) T-beam	$100A_s/b_wh$	0.48	0.26
2) L-beam	$100A_s/b_wh$	0.36	0.20
-	-	-	-

where:
b is the breadth of the section (see Clause 3.4.1.5),
b_w is the breadth or effective breadth of the rib; for a box, T or I section, b_w is taken as the average breadth of the concrete below the flange,
h is the overall depth of the cross-section of a reinforced member.

$b_w / b = (300 / 1200) = 0.25 < 0.4$ ∴ the required area of steel is given by:

$$100A_s / b_wh = 0.18$$

$$b_w h = (300 \times 600) = 180 \times 10^3 \text{ mm}^2$$

$$A_s = \frac{(0.18 \times 180 \times 10^3)}{100} = 324 \text{ mm}^2$$

Note: In flanged beams (see Section 5.10 of this text and Table 3.25 of the code) there is a requirement to provide transverse reinforcement in the flange near the top surface to resist horizontal shear. This is dealt with in Example 5.19 for flanged beams.

(c) Rectangular slab

Extract from Table 3.25

Table 3.25 Minimum percentage of reinforcement			
Situation	Definition of percentage	Minimum percentage	
		$f_y = 250$ N/mm^2 %	$f_y = 460$ N/mm^2 %
-	-	-	-
-	-	-	-
c) rectangular section (in solid slabs this minimum should be provided in both directions)	$100A_s/A_c$	0.24	0.13
-	-	-	-

The required area of steel is given by:

$$100A_s /A_c = 0.13$$

Consider a 1m width of slab:

$$A_c = (1000 \times 275) \quad = 275 \times 10^3 \text{ mm}^2$$

$$A_s = \frac{\left(0.13 \times 275 \times 10^3\right)}{100} = 358 \text{ mm}^2 / \text{ metre width of slab}$$

Note: The minimum area of reinforcement is required in **both** directions in a slab.

In **columns** in addition to minimum percentages of steel there is a requirement for a minimum number and diameter of bars as indicated in Clause 3.1.2.5.3, i.e.

- ◆   in rectangular columns: number of bars   ≥ 4
                                   diameter of bars ≥ 12 mm
- ◆   in circular columns: number of bars   ≥ 6
                                   diameter of bars ≥ 12 mm

The required minimum area of links for beams is as indicated in Table 3.7: see Section 5.4 of this text. In the case of columns, Clause 3.1.2.7.1 specifies that when part or all of the main reinforcement is required to resist compression the links should satisfy the following:

1. link diameter $\geq \dfrac{\text{diameter of the } \textbf{largest} \text{ compression bar}}{4}$

                     $\geq$ 6 mm

2. link spacing $\leq$ 12 × diameter of the **smallest** compression bar

These restrictions limit the *buckling length* of the main compression bars and also prevent the steel from bursting out of the concrete.

5.7.2 Maximum Areas of Steel (Clause 3.12.6)

The maximum percentages of steel given in the code are based on the physical requirements for placing the concrete and fixing the steel. They are:

- in beams (Clause 3.12.6.1):
  $$A_s \leq 4\% A_c$$
  $$A_s' \leq 4\% A_c$$

- in columns (Clause 3.12.6.2):
  a) in vertically-cast columns $\quad\quad\quad\quad\quad A_{sc} \leq 6\% A_c$
  b) in horizontally-cast columns $\quad\quad\quad\quad A_{sc} \leq 8\% A_c$
  c) laps in vertically- or horizontally-cast columns $A_{sc} \leq 10\% A_c$
  (laps are discussed in Section 5.7.6)

  Note: The subscript 'sc' indicates the *total* area of main steel, not necessarily steel in compression. The higher value in b) reflects the better access for the placing and vibrating of concrete in horizontally-cast columns.

- in walls (Clause 3.12.6.3):
  $$A_{sc} \leq 4\% A_c \quad\quad \text{where the subscript 'sc' is as above.}$$

5.7.3 Minimum Spacing of Bars (Clause 3.12.11.1)

Guidance is given in the code for minimum bar spacing to ensure that members can be constructed achieving adequate penetration and compaction of the concrete to enable the reinforcement to perform as designed. It is important that reinforcing bars are surrounded by concrete for two main reasons:

(i) to develop sufficient bond between the concrete and the bars such that the required forces are transferred between the steel and the concrete, and
(ii) to provide protection to the steel against corrosion, fire, etc.

The recommendations for minimum spacing given in Clause 3.12.11.1 are illustrated in Figure 5.58.

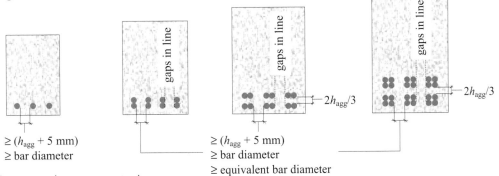

$\geq (h_{agg} + 5 \text{ mm})$
$\geq$ bar diameter

h_{agg} = maximum aggregate size

$\geq (h_{agg} + 5 \text{ mm})$
$\geq$ bar diameter
$\geq$ equivalent bar diameter

Figure 5.58

Where an internal vibrator is to be used for compaction, spacing wider than the minimum should be provided to allow easy flow of the concrete, particularly at the top of the section. In heavily reinforced sections the use of bundled bars can reduce congestion; this does however reduce the bond capacity between the concrete and the steel.

5.7.4 Maximum Spacing of Bars (Clause 3.12.11.2)

The requirement to limit the maximum spacing of reinforcement is to minimize surface cracking. The widths of flexural cracks at a particular point on the surface of a member can be estimated using the method given in Clause 3.8.3 of *BS 8110:Part 2:1985*. This method is designed to give a crack width with an acceptably small probability of being exceeded. It should be recognised that many factors influence the position and width of cracks in concrete and absolute values cannot be predicted. In *BS 8110–1:1997* a number of options are given for determining the maximum permitted clear horizontal distance between bars in tension in situations where crack widths ≤ 0.3 mm are acceptable and the nominal cover ≤ 50 mm.

One of these methods is the use of Table 3.28. The values in this table, which are dependent on the strength of the steel (f_y) and the percentage of redistribution of moment (β_b), are based on the equations given in Part 2 of the code. An Extract from Table 3.28 is given in Figure 5.59.

Extract from Table 3.28

Table 3.28 Clear distance between bars according to percentage redistribution							
f_y	% redistribution to or from section considered						
	−30	−20	−10	0	+10	+20	+30
	mm	mm	mm	mm	mm	mm	mm
250	200	225	255	280	300	300	300
460	110	125	140	155	170	185	200

Figure 5.59

Since very small bars when mixed with larger bars would invalidate the assumptions on which the Table 3.28 values are based, the code specifies that any bar in a section with a diameter < (0.45 × the largest bar in the section) should be ignored except when considering those in the side faces of deep beams.

An alternative to using Table 3.28 is given in Clause 3.12.11.2.4 in which:

$$\text{Clear spacing} \ \leq \ \frac{47{,}000}{f_s} \ \leq \ 300 \text{ mm}$$

Where f_s is the estimated service stress $\ = \ \dfrac{2 f_y A_{s\,\text{req'd}}}{3 A_{s\,\text{prov}}} \times \dfrac{1}{\beta_b}$ as given in Table 3.10.

As indicated in Clause 3.12.11.2.5 the clear distance between the face of a beam and the nearest longitudinal bar in tension should not exceed either:

(Table 3.28 value)/2.0   or   (Clause 3.12.11.2.4 value)/2.0

In deep beams exceeding 750 mm overall depth there is a risk of local yielding of bars in the side faces which can lead to large cracks in the webs. To control cracking in this situation longitudinal bars should be distributed at a spacing ≤ 250 mm near the faces of the beam distributed over a distance of 2/3 of the beam depth measured from the tension face as shown in Figure 5.60.

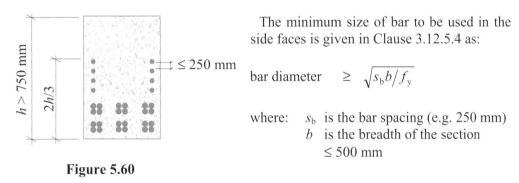

The minimum size of bar to be used in the side faces is given in Clause 3.12.5.4 as:

bar diameter $\geq \sqrt{s_b b / f_y}$

where: s_b is the bar spacing (e.g. 250 mm)
            b is the breadth of the section
            ≤ 500 mm

Figure 5.60

In the case of slabs, rules are given in Clause 3.12.11.2.7 which should ensure adequate control of cracking; these are summarized in Figure 5.61.

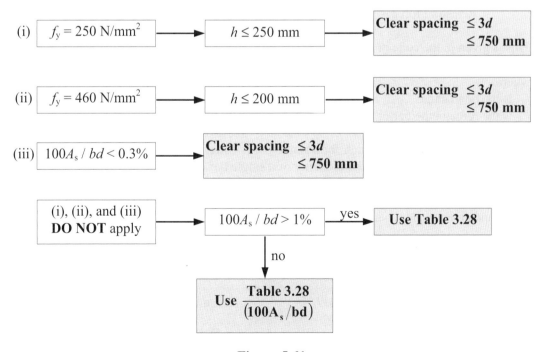

Figure 5.61

5.7.5 *Bond and Anchorage (Clause 3.12.8)*

Reinforced concrete can behave as a composite material only if there is no slip between the reinforcing bars and the surrounding concrete. This condition, which is fundamental to the assumptions made in developing the design equations for strength, is achieved by the development of *bond stresses* at the interface between the two materials. Consider a straight length of steel bar embedded in a block of concrete with one end left projecting from one face as shown in Figure 5.62, and subjected to an axial force *F*.

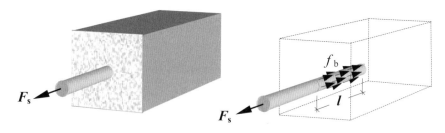

f_b = anchorage bond stress

Figure 5.62

The resistance to pulling this bar out of the concrete is generated by the development of *bond stresses* over the area of contact between the concrete and the steel, i.e.

$$\text{Contact area} = (\text{bar perimeter} \times \text{embedment length})$$
$$= (\pi \times D \times l)$$
$$\text{Resisting force} = F_s = (\text{stress} \times \text{area}) = f_b \times (\pi D l)$$

This is given in the code in '*equation 48*' as:
$$f_b = F_s / (\pi \, \varphi_e \, l)$$

where:

f_b is the bond stress,
F_s is the force in the bar or a group of bars,
l is the anchorage length,
φ_e is the effective bar size which for a single bar is equal to the bar size, and for a group of bars in contact is equal to the diameter of a single bar equal to the total area.

Note: In a group of bars the contact area is less than that for the sum of the individual bars and an equivalent bar diameter is used e.g.

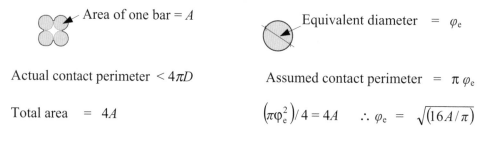

Area of one bar = A

Equivalent diameter   = φ_e

Actual contact perimeter $< 4\pi D$

Assumed contact perimeter = $\pi \, \varphi_e$

Total area = $4A$

$(\pi \varphi_e^2)/4 = 4A$ $\therefore \varphi_e = \sqrt{(16A/\pi)}$

Figure 5.63

The force which can be applied before slip occurs is equal to the maximum force in the

bar, i.e.   Maximum force in the bar F_s = (stress × area) = $\left(\dfrac{f_y}{\gamma_m} \times \dfrac{\pi D^2}{4}\right)$

$$= \left(\dfrac{f_y}{1.05} \times \dfrac{\pi D^2}{4}\right) = 0.238\pi f_y D^2$$

Substituting for F_s in '*equation 48*' gives:

(Let $\varphi_e = D$) $f_b = F_s/(\pi Dl)$ = $(0.238\pi f_y D^2)/(\pi Dl)$
$$= (0.238\,f_y\,D)\,/\,l$$

This can be re-written in terms of the anchorage length l :

$$l = \left(\dfrac{0.238 f_y}{f_b}\right) \times D = C_1 D$$

which gives the minimum required anchorage length as a multiple of the bar diameter. Values for C_1 are given in Table 3.27 of the code for various concrete grades and types of reinforcing bar. An extract from Table 3.27 is given in Figure 5.64.

Extract from Table 3.27

Table 3.27 Ultimate anchorage bond lengths and lap lengths as multiples of bar size					
Reinforcement types	Grade 250 plain	Grade 460			
		Plain	Deformed type 1	Deformed type 2	Fabric
	Concrete cube strength 25				
Tension anchorage and lap length	43	79	55	44	34
-	-	-	-	-	-

Figure 5.64

The value of f_b is normally assumed to be constant over the effective anchorage length and assumed to be equal to the ***design ultimate anchorage bond stress*** as given in Clause 3.1.2.8.4 and Table 3.26 of the code. It is dependent on the type of steel and concrete strength which are related by the *bond coefficient β* given in the table. Types of reinforcing steel are specified in *BS 4449: Specifications for hot rolled steel bars for the reinforcement of concrete*. The bars are identified as Plain, Type 1, Type 2 or Fabric.

- **Plain** bars have a smooth surface and are mostly mild-steel which can be easily bent to form small radius bends for use in links etc.
- **Type 1** are *deformed* bars which are square-twisted to increase the bond resistance. They are better than plain bars but are inferior to Type 2 and are not as readily available.
- **Type 2** are also *deformed* bars having a ribbed surface with enhanced bond characteristics. They are available in both mild and high-yield steel and are the most commonly used type of reinforcement.

♦ **Fabric** is manufactured from bars conforming to BS 4449 or wires conforming to BS 4482. Generally fabric has a higher bond coefficient than individual bars provided that:

(a)   *the fabric is welded in a shear resistance manner conforming to BS 4483, and*

(b)   *the number of welded intersections within the anchorage length is at least equal to (4A$_s$ required)/(A$_s$ provided).*

If condition *(b)* is not satisfied, the anchorage bond stress should be taken as that appropriate to the individual bars or wires in the sheet as indicated in Clause 3.12.8.5.

The design ultimate anchorage bond stress is given by:

$$f_{bu} = \beta \sqrt{f_{cu}}$$

where:
f_{bu} is the bond stress,
β is the bond coefficient,
f_{cu} is the characteristic concrete strength.

5.7.6 Lap Lengths (Clauses 3.12.8.9 to 3.12.8.14)

It is often necessary for practical reasons, e.g. handling long lengths of bar and/or changing bar diameter, to provide reinforcement in several sections rather than in one complete length. When this is necessary, it is important to transfer the force in one section of bar through to the adjacent continuing section. This may be achieved by lapping bars, welding them, or joining them with mechanical devices. The most common practice is to *lap* bars as shown in Figure 5.65 and thereby to transfer the stresses through the concrete. In the case of bars of unequal size the lap length is based on the smaller bar.

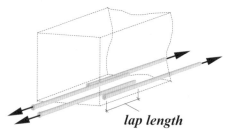

lap length   **Figure 5.65**

The code specifies a number of criteria to accommodate various circumstances as indicated in Table 5.1 of this text.

The information given in Sections 5.75 and 5.76 relate to '*tension anchorages and laps*'. Similar values apply to *compression anchorages and laps* which are also given in Table 3.27 of the code. These values are based on the bond coefficient β for bars in compression as given in Table 3.26 and the requirements of Clause 3.12.8.15 for the design of compression laps.

As indicated in Clause 3.12.8.14, at laps the sum of the reinforcement sizes in a particular layer should not exceed 40% of the breadth of the section at that level.

Condition	Provision (*l*)
Clause 3.12.8.11 In all cases	$l \geq 150$ mm ⎫ for bar ≥ 300 mm ⎭ reinforcement ≥ 250 mm for fabric
Clause 3.12.8.12 Both bars at the lap exceed 20 mm diameter and minimum cover $<$ (1.5 × diameter of the smaller bar)	tension lap length given in Table 3.27 and in addition transverse links should be provided throughout the lap length with: diameter $\geq$ (smaller bar dia.)/4 and spacing of links $\leq$ 200 mm
Clause 3.12.8.13 **(a)** Lap at the top of a section as cast and minimum cover $<$ (2.0 × diameter of the smaller bar)	1.4 × tension lap from Table 3.27
(b) Lap at the corner of a section and minimum cover $<$ (2.0 × diameter of the smaller bar) or where the clear distance between adjacent laps is $<$ 75 mm, or $<$ (6.0 × diameter of the smaller bar) whichever is the greater.	1.4 × tension lap from Table 3.27 1.4 × tension lap from Table 3.27
(c) Where **both** (a) and (b) apply.	2.0 × tension lap from Table 3.27
l is the required lap length	

Table 5.1

5.7.7 Curtailment of Bars and Anchorage at Supports

In the design of a typical simply supported beam subjected to a uniformly distributed load, the ultimate design bending moment occurs at the mid-span. The area of reinforcement calculated to resist this bending moment (e.g. 5/20 mm diameter HYS bars) is theoretically only required at that location. Clearly as the bending moment reduces in value the required area of reinforcement also reduces and individual bars can be cut (i.e. curtailed) leaving only sufficient bars continuing along the span to resist the reduced moment. Eventually a value of bending moment will be reached where only two bars are required and these can be continued to the end of the beam. (**Note:** a minimum of two bars is required to fabricate a reinforcing cage.) This reducing of bars is called *curtailment*.

In practice it is necessary to continue bars a distance beyond where they are theoretically no longer required to resist the bending moment for a number of reasons such as:

♦ to provide a sufficient anchorage length to transfer the force in the bar by bond to the concrete,
♦ to allow for approximations made in the analysis which may result in a bending moment diagram which is not exactly the same as that assumed,
♦ to allow for the misplacement of reinforcement on site,
♦ to control the size of cracking at the cut-off points,
♦ to ensure adequate shear strength.

The code requirements relating to curtailment in all flexural members are as follows:

♦ Except at supports, every bar should extend beyond its theoretical cut-off point for a distance

$\geq$ the effective depth (d) of the member, and
$\geq$ ($12 \times$ bar diameter)

In the case of bars in the tension zone **one** of the additional requirements should also be met: every bar should extend beyond its theoretical cut-off point:

(i)   an anchorage length appropriate to its design strength,
(ii) to the point where the design shear capacity $> (2 \times$ design shear force at that point),

i.e. $v_c b_v d \quad > (2V)$

(iii) to the point where other bars continuing past that point provide double the area required to resist the design bending moment at that section.

In many instances there is insufficient length available to incorporate an additional straight length of bar and an alternative such as a **90° bend** or a **180° hook** is used. These are illustrated in Figure 5.66.

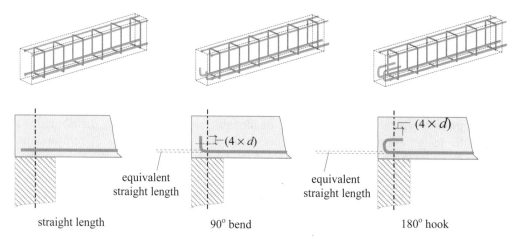

Figure 5.66

The design of bends and hooks should conform to the requirements of *BS 4466:Specification for bending dimensions and scheduling of reinforcement for concrete*, as indicated in Clause 3.12.8.22 of the code. The most common location for 90° bends is at the simply supported end of a member where an effective anchorage length, beyond the centre-line of the support, equivalent to (12 × bar diameter), is required (see Clause 3.12.9.4). The effective anchorage length of a bend or hook is defined in Clause 3.12.8.23 as the **greater** of:

(i)   4r where r is the internal radius of the bend
        (12 × bar diameter)   or   **for 90° bends**
        the actual length of the bar
        (r is equal to 3d in a standard hook or bend)

(ii)   8r where r is the internal radius of the bend
        (24 × bar diameter)   or   **for 180° hooks**
        the actual length of the bar
        (r is normally assumed to be equal to 3d in a standard hook or bend)

In addition, any length of bar in excess of (4 × bar diameter) beyond the end of bend and which lies within the concrete in which the bar is to be anchored may also be included for effective anchorage.

An alternative to calculating precise curtailment positions and anchorage lengths is to adopt the '**Simplified Rules of Curtailment**' which are given in Clause 3.12.10 and Figures 3.24/3.25 of the code.

5.7.7.1 Simplified Detailing Rules for Beams

The rules for beams which are designed for predominantly uniformly distributed loads, and in the case of continuous beams where spans are approximately equal (e.g. approximately 15% difference in length), are illustrated in Figures 5.67, 5.68 and 5.69. For clarity the top and bottom reinforcement in each case is indicated on separate diagrams respectively for the cantilever and continuous spans.

Simply Supported Beams:

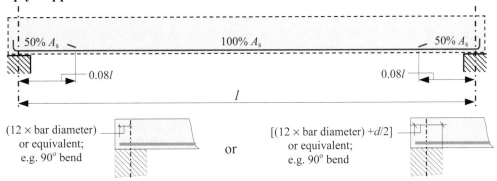

Figure 5.67

Cantilever Beams:

Top Reinforcement

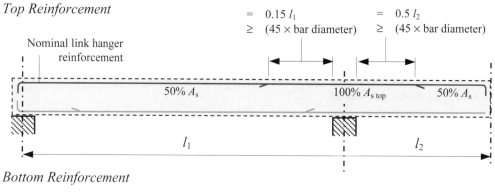

Bottom Reinforcement

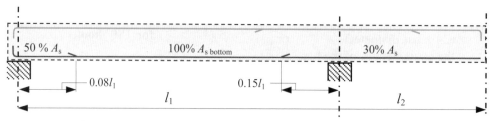

Figure 5.68

Interior Support – Continuous Beams:

Top Reinforcement

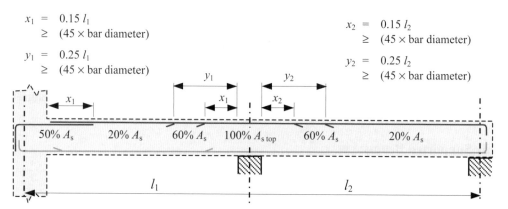

Bottom Reinforcement

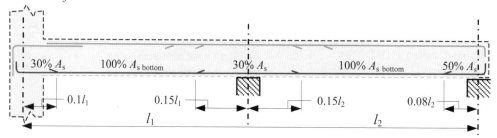

Figure 5.69

Similar curtailment and anchorage rules are given in Clause 3.12.10.3 and Figure 3.25 of the code for slabs. In relatively thin members such as slabs where a simple support has been assumed at the end there is often the possibility of a nominal restraint which induces a negative bending moment. The possibility of cracking in the top of the slab at the support is accommodated by the provision of additional steel as shown in Figure 5.70.

Simply Supported Slabs:
Top Reinforcement

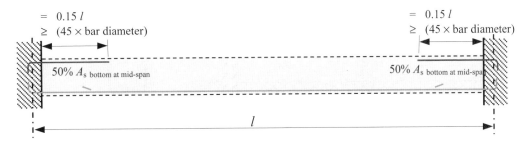

Bottom Reinforcement

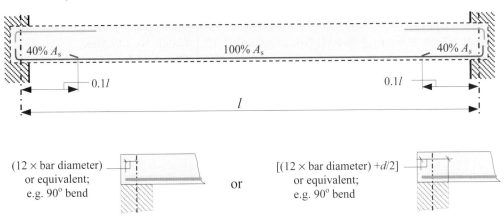

Figure 5.70

5.8   *Example 5.17: Slab and Beam Design*

A simply-supported reinforced concrete beam supports two discontinuous slabs as shown in Figure 5.71. Using the design data given determine:

(i)   the required slab reinforcement
(ii) the required beam reinforcement,   and
(iii) sketch typical reinforcing arrangements indicating the curtailment and anchorage of the steel at the support.

Design Data:
Characteristic dead load (excluding self-weight) 1.0 kN/m^2
Characteristic imposed load 4.0 kN/m^2
Exposure condition   mild

Characteristic strength of concrete (f_{cu}) 40 N/mm^2
Characteristic strength of main reinforcement (f_y) 460 N/mm^2
Characteristic strength of shear reinforcement (f_{yv}) 250 N/mm^2
Nominal maximum aggregate size (h_{agg}) 20 mm
Minimum required fire resistance   1.0 hour
Effective span of main beam   8.0 m

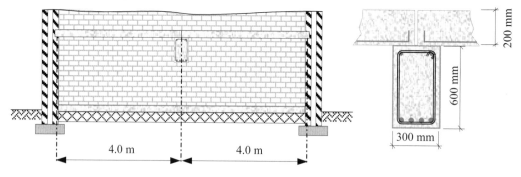

4.0 m   4.0 m

200 mm
600 mm
300 mm

Figure 5.71

5.8.1 Solution to Example 5.17

Contract : Slab & Beam Job Ref. No. : Example 5.17	Calcs. by : W.McK.
Part of Structure : Slab	Checked by :
Calc. Sheet No. : 1 of 8	Date :

References	Calculations	Output
	BS8110:Part 1:1997 **Design Loads:** Characteristic load due to self-weight of slab $= (0.2 \times 24)$ $\qquad = 4.8$ kN/m^2 Characteristic load due to finishes $= 1.0$ kN/m^2 Characteristic load due to imposed loading $= 4.0$ kN/m^2	
Table 2.1	Ultimate design load $= [1.4(4.8 + 1.0) + (1.6 \times 4.0)]$ $\qquad = 14.5$ kN/m^2 Consider a 1.0 m width of slab: Ultimate design load $= 14.5$ kN/m The slab is simply supported over a span of 4.0 m Total design load $\quad = W = (14.5 \times 4.0) = $ 58.0 kN Ultimate design shear $= W/2 = (0.5 \times 58) = $ **29.0 kN** Ultimate design bending moment $= \dfrac{WL}{8} = \dfrac{58 \times 4.0}{8}$ $\qquad\qquad\qquad = $ **29.0 kNm**	

Contract : Slab & Beam Job Ref. No. : Example 5.17	Calcs. by : W.McK.
Part of Structure : Slab **Calc. Sheet No. : 2** of **8**	**Checked by :** **Date :**

References	Calculations	Output
Clause 3.3.7 Clause 3.3.1.2 Table 3.3 Table 3.4	**Cover** Assume 12 mm diameter bars: bar size $\qquad$ = 12 mm exposure condition mild $\qquad \geq$ 20 mm. minimum fire resistance 1 hour $\quad \geq$ 20 mm $\qquad$ **Assume nominal cover to steel = 20 mm**	
	Effective depth d = $(200 - 20 - 6)$ = 174 mm	
Clause 3.3.6 Figure 3.2	Minimum dimensions for 1 hour fire cover: floor thickness $\geq$ 95 mm $\qquad$ **Slab thickness is adequate**	
Clause 3.4.4 Clause 3.4.4.4	**Bending** $K = \dfrac{M}{bd^2 f_{cu}} = \dfrac{29.0 \times 10^6}{1000 \times 174^2 \times 40} = 0.024 \quad < \quad K'\,(=0.156)$ $\qquad$ **Section is singly reinforced**	
	$Z = d\left\{0.5 + \sqrt{\left(0.25 - \dfrac{K}{0.9}\right)}\right\} = d\left\{0.5 + \sqrt{\left(0.25 - \dfrac{0.024}{0.9}\right)}\right\}$ $\qquad\qquad\qquad = 0.97d$ The lever arm is limited to $0.95d$ $A_s = \dfrac{M}{0.95 f_y Z} = \dfrac{29.0 \times 10^6}{0.95 \times 460 \times 0.95 \times 174} = 401 \text{ mm}^2$	**Bottom Reinforcement** Select: **12 mm diameter bars at 250 mm centres providing 452 mm²/m at mid-span**
Clause 3.12.11.1	Minimum spacing of reinforcement is not critical in slabs.	
Clause 3.12.11.2	Maximum spacing of reinforcement: $\leq$ (3 × effective depth) = (3 × 174) = 522 mm $\leq$ 750 mm In addition when using grade 460 steel no further checks are required if the slab is $\leq$ 200 mm thick.	**Bar spacing is adequate**
Clause 3.12.10.3 Figure 3.25	Simplified rules for curtailment of steel in slabs designed for predominantly uniformly distributed loads. 50% of main steel is curtailed a distance $0.1L$ from the support. At support A_s = (0.5×452) = 226 mm² (**Note:** curtail alternate bars) $\dfrac{100 A_s}{bd} = \dfrac{100 \times 226}{1000 \times 174} = 0.13$	

Contract : Slab & Beam Job Ref. No. : Example 5.17	Calcs. by : W.McK.
Part of Structure : Slab	Checked by :
Calc. Sheet No. : 3 of 8	Date :

References	Calculations	Output
Clause 3.5.5	**Shear Resistance**	
Clause 3.5.5.3	Shear stress $\quad v = \dfrac{V}{bd} = \dfrac{29.0\times10^3}{1000\times174} = 0.17\ \text{N/mm}^2$	
	Maximum shear $\leq 0.8\sqrt{f_{cu}} = (0.8\times\sqrt{40})$	
	$\qquad\qquad\qquad = 5.05\ \text{N/mm}^2 \quad$ and	
	$\qquad\qquad\qquad \leq 5.0\ \text{N/mm}^2$	
	$\qquad v < $ maximum permitted value.	
Table 3.8	$v_c \approx (0.41\times1.17) = 0.46\ \text{N/mm}^2$	
	(**Note:** the 1.17 this allows for using C40 concrete)	
Table 3.16	$v < v_c$	**No links are required**
Clause 3.5.7	**Deflection**	
Clause 3.4.6.3	$\dfrac{\text{span}}{\text{effective depth } (d)} \leq$ Table 3.9 value × Table 3.10 value	
	(**Note:** No compression steel is required therefore the Table 3.11 value is not required)	
Table 3.9	Basic (span/d) ratio $\ = 20.0$	
Table 3.10	$\dfrac{M}{bd^2} = K\times f_{cu} = (0.024\times40) = 0.96$	
	Service stress $f_s = \dfrac{2\times f_y\times A_{s,\text{required}}}{3\times A_{s,\text{provided}}} = \dfrac{2\times460\times401}{3\times452}$	
	$\qquad\qquad = 272\ \text{N/mm}^2$	
	Interpolate between values given in Table 3.10	

M/bd^2 / service stress f_s	0.75	1.00
250	1.7	1.55
300	1.44	1.33

	Use conservative estimate of modification factor ≈ 1.33	
Table 3.11	Table 3.9 value × Table 3.10 value = $(20\times1.33) = 26.6$	**Adequate with respect to deflection**
	Actual $\dfrac{\text{span}}{\text{effective depth}} = \dfrac{4000}{174} = 23.0 < 26.6$	
Clause 3.12.5.3 Table 3.25	**Minimum % tension reinforcement** Rectangular sections with $f_y = 460\ \text{N/mm}^2$ $\quad 100A_s/A_c = 0.13$ At least this percentage must be provided in both directions.	

Contract : Slab & Beam **Job Ref. No. :** Example 5.17 **Part of Structure :** Slab **Calc. Sheet No. : 4** of **8**	**Calcs. by : W.McK.** **Checked by :** **Date :**	

References	Calculations	Output
	Minimum A_s required $= \dfrac{0.13 \times 1000 \times 200}{100}$ mm^2 $= 260$ mm^2 < 452 mm^2 **Minimum percentage of reinforcement satisfied**	**Secondary Reinforcement** **Select:** **10 mm diameter bars at 300 centres providing 262 mm^2/m**
Clause 3.12.5.3 Table 3.25	**Secondary Reinforcement:** Minimum % = 0.13 $= 260$ mm^2/m width	
Clause 3.12.10.3 Figure 3.25	**Curtailment:** Curtailment distance of main steel from support $= (0.1 \times 4000)$ $= 400$ mm	
Clause 3.12.10.3.2	Curtailment of bars at end support of slabs (where simple support has been assumed in assessment of moments). A_s provided in the top of the slab at the support: $\geq$ 50% of bottom steel at mid-span $= 200$ mm^2 $\geq$ minimum % given in Table 3.25 $= 260$ mm^2 **Provide the minimum % of steel** This steel should have a full anchorage (e.g. standard 90° bend) into the support and extend: $\geq$ $0.15L$ $= (0.15 \times 4000)$ $= 600$ mm $\geq$ $(45 \times$ bar diameter) $= (45 \times 10)$ $= 450$ mm	**Provide additional steel in the top at the support equal to 10 mm diameter bars at 300 centres.**
Standard 90° bend at support	 600 mm 600 mm 400 mm 400 mm 4000 mm	

Contract : Slab & Beam Job Ref. No. : Example 5.17 Part of Structure : Beam Calc. Sheet No. : 5 of 8	Calcs. by : W.McK. Checked by : Date :

References	Calculations	Output
	Consider the main beam: Characteristic load due to self-weight of the beam: $\quad = \quad (0.3 \times 0.6 \times 24.0)$ $\quad = \quad 4.32$ kN/m length Additional design load due to self-weight $= (1.4 \times 4.32)$ $\quad = \quad 6.0$ kN/m length Area of slab supported / metre length of beam $= \quad 4.0$ m^2 Ultimate design load from both slabs $\quad = \quad (4.0 \times 14.5)$ $\quad = \quad 58.0$ kN/m Total ultimate design load/m $\quad = (6.0 + 58.0) = \quad 64$ kN/m The beam is simply supported over a span of 8.0 m Total design load $\quad = \quad W \; = (64.0 \times 8.0) = \quad 512.0$ kN Ultimate design shear $= W/2 = \quad (0.5 \times 512) = \quad$ **256 kN** Ultimate design bending moment $\; = \; \dfrac{WL}{8} \; = \; \dfrac{512 \times 8.0}{8}$ $\quad = \quad$ **512 kNm**	
Clause 3.3.7 Clause 3.3.1.2 Table 3.3 Table 3.4	**Cover** Assume 25 mm diameter bars for main steel: bar size $\quad\quad\quad\quad\quad\quad\quad = \quad 25$ mm exposure condition mild $\quad\quad \geq \quad 20$ mm minimum fire resistance 1 hour $\quad \geq \quad 20$ mm $\quad\quad\quad$ **Assume nominal cover to steel $\; = \;$ 25 mm** Assume 10 mm diameter bars for links: Effective depth $d \; = \; (600 - 25 - 10 - 12.5) = \quad 552.5$ mm	
Clause 3.3.6 Figure 3.2	Minimum dimensions for 1 hour fire cover: beam width $\quad \geq \quad 200$ mm; $\quad$ actual width $= \quad 300$ mm	
Clause 3.4.4 Clause 3.4.4.4	**Bending** $K = \dfrac{M}{bd^2 f_{cu}} = \dfrac{512 \times 10^6}{300 \times 552.5^2 \times 40} = 0.14 < \; K' \,(= 0.156)$ $\quad\quad\quad\quad\quad\quad\quad$ **Section is singly reinforced** $Z = d \left\{ 0.5 + \sqrt{\left(0.25 - \dfrac{K}{0.9} \right)} \right\} \quad = \quad d \left\{ 0.5 + \sqrt{\left(0.25 - \dfrac{0.14}{0.9} \right)} \right\}$ $\quad\quad\quad\quad\quad = \quad 0.81d \quad < 0.95d$	

References	Calculations	Output
	$A_s = \dfrac{M}{0.95 f_y Z} = \dfrac{512 \times 10^6}{0.95 \times 460 \times 0.81 \times 552.5} = 2618 \text{ mm}^2$	**Bottom Reinforcement**

Contract : Slab & Beam **Job Ref. No. :** Example 5.17
Part of Structure : Beam
Calc. Sheet No. : 6 of 8

Calcs. by : W.McK.
Checked by :
Date :

References	Calculations	Output
	$A_s = \dfrac{M}{0.95 f_y Z} = \dfrac{512 \times 10^6}{0.95 \times 460 \times 0.81 \times 552.5} = 2618 \text{ mm}^2$	**Bottom Reinforcement**
Clause 3.12.11.1	Minimum spacing of reinforcement If the bars are placed horizontally assuming a gap of 25 mm; $h_{agg} = 20$ mm (i.e. [h_{agg} + 5 mm], see Clause 3.12.11.1) The minimum width required = $[(3 \times 25) + (4 \times 25) + (2 \times 10) + (2 \times 20)]$ = 235 mm < actual width = 300 mm	**Select:** **6/25 mm diameter bars at mid-span providing 2950 mm²**
Clause 3.12.11.2 Table 3.28	Maximum spacing of reinforcement: There is no redistribution and f_y = 460 N/mm² The clear distance between bars ≤ 155 mm	
Clause 3.12.10.2 Figure 3.25	Simplified rules for curtailment of steel in beams designed for predominantly uniformly distributed loads. 50% of main steel is curtailed a distance 0.08L from the support. At support $A_s = (0.5 \times 2950) = 1475 \text{ mm}^2$ $\dfrac{100 A_s}{bd} = \dfrac{100 \times 1475}{300 \times 552.5} = 0.89$	
Clause 3.4.5.2	**Shear Resistance**	
Clause 3.5.5.3	Shear stress $\quad v = \dfrac{V}{b_v d} = \dfrac{256 \times 10^3}{300 \times 552.5} = 1.54 \text{ N/mm}^2$	
	Maximum shear $\quad \le \quad 0.8\sqrt{f_{cu}} \quad = (0.8 \times \sqrt{40})$ $= 5.05 \text{ N/mm}^2 \quad$ and $\le 5.0 \text{ N/mm}^2$ $v \quad < \quad$ maximum permitted value.	
Table 3.8	Interpolate between values given in Table 3.8 Effective depth $d > 400$ mm	

$100 A_s / b_v d$	Effective depth d		
	250 mm	300 mm	≥ 400
0.75	-	0.68	0.57
$\overline{}$ 0.89			
1.0	-	0.62	0.63

$v_c \approx (0.6 \times 1.17) = 0.7 \text{ N/mm}^2$
(**Note:** 1.17 – this allows for using C40 concrete)

Contract : Slab & Beam Job Ref. No. : Example 5.17 Part of Structure : Beam Calc. Sheet No. : 7 of 8	Calcs. by : W.McK. Checked by : Date :

References	Calculations	Output
Table 3.7	$0.5v_c = 0.3$ N/mm^2, $(v_c + 0.4) = (0.7 + 0.4) = 1.1$ N/mm^2 $(v_c + 0.4) < v < 0.8\sqrt{f_{cu}}$ or 5N/mm^2 **Design links are required** $A_{sv} \geq b_v s_v (v - v_c)/0.95 f_{yv}$ assuming 10 mm diameter links $s_v \leq \dfrac{0.95 f_{yv} A_{sv}}{(v - v_c) b_v} = \dfrac{0.95 \times 250 \times 157}{(1.54 - 0.7) \times 300} = 148$ mm Check shear stress at 1.0 m from end: Shear force $V_{1m} = 256 - 64 = 192$ kN $v = \dfrac{192 \times 10^3}{300 \times 552.5} = 1.16$ N/mm^2 $> (v_c + 0.4)$ **Design links required** $s_v \leq \dfrac{0.95 \times 250 \times 157}{(1.16 - 0.7) \times 300} = 270$ mm Check shear stress at 2.0 m from end: Shear force $V_{2m} = 256 - 128 = 128$ kN $v = \dfrac{128 \times 10^3}{300 \times 552.5} = 0.77$ N/mm^2 $< (v_c + 0.4)$ **Minimum links required** $A_{sv} \geq 0.4 b_v s_v / 0.95 f_{yv}$ $s_v \leq \dfrac{0.95 \times 250 \times 157}{0.4 \times 300} = 310$ mm	Adopt 10 mm diameter links @ 140 mm centres for the first metre at each end Adopt 10 mm diameter links @ 270 mm centres for the first metre at each end Adopt 10 mm diameter links @ 300 mm centres for the middle 4 m
Clause 3.5.7 Clause 3.4.6.3	**Deflection** $\dfrac{\text{span}}{\text{effective depth }(d)} \leq$ Table 3.9 value $\times$ Table 3.10 value (**Note:** No compression steel is required therefore the Table 3.11 value is not required)	
Table 3.9 Table 3.10	Basic (span/d) ratio = 20.0 $\dfrac{M}{bd^2} = K \times f_{cu} = (0.14 \times 40) = 5.6$ Service stress $f_s = \dfrac{2 \times f_y \times A_{s,\,\text{required}}}{3 \times A_{s,\,\text{provided}}} = \dfrac{2 \times 460 \times 2618}{3 \times 2950}$ $= 272$ N/mm^2	

Contract : Slab & Beam Job Ref. No. : Example 5.17	Calcs. by : W.McK.
Part of Structure : Beam	Checked by :
Calc. Sheet No. : 8 of 8	Date :

References	Calculations	Output
	Interpolate between values given in Table 3.10	

M/bd^2 service stress f_s	5.00	6.00
250	0.87	0.82
300	0.80	0.76

	Use conservative estimate of modification factor $\approx$ 0.76 Table 3.9 value $\times$ Table 3.10 value $= (20 \times 0.76)$ = 15.2 Actual $\dfrac{\text{span}}{\text{effective depth}}$ $=$ $\dfrac{8000}{552.5}$ = 14.5 $<$ 15.2	**Adequate with respect to deflection**
Clause 3.12.5.3 Table 3.25	**Minimum % tension reinforcement** Rectangular sections with f_y = 460 N/mm^2 $\qquad 100A_s/A_c$ = 0.13 Minimum A_s required $=$ $\dfrac{0.13 \times 300 \times 600}{100}$ mm^2 $\qquad\qquad = $ 234mm^2 $\ll 2950$ mm^2 Minimum percentage of reinforcement satisfied	
Clause 3.12.10.2 Figure 3.24	**Curtailment:** Curtailment distance of main steel from support $= (0.08 \times 8000)$ $\qquad\qquad\qquad\qquad\qquad\qquad\qquad\qquad\quad = 640$ mm This steel should have a full anchorage (e.g. standard 90° bend) into the support	

R10 @ 140 c/c   R10 @ 270 c/c   R10 @ 300 c/c   R10 @ 270 c/c   R10 @ 140 c/c

Standard
90° bend
at support

3T25

6T25

3T25

640 mm

640 mm

8000 mm

Cross-section
at mid-span

T25 represents High-Yield steel bars (T) of 25 mm diameter
R10 represents Mild steel bars (R) of 10 mm diameter.

Nominal link hanger bars (e.g. 10 mm or 12 mm diameter) are
used at the top to fabricate the reinforcement cage.

5.9   Example 5.18: Doubly Reinforced Beam

Consider the rectangular beam in Example 5.17 with the overall depth restricted to
550 mm (previously 600 mm) and design suitable reinforcement using the same design
data.

5.9.1 Solution to Example 5.18

Contract : Slab & Beam Job Ref. No. : Example 5.18	Calcs. by : W.McK.
Part of Structure : Beam (300 mm × 550 mm)	Checked by :
Calc. Sheet No. : 1 of 2	Date :

References	Calculations	Output
	Characteristic load due to self-weight of the beam: 　　　　　　$= (0.3 \times 0.55 \times 24.0)$ 　　　　　　$= 3.96$ kN/m length Additional design load due to self-weight $= (1.4 \times 3.96)$ 　　　　　　$= 5.5$ kN/m length Area of slab supported / metre length of beam $= 4.0$ m^2 Ultimate design load from both slabs $= (4.0 \times 14.5)$ 　　　　　　$= 58.0$ kN/m Total ultimate design load/m $= (5.5 + 58.0) = 63.5$ kN/m The beam is simply supported over a span of 8.0 m Total design load $= W = (63.5 \times 8.0) = 508.0$ kN Ultimate design shear $= W/2 = (0.5 \times 508) = $ **254 kN** Ultimate design bending moment $= \dfrac{WL}{8} = \dfrac{508 \times 8.0}{8}$ 　　　　　　$= $ **508 kNm**	
Clause 3.3.7	**Cover** Assume 25 mm diameter bars for main steel:	
Clause 3.3.1.2 Table 3.3 Table 3.4	bar size　　　　　　　$= 25$ mm exposure condition mild　　≥ 20 mm minimum fire resistance 1 hour　≥ 20 mm 　　**Assume nominal cover to steel $= $ 25 mm**	
	Assume 10 mm diameter bars for links: Effective depth $d = (550 - 25 - 10 - 12.5) = 502.5$ mm	
Clause 3.3.6 Figure 3.2	Minimum dimensions for 1 hour fire cover: beam width ≥ 200 mm;　actual width $= 300$ mm 　　　　　**Beam width is adequate**	
Clause 3.4.4	**Bending**	
Clause 3.4.4.4	$K = \dfrac{M}{bd^2 f_{cu}} = \dfrac{508 \times 10^6}{300 \times 502.5^2 \times 40} = 0.168 > K'(=0.156)$ 　　　　　**Section is doubly reinforced**	

Contract : Slab & Beam Job Ref. No. : Example 5.18	Calcs. by : W.McK.
Part of Structure : Beam	Checked by :
Calc. Sheet No. : 2 of 2	Date :

References	Calculations	Output
	$$Z = d\left\{0.5+\sqrt{\left(0.25-\dfrac{K'}{0.9}\right)}\right\} \quad = \quad d\left\{0.5+\sqrt{\left(0.25-\dfrac{0.156}{0.9}\right)}\right\}$$ $$= \quad 0.775d \quad = (0.775 \times 502.5)$$ $$= \quad 389.4 \text{ mm}$$ $$x \quad = \quad (d-z)/0.45 \quad = 0.5d$$ $$= \quad (0.5 \times 389.4) \quad = 194.7 \text{ mm}$$ Assume 20 mm diameter bars for A_s' $$d' \quad = \quad (20+10+10) = \quad 40 \text{ mm}$$ $$d'/x \quad = \quad 40/194.7 \quad = \quad 0.21 \leq 0.37$$ The compression steel will yield and the stress is equal to $0.95f_y$. $$A_s' \quad = \quad (K-K')f_{cu}\,bd^2 / 0.95f_y\,(d-d')$$ $$= \quad \frac{(0.168-0.156)\times 40\times 300\times 502.5^2}{0.95\times 460\times (502.5-40)} \quad = \quad 180 \text{ mm}^2$$ $$A_s \quad = \quad (K'f_{cu}\,bd^2 / 0.95f_y\,z) + A_s'$$ $$= \quad \left(\frac{0.156\times 40\times 300\times 502.5^2}{0.95\times 460\times 389.4}\right) + 180 \quad = \quad 2958 \text{ mm}^2$$ The bars selected are marginally less than that required ($< 0.3\%$ difference) but are acceptable. Shear calculations are carried out as before with minor changes to the values obtained. The deflection calculation should be carried out including an allowance for compression reinforcement using Table 3.11 (as given in Example 5.13). Minimum and maximum areas of steel, curtailment, anchorage and spacing calculations should also be carried out as before. The reader should complete these calculations for this example.	**Top Reinforcement** **Select:** **2/12 mm diameter bars providing 226 mm^2** **Bottom Reinforcement** **Select:** **6/25 mm diameter bars at mid-span providing 2950 mm^2**

5.10   T and L Beams

5.10.1   Introduction

When reinforced concrete slabs are cast integrally with the supporting beams they may contribute to the compressive strength of the beams during flexure. When subject to sagging moments the resulting beam cross-section is either a **T**-section or an **L**-section, as shown in Figures 5.72 and 5.73, where the top surface, i.e. the slab, is in compression. Both types of beam are referred to as *flanged beams*. When subjected to hogging moments the top surface is subject to tension and hence the beams are designed as rectangular sections.

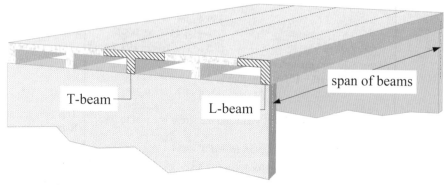

Figure 5.72

The *effective flange width* (b_e) of flanged beams is defined for both cases in Clause 3.4.1.5 of the code as:

for T-beams: web width $+ l_z/5$ or actual flange width if less,
for L-beams: web width $+l_z/10$ or actual flange width if less,

where l_z is the distance between points of zero moment (which, for a continuous beam, may be taken as 0.7 times the effective span).

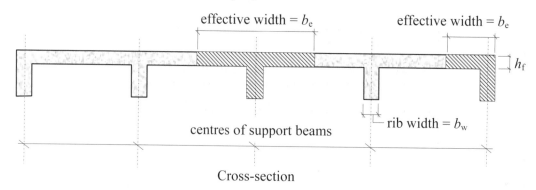

Cross-section

Figure 5.73

Design formulae are given to determine the area of reinforcing steel required, depending on the position of the neutral axis, i.e. either in the flange or below the flange in the rib. In general the neutral axis will lie within the flange. The design process is the same as that used for rectangular beams in which bending, shear and deflection are the main design criteria.

Transverse reinforcement is provided across the top of the slab to prevent cracking due to secondary effects such as shrinkage.

5.10.2   Bending (Clauses 3.4.4.4 and 3.4.4.5)

The area of steel required to resist flexure for beams in which the neutral axis lies within the flange, can be determined using Clause 3.4.4.4 as either:

$$A_s = M/(0.95f_y z)$$ in the case of singly reinforced sections

or

$$A_s' = (K - K')f_{cu}bd^2/(d - d')$$
$$A_s = (K'f_{cu}db^2/0.95f_y z) + A_s'$$ in the case of doubly reinforced sections.

The area of steel required to resist flexure for beams in which the neutral axis lies *below* the flange can be determined using Clause 3.4.4.5. Since this will only occur on very few occasions it is not dealt with in this text.

5.10.3   Shear (Clause 3.4.5.2)

The shear reinforcement is determined using Table 3.7 and Table 3.8 as in ordinary rectangular beams. In the case of flanged beams b_v is defined as the average width of the rib below the flange.

5.10.4   Deflection (3.4.6.3)

The limitations on deflection are governed by satisfying the basic (span/effective depth) ratio from Table 3.9, modified accordingly for tension and compression steel using Table 3.10 and Table 3.11. The basic values for flanged beams given in Table 3.9 apply to ratios of $(b_w/b_e) \le 0.3$. In cases where $(b_w/b_e) > 0.3$, the basic ratio can be found using linear interpolation between the values required for rectangular beams and those required for flanged beams.

5.10.5   Transverse Reinforcement (Clause 3.12.5.3, Table 3.25)

Horizontal shear in the flanges of flanged beams is resisted by providing a minimum area of steel equal to **0.15%** of the flange area extending over the full effective flange breadth b_e as indicated in Table 3.25.

5.10.6   Example 5.19: Single-Span T-Beam Design

A floor system consisting of a solid in-situ reinforced concrete slab cast integrally with the support beams is simply supported over a span of 6.0 m as shown in Figure 5.74.

   (i)   Design suitable reinforcement for a typical **T**-section to satisfy flexure and shear,
   (ii)   check the suitability of a typical **T**-section with respect to deflection,

(iii)   determine the transverse reinforcement required for the flanges of the **T-beams**,

(iv)   prepare a sketch indicating all reinforcement; use the simplified rules indicated in Clause 3.12.10 and Figure 3.24 for curtailment.

Design Data:

Characteristic dead load (excluding self-weight)	g_k =	1.0 kN/m^2
Characteristic imposed load	q_k =	5.0 kN/m^2
Concrete grade	f_{cu} =	40 N/mm^2
Characteristic strength of reinforcing steel	f_y =	460 N/mm^2
Exposure condition	=	mild
Fire resistance	=	1hr minimum
Slab thickness	h_f =	160 mm
Rib width	b =	350 mm
Overall depth	h =	500 mm
Span of main beams	L =	6.0 m
Centres of main beams	=	4.0 m

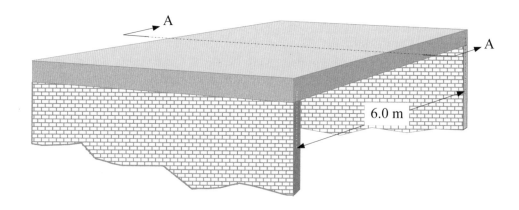

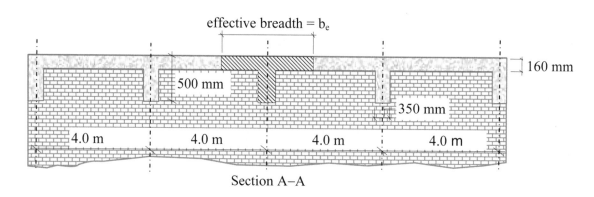

Section A–A

Figure 5.74

5.10.7   Solution to Example 5.19

Contract : Floor Slab Job Ref. No. : Example 5.19	Calcs. by : W.McK.
Part of Structure : T-Beam	Checked by :
Calc. Sheet No. : 1 of 5	Date :

References	Calculations	Output
	BS8110:Part 1:1997 **Design Loads:** Characteristic load due to self-weight of slab: $\quad\quad = \ (0.16 \times 4.0 \times 24) \ = \ 15.36$ kN/m Characteristic load due to self-weight of rib: $\quad\quad = \ (0.35 \times 0.34 \times 24) = \ 2.86$ kN/m Characteristic load due to finishes: $\quad\quad = \ (1.0 \times 4.0) = \ 4.0$ kN/m Characteristic load due to imposed loading: $\quad\quad = \ (5.0 \times 4.0) = \ 20.0$ kN/m Ultimate design load: $\quad = \ [1.4(15.36 + 2.86 + 4.0) + (1.6 \times 20.0)]$ $\quad = \ 63.1$ kN/m Ultimate design shear $= \ (3 \times 63.1) \quad\quad = \ 189.3$ kN Ultimate design bending moment $= \ \dfrac{wL^2}{8} \ = \ \dfrac{63.1 \times 6.0^2}{8}$ $\quad\quad\quad = \ 284$ kNm	
Table 2.1		
Clause 3.4.1.5	**Effective flange width** Effective width of flange beam $= \ b_e \ \le \ (b_w + l_z/5)$ $\quad\quad = \ 350 + \dfrac{6000}{5} \ = \ 1550$ mm Actual flange width $\quad\quad = \ 4000$ mm $\quad\quad\quad\quad \therefore \ \ b_e = \ $**1550 mm**	
	1550 mm 160 mm d 340 mm A_s 350 mm	
Clause 3.3.7 Clause 3.3.1.2 Table 3.3 Table 3.4	**Cover** Assume 25 mm diameter bars for main steel: bar size $\quad\quad\quad\quad = \ 25$ mm exposure condition mild $\quad \ge \ 20$ mm fire resistance $\ \ge 1$ hour $\quad \ge \ 20$ mm $\quad\quad$ **Assume nominal cover to steel 25 mm** Assume 8 mm diameter bars for links. Effective depth $\quad d \ = \ (500 - 25 - 8 - 13) = \ 454$ mm	

References	Calculations	Output
	Contract : Floor Slab **Job Ref. No. : Example 5.19** **Part of Structure :** T-Beam **Calc. Sheet No. : 2 of 5**	**Calcs. by :** W.McK. **Checked by :** **Date :**

References	Calculations	Output
Clause 3.3.6 Figure 3.2	Minimum dimensions for 1 hour fire cover rib width $\geq$ 200 mm, floor thickness $\geq$ 95 mm <div align="right">**Both are satisfied**</div>	**Rib width and floor thickness are adequate for fire resistance**
Clause 3.4.4	**Bending**	
Clause 3.4.4.4	$K = \dfrac{M}{bd^2 f_{cu}} = \dfrac{284 \times 10^6}{1550 \times 454^2 \times 40} = 0.022$ $<$ $K'(= 0.156)$ **Section is singly reinforced** $Z = d\left\{0.5 + \sqrt{\left(0.25 - \dfrac{K}{0.9}\right)}\right\} = d\left\{0.5 + \sqrt{\left(0.25 - \dfrac{0.022}{0.9}\right)}\right\}$ $= 0.97d > 0.95d$ The lever arm is limited to $0.95d$ $x = (d - z)/0.45 = (0.05 \times 454)/0.45 = 50.4$ mm $< h_f (= 160$ mm$)$ The neutral axis lies within the flange. $A_s = \dfrac{M}{0.95 f_y Z} = \dfrac{284 \times 10^6}{0.95 \times 460 \times 0.95 \times 454} = 1507$ mm^2	**Bottom Reinforcement** **Select: 5/25 mm diameter bars at mid-span providing 1570 mm^2**
Clause 3.12.11.1	Minimum spacing of reinforcement If the bars are placed horizontally assuming a gap of 25 mm; h_{agg} = 20 mm (i.e. [h_{agg} + 5 mm], see Clause 3.12.11.1) The minimum width required = $[(4 \times 25) + (5 \times 25) + (2 \times 8) + (2 \times 25)]$ = 291mm $<$ actual width = 350 mm	
Clause 3.12.11.2 Table 3.28	Maximum spacing of reinforcement: There is no redistribution and $f_y = 460$ N/mm^2 The clear distance between bars $\leq$ 155 mm	
Clause 3.12.10.2 Figure 3.24	Simplified rules for curtailment of steel in beams designed for predominantly uniformly distributed loads. (**Note:** curtail two bars) 40% of main steel is curtailed a distance $0.08L$ from the support. At support $A_s = (0.6 \times 1570) = 942$ mm^2 <div align="center">$\dfrac{100 A_s}{b_v d} = \dfrac{100 \times 942}{350 \times 454} = 0.59$</div>	

Contract : Floor Slab Job Ref. No. : Example 5.19 Part of Structure : T-Beam Calc. Sheet No. : 3 of 5	Calcs. by : W.McK. Checked by : Date :

References	Calculations	Output
Clause 3.4.5 Clause 3.4.5.2	**Shear Resistance** Shear stress $\quad v = \dfrac{V}{b_v d} \quad v = \dfrac{189.3 \times 10^3}{350 \times 454} = 1.19 \text{ N/mm}^2$ Maximum shear $\quad \le \quad 0.8\sqrt{f_{cu}} \quad = \quad (0.8 \times \sqrt{40}\,)$ $\qquad\qquad\qquad\qquad\qquad = \quad 5.05 \text{ N/mm}^2$ $\qquad\qquad\qquad\qquad\qquad \le \quad 5.0 \text{ N/mm}^2$ $\qquad\qquad v \quad < \quad$ maximum permitted value.	
Table 3.8	Interpolate between values given in Table 3.8 Effective depth $\quad d > 400$ mm	

$100A_s/b_v d$	Effective depth d		
	250 mm	300 mm	≥ 400
0.50 0.59	-	0.54	0.50
0.75	-	0.68	0.57

References	Calculations	Output
	$v_c \approx (0.52 \times 1.17) = 0.61 \text{ N/mm}^2$ (**Note**: 1.17 – this allows for using C40 concrete)	
Table 3.7	$0.5v_c = 0.33 \text{ N/mm}^2, \quad (v_c + 0.4) = (0.61 + 0.4) = 1.01 \text{ N/mm}^2$ $(v_c + 0.4) \quad < \quad v \quad < \quad 0.8\sqrt{f_{cu}} \;$ or $\; 5\text{N/mm}^2$ $\qquad\qquad\qquad\qquad\qquad$ **Design links are required** $A_{sv} \ge b_v s_v (v - v_c)/0.95 f_{yv} \quad$ assuming 8 mm diameter links $s_v \le \dfrac{0.95 f_{yv} A_{sv}}{(v - v_c)b_v} = \dfrac{0.95 \times 250 \times 101}{(1.19 - 0.61) \times 350} = 118$ mm Check shear stress at 1.0 m from end: Shear force $V_{1m} = 169.3 - 63.1 = 106.2$ kN $\qquad v = \dfrac{106.2 \times 10^3}{350 \times 454} = 0.67 \text{ N/mm}^2 \quad < \quad (v_c + 0.4)$ $\qquad\qquad\qquad\qquad\qquad$ **Minimum links required**	**Adopt 8 mm diameter links @ 100 mm centres for the first metre at each end**
Table 3.7	$A_{sv} \ge 0.4 b_v s_v / 0.95 f_{yv}$ $s_v \le \dfrac{0.95 \times 250 \times 101}{0.4 \times 350} = 171$ mm	**Adopt 8 mm diameter links @ 150 mm centres for the middle 4.0 m**

Contract : Floor Slab Job Ref. No. : Example 5.19	Calcs. by : W.McK.
Part of Structure : T-Beam	Checked by :
Calc. Sheet No. : **4** of **5**	Date :

References	Calculations	Output
	Deflection (Clause 3.4.6)	
	Clause 3.4.6.3	
	$\dfrac{\text{span}}{\text{effective depth } (d)} \quad \leq \quad \text{Table 3.9 value} \times \text{Table 3.10 value}$	
	(**Note:** No compression steel is required therefore the Table 3.11 value is not required)	
Table 3.9	$\dfrac{b_w}{b} = \dfrac{350}{1550} = 0.23 \ \leq \ 0.3$	
	Basic (span/d) ratio $= 16.0$	
Table 3.10	$\dfrac{M}{bd^2} = K \times f_{cu} = (0.022 \times 40) = 0.88$	
	Service stress $f_s = \dfrac{2 \times f_y \times A_{s,\text{required}}}{3 \times A_{s,\text{provided}}} = \dfrac{2 \times 460 \times 1507}{3 \times 1570}$	
	$\qquad\qquad = 294 \text{ N/mm}^2$	
	Interpolate between values given in Table 3.10	

M/bd^2 service stress f_s	0.75	1.00
250	1.70	1.55
300	1.44	1.33

	Use conservative estimate of modification factor $\approx \ 1.33$	
	Table 3.9 value $\times$ Table 3.10 value $= (16 \times 1.33) = 21.28$	Adequate with respect to deflection
	Actual $\dfrac{\text{span}}{\text{effective depth}} = \dfrac{6000}{454} = 13.2 \ < 21.28$	
	Transverse Reinforcement	Adopt 10 mm diameter bars @ 300 mm centres providing 262 mm²/m transverse steel across the full effective flange width of the top flange.
Clause 3.12.5.3	A_s required $= \dfrac{0.15 \times h_f \times 1000}{100}$ mm²/m length	
Table 3.25	$= \dfrac{0.15 \times 160 \times 1000}{100} = 240 \text{ mm}^2/\text{m}$	
Table 3.25	**Minimum % tension reinforcement** Flanged beams with webs in tension and $f_y = 460$ N/mm²	
	$b_w/b = (350/1550) = 0.23 \ < \ 0.4$	
	$\therefore \ 100A_s/b_w h = 0.18$	
	Minimum A_s required $= \dfrac{0.18 \times b_w \times h}{100}$ mm²	

References	Calculations	Output

Contract : Floor Slab   Job Ref. No. : Example 5.19
Part of Structure :   T-Beam
Calc. Sheet No. : 5 of 5

Calcs. by : W.McK.
Checked by :
Date :

References	Calculations	Output

$$\frac{0.18 \times 350 \times 500}{100} = 315 \text{ mm}^2 << 1570 \text{ mm}^2$$

Minimum % reinforcement satisfied

Clause 3.12.10
Figure 3.24:

Curtailment:
Curtailment of main steel = 0.08L = (0.08 × 6000)
                                              =   480 mm
Use 2/12 mm diameter bars as hangers for links

Standard
90° bend at
support

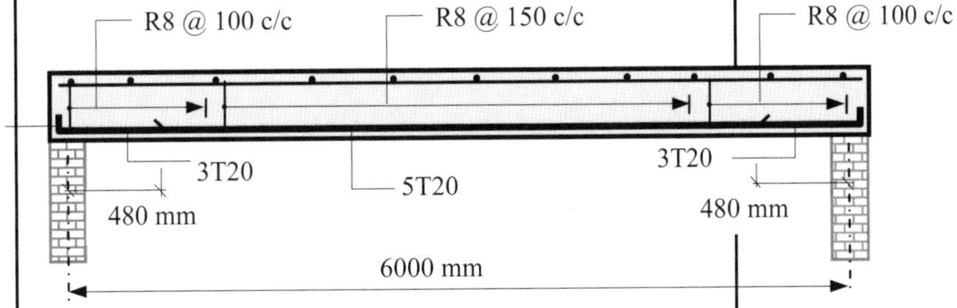

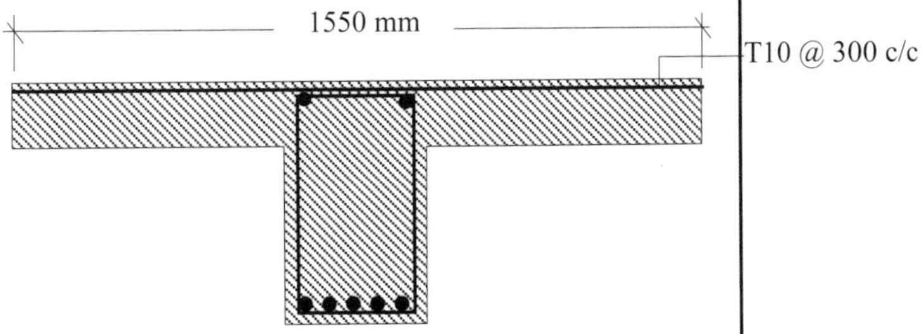

5.11 Multi-Span Beams and Slabs

5.11.1   Analysis

Many reinforced concrete structures are cast in-situ resulting in a load-bearing frame in which the slabs, beam and columns act as a continuum to resist and transfer applied loads to the foundations as shown in Figure 5.75.

In many braced structures, elements such as the shear walls resist the lateral wind loading, whilst the slabs, beams and columns are designed to resist the vertical gravity loading.

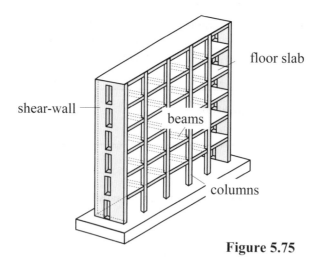

Figure 5.75

The continuity of such structures is maintained by ensuring an adequate provision of reinforcing steel to tie together the various elements at their connections, as indicated in Figure 5.76.

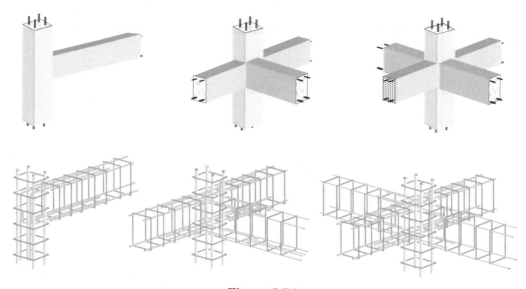

Figure 5.76

The design of continuous sections is based on an analysis to determine maximum sagging and hogging bending moments and the maximum shear forces in the members. The continuity of the structure requires an analysis to be carried out for multi-span beams and/or slabs in addition to multi-storey columns.

The design code BS 8110 permits the use of approximate analysis techniques in which the structure can be considered as a series of sub-frames. The complexity of the sub-frames considered (i.e. the extent to which various columns and beams are included) is given in Clauses 3.2.1.2.1 to 3.2.1.2.5 for monolithic frames not providing lateral stability, e.g. as shown in Figure 5.75 (where the shear walls resist the lateral wind loads).

In Clause 3.2.1.3.1 and Clause 3.2.1.3.2 provisions are given for frames providing lateral stability. Consider the multi-storey frame indicated in Figure 5.77 in which it is assumed that the lateral loading is resisted by separate elements not indicated, such as shear cores. The slabs, beams and columns transfer only vertical loads by rigid-frame action.

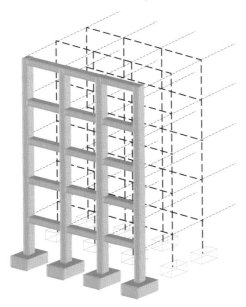

Figure 5.77

Clause 3.2.1.2.1 *(Simplification into sub-frames)*
This Clause states that '*...Each sub-frame may be taken to consist of the beams at one level together with the columns above and below. The ends of the columns remote from the beams may generally be assumed to be fixed unless the assumption of a pinned end is clearly more reasonable (for example, where a foundation detail is considered unable to develop moment restraint.)*'
This is illustrated in Figure 5.78.

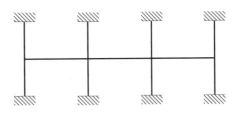

Sub-frame for analysis of beams and columns

Figure 5.78

The critical loading arrangements which should be considered for this sub-frame are defined in Clause 3.2.1.2.2 as:

    a) all spans loaded with the maximum design ultimate load $(1.4G_k + 1.6Q_k)$;

    b) alternate spans loaded with the maximum design ultimate load $(1.4G_k + 1.6Q_k)$ and all other spans loaded with the minimum design ultimate load $(1.0G_k)$.

Clause 3.2.1.2.3: (*Alternative simplification for individual beams and associated columns*)

    This Clause states that '*...the moments and forces in each individual beam may be found by considering a simplified sub-frame consisting only of that beam, the columns attached to the end of that beam and the beams on either side, if any. The column and beam ends remote from the beam under consideration may generally be assumed to be fixed unless the assumption of pinned is clearly more reasonable. The stiffness of the beams on either side of the beam considered should be taken as half their actual values if they are taken to be fixed at their outer ends. The critical loading arrangements should be in accordance with 3.2.1.2.2.*'

    '*The moments in an individual column may also be found from this simplified sub-frame provided that the sub-frame has as its central beam the longer of the two spans framing into the column under consideration.*'

This is illustrated in Figure 5.79.

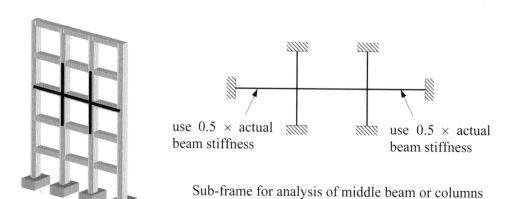

    use 0.5 × actual beam stiffness   use 0.5 × actual beam stiffness

Sub-frame for analysis of middle beam or columns

Figure 5.79

Clause 3.2.1.2.4 (*'Continuous beam' simplification*)

    This Clause states that '*...the moments and forces in the beams at one level may also be obtained by considering the beams as a continuous beam over supports providing no restraint to rotation. The critical loading arrangements should be in accordance with 3.2.1.2.2.*'

This is illustrated in Figure 5.80.

Sub-frame for analysis of beams at any one level

Figure 5.80

Clause 3.2.1.2.5   (Asymmetrically-loaded columns where a beam has been analysed in accordance with 3.2.1.2.4)

> This Clause states that '...*the ultimate moments may be calculated by simple moment distribution procedures, on the assumption that the column and beam ends remote from the junction under consideration are fixed and that the beams possess half their actual stiffness. The arrangement of the design ultimate imposed load should be such as to cause the maximum moment in the column.*'

This is illustrated in Figure 5.81.

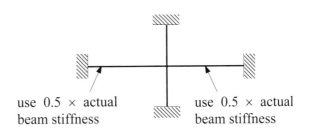

use 0.5 × actual
beam stiffness

use 0.5 × actual
beam stiffness

Alternative sub-frame for analysis of column where beams are analysed using the continuous beam simplification.

Figure 5.81

When using Clauses 3.2.1.2.2, 3.2.1.2.3 and Clause 3.2.1.4 for design, the critical load arrangements should be in accordance with **Clause 3.2.1.2.2:**

'It will normally be sufficient to consider the following arrangements of vertical load:

(a)   all spans loaded with the maximum design ultimate load ($1.4G_k + 1.6Q_k$);
(b)   alternate spans loaded with the maximum design ultimate load ($1.4G_k + 1.6Q_k$) and all other spans loaded with the minimum design ultimate load ($1.0G_k$).'

The shear force and bending moment diagrams can be drawn for each of the load cases required in the patterns of loading. A composite diagram comprising a profile indicating the maximum values including all possible load cases can be drawn; this is known as an ***envelope***. An example of a typical bending moment envelope for a continuous three-span beam is illustrated in Figure 5.82.

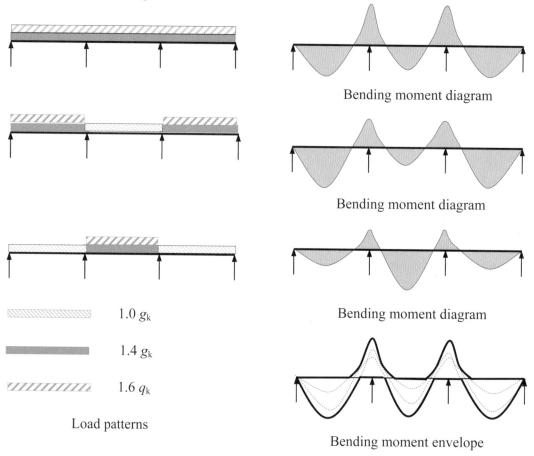

Bending moment diagram

Bending moment diagram

Bending moment diagram

$1.0\,g_k$

$1.4\,g_k$

$1.6\,q_k$

Load patterns

Bending moment envelope

Figure 5.82

This type of analysis is time-consuming and is more conveniently carried out using standard computer techniques. Tables are given in BS 8110 which enable a conservative estimate of shear force and bending moment values to be determined for the design of continuous beams, Table 3.5, and continuous one-way spanning slabs, Table 3.12, with three or more spans.

There are conditions which must be satisfied in each case before these tables can be used. They are:

For beams   (Clause 3.4.3):
*   *the beams should be of approximately equal span,*
*   *the characteristic imposed load Q_k may not exceed the characteristic dead load G_k,*

♦ *loads should be substantially uniformly distributed over three or more spans,*
♦ *variations in span length should not exceed 15% of the longest span.*

For slabs (Clause 3.5.2.4):
♦ *in a one-way spanning slab, the area of each bay exceeds 30 m^2,*
    In this context, a bay means a strip across the full width of a structure bounded on the other two sides by lines of support as shown in Figure 5.83,

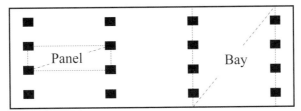

Figure 5.83

♦ *the ratio of the characteristic imposed load to the characteristic dead load does not exceed 1.25,*
♦ *the characteristic imposed load does not exceed 5 kN/m^2 excluding partitions.*

5.11.1.1   Example 5.20: Analysis of a Typical Sub-Frame

A sub-frame from a monolithic, braced frame is shown in Figure 5.84. Using the data given and the simplified analysis methods indicated in Section 3.2 of the code, determine the design moments in columns FJ and FB.

Data:
Characteristic dead load (including self-weight) =   20 kN/m
Characteristic imposed load   =   12 kN/m

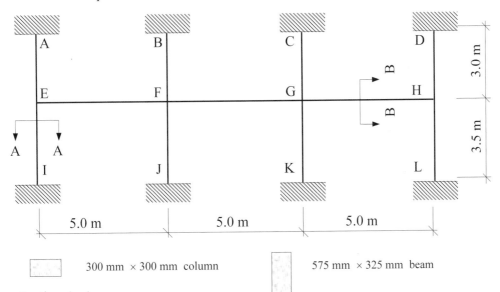

300 mm × 300 mm column

575 mm × 325 mm beam

Section A–A

Section B–B   **Figure 5.84**

Note: The analysis of the sub-frame can be carried out using moment distribution (without sway). Clearly if a computer analysis package is available, the complete frame can be analysed. Since the frame is braced it will be necessary to provide a roller support at each floor level in the computer analysis to prevent sway.

Critical Load Cases   (Clause 3.2.1.2.2):
The critical load cases to be considered are shown in Figure 5.85.

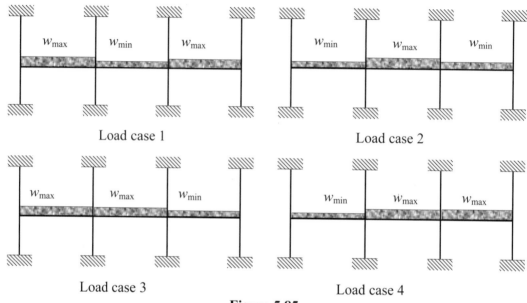

Load case 1   Load case 2

Load case 3   Load case 4

Figure 5.85

w_{max} = $(1.4 \times 20) + (1.6 \times 12)$ = 47.2 kN/m
w_{min} = (1.0×20) = 20.0 kN/m

Using an appropriate elastic analysis, the maximum bending moment in columns FB and FJ are found to be approximately 9.3 kNm and 7.9 kNm respectively.
    These values can be compared with those obtained using other sub-frames, e.g. as shown in Figure 5.86.

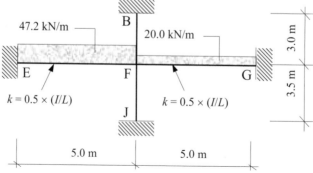

*Alternative sub-frame for analysis of a column where beams
are analysed using the continuous beam simplification.*   **Figure 5.86**

In this simplified frame:

$$I_{beam} = \frac{bd^3}{12} = \frac{0.325 \times 0.575^3}{12} = 5.15 \times 10^{-3} \text{ m}^4$$

$$k_{FE} = \frac{I}{L} = \frac{5.15 \times 10^{-3}}{5.0} = 1.03 \times 10^{-3} \text{ m}^3; \qquad k_{FG} = 1.03 \times 10^{-3} \text{ m}^3$$

$$I_{column} = \frac{bd^3}{12} = \frac{0.3 \times 0.3^3}{12} = 0.68 \times 10^{-3} \text{ m}^4$$

For upper column: $\quad k_{FB} = k_U = \dfrac{I}{L} = \dfrac{0.68 \times 10^{-3}}{3.0} = 0.23 \times 10^{-3} \text{ m}^3$

For lower column: $\quad k_{FJ} = k_L = \dfrac{I}{L} = \dfrac{0.68 \times 10^{-3}}{3.5} = 0.19 \times 10^{-3} \text{ m}^3$

In this sub-frame the beam stiffnesses are assumed to be equal to (0.5 × actual value).
Total stiffness of the joint $\quad k_{total} = [0.5 \times (1.03 + 1.03) + 0.23 + 0.19)] \times 10^{-3}$
$$= 1.45 \times 10^{-3} \text{ m}^3$$

The fixed end moments from the beam loadings are:

$$M_{BA} = \frac{wL^2}{12} = \frac{47.2 \times 5.0^2}{12} = 98.3 \text{ kNm}$$

$$M_{BC} = \frac{wL^2}{12} = \frac{20 \times 5.0^2}{12} = 41.7 \text{ kNm}$$

Completing the moment distribution at the joint B gives:

Moment in the upper column $= M_u = (98.3 - 41.7) \times$ Distribution Factor.
$$= 56.6 \times \frac{0.23}{1.45} = 8.98 \text{ kNm}$$

Moment in the lower column $= M_L = (98.3 - 41.7) \times$ Distribution Factor
$$= 56.6 \times \frac{0.19}{1.45} = 7.42 \text{ kNm}$$

These values compare favourably with the column moments found using the more rigorous analysis (9.35 kNm and 7.95 kNm).

5.11.2 Redistribution of Moments (Clause 3.2.2.1)

When continuous structures approach their failure load there is a redistribution of load as successive plastic hinges develop until failure occurs; this is dependent on the ductility of the material. Advantage can be taken of this behaviour to reduce the maximum moments whilst at the same time increasing others to maintain static equilibrium.

The redistribution of moments in concrete frames should satisfy the conditions indicated in Clause 3.2.2.1 of BS 8110:

> 'redistribution of the moments obtained by means of a rigorous elastic analysis ... may be carried out provided the following conditions are satisfied:
> ♦ equilibrium between internal and external forces is maintained under all appropriate combinations of design ultimate moment,

♦ *where the design ultimate resistance moment of the cross-section subjected to the largest moment within each region of hogging or sagging is reduced, the neutral axis depth should be checked to see that it is not greater than (β_b – 0.4)d where d is the effective depth and β_b is the ratio:*

$$\frac{(moment\ at\ the\ section\ after\ redistribution)}{(moment\ at\ the\ section\ before\ redistribution)} \leq 1$$

from the respective maximum moments diagram,
♦ *the resistance moment at any section should be at least 70% of the moment at that section obtained from an elastic maximum moments diagram covering all appropriate combinations of design ultimate load.*

These conditions limit the redistribution to 30%. In Clause 3.2.2.2 the provisions in Clause 3.2.2.1 are modified for structures over four storeys where the structural frame provides lateral stability such that redistribution is limited to 10% and the 70% criterion should be 90%.

Note: *If Table 3.5 for beams and/or Table 3.12 for slabs is used to estimate the moments in continuous spans then NO redistribution should be carried out on the values from the tables; this has already been allowed for in the coefficients.*

5.11.2.1 Example 5.21: Redistribution of Moments in a Two-span Beam

A two-span beam is required to support an ultimate design load of 150 kN/m as shown in Figure 5.87. Reduce the support moment by 20% and determine the redistributed bending moment diagram.

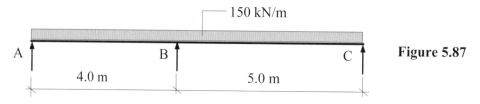

Figure 5.87

Use moment distribution to determine the moments over the supports and in the spans.
Fixed end moments:
Use $wl^2/8$ for FEMs and no carry-over as there are simple supports at both ends.

Span AB $FEM_{AB} = 0$ $FEM_{BA} = + \dfrac{wL^2}{8} = + \dfrac{150 \times 4^2}{8} = + 300$ kNm

Span BC $FEM_{BC} = - \dfrac{wL^2}{8} = - \dfrac{150 \times 5^2}{8} = - 469$ kNm, $FEM_{CB} = 0$

Stiffnesses:

$k_{BA} = \dfrac{I}{4} = 0.25$ $DF_{BA} = \dfrac{0.25}{0.45} = 0.56$

$k_{total} = 0.45$

$k_{BC} = \dfrac{I}{5} = 0.2$ $DF_{Bc} = \dfrac{0.2}{0.45} = 0.44$

Joint	A	B		C
	AB	BA	BC	CB
Distribution Factors (*k*)	1.0	0.56	0.44	1.0
Fixed End Moments (FEMs)	0	+ 300	– 469	0
Balance (no carry over)		+ 94.6	+ 74.4	
Final Moments	0	+ 394.6	– 394.6	0

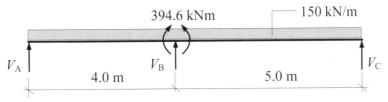

Consider span AB:

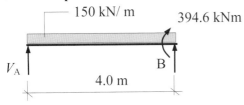

$$\Sigma \text{Moments to the L.H.S.} = 0$$
$$(V_A \times 4.0) + 394.6 - (150 \times 4.0 \times 2.0) = 0$$
$$V_A = 201.4 \text{kN}$$

Consider span BC:

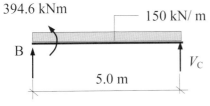

$$\Sigma \text{Moments to the R.H.S.} = 0$$
$$- (V_C \times 5.0) - 394.6 + (150 \times 5.0 \times 2.5) = 0$$
$$V_C = 296.1 \text{kN}$$

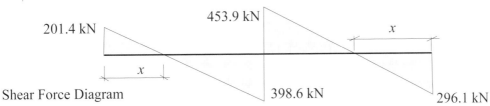

Shear Force Diagram

Span AB: $x = (201.4/150) = 1.34$ m
Maximum bending moment $= (0.5 \times 1.34 \times 201.4) = 134.9$ kNm
Span BC: $x = (296.1/150) = 1.97$ m
Maximum bending moment $= (0.5 \times 1.97 \times 296.1) = 291.7$ kNm

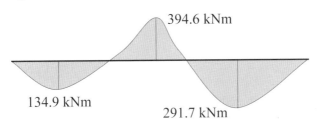

The reduced bending moment at the support $= 0.8 \times 394.6 = 315.7$ kNm

The above calculation must be repeated with the revised value of support moment to determine the redistributed maximum moments in the spans.

Consider span AB:

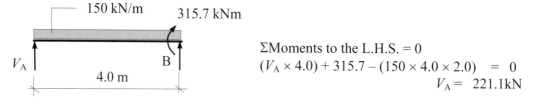

ΣMoments to the L.H.S. = 0
$(V_A \times 4.0) + 315.7 - (150 \times 4.0 \times 2.0) = 0$
$V_A = 221.1$kN

Consider span BC:

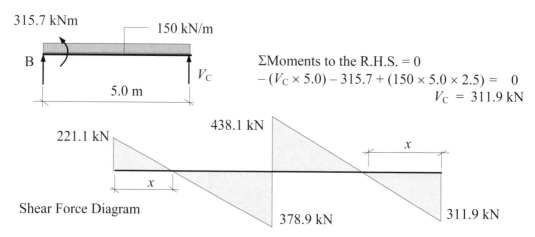

ΣMoments to the R.H.S. = 0
$-(V_C \times 5.0) - 315.7 + (150 \times 5.0 \times 2.5) = 0$
$V_C = 311.9$ kN

Shear Force Diagram

Span AB: $x = (221.1/150) = 1.47$ m
Maximum bending moment $= (0.5 \times 1.47 \times 221.1) = 162.5$ kNm
Span BC: $x = (311.9/150) = 2.08$ m
Maximum bending moment $= (0.5 \times 2.08 \times 311.9) = 324.4$ kNm

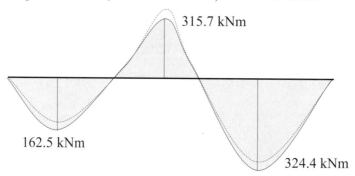

Redistributed Bending Moment Diagram
Figure 5.88

5.11.3   *Example 5.22: Multi-Span Floor System Design*

A floor system consisting of a solid in-situ reinforced concrete slab cast integrally with the support beams is supported over four spans of 6.0 m as shown in Figure 5.89.

(a)   design suitable slab reinforcement,
(b)   check the suitability of the slab with respect to shear and deflection,
(c)   design suitable reinforcement for a typical **T**-section to satisfy flexure and shear,
(d)   check the suitability of a typical **T**-section with respect to deflection,
(e)   determine the transverse reinforcement required for the flanges of the **T**-beams,
(f)   prepare a sketch indicating all reinforcement; use the simplified rules indicated in Clause 3.12.10 and Figure 3.24 for curtailment.

Design Data:

Characteristic dead load (excluding self-weight + finishes)	g_k =	12.0 kN/m^2
Characteristic dead load due to finishes only	g_k =	1.0 kN/m^2
Characteristic imposed load	q_k =	5.0 kN/m^2
Concrete grade	f_{cu} =	40 N/mm^2
Characteristic strength of reinforcing steel	f_y =	460 N/mm^2
Exposure condition	=	severe
Fire resistance	=	1hr minimum
Slab thickness	h_f =	300 mm
Rib width	b_w =	300 mm
Overall depth	h =	600 mm
Span of main beams	L =	6.0 m
Centres of main beams	=	4.0 m

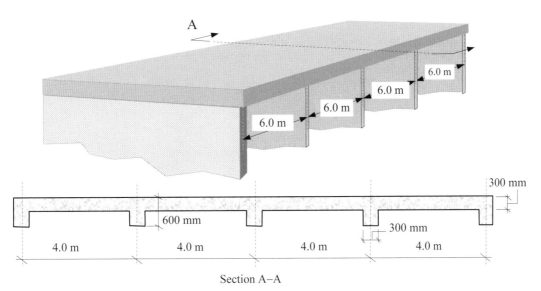

Section A–A

Figure 5.89

5.11.4   Solution to Example 5.22

Contract : Multi-span Floor Job Ref. No. : Example 5.22 Part of Structure : Multi-span Slab Calc. Sheet No. : 1 of 13	Calcs. by : W.McK. Checked by : Date :

References	Calculations	Output
	BS8110:Part 1:1997 **Design Loads:** *Consider the slab and design a 1 metre width strip*: Characteristic load due to self-weight of slab: $\qquad = (0.3 \times 4.0 \times 24) \quad = 28.8 \text{ kN/m}$ Characteristic load due to dead load: $\qquad = (12.0 \times 4.0 \times 1.0) = 48.0 \text{ kN}$ Characteristic load due to finishes: $\qquad = (1.0 \times 4.0 \times 1.0) \quad = 4.0 \text{ kN}$ Characteristic load due to imposed loading: $\qquad = (5.0 \times 4.0) \qquad = 20.0 \text{ kN}$	
Table 2.1	Ultimate design load $= 1.4(28.8 + 4.0 + 48.0) + (1.6 \times 20.0)$ $\qquad\qquad = 145.1 \text{ kN/m width}$	
Clause 3.5.2.3	Area of one bay $\quad = 6.0 \times 16.0 = 96 \text{ m}^2 > 30 \text{ m}^2$ $\qquad \dfrac{q_k}{g_k} \quad = \dfrac{20}{80.8} \; < \; 1.25 \quad$ and $\qquad q_k \quad \leq \; 5.0 \text{ kN/m}^2$	
	Use Table 3.12 to determine the bending moments and shear forces.	
Table 3.12	Near middle of end span ultimate bending moment $= \; 0.086Fl$ At first interior support ultimate bending moment $= \; -0.086Fl$ At outer support ultimate shear force $\qquad\qquad = \; 0.4F$ At first interior support ultimate shear force $\quad = \; 0.6F$ where F is the total design ultimate load $\; = \; (1.4G_k + 1.6Q_k)$ $\qquad\qquad\qquad\qquad\qquad = \; 145.1 \text{ kN}$ Near middle of end span ultimate bending moment: $\qquad = (0.086 \times 145.1 \times 4.0) \qquad = +50 \text{ kNm}$ At first interior support ultimate bending moment: $\qquad = -(0.086 \times 145.1 \times 4.0) \qquad = -50 \text{ kNm}$ At outer support ultimate shear force: $\qquad = (0.4 \times 145.1) \qquad\qquad = 58 \text{ kN}$ At first interior support ultimate shear force: $\qquad = (0.6 \times 145.1) \qquad\qquad = 87 \text{ kN}$	
Clause 3.3.7	**Cover:** Assume 20 mm diameter bars.	
Clause 3.3.1.2 Table 3.3 Table 3.4	bar size $\qquad\qquad\qquad = \; 20 \text{ mm}$ exposure condition severe $\geq \; 40 \text{ mm}$ min. fire resistance 1 hour $\geq \; 20 \text{ mm}$ **Assume minimum cover to main steel = 40 mm**	

References	Calculations	Output

Contract : Multi-span Floor   Job Ref. No. : Example 5.22
Part of Structure :   Multi-span Slab
Calc. Sheet No. : 2 of 13

Calcs. by : W.McK.
Checked by :
Date :

Clause 3.3.6 — Minimum dimensions for 1 hour fire cover
floor thickness ≥ 95 mm

Thickness is adequate

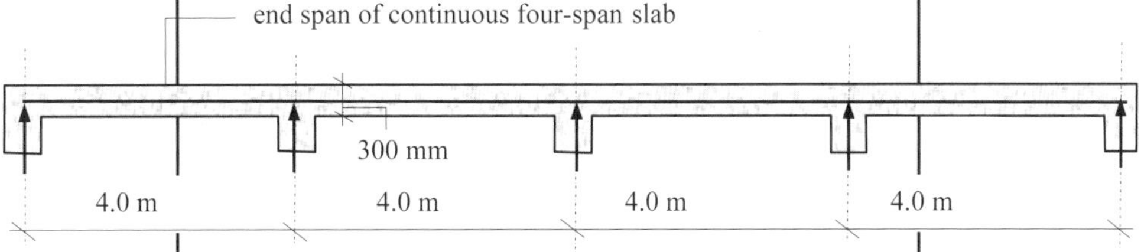

end span of continuous four-span slab

300 mm

4.0 m   4.0 m   4.0 m   4.0 m

Section A–A

Effective depth d = (300 – 40 – 10) = 250 mm

Bending moments and shear forces evaluated using the coefficients from Table 3.12 for an end span with a simple support are:

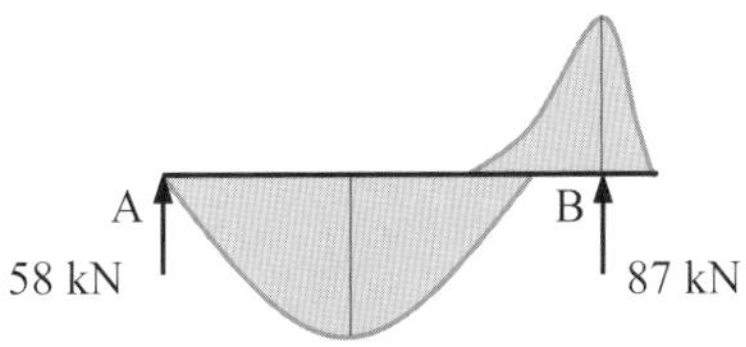

– 50 kNm

A   B
58 kN   87 kN

+ 50 kNm

Note: The area of reinforcing steel required in the bottom of span AB is the same as that required at the top over support B.

Clause 3.4.4 — **Bending:**

Clause 3.4.4.4

$$K = \frac{M}{bd^2 f_{cu}} = \frac{50\times10^6}{1000\times250^2\times40} = 0.02 \quad < \quad K'(=0.156)$$

Section is singly reinforced

$$Z = d\left\{0.5+\sqrt{\left(0.25-\frac{K}{0.9}\right)}\right\} = d\left\{0.5+\sqrt{\left(0.25-\frac{0.02}{0.9}\right)}\right\}$$

$$= 0.97d$$

The lever arm is limited to 0.95d

References	Calculations	Output
Contract : Multi-span Floor Job Ref. No. : Example 5.22		Calcs. by : W.McK.
Part of Structure : Multi-span Slab		Checked by :
Calc. Sheet No. : 3 of 13		Date :

References	Calculations	Output
	$$A_s = \dfrac{M}{0.95 f_y Z} = \dfrac{50 \times 10^6}{0.95 \times 460 \times 0.95 \times 250} = 482 \text{ mm}^2$$	**Bottom Reinforcement** Select: 16 mm diameter bars at 400 mm centres providing 503 mm²/m at mid-span and over first interior support.
Clause 3.12.11.1	Minimum spacing is not critical in slabs.	
Clause 3.12.11.2	Maximum spacing of reinforcement: $\leq \ (3 \times d) = \ (3 \times 250) \qquad = \ 750 \text{ mm}$ $\leq \ 750 \text{ mm}$ % reinforcement $= \dfrac{100 A_s}{bd} = \dfrac{100 \times 482}{1000 \times 250} = \ 0.19\%$ $< 0.3\%$ Since there is less than 0.3% reinforcement no further check is required on spacing.	
Clause 3.12.10.3 Figure 3.25	Simplified rules for curtailment of steel in slabs designed for predominantly uniformly distributed loads. 50% of main steel is curtailed a distance $0.1L$ from the support. At support $A_s = \ (0.5 \times 503) = \ 252 \text{ mm}^2$ (**Note:** curtail alternate bars) $\dfrac{100 A_s}{bd} = \dfrac{100 \times 252}{1000 \times 250} = \ 0.1$	
Clause 3.5.5	**Shear Resistance** First interior support	
Clause 3.5.5.2	Shear stress $\quad v = \dfrac{V}{b_v d} = \dfrac{87 \times 10^3}{1000 \times 250} = \ 0.34 \text{ N/mm}^2$ Maximum shear $\quad \leq \ 0.8\sqrt{f_{cu}} = \ (0.8 \times \sqrt{40})$ $= \ 5.05 \text{ N/mm}^2$ $\leq \ 5.0 \text{ N/mm}^2$ $v \ < \ $ maximum permitted value	
Table 3.8	$v_c \approx \ (0.38 \times 1.17) = \ 0.44 \text{ N/mm}^2$ (**Note:** this allows for using C40 concrete)	
Table 3.16	$v \ < \ v_c$	
	No links are required	
Clause 3.4.6	**Deflection:**	
Clause 3.4.6.3	$\dfrac{\text{span}}{\text{effective depth } (d)} \leq \ $ Table 3.9 value × Table 3.10 value (**Note:** No compression steel/Table 3.11 value is required)	

<table>
<tr><td colspan="2">Contract : Multi-span Floor Job Ref. No. : Example 5.22
Part of Structure : Multi-span Slab
Calc. Sheet No. : 4 of 13</td><td>Calcs. by : W.McK.
Checked by :
Date :</td></tr>
<tr><td>References</td><td>Calculations</td><td>Output</td></tr>
<tr><td>Table 3.9

Table 3.10</td><td>Basic (span/d) ratio = 26.0 (estimate for end span between simple support and continuous span).
$\frac{M}{bd^2} = K \times f_{cu} = 0.02 \times 40 = 0.8$
Service stress $f_s = \frac{2 \times f_y \times A_{s,\,required}}{3 \times A_{s,\,provided}} = \frac{2 \times 460 \times 494}{3 \times 503}$
≈ 301 N/mm^2

Interpolate between values given in Table 3.10:</td><td></td></tr>
</table>

M/bd^2 / service stress f_s	0.75	1.00
300	1.44	1.33
307	1.41	1.30

<table>
<tr><td></td><td>Use conservative estimate of modification factor ≈ 1.30

Table 3.9 value × Table 3.10 value = 26 × 1.30 = 33.8
Actual $\frac{\text{span}}{\text{effective depth}} = \frac{4000}{250} = 16 < 32$</td><td>Adequate with respect to deflection</td></tr>
<tr><td>Table 3.25</td><td>Minimum % tension reinforcement:
For slabs of rectangular section and f_y = 460 N/mm^2
$100A_s/A_c \geq 0.13\%$
At least this percentage should be provided in both directions.
Minimum A_s required $= \frac{0.13 \times b \times h}{100}$ mm^2
$= \frac{0.13 \times 1000 \times 300}{100} = 390$ mm^2
< 494 mm^2
Minimum % reinforcement satisfied</td><td></td></tr>
<tr><td>Clause 3.12.5.3
Table 3.25</td><td>Secondary Reinforcement:
Minimum % = 0.13 = 390 mm^2/m width</td><td rowspan="2">Secondary Reinforcement Select:
16 mm diameter bars at 500 mm centres providing 402 mm^2/ m</td></tr>
<tr><td>Clause 3.12.10
Figure 3.25</td><td>Curtailment:
The steel is curtailed as indicated in Figure 3.25 of the code and is similar to that indicated in Section 5.7.7.1 and Figure 5.69 of this text for beams. Note that the values are slightly different for slabs.</td></tr>
</table>

Contract : Multi-span Floor Job Ref. No. : Example 5.22 **Part of Structure : Multi-span Slab** **Calc. Sheet No. : 5 of 13**		**Calcs. by : W.McK.** **Checked by :** **Date :**

References	Calculations	Output
	At interior support Top reinforcement 100% of steel extends 0.15l = (0.15 × 4000)= 600 mm or 45 bar dia. = (45 × 16) = 720 mm 50% of steel extends 0.3l = (0.3 × 4000) = 1200 mm Curtail bars 720 mm and 1200 mm from the face of the support. Bottom reinforcement 60% of mid-span reinforcement extends to within 0.2l of the centre line of the support. 0.2l = (0.2 × 4000) = 800 mm Curtail bars 800 mm from the centre line of the support. **At end support** (assuming simple support) Top reinforcement	
Clause 3.12.10.3	Provide anti-crack steel equal to 50% of the bottom steel at mid-span and not less than the minimum % as given in Clause 3.12.5.3. 0.15l = (0.15 × 4000) = 600 mm or 45 bar dia. = (45 × 16) = 720 mm This steel should be provided with a full tensile anchorage length. Curtail bars 720 mm from the *face* of the support. Bottom reinforcement 60% of mid-span reinforcement extends to within 0.1l of the centre-line of the support. 0.1 = (0.1 × 4000) = 400 mm Curtail bars 400 mm from the *centre-line* of the support.	
Standard 90° bend at support	Slab detail	

Contract : **Multi-span Floor** **Job Ref. No. : Example 5.22** Part of Structure : **T-Beam** Calc. Sheet No. : **6** of **13**	**Calcs. by : W.McK.** **Checked by :** **Date :**

References	Calculations	Output
	Consider the main beams:	
	Characteristic load due to self-weight of slab:	
	$\quad = \quad (0.3 \times 4.0 \times 24) \quad = \quad 28.8 \text{ kN/m}$	
	Characteristic self-weight of rib:	
	$\quad = \quad (0.3 \times 0.3 \times 24) \quad = \quad 2.16 \text{ kN/m}$	
	Characteristic load due to dead load:	
	$\quad = \quad (12.0 \times 4.0) \quad = \quad 48.0 \text{ kN/m}$	
	Characteristic load due to finishes:	
	$\quad = \quad (1.0 \times 4.0) \quad = \quad 4.0 \text{ kN/m}$	
	Characteristic load due to imposed loading:	
	$\quad = \quad (5.0 \times 4.0) \quad = \quad 20.0 \text{ kN/m}$	
Table 2.1	Ultimate design load $\quad = \quad (1.4g_k + 1.6q_k)$	
	$\quad = \quad 1.4(28.8 + 2.16 + 48.0 + 4.0) + (1.6 \times 20.0)$	
	$\quad = \quad 148.1 \text{ kN/m} \quad = \quad (148.1 \times 6.0) \quad = \quad 888.6 \text{ kN}$	
Clause 3.4.3	Characteristic imposed load $Q_k <$ Characteristic dead load G_k, Loads are substantially uniformly distributed over three or more spans, variations in span lengths do not exceed 15% of the longest span, therefore use Table 3.5.	
Table 3.5	Near middle of end span ultimate bending moment $= \quad 0.09Fl$	
	At first interior support ultimate bending moment $= \quad -0.11Fl$	
	At outer support ultimate shear force $\qquad = \quad 0.45F$	
	At first interior support ultimate shear force $\quad = \quad 0.6F$	
	where F is the total design ultimate load $\quad = \quad (1.4G_k + 1.6Q_k)$	
	$\qquad\qquad = \quad 888.6 \text{ kN}$	
	Near middle of end span ultimate bending moment:	
	$\quad = (0.09 \times 888.6 \times 6.0) \quad = +479.8 \text{ kNm}$	
	At first interior support ultimate bending moment:	
	$\quad = -(0.11 \times 888.6 \times 6.0) \quad = -586.5 \text{ kNm}$	
	At outer support ultimate shear force:	
	$\quad = (0.45 \times 888.6) \quad = 400 \text{ kN}$	
	At first interior support ultimate shear force:	
	$\quad = (0.6 \times 888.6) \quad = 533.2 \text{ kN}$	
	Bending moments and shear forces evaluated using the coefficients from Table 3.5 for an end span with a simple support are:	

References	Calculations	Output
	Contract : Multi-span Floor **Job Ref. No. :** Example 5.22 **Part of Structure :** T-Beam **Calc. Sheet No. : 7 of 13**	**Calcs. by : W.McK.** **Checked by :** **Date :**

References	Calculations	Output
	Consider near the middle of the end span – section can be designed as a **T-beam**, i.e. compression in the top surface.	
Clause 3.4.1.5	Effective width of flange beam = $b_e \leq (b_w + l_z/5)$ $= 300 + \left(\dfrac{0.7 \times 6000}{5} \right)$ $= 1140$ mm Actual flange width $= 4000$ mm $\therefore \quad \boldsymbol{b_e = 1140}$ **mm**	
Clause 3.3.7 Clause 3.3.1.2 Table 3.3 Table 3.4	**Cover** Assume 25 mm diameter bars for main steel: bar size $= 25$ mm exposure condition severe ≥ 40 mm fire resistance ≥ 1 hour ≥ 20 mm **Assume nominal cover to steel 40 mm**	
	Assume 8 mm diameter bars for links. Effective depth $\quad d = (600 - 40 - 8 - 13) = 539$ mm	
Clause 3.3.6	Minimum dimensions for 1 hour fire cover beam width ≥ 200 mm **Beam width is satisfactory**	
Clause 3.4.4 Clause 3.4.4.4	Bending $K = \dfrac{M}{bd^2 f_{cu}} = \dfrac{479.8 \times 10^6}{1140 \times 539^2 \times 40} = 0.04 \quad < \quad K'\,(= 0.156)$ **Section is singly reinforced** $Z = d\left\{0.5 + \sqrt{\left(0.25 - \dfrac{K}{0.9}\right)}\right\} = d\left\{0.5 + \sqrt{\left(0.25 - \dfrac{0.04}{0.9}\right)}\right\}$ $= 0.95d \quad (\leq 0.95d)$	

Contract : Multi-span Floor Job Ref. No. : Example 5.22	Calcs. by : W.McK.
Part of Structure : T-Beam	Checked by :
Calc. Sheet No. : 8 of 13	Date :

References	Calculations	Output
	Depth to the neutral axis: $x = (d-z)/0.45 = (0.05 \times 539)/0.45 = 59.9$ mm $< h_f (= 300$ mm$)$ The neutral axis lies within the flange. $A_s = \dfrac{M}{0.95 f_y Z} = \dfrac{479.8 \times 10^6}{0.95 \times 460 \times 0.95 \times 539} = 2144$ mm^2 The bars selected are 32 mm diameter which are slightly greater than the 25 mm diameter assumed when calculating the effective depth. This slightly reduces the strength but not normally sufficiently to require a re-calculation – the reader should check this.	**Bottom Reinforcement** Select: **3/32 mm diameter bars at mid-span providing 2410 mm^2**
Clause 3.12.11.1	Minimum spacing of reinforcement ≥ 25 mm **Adequate**	
Clause 3.12.11.2 Table 3.28	Maximum spacing of reinforcement: There is no redistribution and $f_y = 460$ N/mm^2 The clear distance between bars ≤ 155 mm Actual spacing $= [300 - (2 \times 40) - (2 \times 8) - (2 \times 32)]/2$ $= 70$ mm **Adequate** Consider at the first interior support: the section must be designed as a **rectangular beam** since the top flange is in tension and does not contribute to the bending strength. Change the value of d to allow for 16 mm diameter slab steel $d = 539 - 16 = 523$ mm 	
Clause 3.4.4 Clause 3.4.4.4	**Bending:** $K = \dfrac{M}{bd^2 f_{cu}} = \dfrac{586.5 \times 10^6}{300 \times 523^2 \times 40} = 0.18 > K' (= 0.156)$ **Section is doubly reinforced**	

References	Calculations	Output
	$$Z = d\left\{0.5 + \sqrt{\left(0.25 - \frac{K'}{0.9}\right)}\right\} = d\left\{0.5 + \sqrt{\left(0.25 - \frac{0.156}{0.9}\right)}\right\}$$ $$= 0.775d = (0.775 \times 523)$$ $$= 405.3 \text{ mm}$$ $$x = (d-z)/0.45 = 0.5d$$ $$= (0.5 \times 405.3) = 202.7 \text{ mm}$$ Assume 16 mm diameter bars for A_s' $$d' = (40 + 8 + 8) = 56 \text{ mm}$$ $$d'/x = 56/202.7 = 0.28 \leq 0.37$$ The compression steel will yield and the stress is equal to $0.95f_y$. $$A_s' = (K - K')f_{cu}bd^2 / 0.95f_y(d - d')$$ $$= \frac{(0.18 - 0.156) \times 40 \times 300 \times 523^2}{0.95 \times 460 \times (523 - 56)} = 386 \text{ mm}^2$$ $$A_s = (K'f_{cu}bd^2 / 0.95f_y z) + A_s'$$ $$= \left(\frac{0.156 \times 40 \times 300 \times 523^2}{0.95 \times 460 \times 405.3}\right) + 386 = 3277 \text{ mm}^2$$ **TOP STEEL**	**Bottom Reinforcement** **Two bars of the mid-span steel are continued through the support and can be used to provide the required area of steel.** **A FULL compression lap is necessary for this steel.** **Top Reinforcement**
Clause 3.12.10.2 Figure 3.24	Simplified rules for curtailment of steel in beams designed for predominantly uniformly distributed loads. A maximum 50% of main bottom steel is curtailed a distance $0.08L$ from the end support. $$A_s = (2/32 \text{ mm diameter bars}) = 1610 \text{ mm}^2$$ $$\frac{100A_s}{b_v d} = \frac{100 \times 1610}{300 \times 539} = 1.0$$ A maximum of 40% of the main top steel at the first interior support is curtailed at a distance equal to the greater of $0.15L$ or $(45 \times$ bar diameters$)$ from the face of the support. $$A_s = (3/32 \text{ mm diameter bars}) = 2410 \text{ mm}^2$$ $$\frac{100A_s}{b_v d} = \frac{100 \times 2410}{300 \times 523} = 1.54$$ For shear at the internal support use 100% of the area: $$A_s = 4020 \text{ mm}^2$$	**Select:** **5/32 mm diameter bars at mid-span providing 4020 mm^2**

References	Calculations	Output
	Contract : Multi-span Floor Job Ref. No. : Example 5.22 **Part of Structure : T-Beam** **Calc. Sheet No. : 10 of 13**	**Calcs. by : W.McK.** **Checked by :** **Date :**

References	Calculations	Output
Clause 3.4.5	**Shear Resistance** At outer support:	
Clause 3.4.5.2	Shear stress $v = \dfrac{V}{b_v d}$ $v = \dfrac{400 \times 10^3}{300 \times 539} = $ 2.47 N/mm^2	
	Maximum shear $\leq$ $0.8\sqrt{f_{cu}}$ $=$ $(0.8 \times \sqrt{40}\,)$ $= $ 5.05 N/mm^2 $\leq$ 5.0 N/mm^2 v $<$ maximum permitted value.	
Table 3.8	Effective depth $d > 400$ mm, $100A_s / b_v d = $ 1.0 $v_c \approx$ $(0.63 \times 1.17) = 0.73$ N/mm^2 (**Note**: 1.17 − this allows for using C40 concrete)	
Table 3.7	$0.5v_c = 0.36$ N/mm^2, $(v_c + 0.4) = (0.73 + 0.4) = 1.13$ N/mm^2 $(v_c + 0.4)$ $<$ v $<$ $0.8\sqrt{f_{cu}}$ or 5N/mm^2 **Design links are required**	
	$A_{sv} \geq b_v s_v(v - v_c)/0.95f_{yv}$ assuming 8 mm diameter links $s_v \leq \dfrac{0.95 f_{yv} A_{sv}}{(v - v_c)b_v} = \dfrac{0.95 \times 250 \times 101}{(2.47 - 0.73) \times 300} = $ 46 mm	**Adopt 8 mm diameter links @ 40 mm centres for the first metre from each end**
	Check shear stress at 1.0 m from end: Shear force V_{1m} $=$ 400 − 148.1 $=$ 251.9 kN $v = \dfrac{251.9 \times 10^3}{300 \times 539} = 1.56$ N/mm^2 $>$ $(v_c + 0.4)$ **Design links required**	
	$s_v \leq \dfrac{0.95 f_{yv} A_{sv}}{(v - v_c)b_v} = \dfrac{0.95 \times 250 \times 101}{(1.56 - 0.73) \times 300} = $ 96 mm	**Adopt 8 mm diameter links @ 80 mm centres 2.0 m from the end**
	Check shear stress at 2.0 m from end: Shear force V_{2m} $=$ 251.9 − 148.1 $=$ 103.8 kN $v = \dfrac{103.8 \times 10^3}{300 \times 539} = 0.64$ N/mm^2 $<$ $(v_c + 0.4)$ **Minimum links required**	
Table 3.7	$A_{sv} \geq 0.4b_v s_v /0.95f_{yv}$ $s_v \leq \dfrac{0.95 \times 250 \times 101}{0.4 \times 300} = $ 200 mm	**Adopt 8 mm diameter links @ 200 mm centres for minimum links**

Contract : Multi-span Floor Job Ref. No. : Example 5.22 Part of Structure : T-Beam Calc. Sheet No. : 11 of 13	Calcs. by : W.McK. Checked by : Date :

References	Calculations	Output
Clause 3.4.5.2	At 1st interior support: Shear stress $\quad v \;=\; \dfrac{V}{b_v d} \quad v \;=\; \dfrac{533.2 \times 10^3}{300 \times 539} = 3.3 \text{ N/mm}^2$ Maximum shear $\quad \leq \quad 0.8\sqrt{f_{cu}} \;=\; (0.8 \times \sqrt{40})$ $\qquad\qquad\qquad\qquad\qquad\qquad = \; 5.05 \text{ N/mm}^2$ $\qquad\qquad\qquad\qquad\qquad\qquad \leq \; 5.0 \text{ N/mm}^2$ $\qquad\qquad v \quad < \quad$ maximum permitted value.	
Table 3.8	Effective depth $\;d > 400$ mm, $\qquad 100A_s / b_v d = \;\; 2.48$ $v_c \;\approx\; (0.85 \times 1.17) = 1.0 \text{ N/mm}^2$ (**Note**: 1.17 – this allows for using C40 concrete)	
Table 3.7	$0.5 v_c = 0.5 \text{ N/mm}^2, \qquad (v_c + 0.4) = (1.0 + 0.4) = 1.4 \text{ N/mm}^2$ $(v_c + 0.4) \;\; < \;\; v \;\; < \;\; 0.8\sqrt{f_{cu}} \;$ or 5N/mm^2 **Design links are required** $s_v \;\leq\; \dfrac{0.95 f_{yv} A_{sv}}{(v - v_c) b_v} \;=\; \dfrac{0.95 \times 250 \times 101}{(3.3 - 1.0) \times 300} = 34 \text{ mm}$ Check shear stress at 1.0 m from 1st interior support: Shear force $V_{1m} \;=\; 533.2 - 148.1 \;=\; 385.1 \text{ kN}$ $\qquad v \;=\; \dfrac{385.1 \times 10^3}{300 \times 539} = 2.38 \text{ N/mm}^2 \quad > \;\; (v_c + 0.4)$ **Design links required** $s_v \;\leq\; \dfrac{0.95 f_{yv} A_{sv}}{(v - v_c) b_v} \;=\; \dfrac{0.95 \times 250 \times 101}{(2.38 - 1.0) \times 300} = \; 58 \text{ mm}$ Check shear stress at 2.0 m from end: Shear force $V_{2m} \;=\; 385.1 - 148.1 \;=\; 237 \text{ kN}$ $\qquad v \;=\; \dfrac{237 \times 10^3}{300 \times 539} = \; 1.47 \text{ N/mm}^2 \quad > \;\; (v_c + 0.4)$ **Design links required** $s_v \;\leq\; \dfrac{0.95 f_{yv} A_{sv}}{(v - v_c) b_v} \;=\; \dfrac{0.95 \times 250 \times 101}{(1.47 - 1.0) \times 300} = \; 170 \text{ mm}$ Check shear stress at 3.0 m from end: Shear force $V_{3m} \;=\; 237 - 148.1 \;=\; 88.9 \text{ kN}$	**Adopt 8 mm diameter links @ 25 mm centres for the first metre from the 1st interior support** **Adopt 8 mm diameter links @ 50 mm centres 1.0 m from the 1st interior support.** **Adopt 8 mm diameter links at 150 mm centres 2.0 m from the 1st interior support.**

Contract : Multi-span Floor Job Ref. No. : Example 5.22 **Part of Structure : T-Beam** **Calc. Sheet No. : 12 of 13**		**Calcs. by : W.McK.** **Checked by :** **Date :**

References	Calculations	Output
	$$v = \frac{88.9 \times 10^3}{300 \times 539} = 0.56 \text{ N/mm}^2 \quad > \quad (v_c + 0.4)$$ <div align="right">Minimum links required as before</div>	
Clause 3.4.6	**Deflection:** $$\frac{\text{span}}{\text{effective depth } (d)}$$ $\leq$ (Table 3.9 value $\times$ Table 3.10 value $\times$ Table 11 value)	
Table 3.9	$\dfrac{b_w}{b} = \dfrac{300}{1140} = 0.26 \leq 0.3$ Basic (span/d) ratio $= 20.8$	
Table 3.10	$\dfrac{M}{bd^2} = K \times f_{cu} = 0.04 \times 40 = 1.6$ Service stress $f_s = \dfrac{2 \times f_y \times A_{s, \text{required}}}{3 \times A_{s, \text{provided}}} = \dfrac{2 \times 460 \times 2144}{3 \times 2410}$ $= 272.8 \text{ N/mm}^2$	
	Interpolate between values given in Table 3.10: 	
		M/bd^2	1.5	2.00
		---	---	---
		service stress f_s		
		250	1.34	1.20
		300	1.16	1.06
	Use conservative estimate of modification factor ≈ 1.06	
Table 3.11	Since no compression reinforcement is required, this does not apply. Table 3.9 value $\times$ Table 3.10 value $= 20.8 \times 1.06 = 22.0$ Actual $\dfrac{\text{span}}{\text{effective depth}} = \dfrac{6000}{539} = 11.1 \; < \; 22.0$ <div align="right">**Adequate with respect to deflection**</div>	
Clause 3.12.5.3 Table 3.25	**Minimum % tension reinforcement:** Flanged beams with webs in tension and $f_y = 460 \text{ N/mm}^2$ $b_w/b = (300/1140) = 0.26 \; < \; 0.4$ $\therefore \; 100 A_s / b_w h = 0.18$ Minimum A_s required $= \dfrac{0.18 \times b_w \times h}{100} \text{ mm}^2$ $= \dfrac{0.18 \times 300 \times 600}{100} = 324 \text{ mm}^2 \; << \; A_s \text{ provided}$ <div align="right">**Minimum % reinforcement satisfied**</div>	

References	Calculations	Output
Contract : Multi-span Floor Job Ref. No. : Example 5.22 **Part of Structure : T-Beam** **Calc. Sheet No. : 13 of 13**		**Calcs. by : W.McK.** **Checked by :** **Date :**

References	Calculations	Output
Clause 3.12.5.3 Table 3.25	**Transverse Reinforcement:** A_s required = $\dfrac{0.15 \times h_f \times 1000}{100}$ mm²/m length = $\dfrac{0.15 \times 300 \times 1000}{100}$ = 450 mm²/m **This is already provided by the slab steel**	
Clause 3.12.10 Figure 3.24	**Curtailment:** At interior support Top reinforcement 100% of steel extends $\geq$ 0.15*l* = 0.15 × 6000 = 900 mm $\geq$ 45 × bar diameter = 45 × 25 = 1125 mm 60% of steel extends $\geq$ 0.25*l* = 0.25 × 6000 = 1500 mm (Two bars will be carried through to support the links)	**Curtailment** **Top Steel :** **5/32 mm diameter bars** **Curtail 2 bars at 1125 mm from the face of the support.** **Curtail 1 bar at 1500 mm from the face of the support.**
	Bottom reinforcement 70% of mid-span reinforcement extends to within 0.15*l* of the *centre-line* of the support. 0.15*l* = 0.15 × 6000 = 900 mm At end support (assuming simple support) Bottom reinforcement 50% of mid-span reinforcement extends to within 0.08*l* of the *centre-line* of the support. 0.08*l* = 0.08 × 6000 = 480 mm	**Curtailment** **Bottom Steel:** **3/32 mm diameter bars** **Curtail 1 bar at 900 mm from the centre-line of the internal support.** **Curtail 1 bar at a distance of 480 mm from the centre-line of the support.**

5.12 Ribbed Slabs (with Solid Blocks, Hollow Blocks or Voids)

5.12.1 Introduction

In long span, solid reinforced concrete slabs, e.g. greater than 5 m, the self-weight becomes excessive when compared to the applied dead and imposed loads, resulting in an uneconomic method of construction. One method of overcoming this problem is to use **ribbed slabs** which are suitable for longer spans supporting light loading, as in residential or commercial properties. They are not suitable for heavy loading such as that found in warehouses, etc.

Ribbed slabs are reinforced concrete slabs in which some of the volume of concrete below the neutral axis (i.e. the area in tension) is removed and replaced with block formers or left as voids, as shown in Figures 5.90 to 5.92. The resulting construction is considerably lighter than a solid cross-section.

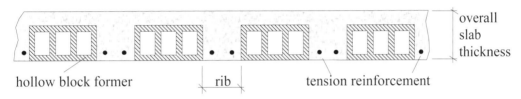

Cross-section through a ribbed slab cast with hollow blocks

Figure 5.90

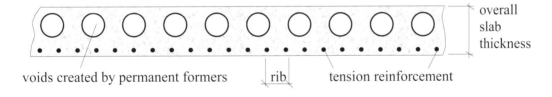

Cross-section through a hollow slab with permanent void formers

Figure 5.91

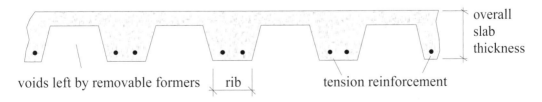

Cross-section through a ribbed slab cast on removable formers

Figure 5.92

*The term ribbed slabs is defined in **Clause 3.6.1.1** in BS 8110:Part 1 as slabs cast in-situ in one of the following ways:*

   a) *Where topping is considered to contribute to the structural strength* (see Table 3.17 for minimum thickness):

Extract from BS 8110:Part 1:1997

Table 3.17 Minimum thickness of structural toppings	
Type of slab	**Minimum thickness of topping (mm)**
Slabs with permanent blocks as described in **3.6.1.1a1** *and* **3.6.1.2**	
a) Clear distance between ribs not more than 500 mm, jointed in cement : sand mortar not weaker than 1 : 3 or 11 N/mm^2	25
b) Clear distance between ribs not more than 500 mm, not jointed in cement : sand mortar	30
c) All other slabs with permanent blocks	40 or one-tenth of clear distance between ribs, whichever is greater
All slabs without permanent blocks As described in **3.6.1.1a2 and 3**	50 or one-tenth of clear distance between ribs, whichever is greater

Figure 5.93

   1) as a series of concrete ribs cast in-situ between blocks which remain part of the completed structure; the tops of the ribs are connected by a topping of concrete of the same strength as that used in the ribs;

   2) as a series of concrete ribs with topping cast on forms which may be removed after the concrete has set;

   3) with a continuous top and bottom face but containing voids of rectangular, oval or other shape.

  b) Where topping is not considered to contribute to structural strength: as a series of concrete ribs cast in-situ between blocks which remain part of the completed structure; the tops of the ribs may be connected by a topping of concrete (not necessarily of the same strength as that used in the ribs).

The requirements for the hollow or solid blocks and formers when required to contribute to the structural strength of a slab are given in Clause 3.6.1.2 :

a) they must be made of concrete or burnt clay;
b) they must have a characteristic strength of at least 14 N/mm², measured on the net section, when axially loaded in the direction of compressive stress in the slab;
c) when made of fired brick, earth, clay or shale, they must conform to BS 3921.

5.12.2   *Spacing and Size of Ribs (Clause 3.6.1.3)*

In-situ ribs should be spaced at centres not exceeding 1.5 m and their depth, excluding any topping, should not exceed four times their width. The minimum width of rib will be determined by considerations of cover, bar spacing and fire. The minimum dimensions for fire resistance for various reinforced concrete members is given in Figure 3.2 of *BS 8110:Part 1:1997.*

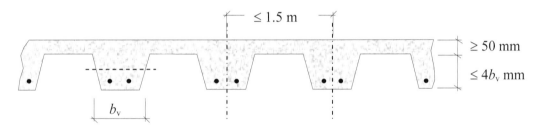

Figure 5.94

5.12.3   *Thickness of Structural Topping (Clause 3.6.1.5)*

The thickness of topping used to contribute to structural strength should not be less than the value given in Table 3.17 of the code.

In hollow block slabs where the topping is *not* used to contribute to structural strength the requirements of **Clause 3.6.1.6** should be satisfied. They are:

'*When a slab is constructed to b) of Table 3.17 the blocks should conform to 3.6.1.2. In addition the thickness of the block material above its void should be not less than 20 mm nor less than one-tenth of the dimension of the void measured transversely to the ribs. The overall thickness of the block and topping (if any) should be not less than one-fifth of the distance between ribs.*'

5.12.4   *Deflection (Clause 3.6.5)*

The deflection of one-way spanning ribbed slabs is checked using the same method as for T-beams, i.e. span/effective depth ratios, except that the rib width may include the walls of the blocks on both sides of the rib.

5.12.5   *Arrangement of Reinforcement (Clause 3.6.6)*

The reinforcement can be curtailed using the simplified procedure given in Clause 3.12.10, as for solid slabs.

In addition to minimum areas of reinforcement as for solid slabs, consideration should be given to providing a single layer of welded steel fabric in the topping. This steel should have a cross-sectional area of not less than 0.12% of the topping, in each direction, and the spacing between the wires should not be greater than half the centre-to-centre distance between ribs.

5.12.6   Links in Ribs (Clause 3.6.6.3)

Provided the geometry of the cross-section satisfies the requirements of Clause 3.6.1.3, ribs reinforced with a single bar do not require links other than to satisfy shear or fire resistance requirements.

Where two or more bars are used in a rib, the use of link reinforcement is recommended. The spacing of links can generally be of the order of 1 m to 1.5 m depending on the size of the main bars.

5.12.7   Design Resistance Moments (Clause 3.6.3)

The same procedure as used for designing beams, i.e. Clause 3.4.4, is used to determine the ultimate moment of resistance. Allowance can be included for the contribution made by the burnt clay or solid blocks in the compression zone, as indicated in Clause 3.6.3.

5.12.8   Design Shear Stress (Clause 3.6.4.2 to Clause 3.6.4.7)

The design shear stress, v, should be calculated using Equation 22, which is the same as Equation 3 given in Clause 3.4.5.2 for beams. The enhancement to the shear strength from hollow blocks, solid blocks and joints between narrow precast units can be evaluated using Clauses 3.6.4.3, 3.6.4.4 and 3.6.4.5 respectively.

It is sometimes necessary to make a section of the slab solid near the end supports or at other locations where high shear forces occur.

5.12.9   Example 5.23: Single Span Hollow-Tile Floor

A hollow-tile floor, in which the permanent blocks are not jointed in a cement:sand mortar, is to be designed as a series of simply supported spans as indicated in Figures 5.95(a) and (b). Using the design data given and assuming that the blocks do not contribute to the strength of the floor, determine:

     i) the suitability of the slab thickness,
    ii) the reinforcement required.

Design Data:

Characteristic dead load (including self-weight)	5.0 kN/m²
Characteristic imposed load	4.0 kN/m²
Characteristic concrete strength (f_{cu})	40 N/mm²
Characteristic of reinforcement (f_y)	460 N/mm²
Exposure condition	mild
Fire rating	Minimum 1 hour
Span of slab	4.0 m
Nominal maximum aggregate size	14 mm

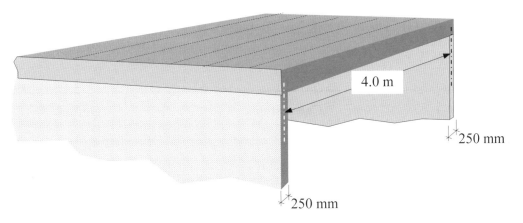

Figure 5.95 (a)

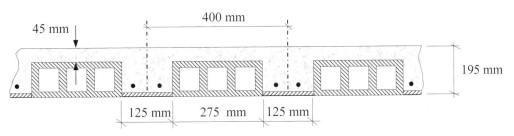

Cross-section through a ribbed slab cast with hollow blocks

Figure 5.95 (b)

5.12.10   Solution to Example 5.23

Contract : Hollow-Tile Floor Job Ref. No. : Example 5.23 Part of Structure : Slab Calc. Sheet No. : **1** of **5**	Calcs. by : W.McK. Checked by : Date :

References	Calculations	Output
	BS8110:Part 1:1997	
Clause 3.6.1.5 Table 3.17	**Thickness of topping:** Since the blocks are not jointed together and the rib spacing is less than 500 mm the minimum thickness of topping should not be less than 30 mm. Actual thickness = 45 mm > 30 mm	**Topping thickness is adequate**
Clause 3.6.1.3	**Spacing and size of ribs:** Maximum spacing of ribs $\leq$ 1.5 m Actual spacing = 400 mm Maximum depth excluding topping $\leq$ $4 \times$ width = (4×125) = 500 mm Actual depth excluding topping = 150 mm	**Spacing and size of ribs is adequate**

Contract :Hollow-Tile Floor Job Ref. No. : Example 5.23 Part of Structure : Slab Calc. Sheet No. : **2** of **5**	Calcs. by : W.McK. Checked by : Date :

References	Calculations	Output
Clause 3.3.6	**Minimum width:** The minimum rib thickness to satisfy fire requirements is given in Figure 3.2 as = 125 mm Actual rib width = 125 mm	**Rib width is adequate**
	Design loads/rib: Characteristic load due to dead load (including self-weight): $\qquad\qquad = \ (5.0 \times 0.4) = \ $ 2.0 kN/m Characteristic load due to imposed loading: $\qquad\qquad = \ (4.0 \times 0.4) = \ $ 1.6 kN/m	
Table 2.1	Ultimate design load $\quad = \ (1.4 \times 2.0) + (1.6 \times 1.6)$ $\qquad\qquad\qquad\qquad = \ $ 5.4 kN/m Ultimate design shear $\ = \ (0.5 \times 3.91 \times 5.4)$ $\qquad\qquad\qquad\qquad = \ $ 10.6 kN	
	Ultimate design bending moment $\ = \ \dfrac{wL^2}{8} \ = \ \dfrac{5.4 \times 3.91^2}{8}$ $\qquad\qquad\qquad\qquad\qquad\qquad\qquad = \ $ 10.3 kNm	
Clause 3.3.7 Clause 3.3.1.2 Clause 3.3.1.3 Table 3.3 Table 3.4	**Cover** Assume 12 mm diameter bars for main steel bar size $\qquad\qquad\qquad\qquad\qquad\qquad = \ $ 12 mm nominal maximum size of aggregate $\quad = \ $ 14 mm exposure condition mild $\qquad\qquad\qquad \geq \ $ 20 mm fire resistance $\ \geq 1$ hour $\qquad\qquad\quad \geq \ $ 20 mm $\qquad\qquad$ **Assume nominal cover to steel 20 mm**	
	Effective depth $\quad d \ = (195 - 10 - 20 - 6) \ = \ $ 159 mm ![T-beam cross-section: 400 mm top flange width, 45 mm top, 195 mm total depth, 10 mm thick slip, 125 mm rib width, effective depth d] 400 mm 45 mm d 195 mm 10 mm thick slip 125 mm	
Clause 3.4.1.2	**Effective span** $\leq$ distance between the centre of the bearings = 4.0 m $\leq$ (clear distance between the supports + the effective depth) $\qquad\qquad\qquad = \ (4.0 - 0.25 + 0.159) \quad = \ $ 3.91 m The hollow blocks are not structural and consequently the cross-section can be designed as a T-beam.	

References	Calculations	Output
	Contract : Hollow-Tile Floor **Job Ref. No. :** Example 5.23 **Part of Structure :** Slab **Calc. Sheet No. : 3 of 5**	**Calcs. by : W.McK.** **Checked by :** **Date :**

References	Calculations	Output
Clause 3.4.1.5	Effective width of flange beam $= b_e \leq (b_w + l_z/5)$ $= \left(125 + \dfrac{3910}{5}\right) = 907$ mm $\leq$ actual flange width $\therefore \; \boldsymbol{b_e = 400}$ **mm** 	
Clause 3.4.4	**Bending**	
Clause 3.4.4.4	$K = \dfrac{M}{bd^2 f_{cu}} = \dfrac{10.3 \times 10^6}{400 \times 159^2 \times 40} = 0.025 < K'\,(=0.156)$ **Section is singly reinforced** $Z = d\left\{0.5 + \sqrt{\left(0.25 - \dfrac{K}{0.9}\right)}\right\} = d\left\{0.5 + \sqrt{\left(0.25 - \dfrac{0.025}{0.9}\right)}\right\}$ $= 0.97d$ the lever arm is limited to $0.95d$ $x = (d-z)/0.45 = (0.05 \times 159)/0.45 = 17.7$ mm $< h_f \;(= 45$ mm) The neutral axis lies within the flange. $A_s = \dfrac{M}{0.95 f_y Z} = \dfrac{10.3 \times 10^6}{0.95 \times 460 \times 0.95 \times 159} = 156$ mm^2	**Bottom Reinforcement** **Select:** **1/16 mm diameter bar providing 201 mm^2** **Alternatively** **2/12 mm diameter bars (226 mm^2) could be used but links are then recommended as in Clause 3.6.6.3**
Clause 3.6.4	**Shear Resistance**	
Clause 3.6.4.7	Shear stress $v = \dfrac{V}{b_v d}$ $\;\; v = \dfrac{10.6 \times 10^3}{125 \times 159} = 0.53$ N/mm^2 Maximum shear $\leq 0.8\sqrt{f_{cu}} = (0.8 \times \sqrt{40})$ $\qquad = 5.05$ N/mm^2 $\qquad \leq 5.0$ N/mm^2 $v < $ maximum permitted value	

References	Calculations	Output

Contract :Hollow-Tile Floor **Job Ref. No. : Example 5.23**
Part of Structure : Slab
Calc. Sheet No. : 4 of 5

Calcs. by : W.McK.
Checked by :
Date :

At support A_s =   201 mm^2

$$\frac{100 A_s}{b_v d} = \frac{100 \times 201}{125 \times 159} = 1.0$$

Table 3.8 v_c > (0.78 × 1.17) = 0.91 N/mm^2

Clause 3.6.4.7 v < v_c

No shear reinforcement required

Clause 3.6.5   **Deflection**

Clause 3.4.6.3 $\dfrac{\text{span}}{\text{effective depth } (d)} \leq$ Table 3.9 value × Table 3.10 value

(**Note:** No compression steel is required therefore the Table 3.11 value is not required)

Table 3.9 $\dfrac{b_w}{b} = \dfrac{125}{400} = 0.32 > 0.3$

(Neglect interpolation between flanged and rectangular beams. i.e. this slightly underestimates the permissible span/d ratio)

Basic (span/d) ratio   =   16.0

Table 3.10 $\dfrac{M}{bd^2} = K \times f_{cu} = (0.025 \times 40) = 1.0$

Service stress $f_s = \dfrac{2 \times f_y \times A_{s,\,required}}{3 \times A_{s,\,provided}} = \dfrac{2 \times 460 \times 156}{3 \times 201}$

                               = 238 N/mm^2

Interpolate between values given in Table 3.10:

service stress f_s	M/bd^2	1.0
200		1.76
250		1.55

Modification factor = 1.6

Table 3.9 value × Table 3.10 value = (16 × 1.6) = 25.6

Actual $\dfrac{\text{span}}{\text{effective depth}} = \dfrac{3910}{159} = 24.5 < 25.6$

Adequate with respect to deflection

References	Calculations	Output
Contract :Hollow-Tile Floor Job Ref. No. : Example 5.23 Part of Structure : Slab Calc. Sheet No. : 5 of 5		Calcs. by : W.McK. Checked by : Date :

References	Calculations	Output
Clause 3.6.6.2	**Topping Reinforcement** A_s required $= \dfrac{0.12 \times h_f \times 1000}{100}$ mm^2/m length $= \dfrac{0.12 \times 45 \times 1000}{100} = 54$ mm^2/m Spacing of wires $\leq \dfrac{\text{rib spacing}}{2} = 200$ mm	**Select mesh from manufacturer's catalogue**

5.13   Two-Way Spanning Slabs

5.13.1   Introduction

When slabs are supported on all four sides, they effectively span in two directions provided that the longer side is no greater than twice the shorter side. It is often more economic to design slabs on this basis and to provide reinforcing steel in both directions to resist the orthogonal bending moments. The magnitude of the bending moment in each direction is dependent on the ratio of the two spans, and the support conditions. In square slabs the load is distributed equally in both directions and in rectangular slabs, the shorter, stiffer span resists a higher percentage of the load than the longer span.

There are two types of slab to be considered:
    (i)   Simply-supported slabs   (Clause 3.5.3.3 and Table 3.13)   and
    (ii) Restrained slabs   (Clause 3.5.3.4 and Table 3.14)

There are nine different types of support condition to be considered which relate to the particular support/restraint provided on each edge of individual slabs; these are illustrated in Figures 5.96 to 5.99.

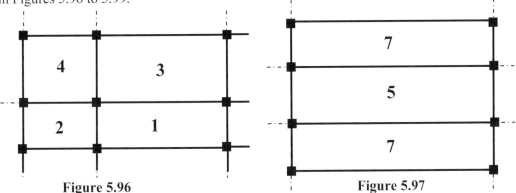

Figure 5.96   Figure 5.97

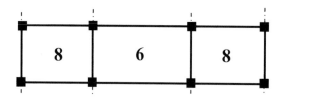

Figure 5.98   **Figure 5.99**

Type 1 *Interior panel*   **Type 2** *One short edge discontinuous*
Type 3 *One long edge discontinuous*   **Type 4** *Two adjacent edges discontinuous*
Type 5 *Two short edges discontinuous* **Type 6** *Two long edges discontinuous*
Type 7 *Three edges discontinuous (one long edge continuous)*
Type 8 *Three edges discontinuous (one short edge continuous)*
Type 9 *Four edges discontinuous*

5.13.2 *Simply-Supported Slabs*

Simply-supported slabs do not have adequate provision to resist torsion at the corners. The maximum design bending moments permitted to prevent lifting and to resist the applied ultimate bending moment can determined using Equations (10) and (11) in the code:

$$m_{sx} = \alpha_{sx} n l_x^2 \qquad \text{Equation (10)}$$

$$m_{sy} = \alpha_{sy} n l_x^2 \qquad \text{Equation (11)}$$

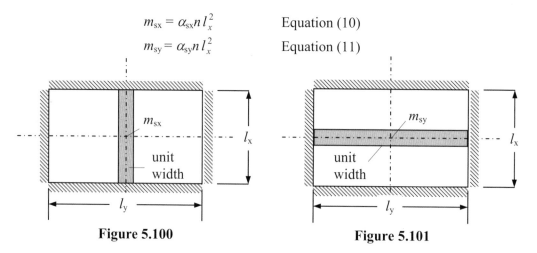

Figure 5.100   **Figure 5.101**

where:
m_{sx} is the maximum design ultimate moment at mid-span on a strip of unit width and span l_x
m_{sy} is the maximum design ultimate moment at mid-span on a strip of unit width and span l_y
n is the total design ultimate load/unit area $= (1.4g_k + 1.6q_k)$
l_x is the length of the shorter side,
l_y is the length of the longer side.

$$\alpha_{sx} \text{ is a moment coefficient} \quad = \quad \frac{(l_y/l_x)^4}{8\{1+(l_y/l_x)^4\}} \qquad \text{Equation (12)}$$

α_{sy} is a moment coefficient $= \dfrac{\left(l_y/l_x\right)^2}{8\left\{1+\left(l_y/l_x\right)^4\right\}}$ Equation (13)

values for α_{sx} and α_{sy} are also given in Table 3.13.

The required areas of reinforcement can be determined using the design charts or the Equations given in Clause 3.4.4.4 for rectangular beams, i.e.

$$A_{sx} = \frac{m_{sx}}{0.95f_yZ_x} \text{ mm}^2/\text{m width of slab in direction } l_x \quad \text{and}$$

$$A_{sy} = \frac{m_{sy}}{0.95f_yZ_y} \text{ mm}^2/\text{m width of slab in direction } l_y$$

Since the shorter span resists a higher bending moment, the main steel will be positioned such that it has a greater effective depth than the longer span and hence $Z_x > Z_y$. Slabs (other than bridge decks etc.) are normally singly reinforced.

5.13.3 Deflection (Clause 3.5.7)
The limit-state of deflection for two-way spanning slabs can be checked using Table 3.9, modified using Table 3.10 as for beams. Only reinforcement at the centre of the shorter span and in that direction should be considered to influence the deflection.

5.13.4 Restrained Slabs
Restrained slabs, in which the corners are prevented from lifting and in which provision for torsion is made at simply supported corners, are designed to resist maximum bending moments given by Equations (14) and (15), i.e.;

$$m_{sx} = \beta_{sx}nl_x^2 \qquad \text{Equation (14)}$$
$$m_{sy} = \beta_{sy}nl_x^2 \qquad \text{Equation (15)}$$

where:
m_{sx} and m_{sy} are the maximum design ultimate moments either over the supports or at the mid-span on strips of unit width as before.
n and l_x are as before
β_{sx} and β_{sy} are bending moment coefficients given in Table 3.14. These coefficients are given for a wide range of support conditions along the edges of panels, to enable the bending moments at the mid-span and over continuous edges to be evaluated.

The use of equations (14) and (15) is dependent on the conditions and rules given in Clause 3.5.3.5 being satisfied. They are given in terms of:

(i) continuous slabs only, and
(ii) in the case of restrained slabs, continuous or discontinuous.

'The conditions in which the equations may be used for continuous slabs only are as follows.

(a) The characteristic dead and imposed loads on adjacent panels are approximately the same as on the panel being considered.

(b) The span of adjacent panels in the direction perpendicular to the line of the common support is approximately the same as the span of the panel considered in that direction.

The rules to be observed when the equations are applied to restrained slabs (continuous or discontinuous) are as follows.

(1) Slabs are considered as divided in each direction into middle strips and edge strips as shown in **figure 3.9** (see Figure 5.102 of this text).

(2) The maximum design moments calculated as above apply only to the middle strips and no redistribution should be made.

(3) Reinforcement in the middle strips should be detailed in accordance with **3.12.10** (simplified rules for curtailment of bars).

(4) Reinforcement in an edge strip, parallel to the edge need not exceed the minimum given in **3.12.5** (minimum areas of tension reinforcement), together with the recommendations for torsion given in (5), (6) and (7).

(5) Torsion reinforcement should be provided at any corner where the slab is simply supported on both edges meeting at that corner. It should consist of top and bottom reinforcement, each with layers of bars placed parallel to the sides of the slab and extending from the edges a minimum distance of one-fifth of the shorter span. The area of reinforcement in each of these four layers should be three-quarters of the area required for the maximum mid-span design moment in the slab.

(6) Torsion reinforcement equal to half that described in the preceding paragraph should be provided at a corner contained by edges over only one of which the slab continues.

(7) Torsion reinforcement need not be provided at any corner contained by edges over both of which the slab is continuous.

5.13.5   Torsion (Clause 3.5.3.5)

The torsion reinforcement provided at corners is not in addition to the main steel included for the edge strips (this is normally the minimum % of reinforcement specified in Table 3.27). If the minimum does not provide sufficient steel then more must be provided to satisfy the requirements of Clause 3.5.3.5.

(**Note:** Four layers of steel are required, i.e. two at both the top and the bottom.)

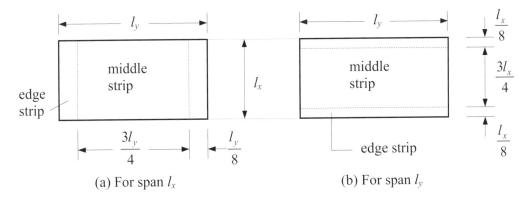

Figure 5.102   Division of slab into middle and edge strips

5.13.6   *Loads on Supporting Beams (Clause 3.5.3.7)*

The design loads on beams which support two-way-spanning solid slabs in which the loads are uniformly distributed can be evaluated using Equations (19) and (20).

$$v_{sy} = \beta_{vy}nl_x \qquad \qquad \text{Equation (19)}$$
$$v_{sx} = \beta_{vx}nl_x \qquad \qquad \text{Equation (20)}$$

where β_{vy} and β_{vx} are shear force coefficients given in Table 3.15 of the code.

5.13.7   *Example 5.24: Simply Supported Two-Way Spanning Slab*

A floor slab in an office building measures 5.0 m × 7.5 m and is simply supported at the edges with no provision to resist torsion at the corners or to hold the corners down. Using the design data given, determine suitable reinforcement.

Design Data:

Characteristic dead load due to finishes and services	g_k	$= 1.25 \text{ kN/m}^2$
Characteristic imposed load	q_k	$= 2.5 \text{ kN/m}^2$
Concrete grade	f_{cu}	$= 40 \text{ N/mm}^2$
Characteristic strength of reinforcing steel	f_y	$= 460 \text{ N/mm}^2$
Self-weight of concrete	γ_{conc}	$= 24 \text{ kN/m}^3$
Exposure condition		mild

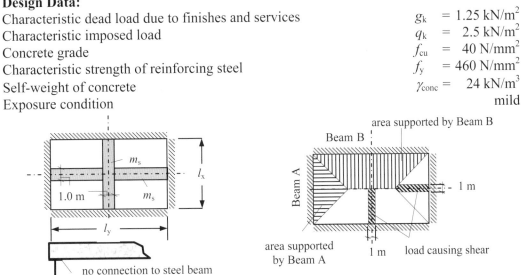

Figure 5.103

5.13.8   Solution to Example 5.24

Contract : Office Block Job Ref. No. : Example 5.24	Calcs. by : W.McK.
Part of Structure : Two-Way Spanning Slab	Checked by :
Calc. Sheet No. : 1 of 4	Date :

References	Calculations	Output
	Initial sizing of the section:	
	The initial trial section depth can be based on deflection i.e.	
	Table 3.9 value $\quad = \quad \dfrac{basic\ span}{effective\ depth} \quad = \quad 20$	
	(This will be modified using the Table 3.10 value as given in Clause 3.4.6.5 for the final deflection check)	
	Estimated effective depth required $\quad d \ \geq \dfrac{5000}{20} = 250$ mm.	
Table 3.3	For C40 concrete with mild exposure steel cover $\ \geq \ $ 20 mm	
	Assume 12 mm diameter bars:	
	The required overall thickness $h \ \geq \ (250 + 6 + 20)$	
	$\qquad\qquad\qquad\qquad\qquad\qquad\ = \ 276$ mm	
	Try a 275 mm thick slab.	
	Actual effective depth for the short span:	
	$d = (275 - 20 - 6) \qquad = 249$ mm	
	Actual effective depth for the long span:	
	$d = (275 - 20 - 12 - 6) = 237$ mm	
	Design loading:	
	Self-weight of slab $\quad = \ (0.275 \times 24) \qquad = \ 6.6$ kN/m^2	
	Finishes and services $\qquad\qquad\qquad\qquad = \ 1.25$ kN/m^2	
	Characteristic dead load $\qquad\qquad g_k \ = \ 7.85$ kN/m^2	
	Characteristic imposed load $\qquad\quad q_k \ = \ 2.5$ kN/m^2	
	Ultimate design load $\ = \ [(1.4 \times g_k) + (1.6 \times q_k)]$	
	$\qquad\qquad\qquad\qquad = \ [(1.4 \times 7.85) + (1.6 \times 2.5)]$	
	$\qquad\qquad\qquad\qquad = \ 15.0$ kN/m^2	
Table 3.13	$l_y / l_x = 7.5/5.0 = \ 1.5 \qquad \alpha_{sx} \ = \ 0.104 \qquad \alpha_{sy} \ = \ 0.046$	
	Design moment in the direction of span l_x	
Clause 3.5.3.3	$m_{sx} = \quad \alpha_{sx} n \, l_x^2 \quad = \quad (0.104 \times 15.0 \times 5^2)$	
	$\qquad\qquad\qquad\qquad = \quad 39.0$ kNm/m width	
	Design moment in the direction of span l_y	
	$m_{sy} = \quad \alpha_{sy} n \, l_x^2 \quad = \quad (0.046 \times 15.0 \times 5^2)$	
	$\qquad\qquad\qquad\qquad = \quad 17.25$ kNm/m width	

Contract : Office Block Job Ref. No. : Example 5.24 Part of Structure : Two-Way Spanning Slab Calc. Sheet No. : 2 of 4	Calcs. by : W.McK. Checked by : Date :

References	Calculations	Output
Clause 3.4.4.4	Reinforcement: short span $K = \dfrac{M}{bd^2 f_{cu}} = \dfrac{39.0 \times 10^6}{1000 \times 249^2 \times 40} = 0.016 \quad < \quad 0.156$ **Section is singly reinforced** $Z = d\left\{0.5 + \sqrt{\left(0.25 - \dfrac{K}{0.9}\right)}\right\} \quad = \quad d\left\{0.5 + \sqrt{\left(0.25 - \dfrac{0.016}{0.9}\right)}\right\}$ $= \quad 0.98d$ The lever arm is limited to $0.95d$ $A_s = \dfrac{M}{0.95 f_y Z} = \dfrac{39.0 \times 10^6}{0.95 \times 460 \times 0.95 \times 249} = 377 \text{ mm}^2/\text{m width}$	**Bottom Reinforcement** **Select:** **12 mm diameter bars** **@ 250 mm centres** **providing 452 mm²/m** **at mid-span**
Table 3.25	For slabs of rectangular section and $f_y = 460 \text{ N/mm}^2$ $100 A_s / A_c \geq 0.13\%$ Minimum A_s required $= \dfrac{0.13 \times b \times h}{100} \text{ mm}^2$ $= \dfrac{0.13 \times 1000 \times 275}{100} = 358 \text{ mm}^2$ $< \quad 452 \text{ mm}^2$ **Minimum % reinforcement satisfied**	
Clause 3.4.4.4	Reinforcement: long span $K = \dfrac{M}{bd^2 f_{cu}} = \dfrac{17.25 \times 10^6}{1000 \times 237^2 \times 40} = 0.008 \quad < \quad 0.156$ **Section is singly reinforced** $Z = d\left\{0.5 + \sqrt{\left(0.25 - \dfrac{K}{0.9}\right)}\right\} \quad = \quad d\left\{0.5 + \sqrt{\left(0.25 - \dfrac{0.008}{0.9}\right)}\right\}$ $= \quad 0.99d$ The lever arm is limited to $0.95d$ $A_s = \dfrac{M}{0.95 f_y Z} = \dfrac{17.25 \times 10^6}{0.95 \times 460 \times 0.95 \times 237} = 175 \text{ mm}^2/\text{m width}$	**Bottom Reinforcement** **Select:** **12 mm diameter bars** **@ 300 mm centres** **providing 377 mm² /m** **at mid-span**
Table 3.25	Minimum area of reinforcement required $A_s = 358 \text{ mm}^2$ (as before) $< \quad 377 \text{ mm}^2$ **Minimum % reinforcement satisfied**	

Contract : **Office Block Job Ref. No. : Example 5.22** Part of Structure : **Two-Way Spanning Slab** Calc. Sheet No. : **3 of 4**	Calcs. by : **W.McK.** Checked by : Date :

References	Calculations	Output
Clause 3.12.11.2	Maximum spacing of reinforcement: $\leq\ (3 \times d) =\ (3 \times 237)\qquad =\ 711$ mm $\leq\ 750$ mm Since there is less than 0.3% reinforcement, no further check is required on spacing.	
Clause 3.12.10	There is no curtailment since the reinforcement would be less than 0.13%.	
Clause 3.5.5	**Shear Resistance**	
Clause 3.5.5.2	Shear stress $\quad v\ =\ \dfrac{V}{b_v d}\ =\ \dfrac{37.5 \times 10^3}{1000 \times 237}\ = 0.16$ N/mm^2 Maximum shear $\quad \leq\quad 0.8\sqrt{f_{cu}}\quad =\quad (0.8 \times \sqrt{40}\ )$ $\qquad\qquad\qquad\qquad\qquad\qquad\quad =\ 5.05$ N/mm^2 $\qquad\qquad\qquad\qquad\qquad\qquad\leq\ 5.0$ N/mm^2 $\qquad\qquad v\ <\ $ maximum permitted value $\dfrac{100 A_s}{bd}\ =\ \dfrac{100 \times 377}{1000 \times 237}\ =\ 0.16; \qquad d\ =\ 237$	
Table 3.8	$v_c\ \approx\ (0.38 \times 1.17)\ =\ 0.44$ N/mm^2 (**Note:** This allows for using C40 concrete)	
Table 3.16	$v\ <\ v_c$ $\qquad\qquad\qquad\qquad\qquad\qquad$ **No links are required**	
Clause 3.5.7	**Deflection:** As for beams in Clause 3.4.6.3 using values for shorter span $\dfrac{\text{span}}{\text{effective depth}}\ \leq\ $ Table 3.9 value $\times$ Table 3.10 value	
Table 3.10	$\dfrac{M}{bd^2}\ =\ K \times f_{cu}\ =\ 0.016 \times 40 = 0.64$ Service stress $\ f_s\ =\ \dfrac{2 \times f_y \times A_{s,\,required}}{3 \times A_{s,\,provided}} = \dfrac{2 \times 460 \times 377}{3 \times 452}$ $\qquad\qquad\quad =\ 256$ N/mm^2 Interpolate between values given in Table 3.10	

M/bd^2 service stress f_s	0.50	0.75
250	1.90	1.70
300	1.60	1.44

Use conservative estimate of modification factor $\quad \approx\quad 1.44$

References	Calculations	Output
	Contract : Office Block Job Ref. No. : Example 5.24 **Part of Structure : Two-Way Spanning Slab** **Calc. Sheet No. : 4** of **4** **Calcs. by : W.McK.** **Checked by :** **Date :**	

References	Calculations	Output
	Table 3.9 value × Table 3.10 value = 20 × 1.44 = 28.8 Actual $\dfrac{\text{span}}{\text{effective depth}} = \dfrac{5000}{249} = 20.1 \; < \; 28.8$	**Adequate with respect to deflection**
Standard 90° bend at support	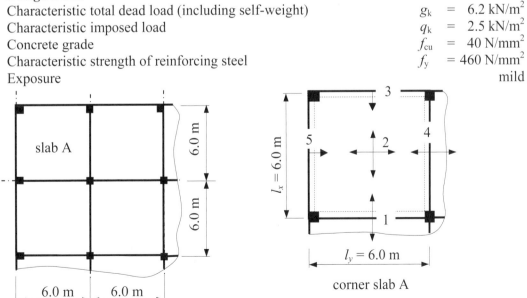	

5.13.9   Example 5.25: Two-Way Spanning Restrained Slab

A part-floor plan for an office building is shown in Figure 5.104. The floor consists of restrained 180 mm thick slabs poured monolithically with edge beams. Using the design data given, design suitable reinforcement for a corner slab.

Design Data:
Characteristic total dead load (including self-weight) $g_k \;=\; 6.2 \text{ kN/m}^2$
Characteristic imposed load $q_k \;=\; 2.5 \text{ kN/m}^2$
Concrete grade $f_{cu} \;=\; 40 \text{ N/mm}^2$
Characteristic strength of reinforcing steel $f_y \;= 460 \text{ N/mm}^2$
Exposure mild

Figure 5.104

5.13.10   Solution to Example 5.25

References	Calculations	Output
	Contract : Office Block Job Ref. No. : Example 5.25 **Part of Structure : Restrained Slab** **Calc. Sheet No. : 1 of 5** **Calcs. by : W.McK.** **Checked by :** **Date :**	

References	Calculations	Output
	Design loading: Characteristic dead load $\quad g_k = 6.2$ kN/m^2 Characteristic imposed load $\quad q_k = 2.5$ kN/m^2 Ultimate design load $\quad = [(1.4 \times g_k) + (1.6 \times q_k)]$ $\quad = [(1.4 \times 6.2) + (1.6 \times 2.5)]$ $\quad = 12.68$ kN/m^2	
Table 3.14	From Figure 5.96 of the text the corner slab is 'type 4'	
Clause 3.5.3.5 Figure 3.9	The slab is divided into middle and edge strips as indicated in Figure 3.9 of the code. $$\frac{l_y}{8} = \frac{l_x}{8} = \frac{6.0}{8} = 0.75 \text{ m}$$	
	(a) For span l_y \qquad\qquad (b) For span l_x	
Table 3.14	$l_x = l_y = 6.0$ m $\quad \dfrac{l_y}{l_x} = 1$	
Table 3.3	For C40 concrete with mild exposure steel cover ≥ 20 mm In most cases of this type the cover requirements regarding 1 hour fire protection is less onerous than exposure requirements. Similarly, strength/deflection thickness requirements normally exceed those required for fire protection. The reader should refer to Table 3.4 and Figure 3.2 of the code. Thickness of slab $= 180$ mm Assume 10 mm diameter bars: The effective depth for the outer layer $d \quad = (180 - 20 - 5)$ $\qquad = 155$ mm the effective depth for the inner layer $d \quad = (180 - 20 - 10 - 5)$ $\qquad = 145$ mm	

Contract : Office Block Job Ref. No. : Example 5.23 Part of Structure : Restrained Slab Calc. Sheet No. : 2 of 5	Calcs. by : W.McK. Checked by : Date :

References	Calculations	Output

References: Table 3.14

Reinforcement for the middle strip of span l_x
$l_y/l_x = $ 1.0

Figure 5.104   position 1 $\beta_{sx} = $ -0.047
              position 2 $\beta_{sx} = $ $+0.036$

Position 1 (continuous edge)
Design moment at position 1:

Clause 3.5.3.4

$m_{sx} = \beta_{sx}nl_x^2$ $= -(0.047 \times 12.68 \times 6^2)$
          $= -21.45$ kNm/m width

Clause 3.4.4.4

$K = \dfrac{M}{bd^2 f_{cu}} = \dfrac{21.45 \times 10^6}{1000 \times 155^2 \times 40} = 0.022$ < 0.156

Section is singly reinforced

$Z = d\left\{0.5 + \sqrt{\left(0.25 - \dfrac{K}{0.9}\right)}\right\} = d\left\{0.5 + \sqrt{\left(0.25 - \dfrac{0.022}{0.9}\right)}\right\}$
$= 0.98d$
The lever arm is limited to $0.95d$

$A_s = \dfrac{M}{0.95f_y Z} = \dfrac{21.45 \times 10^6}{0.95 \times 460 \times 0.95 \times 155} = 333$ mm²/m width

Table 3.25

Actual area of reinforcement $= \dfrac{100 \times 393}{1000 \times 180} = 0.22\% > 0.13\%$

Position 2 (mid-span)
Design moment at position 2:

Clause 3.5.3.4

$m_{sx} = \beta_{sx}l_x^2$ $= (0.036 \times 12.68 \times 6^2)$
          $= 16.43$ kNm/m width

Clause 3.4.4.4

$K = \dfrac{M}{bd^2 f_{cu}} = \dfrac{16.43 \times 10^6}{1000 \times 155^2 \times 40} = 0.017$ < 0.156

Section is singly reinforced

$Z = d\left\{0.5 + \sqrt{\left(0.25 - \dfrac{K}{0.9}\right)}\right\} = d\left\{0.5 + \sqrt{\left(0.25 - \dfrac{0.017}{0.9}\right)}\right\}$
$= 0.98d$
The lever arm is limited to $0.95d$

$A_s = \dfrac{M}{0.95f_y Z} = \dfrac{16.43 \times 10^6}{0.95 \times 460 \times 0.95 \times 155} = 255$ mm²/m width

Output:

Top Reinforcement

Select:
10 mm diameter bars @ 200 mm centres providing 393 mm²/m over continuous edge.

Bottom Reinforcement

Select:
10 mm diameter bars @ 250 mm centres providing 314 mm²/m at mid-span.

Contract : Office Block Job Ref. No. : Example 5.25	Calcs. by : W.McK.
Part of Structure : Restrained Slab	Checked by :
Calc. Sheet No. : 3 of 5	Date :

References	Calculations	Output
Table 3.25	Actual area of reinforcement $= \dfrac{100\times314}{1000\times180} = $ 0.17% $> 0.13\%$	
Clause 3.12.10.3	**Position 3 (discontinuous edge)** At least 50% of bottom steel should be provided in the top at the support and not less than the minimum area required. A_s required $=$ (0.5×255) $=$ 128 mm²/m width Minimum area of reinforcement $= \dfrac{0.13\times1000\times180}{100}$ $= $ 234 mm²/m	**Top Reinforcement** **Select:** **8 mm diameter bars @ 200 mm centres providing 252 mm²/m at mid-span**
Clause 3.5.5 Table 3.15	Shear resistance: $l_y/l_x = $ 1.0 Figure 5.104 position 1 $\beta_{vx} = $ 0.4 position 3 $\beta_{vx} = $ 0.26	
Clause 3.5.3.7	**Position 1 (continuous edge)** Design shear at position 1: $v_{sx} = \beta_{vx}nl_x = (0.4 \times 12.68 \times 6.0) = $ 30.43 kN/m	
Clause 3.5.5	Shear resistance	
Clause 3.5.5.2	Shear stress $v = \dfrac{V}{bd} = \dfrac{30.43\times10^3}{1000\times155} = $ 0.196 N/mm² Maximum shear $\leq 0.8\sqrt{f_{cu}} = (0.8 \times \sqrt{40})$ $= $ 5.05 N/mm² $\leq $ 5.0 N/mm² $v < $ maximum permitted value $\dfrac{100A_s}{bd} = \dfrac{100\times393}{1000\times155} = $ 0.25; $d = $ 155	
Table 3.8	$v_c \approx (0.5 \times 1.17) = $ 0.58 N/mm² (**Note:** 1.17 – allows for using C40 concrete)	
Table 3.16	$v < v_c$	**No links are required**
Clause 3.5.3.7	**Position 3** $v_{sx} = \beta_{vx}nl_x = (0.26 \times 12.68 \times 6.0) = $ 19.78 kN/m	
Clause 3.5.5.2	Shear stress $v = \dfrac{V}{bd} = \dfrac{19.78\times10^3}{1000\times155} = $ 0.136 N/mm²	
Table 3.16	$v < v_c$	**No links are required**

Contract : Office Block Job Ref. No. : Example 5.25 Part of Structure : Restrained Slab Calc. Sheet No. : **4** of **5**	Calcs. by : W.McK. Checked by : Date :

References	Calculations	Output
Clause 3.5.3.5	*Reinforcement for the edge strip of span l_x* Provide minimum reinforcement in accordance with Clause 3.12.10 **The reader should carry out a similar calculation for reinforcement in the l_y direction for positions 2, 4, and 5 using the appropriate β_{sy} and β_{vy} values from Tables 3.14 and 3.15 as necessary.** *Torsion Steel*	**Edge Strips** Select: **8 mm diameter bars** **@ 200 mm centres**

Clause 3.5.3.5	"*Torsion steel consists of top and bottom reinforcement, each with layers of bars placed parallel to the sides of the slab and extending from the edges a minimum distance of one-fifth of the shorter span. The area in each of these four layers should be three-quarters of the area required for the maximum mid-span design moment in the slab.*
	    Torsion reinforcement equal to half that described in the preceding paragraph should be provided at a corner contained by edges over only one of which the slab is continuous.
	    Torsion reinforcement need not be provided at any corner over both of which the slab is continuous."
	
	 $\dfrac{l}{5} \ = \ \dfrac{6000}{5.0} \ = \ 1.2 \text{ m}$
	
	Corner X:
	A_s required = $(0.75 \times 281) = 211$ mm^2/m width
	This is less than the minimum area = 234 mm^2/m
	(**Note:** The 281 mm^2 is based on position 2 in the l_y direction.)
		Torsion steel
		External corner with simple supports
		
		Select:
		8 mm diameter bars
		@ 200 mm centres
		providing 234 mm^2/m

References	Calculations	Output
	Contract : Office Block Job Ref. No. : Example 5.25 **Part of Structure : Restrained Slab** **Calc. Sheet No. : 5 of 5**	**Calcs. by : W.McK.** **Checked by :** **Date :**

References	Calculations	Output
	Corner Y: A_s required $= (0.5 \times 211) = 106 \text{ mm}^2/\text{m width} < 234 \text{ mm}^2$	**Torsion teel corner with one continuous edge**
Clause 3.5.7	**Deflection:** Same as for beams in Clause 3.4.6.3 using values for shorter span $\dfrac{\text{span}}{\text{effective depth}} \leq$ Table 3.9 value $\times$ Table 3.10 value	**Select:** **8 mm diameter bars @ 200 mm centres providing 234 mm²/m**
Table 3.10	$\dfrac{M}{bd^2} = K \times f_{cu} = (0.017 \times 40) = 0.68$ Service stress $f_s = \dfrac{2 \times f_y \times A_{s,\,required}}{3 \times A_{s,\,provided}} = \dfrac{2 \times 460 \times 255}{3 \times 314}$ $\qquad\qquad\quad = 249 \text{ N/mm}^2$ Interpolate between values given in Table 3.10:	

M/bd^2 service stress f_s	0.50	0.75
250	1.90	1.70
300	1.60	1.44

Use conservative estimate of modification factor ≈ 1.44

Table 3.9 value $\times$ Table 3.10 value $= 26 \times 1.44 = 37.44$

$\dfrac{\text{span}}{\text{effective depth}} = \dfrac{6000}{155} = 38.7 > 37.4$

Calculate a more accurate value for the Table 3.10 coefficient.

M/bd^2 service stress f_s	0.50	0.75
284	1.70	1.50

Value for $M/bd^2 = 0.68 = 1.56$

Table 3.9 value $\times$ Table 3.10 value $= 26 \times 1.56 = 40.56$

$\qquad\qquad\qquad\qquad\qquad\qquad > 38.7$

Adequate with respect to deflection

5.14 Design Charts for Bending Moments

As indicated in Clause 3.4.4.2 of the code, design charts have been prepared and are given in *BS 8110:Part 3:1985*. These charts have been derived on the basis of a rectangular-parabolic stress block as described in Section 5.3 with the partial safety factor $\gamma_m = 1.5$ for concrete and $\gamma_m = 1.15$ for reinforcement. The current version of *BS 8110–2:1997* indicates the use of $\gamma_m = 1.05$ for reinforcement and consequently any area of reinforcement determined using the design charts should be modified by multiplying by a factor equal to $(1.05/1.15) = 0.91$.

The design charts are presented for various combinations of f_{cu} and f_y for singly reinforced beams and additionally d'/d values for doubly-reinforced beams. The required percentage areas of reinforcement are given in terms of **100 A_s / bd** and **100 A'_s / bd**. In the case of doubly-reinforced beams graphs are given in for redistribution assumed to be less than 10% (corresponding with $x / d = 0.5$) and with a maximum of 30%.

The charts used in Examples 5.26 to 5.28 are for illustration purposes only.

5.14.1 Example 5.26: Singly-Reinforced Rectangular Beam

Consider the rectangular beam in Example 5.6 which is required to resist a bending moment of 150 kNm and has the following design data:

Characteristic strength of concrete	f_{cu} =	40 N/mm²
Characteristic strength of reinforcing steel	f_y =	460 N/mm²
Breadth of beam	b =	200 mm
Effective depth to the centre of the tension steel	d =	387 mm

Using Design Chart 2 determine the required area of tension reinforcement.

Solution:

In order to use Chart 2 it is necessary to evaluate M/bd^2 $\therefore \dfrac{M}{bd^2} = \dfrac{150 \times 10^6}{200 \times 387^2} = 5.0$

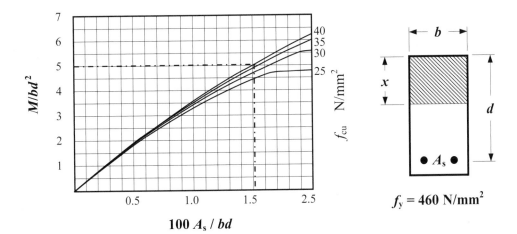

Figure 5.105 – Design Chart 2

From Figure 5.105:

$$\frac{100A_s}{bd} = 1.51 \qquad A_s = \frac{1.51 \times 200 \times 387}{100} = 1169 \text{ mm}^2$$

The modified value allowing for $\gamma_m = 1.05$ $A_s = (1169 \times 0.91) = 1064 \text{ mm}^2$

This value compares with $A_s = 1260 \text{ mm}^2$ calculated using the equations in Clause 3.4.4.4 of the code which were derived using the simplified rectangular stress block.

5.14.2 Example 5.27: Singly-Reinforced Rectangular Slab

Consider the rectangular slab in Example 5.7 which is required to resist a bending moment of 60.02 kNm/metre width and has the following design data:

Characteristic strength of concrete		f_{cu} =	40 N/mm^2
Characteristic strength of reinforcing steel		f_y =	460 N/mm^2
Breadth of slab		b =	1000 mm
Effective depth to the centre of the tension steel		d =	270 mm

Using Design Chart 2 determine the required area of tension reinforcement.

Solution:

$$\frac{M}{bd^2} = \frac{60.02 \times 10^6}{1000 \times 270^2} = 0.82$$

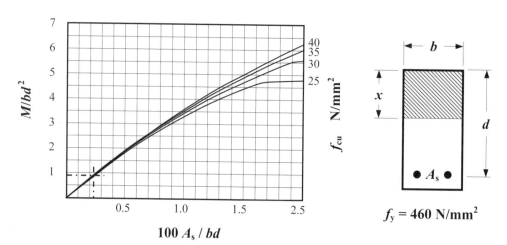

Figure 5.106 – Design Chart 2

From Figure 5.106:

$$\frac{100A_s}{bd} = 0.23 \qquad A_s = \frac{0.23 \times 1000 \times 270}{100} = 621 \text{ mm}^2$$

The modified value allowing for $\gamma_m = 1.05$ $A_s = (621 \times 0.91) = 565 \text{ mm}^2$

This value is the same as calculated using the equations in Clause 3.4.4.4 of the code.

5.14.3   *Example 5.28: Doubly-Reinforced Rectangular Beam*

Consider a rectangular beam which is required to resist a bending moment of 256 kNm and has the following design data:

Characteristic strength of concrete	f_{cu}	=	30 N/mm²
Characteristic strength of reinforcing steel	f_y	=	460 N/mm²
Breadth of beam	b	=	250 mm
Effective depth to the centre of the tension steel	d	=	420 mm
Depth to the centre of the compression steel	d'	=	50 mm

Use Design Charts to determine the required area of tension and compression reinforcement assuming no redistribution of the moments.

Solution:

$$\frac{M}{bd^2} = \frac{256 \times 10^6}{250 \times 420^2} = 5.8; \qquad x/d = 0.5; \qquad d'/d = (50/420) = 0.12$$

Use Design Chart No. 6

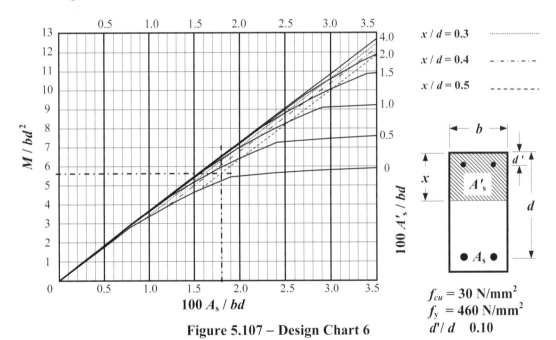

Figure 5.107 – Design Chart 6

From Figure 5.107:

$$\frac{100 A_s}{bd} = 1.8 \quad A_s = \frac{1.8 \times 250 \times 420}{100} \times 0.91^1 = 1720 \text{ mm}^2$$

$$\frac{100 A'_s}{bd} = 0.3 \quad A'_s = \frac{0.3 \times 250 \times 420}{100} \times 0.91 = 300 \text{ mm}^2$$

---

[1] modified value allowing for $\gamma_m = 1.05$

5.15 Columns

Structural frames are constructed from a series of interconnected slabs, beams, walls and columns. The primary purpose of the columns and walls is to transfer the loads in a vertical direction to the foundation. In **braced** frames, i.e. those in which the lateral loading is transferred by structural elements such as shear walls, cores or bracing, the columns are subject to axial loading in addition to moments induced by the dead and imposed loads only. In **unbraced** frames the columns are subject to additional sway moments induced by the lateral wind loading.

In both cases, columns are defined in *BS 8110:Part 1:1997*, Clause 3.8.1.3 as either **short** or **slender**. Slender columns are subject to moments due to the deflection of the columns, which must be added to those calculated for the loading and sway effects.

The definition of 'short' and 'slender' is dependent on the l_{ex}/h and l_{ey}/b ratios of the columns, where:

l_{ex} the effective height in respect of the major axis,
l_{ey} the effective height in respect of the minor axis,
h the depth of the cross-section measured in the plane under consideration,
b the width of a column (dimension of a column perpendicular to h).

The defining slenderness ratios are given in Clause 3.8.1.3 and summarized in Table 5.2.

Defining Slenderness Ratios for **Short** Columns		
	l_{ex}/h	l_{ey}/b
Braced Column	15	15
Unbraced Column	10	10

Table 5.2

Any column which has a slenderness ratio *greater* than the values given in Table 1 should be considered as slender. The effective height (l_e) can be evaluated using the recommendations given in Clause 3.8.1.6, Table 3.19 and Table 3.20 of the code; i.e. $l_e = \beta l_o$ where β is a coefficient which is dependent on the end condition of the column, and l_o is the *clear* height between the end restraints.

The end conditions are graded from 1 to 4, in which 1 corresponds to a significant fixity and 4 represents a free end. These conditions are defined in Clause 3.8.1.6.2 of the code as shown in Figures 5.108 and 5.109.

Extract from *BS 8110:Part 1:1997*

Table 3.19 Values of β for braced columns			
End condition at top	End condition at bottom		
	1	2	3
1	0.75	0.80	0.90
2	0.80	0.85	0.95
3	0.90	0.95	1.00

Figure 5.108

Extract from *BS 8110:Part 1:1997*

End condition at top	End condition at bottom		
	1	2	3
1	1.2	1.3	1.6
2	1.3	1.5	1.8
3	1.6	1.8	—
4	2.2	—	—

Table 3.20 Values of β for unbraced columns

Condition 1:
The end of the column is connected monolithically to beams on either side which are at least as deep as the overall dimension of the column in the plane considered. Where the column is connected to a foundation structure, this should be of a form specifically designed to carry moment.

Condition 2:
The end of the column is connected monolithically to beams or slabs on either side which are shallower than the overall dimension of the column in the plane considered.

Condition 3:
The end of the column is connected to members which, while not specifically designed to provide restraint to rotation of the column will, nevertheless, provide some nominal restraint.

Condition 4:
The end of the column is unrestrained against both lateral movement and rotation (e.g. the free end of a cantilever column in an unbraced structure.

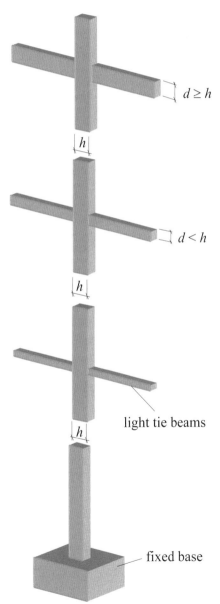

$d \geq h$

$d < h$

light tie beams

fixed base

Figure 5.109

The slenderness limits for columns are given in Clauses 3.8.1.7 and 3.8.1.8 as:

$l_o \leq 60 \times$ minimum thickness generally, and in unbraced columns, one end of which is unrestrained (e.g a cantilever column), its clear height, l_o should not exceed:

$$l_o = \frac{100\,b^2}{h} \leq 60b$$

where h and b are the larger and smaller dimensions of the column respectively.

The design axial forces and moments in columns can be determined according to the requirements of Clauses 3.2 as indicated in Section 5.14.1.

Additional moments induced in slender columns should be added to those evaluated using Clause 3.2.1.3 as indicated in Clause 3.8.2.2. The design of slender columns is not considered in this text.

As indicated in Clause 3.8.2.3, in the case of columns supporting a symmetrical arrangement of approximately equally loaded beams, only the design ultimate axial force need be considered in design, together with a design moment representing a nominal allowance for eccentricity, equal to that recommended in 3.8.2.4.

In Clause 3.8.2.4 a minimum eccentricity, e_{min}, equal to (0.05 × *overall dimension of the column in the plane of bending, but not more than* 20 mm), is defined which must be used to evaluate the nominal design moment.

Note: At no section in a column should the design moment be taken as less than that produced by considering the design ultimate axial load as acting at the minimum eccentricity.

5.15.1   *Design Resistance of Columns*

The rigorous design of columns subject to combined axial loading and bending moments is very laborious and time-consuming. To enable an efficient, economic and rapid design to be undertaken, a series of design charts have been produced for symmetrically-reinforced columns; these are given in *BS 8110:Part 3*. A typical design chart is shown in Figure 5.110. The reinforcement areas determined using the design charts should be modified to reflect the changes to the partial safety factor γ_m as before, i.e. multiplied by 0.91.

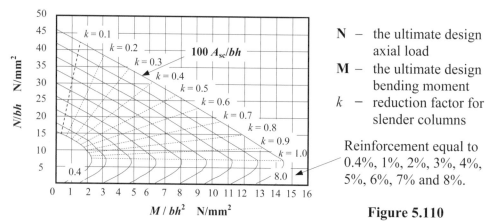

N – the ultimate design axial load

M – the ultimate design bending moment

k – reduction factor for slender columns

Reinforcement equal to 0.4%, 1%, 2%, 3%, 4%, 5%, 6%, 7% and 8%.

Figure 5.110

5.15.1.1 Short Columns Resisting Moments and Axial Forces (Clause 3.8.4.3)

Normally short columns only require to be designed for the maximum moment about one critical axis in addition to the axial load. In the case of a column supporting e.g. a rigid structure or very deep beams, where it cannot be subjected to significant moments, they may be designed in accordance with Clause 3.8.4.3 such that the applied ultimate axial load does not exceed the following:

$$N \ = \ 0.4f_{cu}A_c + 0.8A_{sc}f_y$$

where:
N is the applied ultimate axial load,
A_c is the net cross-sectional area of concrete in the column, i.e. $(bh - A_{sc})$,
A_{sc} is the area of vertical reinforcement, i.e. all main reinforcement,
f_{cu} and f_y are as before.

5.15.1.2 Short Braced Columns Supporting an Approximately Symmetrical Arrangement of Beams (Clause 3.8.4.4)

A reduction from the equation given in Clause 3.8.4.3 is given to allow for moments which will arise from asymmetrical loading on symmetrical beams and is given in Clause 3.8.4.4 in the following equation:

$$N \ = \ 0.35f_{cu}A_c + 0.7A_{sc}f_y$$

where:
a) the beams are designed for uniformly distributed loads; and
b) the beam spans do not differ by more than 15% of the longer.

Both of these equations incorporate an allowance for the partial safety factor γ_m and a reduction of approximately 10% for the required minimum eccentricity specified in Clause 3.8.2.4.

5.15.2 Example 5.29: Axially Loaded Short Column

A short, braced column is subjected to an ultimate applied axial load of 3000 kN and a nominal moment only. Using the design data given:

i) check that the column is short,
ii) determine the required area of main reinforcement,
iii) determine suitable links.

Characteristic strength of concrete	f_{cu} =	40 N/mm²
Characteristic strength of reinforcing steel	f_y =	460 N/mm²
Width of column	b =	350 mm
Depth of column	h =	375 mm
End condition at the top of the column for x–x axis	=	1
End condition at the top of the column for y–y axis	=	2
End condition at the bottom of the column for x–x axis	=	3
End condition at the bottom of the column for y–y axis	=	3
Clear height between the end restraints about both axes	=	5.0 m

5.15.3   Solution to Example 5.29

Contract : Concrete Frame Job Ref. No. : Example 5.29 Part of Structure : Short, Braced Column Calc. Sheet No. : 1 of 1	Calcs. by : W.McK. Checked by : Date :

References	Calculations	Output
Clause 3.8.1.3	The column is braced $\therefore l_{ex}/h$ and l_{ey}/h < 15 if column is short	
Table 3.19	x–x direction: end condition at the top $\quad = \quad 1$ $\qquad\qquad\qquad$ end condition at the bottom $\; = \quad 3$ $\beta \;=\; 0.9 \quad \therefore l_{ex} = (0.9 \times 5000) \qquad = \quad 4500$ mm $\qquad\qquad \therefore l_{ex}/h = (4500/375) \qquad = \quad 12.0 < 15.0$ y–y direction: end condition at the top $\quad = \quad 2$ $\qquad\qquad\qquad$ end condition at the bottom $\; = \quad 3$ $\beta \;=\; 0.95 \quad \therefore l_{ey} = (0.95 \times 5000) \quad = \quad 4750$ mm $\qquad\qquad \therefore l_{ey}/h = (4750/350) \qquad = \quad 13.6 < 15.0$	**Column is short**
Clause 3.8.4.3	$N \;=\; 0.4 f_{cu} A_c + 0.8 A_{sc} f_y$ $(3000 \times 10^3) = 0.4 \times 40 \times [(375 \times 350) - A_{sc}] + (0.8 \times A_{sc} \times 460)$ $A_{sc} = \; 2557$ mm^2	**Select:** **6/25 mm diameter** **bars providing** **2950 mm^2**
Clause 3.12.7.1	Minimum link diameter $\geq$ (0.25 × largest compression bar) $\qquad\qquad\qquad\qquad\qquad = \; 6.25$ mm $\qquad\qquad\qquad\qquad\qquad \geq \; 6$ mm Spacing of links $\qquad\quad \leq \;$ (12 × smallest compression bar) $\qquad\qquad\qquad\qquad\qquad = \; (12 \times 25) \;=\; 300$ mm	**Select:** **8 mm diameter links** **at 300 mm spacing**
	  Typical column/beam/slab intersection details: 	

5.15.4 Example 5.30: Reinforced Concrete Multi-Storey Braced Column

A reinforced concrete industrial frame is shown in Figure 5.111. The structure, of which this frame forms a part, is braced and comprises a series of such frames at 6.0 m centres. Using the design data given determine suitable reinforcement for the internal column at Section *x–x*. Assume the columns bases are fixed.

Design Data:

Characteristic dead load (including self-weight of floor):

Roof level	g_k =	4.0 kN/m^2
1st and 2nd floor levels	g_k =	8.0 kN/m^2
Characteristic imposed load:		
Roof level	q_k =	4.0 kN/m^2
1st and 2nd floor levels	q_k =	12.0 kN/m^2
Characteristic strength of concrete	f_{cu} =	30 N/mm^2
Characteristic strength of reinforcement	f_y =	460 N/mm^2
Exposure condition		mild
Minimum fire resistance		1 hour
All columns		300 mm × 350 mm
All main beams		300 mm wide × 600 mm deep

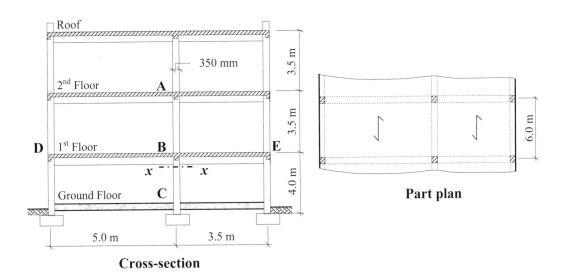

Cross-section   **Part plan**

Figure 5.111

The critical design load case in a braced structure is usually that which induces the largest moment in the column being considered together with the largest compatible axial load. In this example the load case which should be considered is as indicated in Clause 3.2.1.2.5, assuming that the beams DB and BE are to be designed in accordance with Clause 3.2.1.2.4.

5.15.5   Solution to Example 5.30

Contract : Concrete Frame Job Ref. No. : Example 5.30	Calcs. by : W.McK.
Part of Structure : Multi-Storey, Braced Column	Checked by :
Calc. Sheet No. : 1 of 4	Date :

References	Calculations	Output
Clause 3.2.1.2.5	*"The arrangement of the design ultimate imposed load should be such as to cause the maximum moment in the column."* Design roof loading = $[(1.4 \times 4.0 \times 6.0) + (1.6 \times 4.0 \times 6.0)]$ $= 72$ kN/m Design 2nd floor loading $\quad = [(1.4 \times 8.0 \times 6.0) + (1.6 \times 12.0 \times 6.0)]$ $\quad = 182.4$ kN/m Design 1st floor loading $\quad$ Beam DB $\quad = [(1.4 \times 8.0 \times 6.0) + (1.6 \times 12.0 \times 6.0)]$ $\qquad = 182.4$ kN/m $\quad$ Beam BE $\quad = (1.0 \times 8.0 \times 6.0)$ $\qquad = 48.0$ kN/m Design axial load at section x–x: $N = \dfrac{(5.0+3.5)}{2.0} \times (72 + 182.4) + (2.5 \times 182.4) + (1.75 \times 48.0)$ $\quad = 1621.2$ kN The analysis to determine the design bending moment is carried out using a sub-frame as indicated in Figure 5.86.	

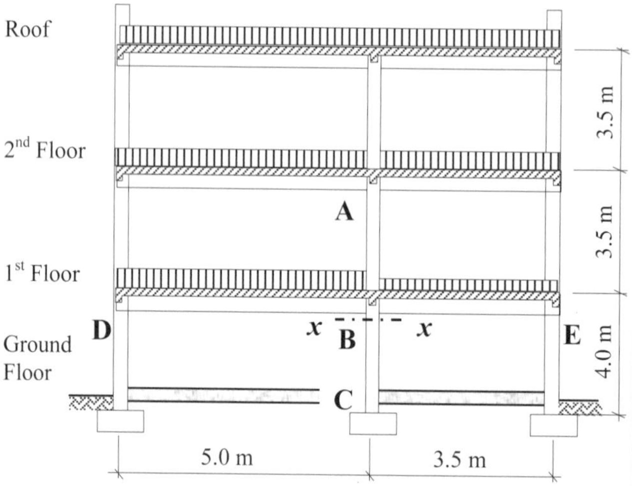

Contract : Concrete Frame Job Ref. No. : Example 5.30 Part of Structure : Multi-Storey, Braced Column Calc. Sheet No. : 2 of 4	Calcs. by : W.McK. Checked by : Date :

References	Calculations	Output

In this simplified frame:

$$I_{BA} = I_{BC} = \frac{bd^3}{12} = \frac{300 \times 350^3}{12} = 1071.9 \times 10^6 \text{ mm}^4$$

$$I_{BD} = I_{BE} = \frac{bd^3}{12} = \frac{300 \times 600^3}{12} = 5400 \times 10^6 \text{ mm}^4$$

$$k_{BA} = \frac{I}{L} = \frac{1071.9 \times 10^6}{3500} = 0.31 \times 10^6 \text{ mm}^3$$

$$k_{BC} = \frac{I}{L} = \frac{1071.9 \times 10^6}{4000} = 0.27 \times 10^6 \text{ mm}^3$$

In the sub-frame the beam stiffnesses are assumed to be equal to (0.5 × actual value)

$$k_{BD} = \left(0.5 \times \frac{I}{L}\right) = \frac{0.5 \times 5400 \times 10^6}{5000} = 0.54 \times 10^6 \text{ mm}^3$$

$$k_{BE} = \left(0.5 \times \frac{I}{L}\right) = \frac{0.5 \times 5400 \times 10^6}{3500} = 0.77 \times 10^6 \text{ mm}^3$$

Total stiffness of the joint k_{total}:
$$= (0.31 + 0.27 + 0.54 + 0.77) \times 10^6 = 1.89 \times 10^6 \text{ m}^3$$

Distribution factors at joint B:

$$DF_{BA} = \frac{0.31}{1.89} = 0.16; \quad DF_{BC} = \frac{0.27}{1.89} = 0.14$$

$$DF_{BD} = \frac{0.54}{1.89} = 0.29; \quad DF_{BA} = \frac{0.77}{1.89} = 0.41$$

Note: The sum of the distribution factors = 1.0

Contract : Concrete Frame Job Ref. No. : **Example 5.30** Part of Structure : **Multi-Storey, Braced Column** Calc. Sheet No. : **3 of 4**	Calcs. by : **W.McK.** Checked by : Date :

References	Calculations	Output

The fixed end moments from the beam loadings are:

$$M_{BD} \text{ and } M_{DB} = \pm\frac{wL^2}{12} = \pm\frac{182.4 \times 5.0^2}{12} = \pm380.0 \text{ kNm}$$

$$M_{BE} \text{ and } M_{EB} = \pm\frac{wL^2}{12} = \pm\frac{48.0 \times 3.5^2}{12} = \pm 49.0 \text{ kNm}$$

Completing the moment distribution at the joint B gives:

Joint	A	D			B		E	C
	AB	DB	BD	BA	BC	BE	EB	CB
DF	0.0	0.0	0.29	0.16	0.14	0.41	0.0	0.0
FEM		− 380	+ 380			− 49.0	+ 49.0	
Bal.			− 96.0	− 53.0	− 46.3	− 135.7		
C.O.	− 26.5	− 48.0					− 67.9	− 23.2
Total	− 26.5	− 428	+ 284	− 53.0	− 46.3	− 184.7	− 18.9	− 23.2

Design moment in the lower column at section x–x:
$$M = 46.3 \text{ kNm}$$
Design axial load at section x–x $N = 1621.2 \text{ kN}$

Clause 3.8.1.3 The column is braced

Assume the effective height of column section BC is equal to 4.0 m about both axes.

Table 3.19 x–x direction: $\therefore l_{ex} / h = (4000 / 300)$ $= 13.3 < 15.0$

y–y direction: $\therefore l_{ey} / h = (4000 / 350)$ $= 11.4 \ < 15.0$

Column is short

Clause 3.3.1.2 Assume bar size $= 25$ mm
Clause 3.3.1.3 Nominal maximum size of aggregate $= 20$ mm
Table 3.3 Exposure condition is mild $= 25$ mm
Table 3.4 Minimum fire resistance 1 hour $= 20$ mm
Assume minimum cover to main steel   = 25 mm

Figure 3.2 Assume all column faces are exposed:
Minimum column dimension $= 200$ mm
Column is adequate with respect to fire resistance

Assume 6 mm diameter links
Effective depth $d = (350 - 25 - 8 - 6)$ $= 311$ mm

Contract : Concrete Frame Job Ref. No. : Example 5.30 Part of Structure : Multi-Storey, Braced Column Calc. Sheet No. : 4 of 4	Calcs. by : W.McK. Checked by : Date :

References	Calculations	Output
	$d/h = (311/350) = 0.89$ Use Design Chart No. 29 $N/bh = \dfrac{1621.2 \times 10^3}{(300 \times 350)} = 15.4$ $M/bh^2 = \dfrac{46.3 \times 10^6}{(300 \times 350^2)} = 1.3$ From the chart: $100 A_{sc}/bh = 1.3 \quad \therefore A_{sc} = \dfrac{(1.3 \times 300 \times 350)}{100} = 1365 \text{ mm}^2$	% of reinforcement ranging from 0.4 to 8.0 **Select:** **4/25 mm diameter** **+ 2/20 mm diameter** **bars providing** **1432 mm²**
Clause 3.12.7.1	Minimum link diameter $\geq$ (0.25 × largest compression bar) $= 6.25 \text{ mm}$ $\geq 8 \text{ mm}$ Spacing of links $\leq$ (12 × smallest compression bar) $= (12 \times 20) = 240 \text{ mm}$	**Select:** **8 mm diameter links** **at 240 mm spacing**
	Similar calculations can be carried out for column sections above and below each floor level and at the ground floor level. Note that this sub-frame cannot be used to determine the beam moments; see Clause 3.2.1.2.4.	

5.16   Foundations

5.16.1   Introduction

The primary function of all structural elements is to transfer the applied dead and imposed loading, from whichever source, to the foundations and subsequently to the ground. The type of foundation required in any particular circumstance is dependent on a number of factors such as:

◆ the magnitude and type of applied loading,
◆ the pressure which the ground can safely support,
◆ the acceptable levels of settlement,
◆ the location and proximity of adjacent structures.

The most common types of foundation currently used are indicated in sections 5.16.2 to 5.16.6.

5.16.2   Pad Foundations

These are normally adopted for single columns (either steel or concrete), and can be either square or rectangular in plan as indicated in Figure 5.112. When only concentric vertical loading is applied, square pads are used, assuming a uniform pressure under the whole base area. If the loading is eccentric or if a moment is applied to the base then it is more efficient to adopt a rectangular base. In this case the pressure under the base is assumed to vary linearly.

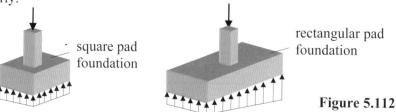

square pad
foundation

rectangular pad
foundation

Figure 5.112

5.16.3   Combined Foundations

These are normally adopted for two columns either when they are relatively close together or when one of the columns is adjacent to an existing structure. The shape of a combined footing is generally rectangular, trapezoidal or a combination of two rectangles, as shown in Figure 5.113. In the last case, a rib-beam the same width as the columns is often incorporated either the whole length of the base or between the columns. This foundation can be designed as a T-section and requires less concrete.

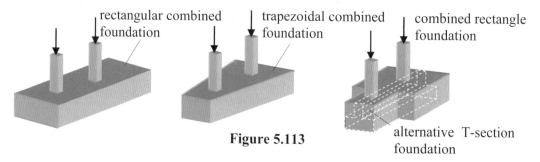

rectangular combined
foundation

trapezoidal combined
foundation

combined rectangle
foundation

Figure 5.113

alternative T-section
foundation

The dimensions of a combined footing can be determined such that the resultant load from the columns passes through the centroid of the base area. In this case the resulting pressure on the soil will be uniform.

A rectangular base does not always provide the most economic arrangement due to the difficulty of making the resultant load pass through the centroid of the base area. The trapezoidal base has the disadvantage of detailing and cutting the transverse reinforcement. It is more suitable when there is a large variation in the column loads and there are limitations on the dimensions of the foundations. The combined rectangular solution can be used in most cases.

The critical section for shear in a combined footing is not specified in BS 8110, but for a wide footing which acts as a thick slab bending in both the longitudinal and transverse directions, it is probably similar to pad foundations and approximately '1.0d' from the column face. In narrow footings the bending is predominantly in the longitudinal direction and the critical section for shear should be taken at the face of the column.

5.16.4   Strip Footings

These are normally adopted for lines of closely spaced columns or under walls, as shown in Figure 5.114. The strips are designed as continuous beams subjected to the ground bearing pressures. In good ground conditions where the soil is firm (and the columns are evenly spaced), the ground pressure can be assumed to be uniform. When columns are unevenly spaced and the soil is firm, the variation in pressure can be assumed to vary linearly. In compressible and/or poor soils the variation in ground pressure will not be linear, resulting in a different distribution of bending moments. In many situations, other than on lightly loaded strip foundations, reinforcement will be necessary.

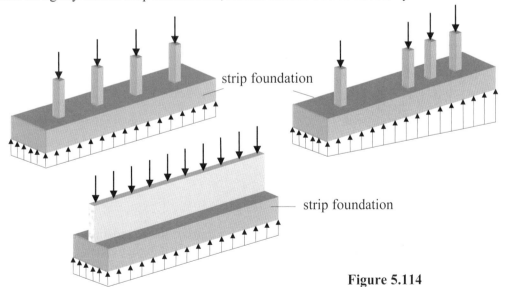

strip foundation

strip foundation

Figure 5.114

5.16.5   Raft Foundations

These are continuous slabs which cover the whole plan area of the structure as shown in Figure 5.115. They are normally used under the following circumstances:

♦ For lightly loaded structures on soft natural ground where it is necessary to spread the load, or where there is variable support due to natural variations such as made-up ground, swallow holes, etc., and the raft is used to bridge the weaker areas.

♦ For heavier structures where the ground conditions are such that there are unlikely to be significant differential settlements. In many such cases the raft can be considered as a nominal one replacing isolated foundations occupying the majority of the available foundation area.

♦ Where differential settlements are likely to be significant. In such cases, the raft will require special design, involving an assessment of the disposition and distribution of loads and contact pressures.

♦ Where mining subsidence is likely to occur. Design of the raft and structure to accommodate mining subsidence requires special consideration, often involving provision of a flexible structure, e.g. avoiding long continuous buildings by creating division of extensive buildings into independent sections of appropriate size each with its own foundations.

The construction of the raft can be a flat slab of continuous thickness, thickened locally around column positions, or can be strengthened by beams to form a ribbed construction.

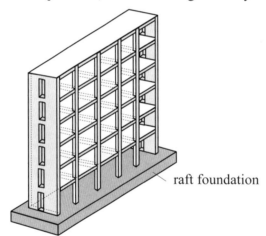

raft foundation

Figure 5.115

5.16.6   *Piled Foundations*

These are used in situations where it is necessary to transfer the foundation loads through strata which have a low bearing capacity to strata, which has a higher capacity, or to rock. In some situations, where it is necessary to resist high uplift forces or to transfer horizontal loads through poor soil, it may also be necessary to introduce piled foundations. Piles are essentially long, slender members, mostly subjected to compression as shown in Figure 5.116. In general, pile groups are subjected to axial load, moments and horizontal loads. The distribution of these loads between individual piles is based on simple elastic analysis.

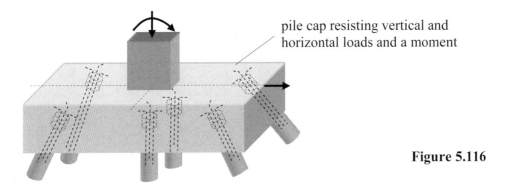

Figure 5.116

5.16.7 *Loading Effects*

The loading effects which occur in foundations, are generally one or more of the following three types:

- ◆ horizontal forces due to lateral loading such as wind on the supported structure or friction between the underside of the base and the ground – horizontal forces are not usually of sufficient magnitude to affect the size of foundations,
- ◆ vertical forces due to columns and/or walls and the bearing pressure from the ground underneath the base,
- ◆ moments due to loading from columns and/or walls etc. which are eccentric to the centroid of the base.

5.16.8 *Base Pressures*

The assumption of a linear pressure distribution under foundations results in one of three possible pressure diagrams under the base. The magnitude of the pressure in each case is determined using elastic analysis.

5.16.8.1 Case 1: Uniform Pressure (compression throughout)

When a base is subject to an axial load only in which the line of action of the applied force passes through the centroid of the base as shown in Figure 5.117, the pressure under the base is assumed to be uniform throughout and is equal to P where:

$$P = \frac{\text{Applied Load}}{\text{Base Area}} = \frac{N}{BD}$$

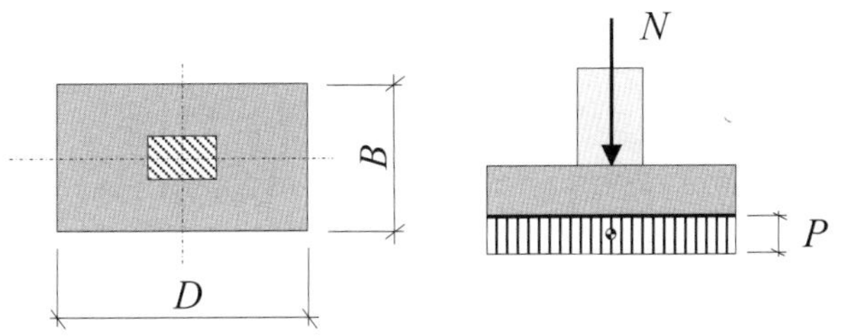

Figure 5.117

5.16.8.2 Case 2: Varying Pressure (compression throughout)

When a foundation is subject to an eccentric load or a central load combined with a moment, the pressure under the base can be either compression throughout or compression and tension. In the case of compression throughout as shown in Figure 5.118, the magnitude can be determined using simple elastic analysis and is equal to the sum of the axial stress and the moment stress, i.e.

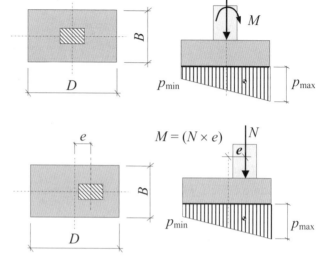

$$p = \frac{N}{BD} \pm \frac{M}{BD^2/6} = \frac{N}{BD} \pm \frac{6M}{BD^2}$$

Figure 5.118

Note: In the second case with the eccentric column, the moment $M = (N \times e)$

5.16.8.3 Case 3: Varying Pressure (compression over part of the base)

In the case of compression over part of the base as shown in Figure 5.119, the magnitude of the maximum pressure can be determined using simple elastic analysis and is equal

to: $$p = \frac{2N}{3B\left(\dfrac{D}{2} - e\right)}$$

Note: In the first case the equivalent eccentricity e can be determined by equating

$$M = (N \times e) \qquad e = \frac{M}{N}$$

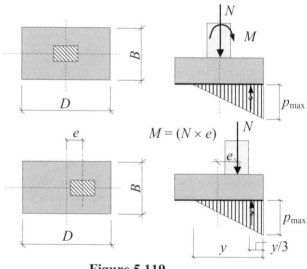

Figure 5.119

Middle Third Rule		
In case 1	$e = 0$	No tension
In case 2	$e \leq D/6$	No tension
In case 3	$e > D/6$	Tension exists

This is normally regarded as the **middle third rule:** i.e. if the eccentricity e of the load lies within the middle third of the base length, then no tension will occur under the base.

5.16.9 *Design of Pad Foundations*

A typical arrangement of the reinforcement in a pad foundation is shown in Figure 5.120. In square bases the reinforcement to resist bending is distributed uniformly across the full width of the foundation. For a rectangular base the reinforcement in the short direction should be distributed with closer spacing in the region under and near the column to allow for the fact that the transverse moments are greater nearer the column. If the foundation is subjected to eccentric loading inducing large moments such that there is only partial bearing, (case 3 above) reinforcement may also be required in the top face.

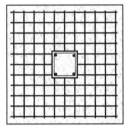 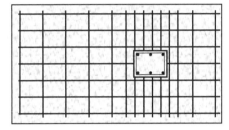

Figure 5.120

Dowel (or **starter**) **bars** provide the continuity of the reinforcement between the column and the base. A length of the column (e.g. 75 mm) is often constructed in the same concrete pour as the foundation to form a '**kicker**' (a support) for the column shutters. In these cases the dowel lap-length should be measured from the top of the kicker as shown in Figure 5.121.

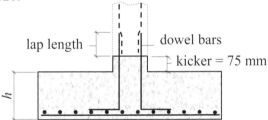

Figure 5.121

5.16.9.1 Critical Section for Bending

The critical section which should be considered for bending is at the face of the column and extending across the full width of the base as shown in Figure 5.122. The area of reinforcement required is calculated in the same manner as for beams using Clause 3.4.4.4. The minimum and maximum area of steel are determined using Table 3.25 and Clause 3.12.6 respectively.

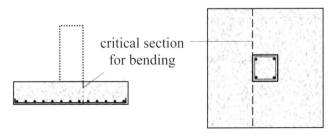

Figure 5.122

5.16.9.2 Critical Sections for Shear

The critical sections which should be considered for shear are as shown in Figure 5.123. The shear stress at the column face should not exceed the maximum values indicated in Clause 3.7.7.2, i.e. $0.8\sqrt{f_{cu}}$ or 5 N/mm^2.

The direct shear stress should be checked at a distance of $1.0d$ from the face of a column, and if it is less than v_c from Table 3.8 no shear reinforcement is required. **Punching failure** can occur on the inclined faces of truncated cones or pyramids, depending on the shape of the loaded area. The possibility of this type of failure can be checked by considering a shear perimeter as indicated in Figure 3.16 of the code. In the case of square/rectangular pad foundations the value of l_p in Figure 3.16 is equal to $1.5d$ as indicated in Clause 3.7.7.6. As in direct shear, when the value of the shear stress is less than v_c from Table 3.8 no shear reinforcement is required. Pad foundations are normally designed such that shear reinforcement is not required.

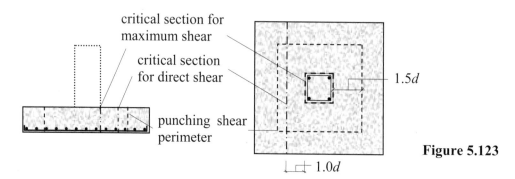

Figure 5.123

5.16.9.3 Design Procedure

Pad foundations should be checked for direct shear, punching shear, and bending. The shearing forces and bending moments are caused by the ultimate design loads from the column and the weight of the base should not be included in these calculations. The thickness of the base is often governed by the requirements for shear resistance. The principal steps in the design calculations are as follows:

- ◆ Calculate the plan size of the foundation using the permissible bearing pressure and the critical loading arrangement for the serviceability limit state.
- ◆ Calculate the bearing pressure associated with the critical loading arrangement at the ultimate limit state.
- ◆ Assume a suitable value of thickness (h) and determine the effective depth (d).
- ◆ Determine the reinforcement required to resist bending.
- ◆ Check that the shear stress at the column face is less than 5 N/mm^2 or $0.8\sqrt{f_{cu}}$, whichever is the smaller.
- ◆ Check that the direct shear stress at a section $1.0d$ from the column face is less than v_c from Table 3.8.
- ◆ Check that the punching shear stress on a perimeter $1.5d$ from the column face is less than v_c from Table 3.8.

Reinforcement to resist bending in the bottom of the base should extend at least a full tension anchorage length beyond the critical section for bending.

5.16.10 Example 5.31: Axially Loaded Pad Foundation

A pad foundation is required to support a single square column transferring an axial load only. Using the data provided:

-   a)   determine a suitable base size,
-   b)   check the base with respect to:
    -   i)   bending,
    -   ii)   direct shear, and
    -   iii)   punching shear,

designing suitable reinforcement where necessary.

Design Data:

Characteristic dead load on column	800 kN
Characteristic imposed load on column	300 kN
Characteristic concrete strength	$f_{cu} = 40 \text{ N/mm}^2$
Characteristic of reinforcement	$f_y = 460 \text{ N/mm}^2$
Net permissible ground bearing pressure	$p_g = 200 \text{ kN/m}^2$
Column dimensions	375 mm × 375 mm
Exposure condition	severe

5.16.11 Solution to Example 5.31

Contract : Foundations Job Ref. No. : Example 5.31 Part of Structure : Axially Loaded Pad Foundation Calc. Sheet No. : 1 of 5	Calcs. by : W.McK. Checked by : Date :

References	Calculations	Output

	Base plan area:
	The **serviceability** limit state loads are used to determine a minimum base area since this is based on the permissible bearing pressure given.
	
	Total design axial load = $(1.0 \, G_k + 1.0 \, Q_k)$
	  = $(800 + 300)$ = 1100 kN
	
	Minimum required base area = $\dfrac{1100}{200}$ = 5.5 m^2
	
	Base length for square base ≥ $\sqrt{5.5}$ = 2.35 m
	
	Base reinforcement (**based on ultimate loads**):
	Total design axial load = $(1.4 \, G_k + 1.6 \, Q_k)$
	  = $(1.4 \times 800) + (1.6 \times 300)$
	  = 1600 kN
		Provide a 2.4 m square base

Contract : Foundations Job Ref. No. : Example 5.31 Part of Structure : **Axially Loaded Pad Foundation** Calc. Sheet No. : **2 of 5**	Calcs. by : **W.McK.** Checked by : Date :

References	Calculations	Output
	Earth pressure due to ultimate loads $= \dfrac{1600}{(2.4 \times 2.4)} = 278 \text{ kN/m}^2$ Base thickness: Assume a base thickness of 550 mm, constructed on a blinding layer of concrete. Cover: Assume 20 mm diameter bars. bar size $\qquad\qquad\qquad\qquad = 20 \text{ mm}$ nominal maximum size of aggregate $= 20 \text{ mm}$ exposure condition is severe $\quad$ cover $\geq 40 \text{ mm}$. $\qquad$ **Assume minimum cover to main steel = 40 mm** Effective depth: $d = (550 - 40 - 20) = 490 \text{ mm}$ (**Note:** This is the mean effective depth since the main reinforcement runs in both directions.) Bending Check the bending at the column face: 1600 kN critical section for bending 375 mm $\times$ 375 mm 278 kN/m^2 $l_c = (1200 - 187.5) = 1013 \text{ mm}$ $\qquad$ 2400 mm Bending moment at the face of the column: $= \; [(278 \times 2.4 \times 1.013) \times 0.51] \;=\; 344.7 \text{ kNm}$ $K = \dfrac{M}{bd^2 f_{cu}} = \dfrac{344.7 \times 10^6}{2400 \times 490^2 \times 40} = 0.015 \; < \; K' \; (= 0.156)$ $\qquad\qquad\qquad\qquad\qquad$ Section is singly reinforced $Z = d\left\{0.5 + \sqrt{\left(0.25 - \dfrac{K}{0.9}\right)}\right\} \;=\; d\left\{0.5 + \sqrt{\left(0.25 - \dfrac{0.015}{0.9}\right)}\right\}$ $\qquad\qquad\qquad\qquad = \; 0.98d$ $\qquad\qquad$ The lever arm is limited to $0.95d$	
Clause 3.3 Clause 3.3.1.2 Clause 3.3.1.3 Table 3.3		
Clause 3.11.3.1 Clause 3.4.4.4		
Clause 3.4.4.4		

References	Calculations	Output
	$$A_s = \frac{M}{0.95 f_y Z} = \frac{344.7 \times 10^6}{0.95 \times 460 \times 0.95 \times 490} = 1694 \text{ mm}^2$$ $$\therefore A_s = (1694/2.4) = 706 \text{ mm}^2/\text{m width}$$	**Bottom Reinforcement**
Table 3.25	Minimum % reinforcement required $= 0.13$ % provided $= \dfrac{100 \times 786}{1000 \times 550} = 0.14$ **Minimum % of steel satisfied**	**Select:** **20 mm diameter bars @ 400 mm centres providing 786 mm²/m in both directions**
Clause 3.12.11.2.7	Maximum spacing: Spacing $\leq (3 \times d) = (3 \times 490) = 1470$ mm ≤ 750 mm $100 A_s / bd = (100 \times 786)/(1000 \times 490) = 0.16\% < 3\%$	**Spacing is adequate**
Clause 3.11.3.2	$[(3c/4) + (9d/4)] = [(0.75 \times 375) + (2.25 \times 490)] = 1383$ mm $l_c = $ distance to the edge of the pad $= 1013$ mm < 1383 mm	**Reinforcement should be distributed uniformly over l_c**
Clause 3.11.3.3 Table 3.8	Shear resistance (Clause 3.5.6 and Clause 3.7.7) Critical shear stress $\dfrac{100 A_s}{b_v d} = \dfrac{100 \times 786}{1000 \times 490} = 0.16 \quad \therefore v_c = 0.35 \text{ N/mm}^2$ (modify to allow for C40 concrete) $v_c = (0.35 \times 1.17) = 0.41 \text{ N/mm}^2$	
Clause 3.7.7.4	Direct shear Check the direct shear at a distance of $1.0d$ from the column face. This location corresponds to a 45° dispersal line from the column face. If the shear stress v is less than v_c then no shear reinforcement is required. At locations closer than this to the column face the load is assumed to distribute through the base in compression.	

1600 kN ┊┈┈ critical section for
maximum shear

1.0d

375 mm × 375 mm

278 kN/m²

$(1200 - 187.5 - 490) = 523$ mm

2400 mm

References	Calculations	Output

At the critical section for direct shear (i.e. $1.0d$ from the column face):

Shear force $= V = (278 \times 2.4 \times 0.523) = 348.9$ kN

Shear stress $= v = \dfrac{V}{bd} = \dfrac{348.9 \times 10^3}{2400 \times 490} = 0.3$ N/mm^2

$$v \; < \; v_c$$

No shear reinforcement required

Punching shear

The possibility of punching shear should be checked at the column face for the maximum shear stress as indicated in Clause 3.7.7.2 and on a perimeter located at $1.5d$ from the column face as indicated in Clause 3.7.7.6. If the calculated shear stress does not exceed v_c on this perimeter then no further checks are required.

Clause 3.7.7.2

Check the maximum shear at the column face:

1600 kN

critical section for maximum shear

375 mm × 375 mm

278 kN/m^2

2400 mm

Clause 3.7.7.2

Shear stress $\quad v = \dfrac{V}{u_o d} = \dfrac{V}{column\ perimeter \times d}$

$v = \dfrac{1600 \times 10^3}{[(4 \times 375) \times 490]} = 2.18$ N/mm^2

Maximum shear $\quad v_{max} \leq 0.8\sqrt{f_{cu}} = (0.8 \times \sqrt{40})$

$= 5.05$ N/mm^2

$v_{max} \leq 5.0$ N/mm^2

$$v \; < \; v_{max}$$

Maximum shear at column face is acceptable

Contract :Foundations   Job Ref. No. : Example 5.31
Part of Structure : Axially Loaded Pad Foundation
Calc. Sheet No. : 4 of 5

Calcs. by : W.McK.
Checked by :
Date :

Contract :Foundations Job Ref. No. : Example 5.31 Part of Structure : Axially Loaded Pad Foundation Calc. Sheet No. : 5 of 5	Calcs. by : W.McK. Checked by : Date :

References	Calculations	Output
Clause 3.7.7.6	Check the shear stress on a perimeter 1.5d from the column face. Critical perimeter $=$ (column perimeter) $+ (8 \times 1.5d)$ $\quad\quad\quad\quad\quad = (4 \times 375) + (8 \times 1.5 \times 490) = 7380$ mm Area within the critical perimeter $= (375 + 3d)^2$ $\quad\quad = [375 + (3 \times 490)]^2 = 3.4 \times 10^6$ mm^2 Shear force outside the critical area $= [278 \times (2.4^2 - 3.4)]$ $\quad\quad\quad\quad = 656.1$ kN Shear stress $\quad v = \dfrac{V}{u_o d} = \dfrac{656.1 \times 10^3}{7380 \times 490} = 0.18$ N/mm^2 $\quad\quad\quad\quad\quad\quad\quad\quad v \ < \ v_c$ **No shear reinforcement required** **Curtailment of reinforcement:** The main reinforcement is not normally curtailed in slab bases. **Anchorage:** The main bars are anchored a standard length beyond the face of the column. In most cases the size of the base will permit a full anchorage to be obtained. In situations where this is not the case then a standard 90° bend can be provided.	
Clause 3.7.7.2		

5.16.12   Example 5.32: *Pad Foundation with Axial Load and Moment*

A rectangular pad foundation is required to support a single square column transferring an axial load and a moment as shown in Figure 5.124. Using the data provided check the base with respect to:

        i)   bending,
        ii) direct shear, and
        iii) punching shear,

designing suitable reinforcement where necessary.

Design Data:

Characteristic dead axial load on column	250 kN
Characteristic imposed axial load on column	350 kN
Characteristic dead moment on column	125 kN
Characteristic imposed moment on column	175 kN
Characteristic concrete strength	$f_{cu} = 40\ \text{N/mm}^2$
Characteristic of reinforcement	$f_y = 460\ \text{N/mm}^2$
Net permissible ground bearing pressure	$p_g = 300\ \text{kN/m}^2$
Column dimensions	375 mm × 375 mm
Exposure condition	moderate

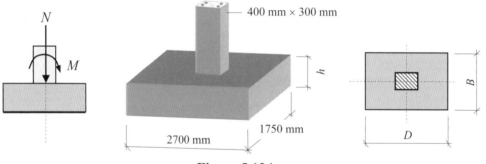

Figure 5.124

5.16.13   Solution to Example 5.32

Contract : Foundations Job Ref. No. : Example 5.32	Calcs. by : W.McK.
Part of Structure : Pad Foundation with Axial load and Moment	**Checked by :**
Calc. Sheet No. : 1 of 7	**Date :**

References	Calculations	Output
	It is necessary to establish if tension exists under the base under the action of the bending moment. Refer to Section 5.19.8, and the 'middle third rule'.	

Contract : Foundations Job Ref. No. : Example 5.32 Part of Structure :Pad Foundation with Axial Load and Moment Calc. Sheet No. : 2 of 7	Calcs. by : W.McK. Checked by : Date :

References	Calculations	Output

Characteristic design axial load   =   600 kN

Characteristic design bending moment   =   300 kNm

The equivalent eccentricity $e = \dfrac{M}{N} = \dfrac{300 \times 10^3}{600}$

$= \ 500$ mm

Equivalent load system

$D/6 = (2700/6) = 450$ mm   eccentricity $e > D/6$

Case 3 applies (see 5.19.8.3)

The length of the pressure diagram can be found from the equivalent load diagram as follows:

$y/3 = \ [(0.5 \times 2700) - 500] \ = \ 850$ mm $\therefore \ y = 2550$ mm

In the case of compression over part of the base the magnitude of the maximum earth pressure can be determined using simple elastic analysis.

$$p_{max} = \frac{2N}{3B\left(\dfrac{D}{2} - e\right)} = \frac{2 \times 600}{3 \times 1.75\left(\dfrac{2.7}{2} - 0.5\right)} = \ 269 \ \text{kN/m}^2$$

$< \ 300 \ \text{kN/m}^2$

Ground bearing pressure satisfactory

Assume an average partial safety factor equal to 1.5 to determine an *estimate* of the ultimate design pressure under the base. (**Note:** The values of e and y are slightly different from those determined when using the factored values for axial load and moment.)

References	Calculations	Output

Contract : Foundations Job Ref. No. : Example 5.32
Part of Structure : Pad Foundation with Axial Load and Moment
Calc. Sheet No. : 3 of 7

Calcs. by : W.McK.
Checked by :
Date :

$p_{max} \approx (1.5 \times 269) = 403.5$ kN/m^2 (net **ultimate** ground bearing pressure)

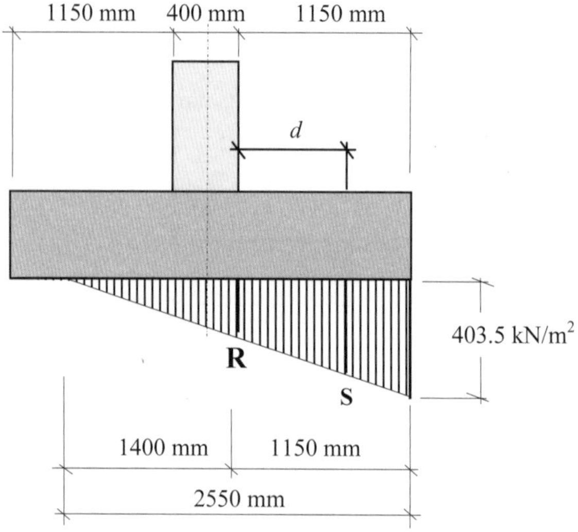

Base thickness
Assume a base thickness of equal to $(0.5 \times$ cantilever length)
$h = (0.5 \times 1150) = 575$ mm
Assume a 600 mm thick base
(**Note:** This depth must be sufficient to accommodate a 'starter bar' bond length.)

Clause 3.3

Cover:
Assume 32 mm diameter bars.

Clause 3.3.1.2
Clause 3.3.1.3
Table 3.3

bar size	=	32 mm
nominal maximum size of aggregate	=	20 mm
exposure condition is moderate cover	$\geq$	30 mm

Assume minimum cover to main steel = 32 mm

Effective depth:
$d = (600 - 32 - 32) = 536$ mm
(**Note:** This is the mean effective depth since the main reinforcement runs in both directions.)

The plane for direct shear occurs at a distance of 536 mm from the column face, i.e. $(1400 + 536) = 1936$ mm from the end of the pressure diagram.

Contract : Foundations Job Ref. No. : Example 5.32	Calcs. by : W.McK.
Part of Structure :Pad Foundation with Axial Load and Moment	Checked by :
Calc. Sheet No. : 4 of 7	Date :

References	Calculations	Output
	Bearing pressure at R $= \dfrac{403.5 \times 1400}{2550} = 221.5 \text{ kN/m}^2$ Bearing pressure at S $= \dfrac{403.5 \times 1936}{2550} = 306.3 \text{ kN/m}^2$ **Clause 3.11.3.1** **Clause 3.4.4.4** Bending: Check the bending at the column face Bending moment at R $=$ $[(221.5 \times 1.15 \times 1.75) \times (0.575) +$ $\qquad (403.5 - 221.5) \times (0.5 \times 1.15 \times 1.75) \times (0.767)]$ $\qquad = (195.64 + 152.2) \approx 397 \text{ kNm}$ $K = \dfrac{M}{bd^2 f_{cu}} = \dfrac{397 \times 10^6}{1750 \times 536^2 \times 40} = 0.02 < K' (= 0.156)$ **Section is singly reinforced** $Z = d\left\{0.5 + \sqrt{\left(0.25 - \dfrac{K}{0.9}\right)}\right\} = d\left\{0.5 + \sqrt{\left(0.25 - \dfrac{0.02}{0.9}\right)}\right\}$ $\qquad = 0.98d$ $\qquad$ The lever arm is limited to $0.95d$ $A_s = \dfrac{M}{0.95 f_y Z} = \dfrac{397 \times 10^6}{0.95 \times 460 \times 0.95 \times 536} = 1784 \text{ mm}^2$ $\therefore A_s = (1784/1.75) = 1019 \text{ mm}^2/\text{m width}$ Minimum % reinforcement required $= 0.13$ $\quad$ % provided $= \dfrac{100 \times 1019}{1000 \times 600} = 0.17$ **Minimum % of steel satisfied**	

(References column, continued: Clause 3.4.4.4; Table 3.25)

(Output column: **Bottom Reinforcement** / **Select:** / **20 mm diameter bars @ 300 mm centres providing 1050 mm²/m in both directions**)

Contract : Foundations Job Ref. No. : Example 5.32	Calcs. by : W.McK.
Part of Structure :Pad Foundation with Axial Load and Moment	Checked by :
Calc. Sheet No. : 5 of 7	Date :

References	Calculations	Output
Clause 3.12.11.2.7	**Maximum spacing:**	
	Spacing $\leq (3 \times d) = (3 \times 536) = 1608$ mm	
	≤ 750 mm	**Spacing is adequate**
	$100 A_s /bd = (100 \times 1050)/(1000 \times 536) = 0.2\% < 3\%$	
Clause 3.11.3.2	$[(3c/4) + (9d/4)] = [(0.75 \times 400) + (2.25 \times 536)] = 1506$ mm	
	$l_c =$ distance to the edge of the pad $= 1150$ mm < 1506 mm	**Reinforcement should be distributed uniformly over l_c**
	Shear force at section R:	
	$$= \left(\frac{221.5 + 403.5}{2} \right) \times 1.15 \times 1.75$$	
	$$= 628.9 \text{ kN}$$	
	Shear force at section S:	
	$$= \left(\frac{306.3 + 403.5}{2} \right) \times (1.15 - 0.536) \times 1.75$$	
	$$= 381.3 \text{ kN}$$	
Clause 3.11.3.3 Table 3.8	Shear resistance (Clause 3.5.6 and Clause 3.7.7) Critical shear stress	
	$\dfrac{100 A_s}{b_v d} = 0.2 \quad \therefore v_c = 0.37 \text{ N/mm}^2$	
	(modify to allow for C40 concrete)	
	$v_c = (0.37 \times 1.17) = 0.43 \text{ N/mm}^2$	
Clause 3.7.7.4	Direct shear	
	Check the direct shear at a distance of $1.0d$ from the column face.	
	This location corresponds to a $45°$ dispersal line from the column face. If the shear stress v is less than v_c then no shear reinforcement is required. At locations closer than this to the column face the load is assumed to distribute through the base in compression.	
	At the critical section for direct shear (i.e. $1.0d$ from the column face):	
	536 mm	
	381.3 kN	

Contract : Foundations Job Ref. No. : Example 5.32	Calcs. by : W.McK.
Part of Structure : Pad Foundation with Axial load and Moment	Checked by :
Calc. Sheet No. : 6 of 7	Date :

References	Calculations	Output
	Shear stress $= v = \dfrac{V}{bd} = \dfrac{381.3 \times 10^3}{1750 \times 536} = 0.41$ N/mm^2	

$$v < v_c$$

No shear reinforcement required

Punching shear
The possibility of punching shear should be checked at the column face for the maximum shear stress as indicated in Clause 3.7.7.2 and on a perimeter located at 1.5d from the column face. If the calculated shear stress does not exceed v_c on this perimeter then no further checks are required.

Clause 3.7.7.2 — Check the maximum shear at the column face:

Design vertical load on column $= (1.4 \times 250) + (1.6 \times 350)$
$= 910$ kN

Clause 3.7.7.2 — Shear stress $v = \dfrac{V}{u_o d} = \dfrac{V}{column\ perimeter \times d}$

$$v = \dfrac{910 \times 10^3}{[2(400+300) \times 536]} = 1.21 \text{ N/mm}^2$$

Maximum shear $v_{max} \leq 0.8\sqrt{f_{cu}} = (0.8 \times \sqrt{40})$
$= 5.05$ N/mm^2
$v_{max} \leq 5.0$ N/mm^2

$$v < v_{max}$$

Maximum shear at column face is acceptable

Clause 3.7.7.6 — Check the shear stress on a perimeter 1.5d from the column face. In this case, punching shear is more complex than for a concentrically axially loaded base. A procedure often adopted is to consider the total load due to the pressure under a reduced area (shown shaded) acting along the plane 'n–n' indicated.

Contract : Foundations Job Ref. No. : Example 5.32	Calcs. by : W.McK.
Part of Structure :Pad Foundation with Axial Load and Moment	Checked by :
Calc. Sheet No. : 7 of 7	Date :

References	Calculations	Output

$$1.5d = (1.5 \times 536) = 804 \text{ mm}$$
$$x = (1750 - 300)/2 = 725 \text{ mm}$$

In this problem the 1.5d line lies further from the column face than the intersection between the 45° line and the edge of the base. This is clearly less critical than the direct shear case considered previously in which '*n-n*' is the full width equal to B.

Punching shear is satisfactory

Anchorage:
The main bars are anchored a standard length beyond the face of the column. In most cases the size of the base will permit a full anchorage to be obtained. In situations where this is not the case then a standard 90° bend can be provided.

Transverse bending:
The critical section for transverse bending is at the column face. This can be checked in the same manner as longitudinal bending and reinforcement designed accordingly.

5.16.14 Example 5.33: Inverted T-Beam Combined Foundation

An inverted T-beam combined foundation is required to support two square columns transferring axial loads as shown in Figure 5.125. Using the data provided, design suitable reinforcement for the base.

Design Data:

Characteristic dead load on column A	450 kN
Characteristic imposed load on column A	450 kN
Characteristic dead load on column B	750 kN
Characteristic imposed load on column B	750 kN
Characteristic concrete strength	f_{cu} = 40 N/mm^2
Characteristic of reinforcement	f_y = 460 N/mm^2
Net permissible ground bearing pressure	p_g = 175 kN/m^2
Nominal cover to centre of main reinforcement	40 mm
Column A dimensions	350 mm × 350 mm
Column B dimensions	350 mm × 350 mm

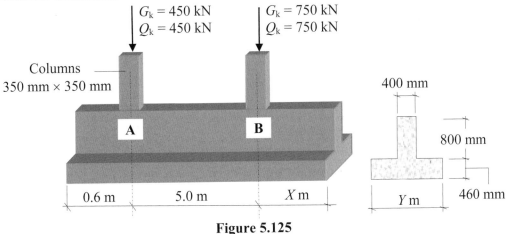

Figure 5.125

5.16.15 Solution to Example 5.33

Contract : Foundations Job Ref. No. : Example 5.33 Part of Structure : Inverted T-Beam, Combined Foundation Calc. Sheet No. : 1 of 7	Calcs. by : W.McK. Checked by : Date :

References	Calculations	Output
	The foundation should be designed as a T-beam between the columns where the rib is in tension and as a rectangular beam at the column positions where the bottom slab is in tension. Using the service loads it is first necessary to determine a suitable base area and dimensions X and Y such that the centre-line of the base coincides with the centre-of-gravity of the column loads. This ensures a uniform earth pressure under the base.	

References	Calculations	Output
	Base Area:	

Base Area:
Total service load $= (450 + 450 + 750 + 750) = 2400$ kN

Minimum area required $= \dfrac{service\ load}{permissible\ bearing\ pressure}$

$$= \dfrac{2400}{175} = 13.71 \text{ m}^2$$

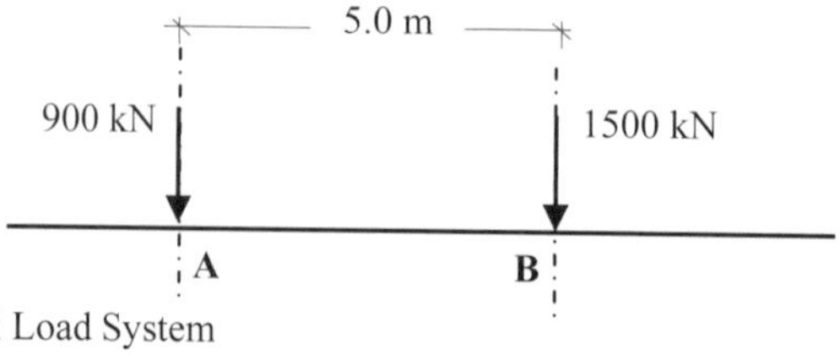

Actual Load System

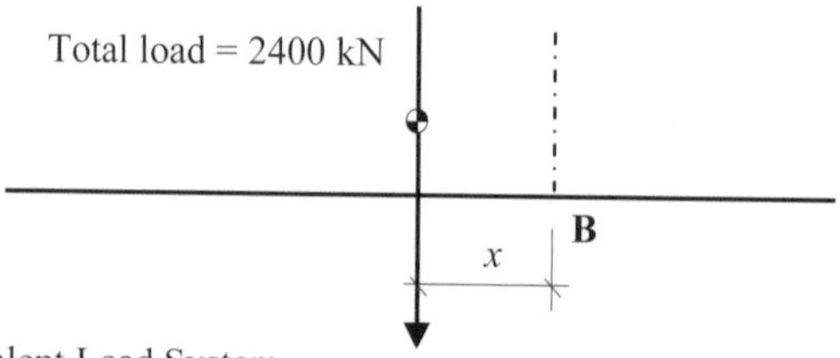

Equivalent Load System

Equate the moments of the force systems about the centre-line of column B:

$M_{\text{centre-line column B}} = (900 \times 5.0) = (2400 \times x) \quad \therefore x = 1.875$ m

The centre-line of the base should coincide with this position.

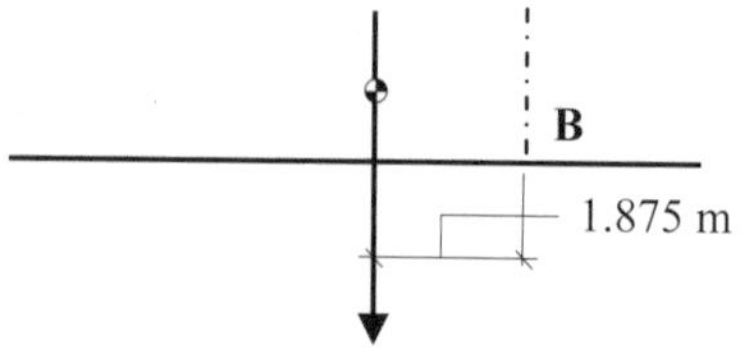

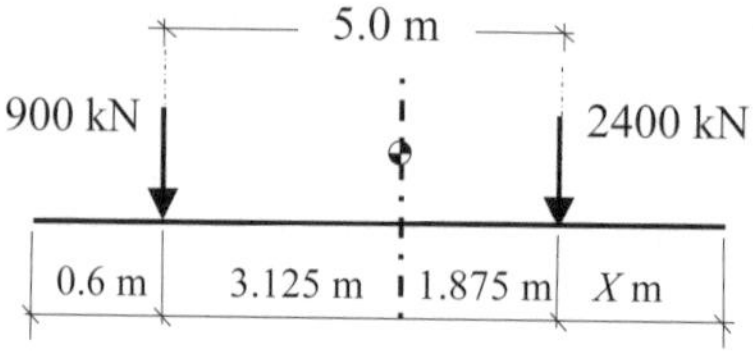

References	Calculations	Output
	Contract : Foundations **Job Ref. No. :** Example 5.33 **Calcs. by : W.McK.** **Part of Structure :** Inverted T-Beam, Combined Foundation **Checked by :** **Calc. Sheet No. : 3 of 7** **Date :**	

References	Calculations	Output
	Length of the base $= [2 \times (3.125 + 0.6)] = 7.45$ m $\therefore X = (7.45 - 5.6) = 1.85$ m	

Length of the base $= [2 \times (3.125 + 0.6)] = 7.45$ m
$$\therefore X = (7.45 - 5.6) = 1.85 \text{ m}$$

Width of the base $Y = (13.71 / 7.45) = 1.84$ m

Ultimate bearing pressure:

Ultimate load on column A $= (1.4 \times 450) + (1.6 \times 450)$
$$= 1350 \text{ kN}$$

Ultimate load on column B $= (1.4 \times 750) + (1.6 \times 750)$
$$= 2250 \text{ kN}$$

Ultimate design pressure under the base $= \dfrac{(1350 + 2250)}{(7.45 \times 1.84)}$
$$= 262.6 \text{ kN/m}^2$$

The combined base can be regarded as a beam 1.84 m wide.
Longitudinal load/m $= (262.6 \times 1.84) = 483.2$ kN/m

Shear force at Q $= (483.2 \times 0.6)$ $= +289.9$ kN
and $= (289.9 - 1350)$ $= -1060$ kN

Shear force at R $= -1060 + (483.2 \times 5.0) = +1356$ kN
and $= (1356 - 2250)$ $= -894.1$ kN

Position of zero shear $= \dfrac{1060}{483.2} = 2.19$ m

Shear force diagram

Bending moment at Q $=$ shaded area
$$= \left(\frac{0.6}{2} \times 289.9\right) = 86.8 \text{ kNm}$$

Contract : Foundations Job Ref. No. : Example 5.33 Part of Structure :Inverted T-Beam, Combined Foundation Calc. Sheet No. : 4 of 7	Calcs. by : W.McK. Checked by : Date :

References	Calculations	Output

Bending moment at R = shaded area

$$= \left(\frac{1.85}{2} \times 894 \right) = 827 \text{ kNm}$$

Maximum bending moment at point of zero shear:

$$= \text{resultant shaded area} = 87 - \left(\frac{2.19}{2} \times 1060 \right) = -1074 \text{ kNm}$$

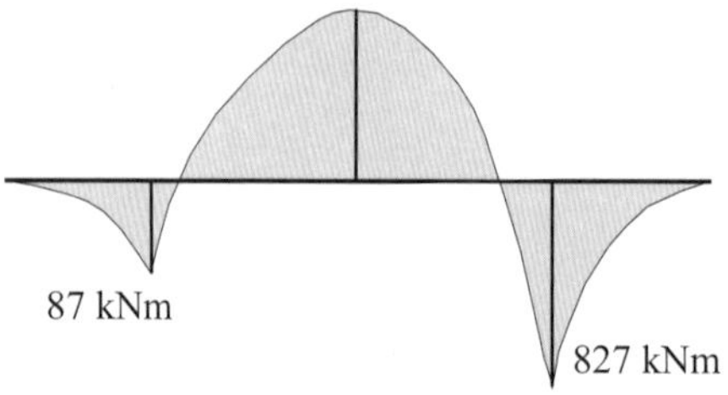

1074 kNm

87 kNm

827 kNm

Design the base as a T-beam between A and B and as a rectangular beam at the column locations.

Assume 32 mm diameter bars for the main steel.

Effective depth:
$d = (1260 - 40 - 16) = 1204 \text{ mm}$
(**Note:** This is the mean effective depth)

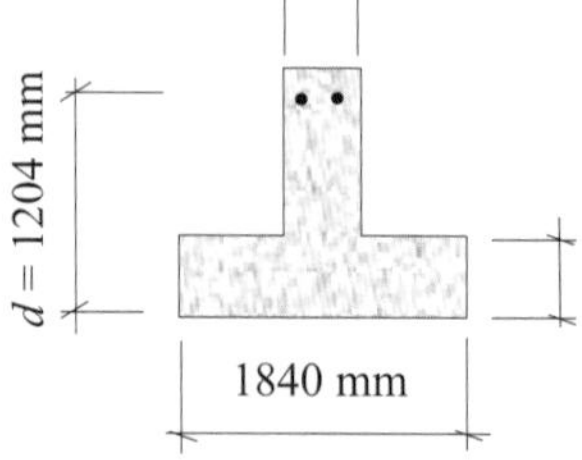

400 mm

$d = 1204$ mm

460 mm

1840 mm

Bending between the columns A and B   (T-beam):

Design bending moment between the columns = 1074 kNm

Clause 3.4.4.4
$$K = \frac{M}{bd^2 f_{cu}} = \frac{1074 \times 10^6}{1840 \times 1204^2 \times 40} = 0.01 \quad < \quad K'(=0.156)$$

Section is singly reinforced

Contract : Foundations Job Ref. No. : Example 5.33 Part of Structure :Inverted T-Beam, Combined Foundation Calc. Sheet No. : 5 of 7	Calcs. by : W.McK. Checked by : Date :

References	Calculations	Output
	$x = (d - z)/0.45 = (0.05 \times 1204)/0.45 = 133.8$ mm < 460 mm $\qquad$ The neutral axis lies in the flange $Z = d\left\{0.5 + \sqrt{\left(0.25 - \dfrac{K}{0.9}\right)}\right\} = d\left\{0.5 + \sqrt{\left(0.25 - \dfrac{0.01}{0.9}\right)}\right\}$ $\qquad\qquad = 0.98d$ $\qquad$ The lever arm is limited to $0.95d$ $A_s = \dfrac{M}{0.95 f_y Z} = \dfrac{1074 \times 10^6}{0.95 \times 460 \times 0.95 \times 1204} = 2149$ mm^2 Design bending moment at column B $\quad$ (Rectangular section): $= \left[483.2 \times \dfrac{(1.85 - 0.175)^2}{2}\right] = 678$ kNm	**Top Reinforcement** Select: **3/32 mm diameter bars providing 2410 mm^2**
	(diagram: T-section, 400 mm top width, $d = 1204$ mm, 460 mm, 1840 mm)	
Clause 3.4.4.4	$K = \dfrac{M}{bd^2 f_{cu}} = \dfrac{678 \times 10^6}{400 \times 1204^2 \times 40} = 0.03 \; < \; K'\,(= 0.156)$ $\qquad\qquad$ **Section is singly reinforced** $Z = d\left\{0.5 + \sqrt{\left(0.25 - \dfrac{0.03}{0.9}\right)}\right\} = 0.96d$ $\qquad$ The lever arm is limited to $0.95d$ $A_s = \dfrac{M}{0.95 f_y Z} = \dfrac{678 \times 10^6}{0.95 \times 460 \times 0.95 \times 1204} = 1356$ mm^2	**Bottom Reinforcement** Select: **3/25 mm diameter bars providing 1470 mm^2**
Table 3.25	Minimum area of steel $= \dfrac{100 A_s}{b_w h} = 0.26\%$ $A_s \geq \dfrac{0.26 \times 400 \times 1260}{100} = 1310$ mm^2 < 1356 mm^2 *The bottom reinforcement required at column A should be determined by the reader.*	

Contract : **Foundations Job Ref. No. : Example 5.33** Part of Structure :**Inverted T-Beam, Combined Foundation** Calc. Sheet No. : **6 of 7**	Calcs. by : **W.McK.** Checked by : Date :

References	Calculations	Output
	The reinforcement required to resist transverse bending can be determined in a similar manner. Consider a 1m width strip: 720 mm 1.0 m 483.2 kN/m^2 Assume 25 mm diameter bars: $d = (460 - 40 - 25 - 13) = 382$ mm Transverse bending moment $= \left(483.2 \times \dfrac{0.72^2}{2}\right)$ $= 125.2$ kNm/m width	
Clause 3.4.4.4	$K = \dfrac{M}{bd^2 f_{cu}} = \dfrac{125.2 \times 10^6}{1000 \times 382^2 \times 40} = 0.02 \; < \; K'\,(=0.156)$ **Section is singly reinforced** $Z = d\left\{0.5 + \sqrt{\left(0.25 - \dfrac{0.02}{0.9}\right)}\right\} = 0.98d$ The lever arm is limited to $0.95d$ $A_s = \dfrac{M}{0.95 f_y Z} = \dfrac{125.2 \times 10^6}{0.95 \times 460 \times 0.95 \times 382} = 789$ mm^2/m	**Bottom Reinforcement**
Table 3.25	Minimum area of steel $= 0.15\%$ $A_s \geq \dfrac{0.15 \times 1000 \times 460}{100} = 690$ mm^2/m $\quad < \; 789$ mm^2/m	**Select:** **20 mm diameter bars** **@ 400 mm centres** **providing 786 mm^2/m**
Clause 3.4.5 Table 3.8	Shear resistance of the T-beam: Consider shear at the face of Column B Critical shear stress $\dfrac{100 A_s}{b_v d} = \dfrac{100 \times 2410}{400 \times 1204} = 0.5 \; \therefore v_c = \; 0.5$ N/mm^2 Modify to allow for using C40 concrete $v_c = (0.5 \times 1.17) = 0.59$ N/mm^2	
Table 3.7	$(v_c + 0.4) = 0.99$ N/mm^2	

Contract : **Foundations Job Ref. No. : Example 5.33** Part of Structure :**Inverted T-Beam, Combined Foundation** Calc. Sheet No. : **7 of 7**	Calcs. by : W.McK. Checked by : Date :

References	Calculations	Output
	Shear force at the face of the column:	

Shear force at the face of the column:

$V = (1356 - (483.2 \times 0.175)) = 1271.4 \text{ kN}$

$v = \dfrac{V}{b_v d} = \dfrac{1271.4 \times 10^3}{400 \times 1204} = 2.64 \text{ N/mm}^2$

Since $(v_c + 0.4) < v < 0.8\sqrt{f_{cu}}$

$< 5.0 \text{ N/mm}^2$

Design links are required

$A_{sv} \geq \dfrac{b_{sv} s_v (v - v_c)}{0.95 f_{yv}}$

Assume two-legged 10 mm diameter links $\therefore A_{sv} = 157 \text{ mm}^2$

$S_v \leq \left[\dfrac{157 \times 0.95 \times 250}{400 \times (2.64 - 0.59)} \right] = 45.5 \text{ mm}$

A similar calculation should be carried out by the reader to determine the links required over the full length of the T-beam.

Output: Adopt double 10 mm diameter two-legged links @ 100 mm centres for the 1st metre from column B

Clause 3.4.5 — Shear resistance of the flange: Consider shear at the face of rib

720 mm

483.2 kN/m^2

Table 3.8 — Critical shear stress

$\dfrac{100 A_s}{b_v d} = \dfrac{100 \times 786}{1000 \times 382} = 0.2 ; \quad d = 382 \text{ mm}$

$v_c = 0.37 \text{ N/mm}^2 ;$ modify to allow for using C40 concrete

$v_c = (0.37 \times 1.17) = 0.43 \text{ N/mm}^2$

$(v_c + 0.4) = 0.83 \text{ N/mm}^2$

Shear force $V = (483.2 \times 1.0 \times 0.72) = 348 \text{ kN}$

$v = \dfrac{V}{b_v d} = \dfrac{348 \times 10^3}{1000 \times 382} = 0.91 \text{ N/mm}^2 < 0.8\sqrt{f_{cu}}$

$< 5.0 \text{ N/mm}^2$

Table 3.16 — Since the shear stress v is greater than v_c shear reinforcement is required. Shear links are not normally provided in slabs. Alternatives to providing links are to increase the % of reinforcement A_s and hence increase the value of v_c sufficiently, or to increase the thickness of the slab, which will reduce the value of v.

6. Design of Structural Steelwork Elements (BS 5950)

Objective: *to illustrate the process of design for structural steelwork elements.*

6.1 Introduction

The origins of modern building materials such as structural steelwork can be traced back to the birth of the Industrial Revolution in the latter part of the 18th century. The construction of the Iron Bridge (manufactured from cast iron) across the River Severn near Coalbrookdale in 1779 marked the end of an era in which timber and masonry were the dominant materials of building.

Initially cast iron was used to replace timber columns and beams in buildings but masonry load-bearing walls were still used as an external envelope well into the 1800's. The development of extensive railway networks and their associated infrastructure throughout the U.K. in the 19th century resulted in the widespread use of wrought iron, a purer material which was much more reliable in both tension and compression than cast iron. The Bessemer process for producing steel which was strong, ductile and economically viable virtually eclipsed the use of cast and wrought iron by the end of the 19th century.

During this period the improved scientific principles and model testing on which structural design was based resulted in the extensive use of pre-fabricated units in structures such as the Crystal Palace in London, designed by Joseph Paxton, and the 1000-bed hospital, designed by Isambard Kingdom Brunel and shipped out to Crimea.

The major improvements in material production, analysis and design techniques were reflected in major steel structures throughout the world such as the Forth Rail Bridge–Scotland (1890), the Guaranty Building, in Buffalo New York (1894) and the Eiffel Tower–Paris (1899).

Steel had become a very important construction material by the end of the 19th century. Hot-rolled steel sections were available in quantity at affordable prices and methods of connecting elements together by rivets and later bolts were well established.

During the 1st World War the method of joining steel members together by metal arc welding was established. This is now used extensively in the construction of modern steel structures such as multi-storey frames, bridges, and oil production platforms. During the welding process an electrical arc is struck between a metal rod (the electrode) and the two steel members to be welded. The metal is fused at both ends of the arc and the fused electrode is deposited in the joint in a series of layers until it is filled. The resulting joint is smaller and more efficient than bolts but does require the use of highly trained personnel and sophisticated examination techniques to ensure the integrity of the connection.

The design of modern steelwork structures is undertaken in the U.K. to comply with the requirements of BS 5950-1:2000, *Structural use of steelwork in building–Part 1: Code of practice for design-rolled and welded sections*. This text relates to the use of Part 1 of this code.

6.2 Material Properties

6.2.1 Stress-Strain Characteristics

The stress-strain characteristics for a typical structural steel as shown in Figure 5.11 of Chapter 5 indicate a ductile material which exhibits linearly elastic behaviour followed by significant plasticity before failure occurs. The minimum yield stress and the maximum tensile stress indicated in Figure 5.11 are represented in Clause 3.1.1 of BS 5950-1:2000 by the symbols Y_s and U_s respectively. The assumed strength (p_y) for design purposes is the smaller of $1.0Y_s$ and $U_s/1.2$, and is dependent on the thickness of the material being used. Values of p_y for the more commonly used grades and thicknesses of steel are given in Table 9 of the code, as shown in Figure 6.1.

Steel Grade	Thickness* less than or equal to (mm)	Design strength (N/mm^2)
S 275	16	275
	40	265
	63	255
	80	245
	100	235
	150	225
S 355	16	355
	40	345
	63	335
	80	325
	100	315
	150	295
S 460	16	460
	40	440
	63	430
	80	410
	100	400
* For rolled sections, use the specified thickness of the thickest element of the cross-section.		

Figure 6.1 **(BS 5950-1:2000 – Table 9: Design Strength p_y)**

Steel grades are specified in accordance with *BS 5950-2:2000* in which reference is made to the following European Standards:

- BS EN 10025   *Hot rolled products of non-alloy structural steels – Technical delivery conditions.*
- BS EN 10113-2   *Hot rolled products in weldable fine grain structural steels – Part 2: Delivery conditions for normalized/normalized rolled steels.*
- BS EN 10113-3   *Hot rolled products in weldable fine grain structural steels – Part 3: Delivery conditions for thermomechanical rolled steels.*
- BS EN 10137-2   *Plates and wide flats made of high yield strength structural steels in the quenched and tempered or precipitation*

> *hardened conditions – Part 2: Delivery conditions for quenched and tempered steels.*

♦ BS EN 10155   *Structural steels with improved atmospheric corrosion resistance – Technical delivery conditions.* (i.e. weathering steels)

♦ BS EN 10210-1 *Hot finished structural hollow sections of non-alloy and fine grain structural steels – Part 1: Technical delivery requirements.*

In each code a designation system is used to describe the steel:

BS EN X – Y Z

where:

X identifies the standard,
Y identifies the type of steel i.e. **S** for structural steel, **E** for engineering steel
Z identifies the minimum yield strength e.g. 235 N/mm^2, 275 N/mm^2 etc.

The following example represents the designation for non-alloy steel with a minimum yield strength of 255 N/mm^2 used to manufacture hot-rolled sections and plates:

BS EN 10025 – S 255

6.2.2 Ductility (Clause 2.4.4)

It is important to ensure that steel has sufficient ductility, particularly at low temperatures, to avoid brittle fracture. The ductility is measured in terms of the *notch toughness* and the *Charpy value* from the Charpy V-notch test.

The Charpy test is a notched-bar impact test in which a notched specimen, fixed at both ends, is struck behind the notch by a striker carried on a pendulum. The energy absorbed in the fracture is measured by the height to which the pendulum rises.

In addition to specifying the grade of the steel, e.g. S 275, it is necessary to identify the appropriate quality by specifying a sub-grade. Sub-grades are defined in the appropriate British Standards for steels and used in Tables 4, 5, 6 and 7 of *BS 5950-1:2000*; e.g.

♦ JR   represents an impact resistance value of 27 Joules at room temperature
♦ J0   represents an impact resistance value of 27 Joules at   0°C
♦ J2   represents an impact resistance value of 27 Joules at –20°C
♦ K2   represents an impact resistance value of 27 Joules at –30°C

The full designation for a steel includes the sub-grade, e.g. **BS EN 10025 – S 255J2.**

When determining the required sub-grade for a particular situation consideration must be given to a number of factors:

♦ the minimum service temperature,
♦ the thickness,
♦ the steel grade,
♦ the type of detail,

♦ the stress level,
♦ the strain level or strain rate.

as indicated in Clause 2.4.4 of the code. The appropriate steel sub-grade is selected such that the thickness t of each component satisfies the following criteria:

(i) $\quad t \leq Kt_1$

where:
K is a factor that depends on the type of detail, the general stress level, the stress concentration effects and the strain conditions and is given in Table 3 of the code
t_1 is the limiting thickness at the appropriate minimum service temperature T_{min} for a given steel grade and quality, when the factor $K = 1$. Values of t_1 are given in Tables 4 and 5 of the code for plates, flats and rolled sections and for structural hollow sections respectively.

and

(ii) $\quad t \leq t_2$

where:
t_2 is the thickness at which the full Charpy impact value applies to the selected steel quality for that product type and steel grade, according to the relevant product standard. Values of t_2 are given in Table 6 of the code.

For rolled sections t and t_1 should be related to the same element of the cross-section as the factor K, but t_2 should be related to the thickest element of the cross-section.

The value of t_1 can also be determined from equations given in the code. The application of Clause 2.4.4, is illustrated in Examples 6.1 and 6.2.

6.2.2.1 Example 6.1
Consider an internal steel structure fabricated from rolled steel sections in which both welding and bolting are used. Assuming the design data given, check the suitability of the proposed steel designation.

Design Data:
Maximum steel stress $\qquad > 150\ \text{N/mm}^2$
Maximum thickness of element $\qquad$ (i) $t = 14\ \text{mm}$ and (ii) $t = 35\ \text{mm}$
Proposed steel grade $\qquad$ BS EN 10025 S 275
Assume all connections are bolted using punched holes and welded end plates/cleats.

Clause 2.4.4 $\quad t \leq Kt_1$
Table 3 $\qquad K$ requires the type of steel and stress level and the value of Y_{nom} which is the nominal yield strength as in the steel grade designation.

Case (i) $t = 14$ mm ; $Y_{nom} = 275$ N/mm^2
Table 3 $0.3Y_{nom} = (0.3 \times 275) = 82.5$ N/mm^2
             Stress level > 150 N/mm$^2 > 0.3Y_{nom}$ $\left.\rule{0cm}{1.3cm}\right\}$ $\therefore K = 1.0$
             Welded generally and punched holes

Table 4   Normal temperatures / Internal $(-5^\circ C)$
             BS EN 10025 S 275 gives $t_1 = 25$ mm ≥ 14 mm

Table 6   Maximum thickness based on Charpy value $= 100$ mm for sections.
                    Proposed steel designation BS EN 10025 S 275 is adequate

Case (ii) $t = 35$ mm ; $Y_{nom} = 275$ N/mm^2
Table 3 $0.3Y_{nom} = (0.3 \times 275) = 82.5$ N/mm^2
             Stress level > 150 N/mm$^2 > 0.3Y_{nom}$ $\left.\rule{0cm}{1.3cm}\right\}$ $\therefore K = 1.0$
             Welded generally and punched holes

Table 4   Normal temperatures / Internal $(-5^\circ C)$
             BS EN 10025 S 275J0 gives $t_1 = 65$ mm ≥ 35 mm

Table 6   Maximum thickness based on Charpy value $= 100$ mm for sections.
                    Required steel designation is BS EN 10025 S 275J0

6.2.2.2   Example 6.2

Using the design data given, check the suitability of the proposed steel designation for a structure which is exposed to a temperature of $- 30^\circ C$.

Design Data:
Maximum steel stress > 240 N/mm^2
Maximum thickness of element $t = 20$ mm
Proposed steel grade   BS EN 10025 S 355J2
Assume all connections are welded to unstiffened flanges.

Clause 2.4.4 $t \leq Kt_1$
Table 3 $t = 20$ mm ; $Y_{nom} = 355$ N/mm^2;
             $0.3Y_{nom} = (0.3 \times 355) = 106.5$ N/mm^2
             Stress level > 150 N/mm$^2 > 0.3Y_{nom}$ $\left.\rule{0cm}{1.6cm}\right\}$ $\therefore K = 0.5$
             Welded connections to unstiffened flanges
             $t \leq Kt_1$ $\therefore 20 \leq 0.5t_1$ $\therefore$ Required $t_1 \geq 40$ mm

Table 4   Lower temperatures $(-35^\circ C)$
             BS EN 10025 S 355J2 gives $t_1 = 38$ mm < 40 mm

             Change the steel designation:
             BS EN 10025 S 355K2 gives $t_1 = 46$ mm ≥ 40 mm
Table 6   Maximum thickness based on Charpy value $= 100$ mm for sections.
                    Required steel designation is BS EN 10025 S 355K2

A more precise value of t_1 for ($-30°C$) may be calculated from the equations given in the code as follows:

Table 7 $\quad T_{27J} = -20°C \quad$ (assuming BS EN 10025 S 355J2)

$\quad\quad\quad\quad\quad T_{min} =$ the minimum service temperature (in $°C$) expected to occur in the steel within the intended design life of the part

$\quad\quad\quad\quad\quad\quad\quad = -30°C$

If $T_{27J} \leq T_{min} + 20°C$ then $\quad t_1 \leq 50(1.2)^N \left[\dfrac{355}{Y_{nom}}\right]^{1.4}$

where $\quad N = \left[\dfrac{T_{min} - T_{27J}}{10}\right] = \left[\dfrac{-30 + 20}{10}\right] = -1.0$

$\quad\quad \therefore \quad t_1 \leq 50 \times (1.2)^{-1} \times \left[\dfrac{355}{355}\right]^{1.4} = 41.7 \text{ mm} \geq 40 \text{ mm}$

In this instance BS EN 10025 S 355J2 is satisfactory

6.2.3 Fatigue (Clause 2.4.3)

Metals which are subject to continuously varying or alternating loads can fracture at values of stress considerably less than the ultimate value found during static tests. Experimental evidence has indicated that fluctuating stresses, in some cases smaller than the elastic limit, will induce fracture if repeated a sufficient number of times. This type of failure is called *fatigue failure* and is dependent on the *number of cycles* and the *range of stress* to which an element is subjected.

In general, it is not necessary to consider fatigue in design unless a structural element is subjected to numerous significant fluctuations, e.g. in members supporting heavy vibrating machinery and certain classes of crane supporting structures as indicated in Clause 2.4.3 of the code. When it is considered necessary to evaluate resistance to fatigue, reference should be made to BS 7608 *Code of practice for fatigue design and assessment of steel structures*.

6.2.4 Elastic Properties (Clause 3.1.3)

The most commonly required material elastic properties for steel are given in Clause 3.1.3 as:

- Modulus of Elasticity $\quad\quad\quad\quad\quad\quad\quad E = 205 \times 10^3 \text{ N/mm}^2$
- Poisson's Ratio $\quad\quad\quad\quad\quad\quad\quad\quad\quad v = 0.3$
- Shear Modulus $\quad\quad\quad\quad\quad\quad\quad\quad\quad G = E/[2(1+v)] = (78.85 \times 10^3 \text{ N/mm}^2)$
- Coefficient of Thermal Expansion $\quad \alpha = 12 \times 10^{-6}/°C$

6.2.5 Section Classification (Clause 3.5)

In Clause 3.5 of BS 5950-1:2000, the compression elements of structural members are classified into four categories depending upon their resistance to local buckling effects which may influence their load carrying capacity. The compression may be due to direct axial forces, bending moments, or a combination of both. There are two distinct types of element in a cross-section identified in the code:

1. *Outstand elements* – elements which are attached to an adjacent element at one edge only, the other edge being free, e.g. the flange of an **I**-section.

2. *Internal elements* – elements which are attached to other elements on both longitudinal edges, including:
    –   webs comprising the internal elements perpendicular to the axis of bending
    –   flanges comprising the internal elements parallel to the axis of bending
    e.g. the webs and flanges of a rectangular hollow section.

The classifications specified in the code are:

*   Class 1   **Plastic** *Sections*
*   Class 2   **Compact** *Sections*
*   Class 3   **Semi-compact** *Sections*
*   Class 4   **Slender** *Sections*

and are determined by consideration of the limiting values given in Tables 11 and 12. The classifications are based on a number of criteria.

6.2.5.1   Aspect Ratio

The aspect ratio for various types of element can be determined using the variables indicated in Figure 5 of the code for a wide range of cross-sections. A typical example is the **I**-section indicated in Figure 6.2.

Element	Aspect ratio
outstand of compression flange	b/T
web	d/t

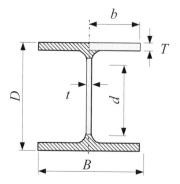

Figure 6.2

The limiting aspect ratios given must be modified to allow for the design strength p_y. This is done by multiplying each limiting ratio by ε which is defined in each Table as:

$$\varepsilon = \left(\frac{275}{p_y}\right)^{0.5}$$. In the case of the web of a hybrid section ε should be based on the

design strength p_{yf} of the flanges.

In addition to ε, some limiting values in Tables 11 and 12 also include parameters r_1 and r_2 which are stress ratios, defined in Clause 3.5.5 as:

$$r_1 = \frac{F_c}{dtp_{yw}} \qquad \text{but } -1 < r_1 \le 1$$

$$r_2 = \frac{F_c}{A_g p_{yw}}$$

For **I** and **H** sections with *equal* flanges

$$r_1 = \frac{F_c}{dtp_{yw}} + \frac{(B_t T_t - B_c T_c)p_{yf}}{dtp_{yw}} \qquad \text{but } -1 < r_1 \le 1$$

$$r_2 = \frac{f_1 + f_2}{2p_{yw}}$$

For **I** and **H** sections with *unequal* flanges

$$r_1 = \frac{F_c}{2dtp_{yw}} \qquad \text{but } -1 < r_1 \le 1$$

$$r_2 = \frac{F_c}{A_g p_{yw}}$$

For **RHS** or **welded box** sections with *equal* flanges

where:
A_g is the gross cross-sectional area,
B_c is the width of the compression flange,
B_t is the width of the tension flange,
d is the web depth,
F_c is the axial compression (*negative for tension*),
f_1 is the maximum compressive stress in the web (as indicated in Figure 7 of the code),
f_2 is the minimum compressive stress in the web (*negative for tension*, see Figure 7),
p_{yf} is the design strength of the flanges,
p_{yw} is the design strength of the web (but $p_{yw} \le p_{yf}$),
T_c is the thickness of the compression flange,
T_t is the thickness of the tension flange,
t is the web thickness.

6.2.5.2 Type of Section

Circular hollow sections[*]   – including welded tubes
Hot finished rectangular hollow sections – including square sections   } Table 12
Cold formed rectangular hollow sections

All other sections   Table 11

* Note: In the case of circular hollow sections, classification is defined seperately for axial compression and for bending as indicated in Clause 3.5.1 of the code.

The classifications given in Tables 11 and 12 indicate the moment/rotation characteristics of a section, as shown in Figure 6.3.

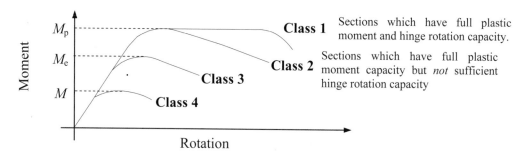

Figure 6.3

where:

M_p = plastic moment of resistance
M_e = limiting elastic moment of resistance
M = elastic moment of resistance

These characteristics determine whether or not a fully plastic moment can develop within a section and whether or not the section possesses sufficient rotational capacity to permit redistribution of the moments in a structure.

Consider a section subject to an increasing bending moment; the bending stress diagram changes from a linearly elastic condition with extreme fibre stresses less than the design strength (p_y), to one in which all of the fibres can be considered to have reached the design strength, as shown in Figure 6.4.

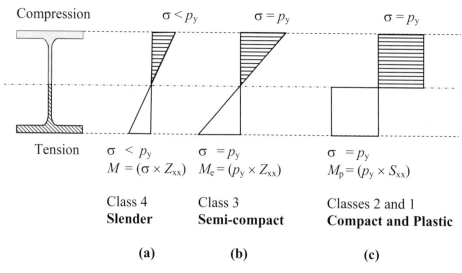

where:

Z_{xx} = elastic section modulus
S_{xx} = plastic section modulus
σ = elastic stress
p_y = design strength

Figure 6.4

Note: The **shape factor** of a section is defined as:

$$\nu = \frac{\text{plastic mod ulus}}{\text{elastic mod ulus}} = \frac{S_{xx}}{Z_{xx}} \quad \text{The value of } \nu \text{ for most I-sections} \approx 1.15.$$

6.2.5.3   Plastic Sections
The failure of a structure such that plastic collapse occurs is dependent on a sufficient number of plastic hinges developing within the cross-sections of the members (i.e. value of internal bending moment reaching M_p), to produce a mechanism. For full collapse this requires one more than the number of redundancies in the structure, as illustrated in the rigid-jointed rectangular portal frame in Figure 6.5.

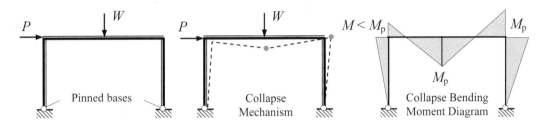

Number of redundancies = 1   Number of hinges = 2

Figure 6.5

The required number of hinges will only develop if there is sufficient rotational capacity in the cross-section to permit the necessary redistribution of the moments within the structure. When this occurs, the stress diagram at the location of the hinge is as shown in Figure 6.4(c), and the aspect ratios of the elements of the cross-section are low enough to prevent local buckling from occurring. Such cross-sections are defined as plastic sections and classified as **Class 1**. *Full plastic analysis and design can only be carried out using Class 1 sections.*

6.2.5.4   Compact Sections
When cross-sections can still develop the full plastic moment as in Figure 6.4(c), but are prevented by the possibility of local buckling from undergoing enough rotation to permit redistribution of the moments, the section is considered to be **compact** and is classified as **Class 2**. *Compact sections can be used without restricting their capacity, except at plastic hinge positions.*

6.2.5.5   Semi-compact Sections
Semi-compact sections may be prevented from reaching their full plastic moment capacity by local buckling of one or more of the elements of the cross-section. The aspect ratios may be such that only the extreme fibre stress can attain the design strength before local buckling occurs. *Such sections are classified as* **Class 3** *and their capacity is therefore based on the limiting elastic moment* as indicated in Figure 6.4(b).

There is provision in Clause 3.6.5 in the case of Class 3, semi-compact cross-sections for adopting an alternative method of determining a reduced capacity. In this case the design strength p_y is modified to produce a reduced design strength p_{ry}, defined as:

$$p_y = (\beta_3/\beta)^2 p_y$$

where β is the value of b/T, b/t, D/t or d/t that exceeds the limiting value (i.e. β_3) given in Table 11 or Table 12 for a Class 3, semi-compact section. The reduced design strength is subsequently used in strength calculations for members subject to bending, lateral-torsional buckling, axial compression and combined bending and compression.

6.2.5.6   Slender Sections

When the aspect ratio is relatively high, then local buckling may prevent any part of the cross-section from reaching the design strength. *Such sections are called* **slender** *sections and are classified as* **Class 4** *sections; their reduced capacity is based on effective cross-section properties* as specified in Clause 3.6 of BS 5950-1:2000.

The classification of cross-sections is illustrated in Example 6.3.

6.2.5.7   Example 6.3: Classification of Sections

Determine the section classification of the sections indicated in (1) to (3), considering the loading conditions given:

1. Section: 610 × 229 × 125 UB S 275
   a) pure bending
   b) pure compression
      i) axial load   = 3200 kN
      ii) axial load   = 4350 kN
   c) combined bending and compression
2. Section: Double angle strut composed of two 100 × 75 × 10 S 355 angles connected to gusset plates by the long legs back-to-back, and subjected to axial compression.
3. Section: Hot-rolled circular section 273.0 × 6.3 CHS   S 275, subject to compression due to bending.

Solution:
1. Section: 610 × 229 × 125 UB   S 275
Section properties:
A_g = 159.0 cm², $b/T = 5.84$, $d/t = 46.0$, $d = 547.6$ mm, $t = 11.9$ mm, $T = 19.6$ mm

(a) Pure bending:
The section is symmetrical about the major axis (i.e. the axis of bending) and consequently the neutral axis is at mid-height.

Tables 9/11 p_y = 265 N/mm² $\therefore \varepsilon = \left(\dfrac{275}{p_y}\right)^{0.5} = 1.02$

Consider the outstand element of the compression flange for rolled sections:

The limiting value of b/T for Class 1 sections $= 9\varepsilon = 9.18 \quad > 5.84$ (the actual value)

The flanges are plastic

Consider the web of an **I**-beam with the neutral axis at mid-depth $(p_y = 275 \text{ N/mm}^2)$:

The limiting value of d/t for Class 1 sections $= 80\varepsilon = 80.0 \quad > 46.0$ (the actual value)

The web is plastic

Clause 3.5.*1*

The cross-section is plastic

(b) Pure compression:

(i) Axial load $= 3200$ kN

Table 11 $\quad p_y = 265 \text{ N/mm}^2 \quad \therefore \varepsilon = \left(\dfrac{275}{p_y}\right)^{0.5} = 1.02$

The flanges are plastic as before

Consider the web of an **I**-beam subject to axial compression $(p_y = 275 \text{ N/mm}^2)$:

If $d/t \geq \dfrac{120\varepsilon}{(1+2r_2)}$ or $\geq 40\varepsilon$ then the section is considered to be slender otherwise it is

considered to be semi-compact, i.e. Class 3. The value of r_2 is defined in Clause 3.5.5(a).

Clause 3.5.5(a): $\quad r_2 = \dfrac{F_c}{A_g p_{wy}} = \dfrac{3200 \times 10^3}{159 \times 10^2 \times 275} = 0.73$

$\therefore$ limiting value of $d/t = \dfrac{120\varepsilon}{(1+2r_2)} = \dfrac{120 \times 1.0}{\left[1+(2 \times 0.73)\right]} = 48.78$

> 46.0 (the actual value)

The web is semi-compact

Clause 3.5.1

The cross-section is semi-compact

(ii) Axial load $= 4350$ kN

Table 11 $\quad p_y = 265 \text{ N/mm}^2 \quad \therefore \varepsilon = \left(\dfrac{275}{p_y}\right)^{0.5} = 1.02$

The flanges are plastic as before

Consider the web of an **I**-beam subject to axial compression $(p_y = 275 \text{ N/mm}^2)$:

Clause 3.5.5(a): $\quad r_2 = \dfrac{F_c}{A_g p_{wy}} = \dfrac{4350 \times 10^3}{159 \times 10^2 \times 275} = 0.99$

$\therefore$ limiting value of $d/t = \dfrac{120\varepsilon}{(1+2r_2)} = \dfrac{120 \times 1.0}{\left[1+(2 \times 0.99)\right]} = 40.26$

< 46.0 (the actual value)

The web is slender

Clause 3.5.1

The cross-section is slender

(c) Combined bending and compression:
     Axial load = 2500 kN

Table 11 $p_y = 265 \text{ N/mm}^2$ $\therefore \varepsilon = \left(\dfrac{275}{p_y}\right)^{0.5} = 1.02$

The flanges are plastic as before

Consider the web of an **I**-beam 'generally':

Clause 3.5.5(a): $r_1 = \dfrac{F_c}{dtp_{wy}} = \dfrac{2500\times10^3}{547.6\times11.9\times275} = 1.4$

Clause 3.5.5(a): $r_2 = \dfrac{F_c}{A_g P_{wy}} = \dfrac{2500\times10^3}{159\times10^2 \times 275} = 0.57$

Table 11:

Class 1 limiting value of $d/t = \dfrac{80\varepsilon}{(1+r_1)} = \dfrac{80\times1.0}{(1+1.4)} = 33.3$ ⎫
                          $\geq 40\varepsilon$ ⎬ $= 40\varepsilon$
                                                                                            ⎭

Class 2 limiting value of $d/t = \dfrac{100\varepsilon}{(1+1.5r_1)} = \dfrac{100\times1.0}{[1+(1.5\times1.4)]} = 32.3$ ⎫
                          $\geq 40\varepsilon$ ⎬ $= 40\varepsilon$
                                                                                                                   ⎭

Class 3 limiting value of $d/t = \dfrac{120\varepsilon}{(1+2r_2)} = \dfrac{120\times1.0}{[1+(2\times0.57)]} = 56.1$ ⎫
                          $\geq 40\varepsilon$ ⎬ $= 56.1\varepsilon$
                                                                                                                 ⎭

Actual $d/t = 46.0$
The web is semi-compact (non-slender)
Clause 3.5.1   **The cross-section is semi-compact**

2. Section: *Double angle strut composed of two 100 × 75 × 10 S 355 angles connected to gusset plates by the long legs back-to-back and subjected to axial compression.*

Section properties:
$b = 75$ mm, $d = 100$ mm, $t = 10.0$ mm (see Figure 5 of the code)
$b/t = (75/10) = 7.5$, $d/t = (100/10) = 10.0$, $(b+d)/t = (175/10) = 17.5$

Table 11 $p_y = 355 \text{ N/mm}^2$ $\therefore \varepsilon = \left(\dfrac{275}{p_y}\right)^{0.5} = \left(\dfrac{275}{355}\right)^{0.5} = 0.88$

Classes 1 and 2 are not applicable. Three criteria must be satisfied to comply with a semi-compact classification:
Class 3   limiting value of b/t $= 15\varepsilon$ $= (15\times0.88)$ $= 13.2 > 7.5$
            limiting value of d/t $= 15\varepsilon$ $= 13.2 > 10.0$
            limiting value of $(b+d)/t = 24\varepsilon$ $= (24\times0.88)$ $= 21.12 > 17.5$
Clause 3.5.1   **The cross-section is semi-compact**

3. Section: *Hot-rolled circular section 273.0 × 6.3 CHS S 275 subject to compression due to bending.*

Section properties: $D/t = 43.3$

Table 12 $p_y = 275 \text{ N/mm}^2$ $\varepsilon = \left(\dfrac{275}{p_y}\right)^{0.5} = 1.0$

Class 1 limiting value of $D/t = 40\varepsilon = 40$ < 43.3
Class 2 limiting value of $D/t = 50\varepsilon = 50$ > 40.3
Clause 3.5.1   ***The cross-section is compact***

6.2.6 Cross-Section Properties (Clauses 3.4, 3.5.6 and 3.6)

The cross-sectional properties used in design are dependent on the classification of the cross-section and are defined in the following Clauses:

- Clause 3.4.1   *gross* cross-sectional properties,
- Clause 3.4.2   *net* cross-sectional area allowing for the reduced area due to bolt holes,
- Clause 3.4.3   *effective net* areas which are dependent on the grade of steel,
- Clause 3.5.6   *effective plastic section modulus* for Class 3, **semi-compact sections**,
- Clause 3.6   *effective cross-sectional area and section modulus* for Class 4 **slender sections.**

These are considered separately in Sections 6.2.6.1 to 6.2.6.5.

6.2.6.1   Gross Cross-sectional Properties (Clause 3.4.1)

The gross cross-sectional properties for elements is calculated on the basis of the nominal dimensions without any allowance being made for bolt holes. Larger holes, e.g. to accommodate services, should be allowed for.

6.2.6.2   Net Cross-sectional Area (Clauses 3.4.2 and 3.4.4)

The reduction in the cross-sectional area due to bolt holes is based on the holes due allowance being made for the clearance dimensions, e.g. a 20 mm diameter bolt requires a 22 mm diameter hole.

If holes are *not staggered* across the width of a member, then the area to be deducted is the maximum sum of the sectional areas of the holes in any cross-section perpendicular to the direction of the applied stress in the member.

For example, consider the flat plate tie member shown in Figure 6.6.

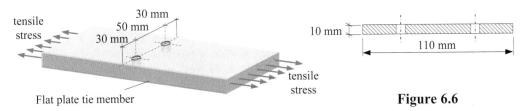

Flat plate tie member

Figure 6.6

Assuming the bolt diameter is equal to 20 mm then the hole diameter = (20 + 2) = 22 mm
Area to be deducted   =   [(22 × 10) × 2]   =   440 mm²
Net area a_n =   [(110 × 10) − 440] =   660 mm²

If holes are *staggered* across the width of a member then it is necessary to consider all possible failure paths extending progressively across the member. The area to be deducted is equal to the greater of the sum of the sectional areas of all holes in the path less an allowance equal to $(0.25s^2 t/g)$ for each gauge space which traverses diagonally in the path as shown in Figure 6.7.

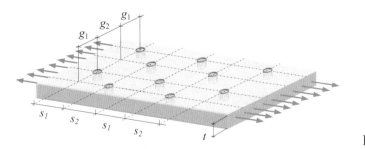

where:
s   is the staggered pitch
g   is the gauge
t   is the thickness

Figure 6.7

Consider Example 6.4 shown in Figure 6.8 and determine the net area.

6.2.6.3   Example 6.4: Net Cross-sectional Area
Determine the net cross-sectional area of the axially loaded plate shown in Figure 6.8.

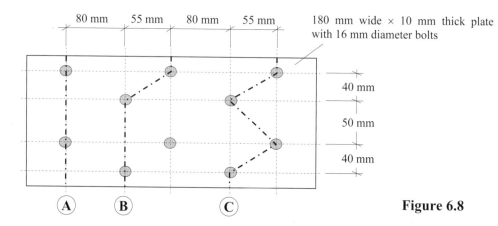

180 mm wide × 10 mm thick plate with 16 mm diameter bolts

Figure 6.8

Solution:
Path A   Area to be deducted   =   [2(18 × 10)]   =   360 mm²
Path B   Area to be deducted   =   [3(18 × 10) − (0.25 × 55² × 10)/40] = (540 − 189)
                                                                          =   351 mm²

Path C   Area to be deducted

          =   [4(18 × 10) − 2(0.25 × 55² × 10)/40 − (0.25 × 55² × 10)/50]
          =   (720 − 378.13 − 151.25)   =   190.62 mm²
Path A is the most severe with the highest area to be deducted and therefore:
          Net area a_n = (180 × 10) − 360 = 1440 mm²

6.2.6.4 Effective Net Cross-sectional Area (Clauses 3.4.3)

Experimental testing has indicated that the effective capacity of a member in tension is not reduced by the presence of holes provided that the ratio:

$$\frac{\text{net cross - sectional area}}{\text{gross cross - sectional area}} \quad > \quad \frac{\text{yield strength}}{\text{ultimate strength}} \quad \text{by a suitable margin.}$$

The effective net area a_e of a cross-section with bolt holes is determined by multiplying a_n by a modification factor K_e to allow for this, and hence:

$$a_e = K_e a_n \quad \leq a_g$$

where:

a_g is the gross cross-sectional area,

a_n is the net cross-sectional area,

K_e is the effective net area coefficient defined in Clause 3.4.3 of the code as indicated in Figure 6.9.

Grade of Steel	Effective Net Area Coefficient K_e
S 275	1.2
S 355	1.1
S 460	1.0
All other grades	$(U_s/1.2)/p_y$
where: U_s is the specified minimum tensile strength p_y is the design strength	

Figure 6.9

6.2.6.5 Example 6.5: Effective Net Cross-sectional Area

Determine the effective net cross-sectional area for the plate given in Example 6.4, assuming the grade of steel to be S 355.

Solution:

Effective net area $a_e = K_e a_n \quad \leq a_g$

For grade S 355 steel $K_e = 1.1, \quad a_g = (180 \times 10) = 1800 \text{ mm}^2, \, a_n = 1440 \text{ mm}^2$

$$\therefore \quad a_e = (1.1 \times 1440) = 1584 \text{ mm}^2 < a_g$$

6.2.6.6 Effective Plastic Section Modulus for Class 3 Sections (Clause 3.5.6)

The design of Class 3, semi-compact sections subject to bending can be carried out conservatively, using the elastic section modulus (Z), or alternatively an effective plastic section modulus (S_{eff}) can be adopted using the equations given in the code. The general form of these equations is:

$$S_{eff} = Z + [(S - Z) \times \text{'formula'}]$$

The 'formula' is dependent on several factors, i.e. the type of section and the limiting d/t and b/T ratios for a given section, and is given in the code as:

$$\frac{\left(\dfrac{\beta_{3w}}{d/t}\right)^2 - 1}{\left(\dfrac{\beta_{3w}}{\beta_{2w}}\right)^2 - 1} \quad \text{but} \quad \leq \quad \left[\frac{\dfrac{\beta_{3f}}{b/T} - 1}{\dfrac{\beta_{3f}}{\beta_{2f}} - 1}\right]$$

For **I** and **H** sections with equal flanges considering bending about the **major axis**, (i.e. the x–x axis).

$$\left[\frac{\dfrac{\beta_{3f}}{b/T} - 1}{\dfrac{\beta_{3f}}{\beta_{2f}} - 1}\right]$$

For **I** and **H** sections with equal flanges considering bending about the **minor axis**, (i.e. the y–y axis)

where:
b is the flange outstand as given in Figure 5 of the code,
d is the web depth,
T is the flange thickness,
t is the web thickness,
β_{2f} is the limiting value of b/T from Table 11 for a Class 2 compact flange,
β_{2w} is the limiting value of d/t from Table 11 for a Class 2 compact web,
β_{3f} is the limiting value of b/T from Table 11 for a Class 3 semi-compact flange,
β_{3w} is the limiting value of d/t from Table 11 for a Class 3 semi-compact web.

$$\left[\frac{\dfrac{\beta_{3w}}{d/t} - 1}{\dfrac{\beta_{3w}}{\beta_{2w}} - 1}\right] \quad \text{but} \quad \leq \quad \left[\frac{\dfrac{\beta_{3f}}{b/t} - 1}{\dfrac{\beta_{3f}}{\beta_{2f}} - 1}\right]$$

For **rectangular hollow** sections considering bending about either the **major axis** (i.e. the x–x axis) or the **minor axis** (i.e. the y–y axis).

where:
β_{2f} is the limiting value of b/t from Table 12 for a Class 2 compact flange,
β_{2w} is the limiting value of d/t from Table 12 for a Class 2 compact web,
β_{3f} is the limiting value of b/t from Table 12 for a Class 3 semi-compact flange,
β_{3w} is the limiting value of d/t from Table 12 for a Class 3 semi-compact web.

$$1.485\left\{\left[\left(\frac{140}{D/t}\right)\left(\frac{275}{p_y}\right)\right]^{0.5} - 1\right\} \quad \text{For \textbf{circular hollow} sections.}$$

In all cases, $S_{\text{eff}} \leq S$.

6.2.6.7   Example 6.6: Effective Plastic Section Modulus for Class 3 Sections

Determine the effective plastic section modulus for a $457 \times 191 \times 67$ UB S 275 section subject to bending about the major axis and an axial compressive load equal to 950 kN.

Solution:
Section properties:
$d = 407.6$ mm, $t = 8.5$ mm, $b/T = 7.48$, $d/t = 48.0$,
$A = 85.5$ cm^2, $Z_x = 1300$ cm^2, $S_x = 1470$ cm^2

Clause 3.5.6.3

$$S_{x,eff} = Z_x + (S_x - Z_x) \frac{\left(\dfrac{\beta_{3w}}{d/t}\right)^2 - 1}{\left(\dfrac{\beta_{3w}}{\beta_{2w}}\right)^2 - 1} \quad \text{but} \quad \leq Z_x + (S_x - Z_x) \frac{\dfrac{\beta_{3f}}{b/T} - 1}{\dfrac{\beta_{3f}}{\beta_{2f}} - 1}$$

$$\leq S_x$$

Table 11 $\beta_{3w} = \left(\dfrac{120\varepsilon}{1+2r_2}\right) \geq 40\varepsilon$ where $\varepsilon = (275/p_y)^{0.5} = 1.0$

$$r_2 = \frac{F_c}{A_g p_{yw}} = \left(\frac{950 \times 10^3}{85.5 \times 10^2 \times 275}\right) = 0.404$$

$$\beta_{3w} = \left[\frac{120}{1+(2 \times 0.404)}\right] = 66.37 \qquad \geq 40\varepsilon \quad \therefore \beta_{3w} = \mathbf{66.37}$$

$$\beta_{2w} = \left(\frac{100\varepsilon}{1+1.5r_1}\right) \geq 40\varepsilon \quad \text{where}$$

$$r_1 = \frac{F_c}{dtp_{yw}} = \left(\frac{950 \times 10^3}{407.6 \times 8.5 \times 275}\right) = 0.997$$

$$\beta_{2w} = \left[\frac{100}{1+(1.5 \times 0.997)}\right] = 40.07 \qquad \geq 40\varepsilon \quad \therefore \beta_{3w} = \mathbf{40.07}$$

$$\beta_{3f} = 15\varepsilon = 15.0$$
$$\beta_{2f} = 10\varepsilon = 10.0$$

$$\frac{\left(\dfrac{\beta_{3w}}{d/t}\right)^2 - 1}{\left(\dfrac{\beta_{3w}}{\beta_{2w}}\right)^2 - 1} = \frac{\left(\dfrac{66.37}{48.0}\right)^2 - 1}{\left(\dfrac{66.37}{40.07}\right)^2 - 1} = 0.523$$

                                                      Use smaller value

Check $\dfrac{\dfrac{\beta_{3f}}{b/T} - 1}{\dfrac{\beta_{3f}}{\beta_{2f}} - 1} = \dfrac{\left(\dfrac{15.0}{7.48}\right) - 1}{\left(\dfrac{15.0}{10.0}\right) - 1} = 0.201$

$$S_{x,eff} = \{1300 + [(1470 - 1300) \times 0.523]\} = 1386 \quad \leq S_x$$
$$\therefore S_{x,eff} = \mathbf{1386 \ mm^3}$$

6.2.6.8 Effective Cross-sectional Area and Section Modulus for Class 4 Sections (Clause 3.5.6)

In the case of slender cross-sections, the reduced capacity due to local buckling effects is determined using an effective cross-sectional area and section modulus as indicated in Figures 8(a) and (b) of the code, which relate to doubly symmetric slender sections as follows:

Figure 8(a) of the code: Effective cross-sectional (A_{eff}) area when considering *pure compression*.

Consider two typical cross-sections, i.e. a rolled **I**-section and a hot finished rectangular hollow section (RHS).

The effective width of a slender web element or internal flange element is considered to be equal to 40ε comprising two equal portions with a central non-effective zone. Note that this does not change the position of the centroidal axis of the effective cross-section.

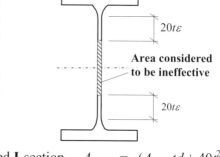

Rolled **I**-section $A_{eff} = (A_g - td + 40t^2\varepsilon)$

Figure 6.10

Shaded areas are considered ineffective

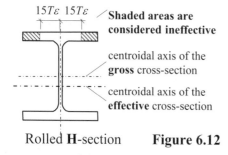

1.5*t* 20*tε* 20*tε* 1.5*t*

Figure 6.11

Rectangular hollow section (RHS)
Web only slender:
$$A_{eff} = A_g - 2t(D - 3t - 40t\varepsilon)$$
Web and flanges slender:
$$A_{eff} = A_g - 2t\{[D - (3t + 40t\varepsilon)] + [B - (3t + 40t\varepsilon)]\}$$

Similar information relating to rolled **H**-sections, welded and cold formed sections is also provided in Figure 8(a).

Figure 8(b) of the code: Effective section modulus (Z_{eff}) considering *pure bending* where the web is not slender.

15*Tε* 15*Tε*   **Shaded areas are considered ineffective**

centroidal axis of the gross cross-section

centroidal axis of the effective cross-section

Rolled **H**-section   **Figure 6.12**

The Z_{eff} used should be the minimum value of elastic section modulus considering the centroidal axis of the effective section and ignoring the shaded areas.

If neither the web nor the flanges are slender then $Z_{eff} = Z$.

Figure 9 of the code: Effective Section Modulus (Z_{eff}) considering *pure bending* where the web is slender.

This Clause is not applicable to hot-rolled sections since none of them have slender webs when the neutral axis is at mid-depth. It is used when considering welded sections such as plate girders and is not included in this text.

For slender hot-rolled angle sections, either the method given in Clause 3.6.3 can be used or an alternative, conservative method can be adopted in which:

The effective cross-sectional area: $\dfrac{A_{eff}}{A} = \dfrac{12\varepsilon}{b/t} \qquad \therefore \quad A_{eff} = \dfrac{12\varepsilon\, A}{b/t}$

The effective section modulus: $\dfrac{Z_{eff}}{Z} = \dfrac{15\varepsilon}{b/t} \qquad \therefore \quad Z_{eff} = \dfrac{15\varepsilon\, Z}{b/t}$

where *b* is the leg length and *t* is the thickness.

An alternative conservative method for all slender section properties is indicated in Clause 3.6.5 in which the sections are considered to be Class 3 semi-compact and designed on the basis of a reduced design strength p_{yr} obtained from:

$$p_{yr} = (\beta_3/\beta)^2 p_y$$

in which β is the value of *b/T*, *b/t*, *D/t* or *d/t* which exceeds the limiting value β_3 given in Table 11 or Table 12 for a Class 3 semi-compact section.

The provisions of Clauses 3.6.2.1 to 3.6.2.3 and Figures 8(a) and (b) are illustrated in Example 6.6.

6.2.6.9   Example 6.7: Effective Plastic Section Properties for Class 4 Sections

Determine:

(i)   the effective cross-sectional area for a 610 × 229 × 125 UB S 275 subjected to pure compression from an axial load of 4350 kN,

(ii)   the effective plastic section modulus for a 300 × 300 × 6.3 square hollow section (SHS) S 275 section subject to pure bending about the major axis.

Solution:
(i) 610 × 229 × 125 UB S 275
Section properties:
$d = 547.6$ mm, $t = 11.9$ mm, $A = 159$ cm^2
In Example 6.3 this section was shown to be slender when subjected to an axial load equal to 4350 kN and $\varepsilon = 1.0$
Figure 8(a)   (see Figure 6.10 of this text)

$$\begin{aligned}
\text{Rolled I-section} \quad A_{eff} &= (A_g - td + 40t^2\varepsilon) \\
&= [(159 \times 10^2) - (11.9 \times 547.6) + (40 \times 11.9^2)] \\
&= 15048 \text{ mm}^2
\end{aligned}$$

(ii)   300 × 300 × 6.3 SHS S 275
Section properties:
$d/t = b/t = 44.6$, $t = 6.3$ mm, $A = 73.6$ cm^2, $I_{xx} = 10500$ cm^4

Table 12: $b/t > 40\varepsilon$ ∴ the flanges are slender
                $d/t < 64\varepsilon$ ∴ the webs are plastic   (neutral axis at mid-height)

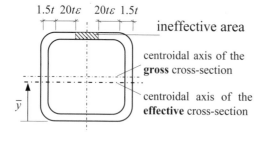

ineffective area

Width of ineffective area:
= $[b - 2(1.5t + 20t\varepsilon)]$
= $[300 - 2(21.5 \times 6.3)]$ = 29.1 mm^2

Ineffective area:
= (29.1×6.3) = 183.33 mm^2

$$\bar{y} = \frac{[(7360 \times 150) - (183.33 \times 296.85)]}{(7360 - 183.33)} = 146.25 \text{ mm}$$

$$I_{xx} = \sum\left[I_{\text{centroidal axis}} + Ay^2\right]$$
$$= \{(10500 \times 10^4) + [7360 \times (150 - 146.25)^2]\}$$
$$- \left\{\left(\frac{29.1 \times 6.3^3}{12}\right) + \left[183.33 \times (296.85 - 146.25)^2\right]\right\}$$
$$= 100.94 \times 10^6 \text{ mm}^4$$

$$Z_{\text{eff}} = \frac{I_{xx}}{(D - \bar{y})} = \frac{100.94 \times 10^6}{(300 - 146.25)} = 656.5 \times 10^3 \text{ mm}^3$$

6.3   Axially Loaded Elements

6.3.1 Introduction

The design of axially loaded members considers any member where the applied loading induces either axial tension or axial compression. Members subject to axial forces most frequently occur in bracing systems, pin-jointed trusses, lattice girders or suspension systems, as shown in Figure 6.13.

Frequently, in structural frames, sections are subjected to combined axial and bending effects which may be caused by eccentric connections, wind loading or rigid-frame action. In most cases in which UB and UC sections are used as columns in buildings, they are subjected to combined axial and bending effects. The design of such members is discussed and illustrated in Section 6.3.5.

The types of section used for axially loaded members range from rolled uniform beams, columns and hollow sections to threaded bars, flat plates and wire ropes.

The following discussion relates primarily to pin-jointed structures, which comprise the majority of structures with members subject to axial loads only.

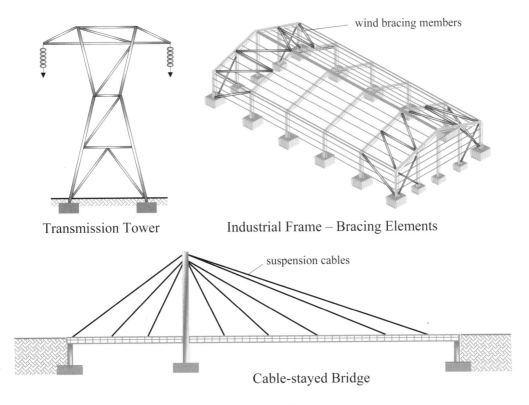

Transmission Tower   Industrial Frame – Bracing Elements

Cable-stayed Bridge

Figure 6.13

The use of beams and plate girders does not always provide the most economic or suitable structural solution when spanning large openings. In buildings which have lightly loaded long span roofs, when large voids are required within the depth of roof structures for services, when plate girders are impractical, or for aesthetic/architectural reasons, the use of roof trusses or lattice girders may be more appropriate.

Trusses are frequently used as secondary structural elements to distribute wind loading to the foundations, as temporary bracing during construction and for torsional and lateral stability.

Roof trusses and lattice girders are open-web flexural members which transmit the effects of loads applied within their spans to support points by means of bending and shear. In the case of beams, the bending and shear is transmitted by inducing bending moments and shear forces in the cross-sections of structural members. Trusses and lattice girders, however, generally transfer their loads by inducing axial tensile or compressive forces in the individual members. The form of a truss is most economic when the arrangement is such that most members are in tension.

The magnitude and sense of these forces can be determined using standard methods of analysis such as *the method of sections*, *joint resolution*, *tension coefficients*, *graphical techniques* or the use of *computer software*.

The arrangement of the internal framing of a roof truss depends upon its span. Rafters are normally divided into equal panel lengths and ideally the loads are applied at the node points by roof purlins. Purlin spacing is dependent on the form of roof cladding that is used and is usually based on manufacturers' data sheets. In instances where the purlins do not coincide with the node points, the main members (i.e. the rafters, or the top and bottom booms of lattice girders) are also subjected to local bending which must be allowed for in the design.

The internal structure of trusses should be such that, where possible, the long members are **ties** (in tension), while the short members are **struts** (in compression). In long span trusses the main ties are usually cambered to offset the visual sagging effects of the deflection.

In very long span trusses, e.g. 60 metres, it is not usually possible to maintain a constant slope in the rafter owing to problems such as additional heating requirements caused by the very high ridge height. This problem can be overcome by changing the slope to provide a *mansard-type* truss in which the slope near the end of the truss is very steep, while it is shallower over the rest of the span.

A few examples of typical pitched roof trusses are illustrated in Figure 6.14. Lattice girders are generally trusses with parallel top and bottom chords (known as booms) with internal web bracing members. In long span construction they are very useful since their relatively small span/depth ratio (typically 1/10 to 1/14) gives them an advantage over pitched roof trusses.

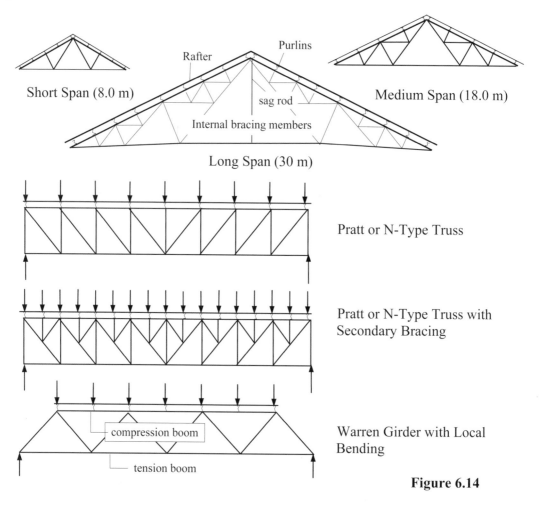

Figure 6.14

As with roof trusses, the framing should be triangulated, considering the span and the spacing of the applied loads. If purlins do not coincide with the panel points then secondary bracing as shown in Figure 6.14 can be adopted as an alternative to designing for combined axial and local bending effects.

Generally, the four main assumptions made when analysing trusses are:

♦ *Truss members are connected together at their ends only.*

In practice, the top and bottom chords are normally continuous and span several joints rather than being a series of discontinuous, short members. Since truss members are usually long and slender and do not support significant bending moments, this assumption in the analysis is acceptable.

♦ *Truss members are connected together by frictionless pins.*

In real trusses the members are connected at the joints using bolted or welded gusset plates or end plates, as shown in Figure 6.15, rather than frictionless pins.

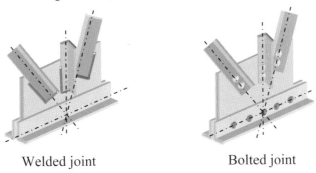

Welded joint   Bolted joint

Figure 6.15

Provided that the setting-out lines of the bolts or the centroidal axes of the members intersect at the assumed joint locations, experience has shown that this idealisation is acceptable.

♦ *The truss structure is loaded only at the joints.*

Often the exact location of purlins relative to the joints on the top of the compression chord/rafters is unknown at the design stage of a truss. In these circumstances, assuming that the purlins do not coincide with the position of the joints, a local bending moment in addition to the axial load is assumed in the truss members.

A number of empirical rules are given in Clause 4.10 for this situation. The design of such members is carried out assuming a combined axial load and bending moments, in accordance with the requirements of Clause 4.8 in which the bending moment is assumed to be equal to $wL^2/6$ in which:

w is the total load/unit length applied perpendicular to the rafters, and
L is the length of the rafter between the joints.

♦ *The self-weight of the members may be neglected or assumed to act at the adjacent nodes.*

Frequently, in the analysis of small trusses it is reasonable to neglect the self-weight of the members. This may not be acceptable for large trusses, particularly those used in bridge construction. Common practice is to assume that half of the weight of each member acts at each of the two joints that it connects.

6.3.2 Tension Members (Clause 4.6)

As indicated in Section 6.3.1, the types of element used in the design of tension members are numerous and varied; open and closed single-rolled sections being adopted for light trusses and lattice girders, compound sections comprising either multiple-rolled sections or welded plates for heavy trusses with ropes and cables being used in suspension structures such as bridges and roofs.

There are a number of potential problems that may arise from using light, slender sections such as bars, flats, rolled angle and channel sections, e.g.

♦ excessive sag under self-weight,
♦ vibration during dynamic loading   and
♦ damage during transportation to site.

The introduction of sag rods as indicated in Figure 6.14 and the use of intermediate packing in double angle or channel members will assist in minimizing the first two of these problems.

In general, if the leg length of an angle tie is at least equal to $\dfrac{\text{member length}}{60}$, the member will have sufficient stiffness to prevent damage during transport.

The tension capacity P_t of a member should be determined from

$$P_t = p_y \times A_e$$

where:
p_y is the design strength of the steel as given in Table 9   and
A_e is the sum of the effective net areas (a_e) of all the elements of the cross-section as defined in Clause 3.4.3 but $\leq (1.2 \times \text{Total net area } A_n)$.

Note: When designing bolted joints it is also necessary to consider the possibility of failure by *block shear* as indicated in Clause 6.2.4 (see Section 6.3.6.4.5 on connections).

6.3.2.1   Effects of Eccentric Connections (Clause 4.6.2)

Although theoretically tension members are inherently stable and the most economic structural elements, the introduction of secondary effects such as bending due to eccentricities at connections reduces their efficiency. With the exception of angles, channels and T-sections, the secondary effects should be allowed for in the design by considering members subject to combined axial and bending load effects, as indicated in Clause 4.8.2.

6.3.2.2 Single Angles, Channels and T-Sections (Clause 4.6.3.1)

For asymmetric connections where secondary effects will occur, the tension capacity should be calculated based on a reduced effective net area as indicated in Figure 6.16.

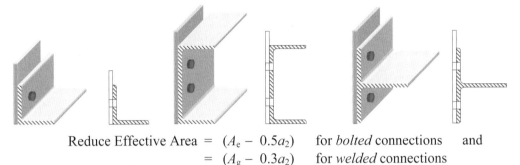

$$\text{Reduce Effective Area} = (A_e - 0.5a_2) \quad \text{for } \textit{bolted} \text{ connections} \quad \text{and}$$
$$= (A_g - 0.3a_2) \quad \text{for } \textit{welded} \text{ connections}$$

where :
A_g is the gross cross-sectional area,
a_1 is the gross cross-sectional area of the connected element[*],
a_2 is equal to $(A_g - a_1)$.

Figure 6.16

* **Note:** the gross cross-sectional area of the connected element of an angle is defined in Clause 4.6.3.1 as '... *the product of its thickness and the overall leg width, the overall depth for a channel or the flange width for a T-section.*'

The above reduced area also applies to double angles, channels and T-sections which are connected to the *same* side of a gusset plate as shown in Figure 6.17. The total capacity is equal to the sum of the capacities of each component.

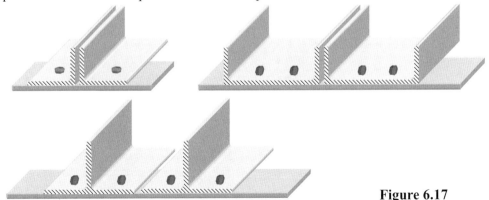

Figure 6.17

6.3.2.3 Double Angles, Channels and T-Sections Connected, to *Both* Sides of a Gusset Plate (Clause 4.6.3.2)

For double angles connected back-to-back and to the same side of a gusset plate or section as shown in Figure 6.18, a similar calculation is made to determine the effective area.

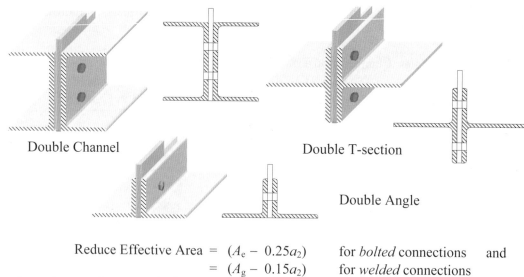

Double Channel   Double T-section

Double Angle

Reduce Effective Area $= (A_e - 0.25a_2)$ for *bolted* connections   and

$= (A_g - 0.15a_2)$ for *welded* connections

where A_g, a_1 and a_2 are as before.

Figure 6.18

In double ties as indicated in Figure 6.18 it is assumed that the components are interconnected by bolts or welds and are held apart and longitudinally parallel by battens or solid packing pieces in at least two locations within their length. In the case of angles the outermost connection should be within a distance from each end equal to ($10 \times$ smaller leg length); in the case of channels and T-sections to ($10 \times$ the smaller overall dimension).

6.3.2.4   Other Simple Ties (Clause 4.6.3.3)

In simple connections such as single angles connected through both legs, channels connected by both flanges or T-sections connected either through the flange and the stem or the stem only as in Figure 6.19 secondary bending effects can be ignored and the sections designed assuming the effective net area as calculated using Clause 3.4.3.

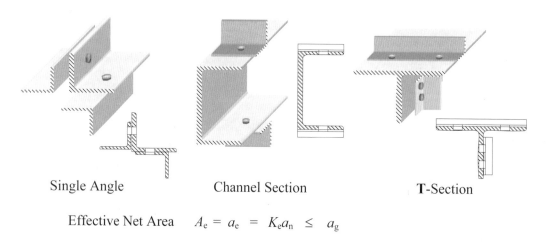

Single Angle   Channel Section   T-Section

Effective Net Area $A_e = a_e = K_e a_n \leq a_g$

Figure 6.19

6.3.2.5 Example 6.8: Plate with Staggered Holes and Single and Double Angle Sections

Determine the tension capacity of the plate and the angle sections shown in Figures 6.20(a), (b) and (c), considering the strength of the sections only (i.e. ignoring block shear and bolt strength).

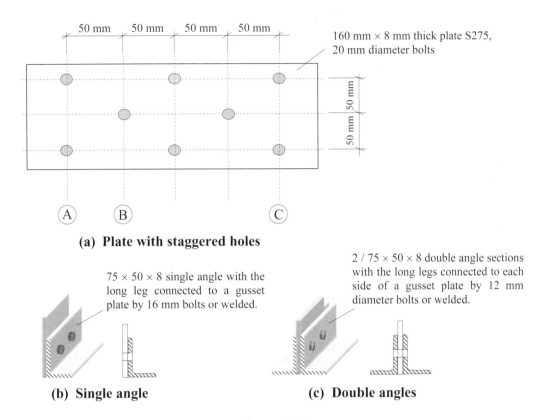

(a) **Plate with staggered holes**

(b) **Single angle**

(c) **Double angles**

Figure 6.20

6.3.2.6 Solution to Example 6.8

Contract : Ties Job Ref. No. : Example 6.8	Calcs. by : W.McK.
Part of Structure : Plate and Angle Sections	Checked by :
Calc. Sheet No. : 1 of 3	Date :

References	Calculations	Output
Clause 3.4.3	**(a) Plate:** $A_e = a_e = K_e a_n \leq a_g$ $\leq 1.2\,(a_{n1} + a_{n2})$	
Clause 3.4.4.3	A series of staggered failure patterns must be considered to determine the critical case.	

References	Calculations	Output
	Contract : Ties Job Ref. No. : Example 6.8	

Contract :   Ties   Job Ref. No. : Example 6.8
Part of Structure :   Plate and Angle Sections
Calc. Sheet No. : 2 of 3

Calcs. by : W.McK.
Checked by :
Date :

References	Calculations	Output
	(**Note:** The hole diameter is equal to the bolt diameter + 2 mm)	
	Path A: (two holes)	
	Area to be deducted $\quad = \quad 2\,(t \times D) \quad = \quad (2 \times 8 \times 22)$	
	$\qquad\qquad\qquad\qquad\quad = \quad 352 \text{ mm}^2$	
	Path B: (two holes and one diagonal path)	
	Area to be deducted $\quad = \quad [2\,(t \times D) - 0.25s^2t/g]$	
	$\qquad\qquad\qquad\qquad\quad = \quad [352 - (0.25 \times 50^2 \times 8)/50]$	
	$\qquad\qquad\qquad\qquad\quad = \quad 252 \text{ mm}^2$	
	Path C: (three holes and two diagonal paths)	
	Area to be deducted $\quad = \quad [3\,(t \times D) - 2\,(0.25s^2t/g)]$	
	$\qquad\qquad\qquad\quad = \quad [(3 \times 8 \times 22) - (2 \times 0.25 \times 50^2 \times 8)/50]$	
	$\qquad\qquad\qquad\quad = \quad 328 \text{ mm}^2$	
	Path A is the most critical and hence:	
	$A_n = [(160 \times 8) - 352)] \qquad = \quad 928 \text{ mm}^2$	
	$A_g = (160 \times 8) \qquad\qquad\quad = \quad 1280 \text{ mm}^2$	
Clause 3.4.3	$K_e = 1.2$ for S 275 steel	
	$A_e = K_e\,a_n = (1.2 \times 928) \quad = \quad 1113.6 \text{ mm}^2$	
		Plate Capacity:
Clause 4.6.1	$P_t = (p_y \times A_e) \quad = \quad (275 \times 1113.6)/10^3 = \quad 306.2 \text{ kN}$	**306.2 kN**
	(b) Single angle:	
	(i) Consider the connection to be bolted	
Clause 4.6.3.1	$P_t = p_y \times (A_e - 0.5a_2)$	
Clause 3.4.3	$A_e = (a_{e1} + a_{e2}) \leq 1.2\,(a_{n1} + a_{n2})$	
	$a_{e1} =$ effective area of the connected leg $\qquad = \quad K_e\,a_{n1} \leq a_1$	
	$a_{e2} =$ effective area of the unconnected leg $\quad = \quad K_e\,a_{n2} \leq a_2$	
	$a_{n1} = (a_1 - \text{area of bolt holes})$	
	$a_{n2} = (a_2 - \text{area of bolt holes})$	
Section Tables	$A_g = 9.41 \text{cm}^2$	

Contract : Ties Job Ref. No. : Example 6.8	Calcs. by : W.McK.
Part of Structure : Plate and Angle Sections	Checked by :
Calc. Sheet No. : 3 of 3	Date :

References	Calculations	Output
Clause 4.6.3.1	$a_1 = (75 \times 8)$ $= 600 \text{ mm}^2$ $a_2 = (A_g - a_1) = (941 - 600)$ $= 341 \text{ mm}^2$ $a_{n1} = [600 - (8 \times 18)]$ $= 456 \text{ mm}^2$ $a_{n2} = 341 \text{ mm}^2$	
Clause 3.4.3	$K_e = 1.2$ $a_{e1} = (1.2 \times 456) = 547.2 \text{ mm}^2 \quad \leq \quad a_1$ $a_{e2} = 341 \text{ mm}^2 \quad\quad\quad\quad\quad\quad \leq \quad a_2$ $A_e = (a_{e1} + a_{e2}) = (547.2 + 341) = 888.2 \text{ mm}^2$ $1.2 (a_{n1} + a_{n2}) \quad = 1.2 (456 + 341) = 956.4 \text{ mm}^2$ $\therefore A_e \quad \leq \quad 1.2 (a_{n1} + a_{n2})$	
Clause 4.6.3.1	$P_t = p_y \times (A_e - 0.5a_2) = \{275 \times [888.2 - (0.5 \times 341)]\}/10^3$ $\quad\quad = 197.4 \text{ kN}$	**Bolted Single Angle Capacity:** **197.4 kN**
Clause 4.6.3.1	(ii) Consider the connection to be welded $P_t = p_y \times (A_g - 0.3a_2) = \{275 \times [941.0 - (0.3 \times 341)]\}/10^3$ $\quad\quad = 230.6 \text{ kN}$	**Welded Single Angle Capacity:** **230.6 kN**
Clause 4.6.3.2 Section Tables Clause 4.6.3.2	**(c) Double angle:** (i) Consider the connection to be bolted $P_t = p_y \times (A_e - 0.25a_2)$ $A_g = 18.8 \text{ cm}^2$ $(= 9.41 \text{ cm}^2/\text{angle})$ Evaluate area/angle: $a_1 = (75 \times 8)$ $= 600 \text{ mm}^2$ $a_2 = (A_g - a_1) = (941 - 600)$ $= 341 \text{ mm}^2$ $a_{n1} = [600 - (8 \times 14)]$ $= 488 \text{ mm}^2$ $a_{n2} = 341 \text{ mm}^2$	
Clause 3.4.3	$K_e = 1.2$ $a_{e1} = (1.2 \times 488) = 585.6 \text{ mm}^2 \quad \leq \quad a_1$ $a_{e2} = 341 \text{ mm}^2 \quad\quad\quad\quad\quad\quad \leq \quad a_2$ $A_e = (a_{e1} + a_{e2}) = (585.6 + 341) = 926.6 \text{ mm}^2$ $1.2 (a_{n1} + a_{n2}) \quad = 1.2 (488 + 341) = 994.8 \text{ mm}^2$ $\therefore A_e \quad \leq \quad 1.2 (a_{n1} + a_{n2})$	
Clause 4.6.3.2	Tension capacity for both angles: $P_t = 2 [p_y \times (A_e - 0.25a_2)]$ $\quad = 2 \{275 \times [926.6 - (0.25 \times 341)]\}/10^3 \quad = 462.7 \text{ kN}$	**Bolted Double Angle Capacity:** **462.7 kN**
Clause 4.6.3.2	(ii) Consider the connection to be welded $P_t = 2 [p_y \times (A_g - 0.15a_2)]$ $\quad = 2 \{275 \times [941.0 - (0.15 \times 341)]\}/10^3 \quad = 489.0 \text{ kN}$	**Welded Double Angle Capacity:** **489.0 kN**

6.3.3 *Compression Members (Clause 4.7)*

The design of compression members is more complex than that of tension members and encompasses the design of structural elements referred to as **columns, stanchions** or **struts**. The term 'strut' is usually used when referring to members in lattice/truss frameworks, while the other two generally refer to vertical or inclined members supporting floors and/or roofs in structural frames.

As with tension members, in many cases such members are subjected to both axial and bending effects. This section deals primarily with those members on lattice/truss frameworks in which it is assumed that all members are subjected to concentric axial loading. Column/stanchion design in which combined axial compression and bending are present is discussed in Section 6.3.5.

6.3.3.1 Compressive Strength p_c (Clause 4.7.5)

The dominant mode of failure to be considered when designing struts is axial buckling, and consequently the compressive strength (p_c) is dependent on the factors which determine the buckling strength. Buckling failure is caused by secondary bending effects induced by factors such as:

♦ the inherent eccentricity of applied loads due to asymmetric connection details,
♦ imperfections present in the cross-section and/or profile of a member throughout its length,
♦ non-uniformity of material properties throughout a member.

The first two of these factors are the most significant and their effect is to introduce initial curvature, secondary bending and consequently premature failure by buckling before the stress in the material reaches the yield value. The stress at which failure will occur is influenced by several variables, e.g.

♦ the cross-sectional shape of the member,
♦ the slenderness of the member,
♦ the yield strength of the material,
♦ the pattern of residual stresses induced by differential cooling of the member after the rolling process.

The effects of these variables are reflected in the various compressive strength (p_c) strut curves (see Chapter 4, Section 4.2) as indicated in Tables 23 and 24 of the code. The values in Table 24 are presented in terms of the slenderness λ and the material strengths S 275, S 355 and S 460. The contents of Table 24 are summarized in Figure 6.21.

If, as indicated in Figure 14, additional flange plates have been welded to rolled **I** or **H** sections, it is necessary to allow for the residual stresses induced by the welding process. These stresses differ in both magnitude and distribution from those caused during the rolling/cooling processes, and effectively further reduce the stress at which buckling may occur. The appropriate strut curve for such composite sections is indicated in Table 23.

In the case of welded **I, H** or box sections, the value of compressive strength (p_c) is determined from Table 24 by using a p_y value 20 N/mm^2 less than is obtained from Table 9.

Table 24	Slenderness λ	Strut Curve	Robertson Constant
1	< 110	a	2.0
2	≥ 110		
3	< 110	b	3.5
4	≥ 110		
5	< 110	c	5.5
6	≥ 110		
7	< 110	d	8.0
8	≥ 110		

Figure 6.21

6.3.3.2 Compressive Resistance (Clause 4.7.4)

In Clause 4.7.4, the compressive resistance P_c of a member is given by:

$$P_c = A_g p_c \quad \text{for plastic, compact and semi-compact sections} \quad \text{and}$$
$$P_c = A_{eff} p_{cs} \quad \text{for slender sections}$$

where:
A_g is the gross cross-sectional area as defined in Clause 3.4.1,
A_{eff} is the effective cross-sectional area as defined in Clause 3.6,
p_{cs} is the value of p_c as defined in Clause 4.7.5 for a reduced slenderness of $\lambda(A_{eff}/A_g)^{0.5}$ in which λ is based on the radius of gyration r of the gross cross-section,
p_c is the compressive strength as defined in Clause 4.7.5.

The value of compressive strength to be used in any particular circumstance is determined by identifying first the appropriate strut curve from Table 23 and subsequently p_c from Table 24 (1) to (8). In order to obtain a value from Table 24 it is necessary to evaluate the slenderness.

A practical and realistic assessment of the critical slenderness (λ) of a strut is the most important criterion in determining the compressive strength.

6.3.3.3 Slenderness (Clause 4.7.2)

Slenderness is generally evaluated using:

$$\lambda = \frac{L_E}{r}$$

where:
λ is the slenderness ratio,
L_E is the effective length with respect to the axis of buckling being considered,
r is the radius of gyration with respect to the axis of buckling being considered.

In most cases the value of L_E for a compression member can be assessed in accordance with Table 22 which provides coefficients (e.g. $07L$, $0.85L$) for both non-sway and sway structures. There are a number of sections for which the effective lengths are specified separately, as indicated in Figure 6.22.

Type of Strut		Relevant Clause
Laced	triangular lacing	Clause 4.7.8
Battened	battens	Clause 4.7.9
Single angle		Clause 4.7.10.2
Double angles		Clause 4.7.10.3
Single channels		Clause 4.7.10.4
Single T-sections		Clause 4.7.10.5
Starred angle		Clause 4.7.11
Battened Parallel Angles		Clause 4.7.12
All Others		Table 22

Figure 6.22

In addition, Appendix D of the code gives the appropriate coefficients to be used when assessing the effective lengths for columns in single-storey buildings using simple construction and for columns supporting an internal platform floor of simple design. Effective lengths for continuous structures with moment-resisting joints should be assessed in accordance with the provisions of Appendix E.

6.3.3.4   Example 6.9: Single Angle Strut (Clause 4.7.10.2)

A typical joint from a lattice girder is shown in Figure 6.23. Using the data given, determine the maximum compressive load which can be carried by member AB.

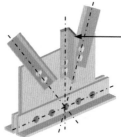

Member AB:
$1/100 \times 75 \times 10$ S 275 single angle section with the long leg connected to a 12 mm gusset plate.
Double bolted in standard clearance holes in line at each end.
Length between the intersection of the setting out lines of the bolts = 3.0 m

Figure 6.23

6.3.3.5 Solution to Example 6.9

References	Calculations	Output
Contract : Struts	**Job Ref. No. : Example 6.9**	**Calcs. by : W.McK.**

Contract : Struts   Job Ref. No. : Example 6.9
Part of Structure : Single Angle Section
Calc. Sheet No. : 1 of 1

Calcs. by : W.McK.
Checked by :
Date :

References	Calculations	Output
	Member AB: 1 / 100 × 75 × 10 Single Angle S 275	

A_g = 1660 mm^2
r_a = 21.6 mm
r_b = 31.2 mm
r_v = 15.9 mm
b = 75 mm
d = 100 mm
t = 10 mm

References	Calculations	Output
Table 9 Table 11	Design strength: p_y = 275 N/mm^2 ; ε = 1.0 Section Classification: Single angle subject to axial compression $\quad b/t \quad = \quad (75/10) \qquad = \quad 7.5$ $\quad d/t \quad = \quad (100/10) \qquad = \quad 10$ $(b+d)/t \quad = \quad (75+100)/10 \quad = \quad 17.5$	
Figure 5	**Note:** For an angle d is the width of the connected leg.	
Table 11	Since $\qquad b/t \; < \quad 15\varepsilon$ $\qquad\qquad d/t \; < \quad 15\varepsilon$ $\qquad (b+d)/t \; < \quad 24\varepsilon$ The section is not slender.	
Clause 4.7.10.1	Effective length $L_E =$ 3.0 m for all axes in this case	
Clause 4.7.2 Clause 4.7.10.2 (a)	Slenderness $\quad \lambda \quad = \quad L_E/r$ $0.85L_v/r_v \qquad\qquad = \quad [(0.85\times3000)/15.9] \qquad = \quad 160$ $(0.7L_v/r_v + 15) \quad = \quad [(0.7\times3000)/15.9 + 15] \quad = \quad 147$ $1.0L_a/r_a \qquad\qquad = \quad [(1.0\times3000)/21.6] \qquad = \quad 139$ $(0.7L_a/r_a + 30) \quad = \quad [(0.7\times3000)/21.6 + 30] \quad = \quad 127$ $0.85L_b/r_b \qquad\quad = \quad [(0.85\times3000)/31.2] \qquad = \quad 81.7$ $(0.7L_b/r_b + 30) \quad = \quad [(0.7\times3000)/31.2 + 30] \quad = \quad 97.3$ The critical slenderness $\quad = \quad \lambda_v \; = \quad 160$	
Table 23 Table 24(6)	Use strut curve (c) for all axes and since $\lambda \geq 110$ use Table 24(6) Using S 275 steel $\qquad p_y \; = \quad 275$ N/mm^2 $\qquad\qquad\qquad \therefore \; p_c \; = \quad 61$ N/mm^2	**The maximum compressive load which can be supported = 101 kN**
Clause 4.7.4	$\qquad\qquad P_c \; = \quad A_g p_c$ $\qquad\qquad\quad = \quad (1660 \times 61)/10^3 \quad = \quad 101$ kN	

6.3.3.6   Example 6.10: Double Angle Strut (Clause 4.7.10.3)

A compression member of a lattice girder comprises 2/150 × 75 × 15 S 275 angle sections with the long legs connected back-to-back and double bolted at the ends to 12 mm thick gusset plates as indicated in Figure 6 24. Using the data given determine the maximum compressive load which can be carried by the member.

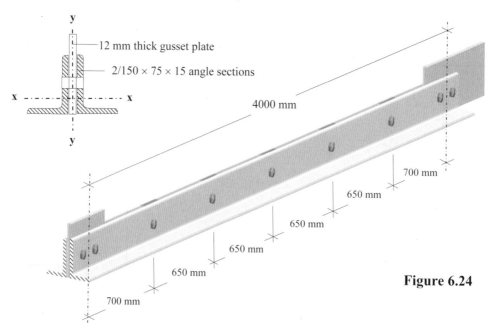

Figure 6.24

6.3.3.7   Solution to Example 6.10

References	Calculations	Output
	Contract : Struts Job Ref. No. : Example 6.10 **Part of Structure : Double Angle Section** **Calc. Sheet No. : 1 of 2**	**Calcs. by : W.McK.** **Checked by :** **Date :**

References	Calculations	Output
	Member: 2/150 × 75 × 15 Double Angles S 275 Double Angle Section Properties: A_g = 6340 mm^2 r_x = 47.5 mm r_y = 30.9 mm Single Angle Section Properties: r_x = 47.5 mm r_y = 19.4 mm r_v = 15.8 mm r_u = 48.8 mm b = 75 mm d = 150 mm t = 15 mm	

Contract : Struts Job Ref. No. : Example 6.10 Part of Structure : Double Angle Section Calc. Sheet No. : 2 of 2	Calcs. by : W.McK. Checked by : Date :

References	Calculations	Output
Table 9 Table 11	Design strength: $p_y = 275 \text{ N/mm}^2$; $\varepsilon = 1.0$ Section Classification: Double angle subject to axial compression $\begin{aligned} b/t &= (75/15) &= 5 \\ d/t &= (150/15) &= 10 \\ (b+d)/t &= (75+150)/15 &= 15.0 \end{aligned}$	

Table 11	Since $b/t < 15\varepsilon$
	       $d/t < 15\varepsilon$
	   $(b+d)/t < 24\varepsilon$
	The section is not slender.	

Clause 4.7.10.1	Effective length $L_E = 4.0$ m for all axes in this case	

Clause 4.7.2
Clause 4.7.10.3(c)
and Table 25
	Slenderness $\lambda = L_E/r$
	$\lambda \geq 0.85L_x/r_x$
	   $\geq (0.7L_x/r_x + 30)$
	   $\geq [(L_y/r_y)^2 + \lambda_c^2]^{0.5}$
	   $\geq 1.4\lambda_c$
	where:
	L_x and L_y are taken as the length L between the intersections of the centroidal axes or the setting out lines of the bolts.	

Clause 4.7.9(c)	$\lambda_c = \dfrac{L_v}{r_v} \leq 50$ and L_v is the distance between intermediate packing pieces r_v is the minimum radius of gyration for a single angle, i.e. about the v–v axis $L_x = L_y = 4000$ mm For a $150 \times 75 \times 15$ double angle $\lambda_c = \dfrac{700}{15.8} = 44.3$	
Clause 4.7.10.3(c)	$\begin{aligned} 0.85L_x/r_x &= [(0.85 \times 4000)/47.5] &= 71.6 \\ (0.7L_x/r_x + 30) &= [(0.7 \times 4000)/47.5 + 30] &= 88.9 \\ [(L_y/r_y)^2 + \lambda_c^2]^{0.5} &= [(4000/30.90)^2 + 44.3^2]^{0.5} &= 136.8 \\ 1.4\lambda_c &= (1.4 \times 44.3) &= 62.0 \end{aligned}$ The critical slenderness $= \lambda = 136.8$	

Table 23
Table 24(6)
 	
Clause 4.7.4
	Use strut curve (c) for all axes and since $\lambda \geq 110$ use Table 24(6)
	Using S 275 steel; $p_y = 275 \text{ N/mm}^2$
	Interpolate $p_c = \{81 - [(81 - 76) \times 1.8/(140 - 135)]\} \text{ N/mm}^2$
	   $\therefore p_c = 79.2 \text{ N/mm}^2$
	   $P_c = A_g p_c$
	   $= (6340 \times 79.2)/10^3 = 502 \text{ kN}$
		The maximum compressive load which can be supported = 502 kN

6.3.3.8   Example 6.11: Side Column with Lateral Restraint to Flanges

The side column of a single storey building in which the roof truss is supported on a cap plate and simply connected such that it does not develop any significant moment is shown in Figure 6.25. Determine the maximum compressive load which the column can support.

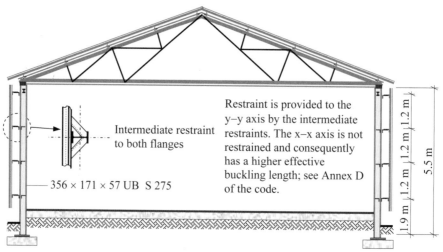

Intermediate restraint
to both flanges

Restraint is provided to the y–y axis by the intermediate restraints. The x–x axis is not restrained and consequently has a higher effective buckling length; see Annex D of the code.

356 × 171 × 57 UB S 275

1.9 m 1.2 m 1.2 m 1.2 m 1.2 m

5.5 m

Figure 6.25

6.3.3.9   Solution to Example 6.11

Contract : Struts Job Ref. No. : Example 6.11	Calcs. by : W.McK.
Part of Structure : Side Column with Restraint Calc. Sheet No. : 1 of 2	Checked by : Date :

References	Calculations	Output
	Member: 356 × 171 × 57 UB S 275 Section properties: A_g = 7260 mm^2 r_x = 149 mm r_y = 39.1 mm t = 8.1 mm T = 13.0 mm b/T = 6.62 d/t = 38.5	
Table 9	Design strength: p_y = 275 N/mm^2 ; ε = 1.0	
Table 11	Section Classification: Outstand element of a compression flange – Rolled sections $\quad b/T$ = 6.62 < 9ε $\quad d/t$ = 38.5 < 40ε	**Flanges are plastic** **Web is non-slender**

References	Calculations	Output
Contract : Struts Job Ref. No. : Example 6.11 Part of Structure : Side Column with Restraint Calc. Sheet No. : 2 of 2		Calcs. by : W.McK. Checked by : Date :

References	Calculations	Output
Clause 4.7.2 Clause 4.7.3 Annex D	Slenderness $\lambda = L_E/r$ For single-storey columns with intermediate restraint to both flanges the effective lengths are indicated in Figure D2 of Annex D1. $L_x = 1.5L = (1.5 \times 5.5) = 8.25$ m $\lambda_x = \dfrac{L_x}{r_x} = \dfrac{8250}{149} = 55.4$ $L_y \geq 0.85L_1 = (0.85 \times 1.9) = 1.615$ m * critical value $ \geq 1.0\,L_2 = (1.0 \times 1.2) = 1.2$ m $\lambda_y = \dfrac{L_y}{r_y} = \dfrac{1615}{39.1} = 41.3$	
Table 23	$T = 13.0$ mm $<$ 40 mm For Rolled **I** Sections ∴ Use strut curve (**a**) for the x–x axis ∴ Use strut curve (**b**) for the y–y axis	
Table 24(1) Table 24(3) Clause 4.7.4	$\lambda_x = 55.4,\;\; p_y = 275$ N/mm^2 ∴ $p_c \approx 245$ N/mm^2 $\lambda_y = 41.3,\;\; p_y = 275$ N/mm^2 ∴ $p_c \approx 248.5$ N/mm^2 ∴ Critical value of $p_c = 245$ N/mm^2 $\qquad\qquad P_c = A_g p_c$ $\qquad\qquad\;\;\; = (7260 \times 245)/10^3 = 1779$ kN	**The maximum compressive load which can be supported = 1779 kN**

6.3.3.10 Example 6.12: Concentrically Loaded Column

A column in a braced building supports a symmetrical arrangement of beams in addition to a vertical load from above, as shown in Figure 6.26. Using the characteristic loads indicated, check the suitability of a $203 \times 203 \times 60$ UC S 355 section.

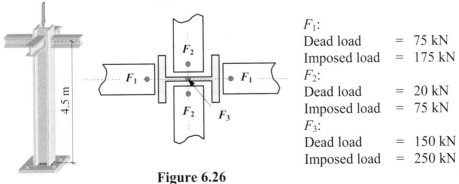

F_1:		
Dead load	=	75 kN
Imposed load	=	175 kN

F_2:		
Dead load	=	20 kN
Imposed load	=	75 kN

F_3:		
Dead load	=	150 kN
Imposed load	=	250 kN

Figure 6.26

6.3.3.11 Solution to Example 6.12

References	Calculations	Output
	Contract : Struts Job Ref. No. : Example 6.12 **Calcs. by : W.McK.** **Part of Structure : Concentrically Loaded Column** **Checked by :** **Calc. Sheet No. : 1 of 2** **Date :**	

References	Calculations	Output
	Member AB: $203 \times 203 \times 60$ UC S 355 Section properties: A_g = 7640 mm^2 r_x = 89.6 mm r_y = 52.0 mm t = 9.4 mm T = 14.2 mm b/T = 7.25 d/t = 17.1	
	Total characteristic dead load $G_k = [2 \times (75 + 20) + 150]$ $= 340$ kN Total characteristic imposed load $Q_k = [2 \times (175 + 75) + 250]$ $= 750$ kN	
Table 2	Partial safety factors: $\gamma_{m, \text{dead load}}$ = 1.4 $\gamma_{m, \text{imposed load}}$ = 1.6 Design load = $[(1.4 \times G_k) + (1.6 \times Q_k)]$ = $[(1.4 \times 340) + (1.6 \times 750)]$ = 1676 kN	
Table 9	Design strength: p_y = 355 N/mm^2 ε = $(275/355)^{0.5}$ = 0.88	
Table 11	Section Classification: Outstand element of a compression flange – Rolled sections 9ε = (9×0.88) = 7.92 b/T = 7.25; $b/T < 9\varepsilon$	**Flanges are plastic**
Clause 3.5.5	Webs with axial compression $r_2 = \dfrac{F_c}{A_g p_{yw}}$ $= \left(\dfrac{1676 \times 10^3}{7640 \times 355}\right)$ = 0.618 $\therefore \dfrac{120\varepsilon}{1 + 2r_2} = \left[\dfrac{(120 \times 0.88)}{1 + (2 \times 0.618)}\right]$ = 47.2 $\geq$ 40ε d/t = 17.1 $< \dfrac{120\varepsilon}{1 + 2r_2}$ **Note:** In this case it is not necessary to evaluate the full expression as above since clearly $d/t < 40\varepsilon$	**Web is non-slender**

References	Calculations	Output
	Contract : Struts **Job Ref. No. : Example 6.12** **Calcs. by : W.McK.** **Part of Structure : Concentrically Loaded Column** **Checked by :** **Calc. Sheet No. : 2 of 2** **Date :**	

References	Calculations	Output
Clause 4.7.2	Slenderness $\lambda = L_E/r$	
Table 22	(a) No-sway mode. Assume the column to be effectively held in position at both ends and not restrained in direction at either end. Effective length $L_E = 1.0L = (1.0 \times 4.5) = 4.5$ m $\lambda_x = \dfrac{L_x}{r_x} = \dfrac{4500}{89.6} = 50.2$ $\lambda_y = \dfrac{L_y}{r_y} = \dfrac{4500}{52.0} = 86.5$	
Table 23	$T = 14.2$ mm < 40 mm For Rolled **H** Sections $\therefore$ Use strut curve (**b**) for the x–x axis $\therefore$ Use strut curve (**c**) for the y–y axis	
Table 24(3) Table 24(5)	$\lambda_x = 50.2, \quad p_y = 355$ N/mm^2 $\therefore p_c \approx 297.5$ N/mm^2 $\lambda_y = 86.5, \quad p_y = 355$ N/mm^2 $\therefore p_c \approx 169$ N/mm^2	
Clause 4.7.4	$\therefore$ Critical value of $p_c = 149$ N/mm^2 $\quad\quad P_c = A_g p_c$ $\quad\quad\quad = (7640 \times 169)/10^3 = 1291$ kN $\quad\quad\quad\quad\quad < \text{Design load}$	**Increase the column size**

6.3.3.12 Column Base Plates

Columns which are assumed to be nominally pinned at their bases are provided with a slab base comprising a single plate fillet welded to the end of the column and bolted to the foundation with four **holding down (H.D.) bolts**. The base plate, welds and bolts must be of adequate size, stiffness and strength to transfer the axial compressive force and shear at the support without exceeding the bearing strength of the bedding material and concrete base, as shown in Figure 6.27.

The dimensions of the plate must be sufficient to distribute the axial compressive load to the foundations and to accommodate the holding down bolts. It is usual to calculate the thickness as indicated in Example 6.12 and to ensure that it is ≥ to the flange thickness.

The purpose of the welds is to transfer the shear force at the base and to securely attach the plate to the column. In most cases either 6 mm or 8 mm fillet welds run along the flanges and for a short distance on either side of the web will be adequate.

The holding down bolts are generally cast within location cones in a concrete base and fitted with an anchor plate to prevent pull-out. The purpose of the location cone is to allow for movement before final grouting and hence to permit site adjustment during construction. The diameter at the top of the location cones is usually at least 100 mm or 3 × bolt diameter. The recommended size of H.D. bolts is M20 for light construction, M24 for bases up to 50 mm thick, increasing to M36 for heavier plates. Clearance holes in the base plates should be 6 mm larger than the bolt diameter. The bolts are embedded in the concrete base to a length equal to approximately 16 to 18 × bolt diameter with a threaded length at least equal to the bolt diameter plus 100 mm. In cases where a moment is applied to the base and tension may develop in the bolts, reference should be made to Clause 4.13.2.4.

Bedding material can be mortar, fine concrete or a proprietary, non-shrink grout. In lightly loaded bases a gap of 25 mm to 50 mm is normally provided; this allows access for grouting the H.D. bolt pockets and ensuring that the gap under the base plate is completely filled. It is normal for the strength of the bedding material to be at least equal to that of the concrete base. In BS 5950-1:2000 Clause 4.13.1.2, the bearing strength for concrete foundations is given as $0.6f_{cu}$, where f_{cu} is the characteristic concrete cube strength at 28 days.

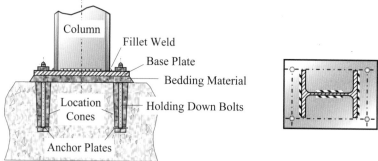

Figure 6.27

In BS 5950-1:2000 Clause 4.13.2.2, the formula for determining the minimum thickness of a rectangular base plate supporting a concentrically loaded column is given as:

$$t_p = c[3w/p_{yp}]^{0.5}$$
$$\geq T$$

where:

c is the largest perpendicular distance from the edge of the effective portion (see Figure 6.28) of the baseplate to the face of the column cross-section,

p_{yp} is the design strength of the baseplate,

T is the flange thickness (or maximum thickness) of the column,

w is the pressure under the baseplate, based on an assumed uniform distribution of pressure throughout the effective portion.

This equation is based on assuming an effective area as defined in Figure 15 of the code, in which a portion of the area of any baseplate which is larger than is required to limit the bearing pressure to $0.6f_{cu}$, is considered ineffective. This situation frequently occurs, for example, when the minimum dimensions are governed by the size required to accommodate the column section, welds and H.D. bolts. The effective areas indicated for various cases in the code are shown in Figure 6.28.

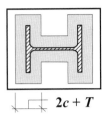 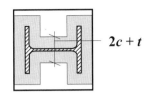 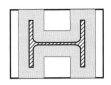

$2c + t$

$2c + T$

Shaded areas are considered to be the effective areas
when evaluating the pressure under the baseplates

$2c + t$ stiffener

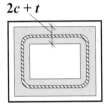 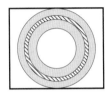 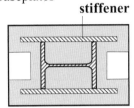

Figure 6.28

6.3.3.13 Example 6.13: Column Base Plate

Design a suitable base plate for a 203 × 203 × 60 UC S 275 column section considering
the following two design loads:

(i)   750 kN   and
(ii)   1900 kN

6.3.3.14 Solution to Example 6.13

Contract : **Struts** **Job Ref. No. : Example 6.13** **Part of Structure : Column Base Plate** **Calc. Sheet No. : 1 of 3**	**Calcs. by : W.McK.** **Checked by :** **Date :**

References	Calculations	Output
Section Tables Table 9	Design Data: 203 × 203 × 60 UC S 275 Section Properties Surface Area = 1.21 m²/ metre length ∴ Perimeter = 1210 mm A_g = 7640 mm²; D = 209.6 mm; B = 205.8 mm; T = 14.2 mm; t = 9.4 mm f_{cu} = 20 N/mm² Assume the base plate is ≤ 16 mm thick Design strength: p_{yp} = 275 N/mm² ; ε = 1.0	

References	Calculations	Output
	Contract : Struts Job Ref. No. : Example 6.13 **Part of Structure : Column Base Plate** **Calc. Sheet No. : 2 of 3**	**Calcs. by : W.McK.** **Checked by :** **Date :**

References	Calculations	Output
Figure 15	Effective area of baseplate: Shaded area = Cross-sectional area $+ 4c^2$ + (Perimeter $\times$ c) Effective area = $[A_g + 4c^2 +$ (Perimeter $\times$ c)] = $(7640 + 4c^2 + 1210c)$	
	(i) Design Load = 750 kN	

Clause 4.13.1	Effective area required $= \dfrac{750 \times 10^3}{(0.6 \times 20)} = 62{,}500$ mm^2
	
	$\therefore$ 62,500 = $4c^2 + 1210c + 7640$
	            0 = $4c^2 + 1210c - 54860$
	            0 = $c^2 + 302.5c - 13715$
	
	$c = \dfrac{-302.5 \pm \sqrt{302.5^2 + (4 \times 13715)}}{2} = \dfrac{(-302.5 + 382.6)}{2}$
	
	   = 40 mm
	
	$(D - 2T - 2c)$ = $[209.6 - (2 \times 14.2) - (2 \times 40)]$ = 101.2 mm
	Since $(D - 2T - 2c)$ is +ve there is no overlap of the c values between the flanges.
	$(D + 2c)$ = $[209.6 + (2 \times 40)]$ = 289.6 mm
	$(B + 2c)$ = $[205.8 + (2 \times 40)]$ = 285.8 mm	

Clause 4.13.2.2	t_p = $c[3w/p_{yp}]^{0.5}$ = $\{40 \times [(3 \times 0.6 \times 20)/275]^{0.5}\}$ = 14.5 mm $\le$ 16 mm $(\ge T)$	**Adopt a base plate:** **300 mm $\times$ 300 mm $\times$** **15 mm thick.**

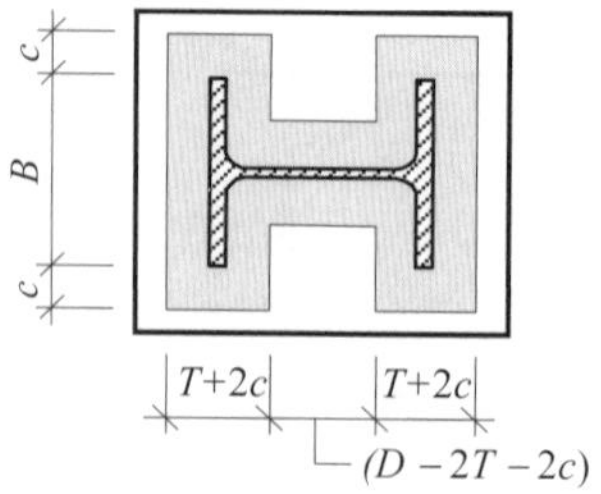

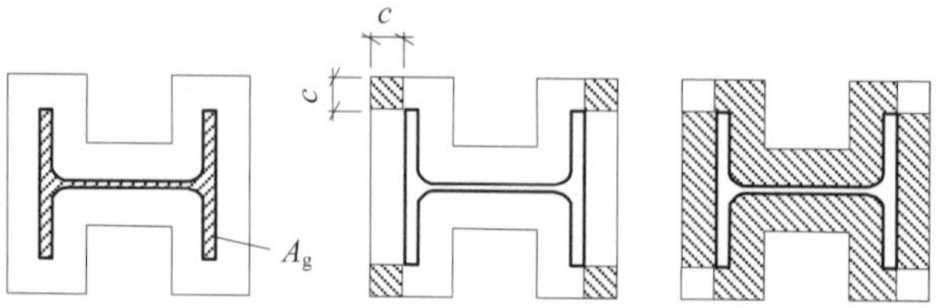

References	Calculations	Output
	Contract : Struts Job Ref. No. : Example 6.13 **Part of Structure : Column Base Plate** **Calc. Sheet No. : 3 of 3** **Calcs. by : W.McK.** **Checked by :** **Date :**	

References	Calculations	Output
	(i) Design Load = 1900 kN	
	Effective area $= (7640 + 4c^2 + 1210c)$	
Clause 4.13.1	Effective area required $= \dfrac{1900 \times 10^3}{(0.6 \times 20)} = 158{,}333 \text{ mm}^2$	
	$\therefore 158{,}333 = 4c^2 + 1210c + 7640$	
	$0 = 4c^2 + 1210c - 150{,}693$	
	$0 = c^2 + 302.5c - 37673$	
	$c = \dfrac{-302.5 \pm \sqrt{302.5^2 + (4 \times 37673)}}{2} = \dfrac{(-302.5 + 492.1)}{2}$	
	$= 94.8 \text{ mm}$	
	$(D - 2T - 2c) = [209.6 - (2 \times 14.2) - (2 \times 94.8)] = -8.4 \text{ mm}$	
	Since $(D - 2T - 2c)$ is −ve there is insufficient width available between the flanges. Assume the effective area comprises a complete rectangle.	
	Effective area $= [(B + 2c) \times (D + 2c)]$ $= [BD + 2c(B + D) + 4c^2]$	
	$158{,}333 = [4c^2 + 830.8c + 43{,}136]$	
	$0 = 4c^2 + 830.8c - 115197$	
	$0 = c^2 + 207.7c - 28799$	
	$c = \dfrac{-207.7 \pm \sqrt{207.7^2 + (4 \times 28799)}}{2} = \dfrac{(-207.7 + 397.9)}{2}$	
	$= 95.1 \text{ mm}$	
	Plate size $\geq [(D + 2c) \times (B + 2c)]$ $= [209.6 + (2 \times 95.1)] \times [205.8 + (2 \times 95.1)]$ $= 400 \text{ mm} \times 396 \text{ mm}$	
Table 9	Assume $16 \text{ mm} < t_p \leq 63 \text{ mm} \therefore p_y = 265 \text{ N/mm}^2$	
Clause 4.13.2.2	$t_p = c[3w/p_{yp}]^{0.5} = \{95.1 \times [(3 \times 0.6 \times 20)/265]^{0.5}\}$	**Adopt a base plate:** **400 mm × 400 mm ×**
	$= 35 \text{ mm} \leq 63 \text{ mm} \quad (\geq T)$	**35 mm thick**
	H.D. Bolts Assume M24's threaded length $\approx 24 + 100 = 124 \text{ mm}$ embedded length $\approx 16 \times 24 = 384 \text{ mm}$ thickness of bedding material $= 30 \text{ mm}$ Total length of bolt $= (124 + 35 + 30 + 384) = 573 \text{ mm}$	**6 mm fillet welds to flanges and web, M24 mm H.D. bolts 575 mm long with 125 mm threaded length.**

6.3.4 Flexural Elements

The most frequently used, and possibly the earliest used, structural element is the beam. The primary function of a beam is to transfer vertical loading to adjacent structural elements such that the load can continue its path through the structure to the foundations. Loading can be imposed on a beam from one or several of a number of sources, e.g. other secondary beams, columns, walls, floor systems or directly from installed plant and/or equipment. In most cases static loading will be considered the most appropriate for design purposes, but dynamic and fatigue loading may be more critical in certain circumstances.

The structural action of a beam is predominantly **bending**, with other effects such as **shear**, **bearing** and **buckling** also being present. In addition to ensuring that beams have sufficient strength capacities to resist these effects, it is important that the stiffness properties are adequate to avoid excessive **deflection** or **local buckling** of the cross-section.

A large variety of cross-sections are available when selecting a beam for use in any one of a wide range of applications. The most common types of beam, with an indication of the span range for which they may be appropriate, are given in Figure 6.29.

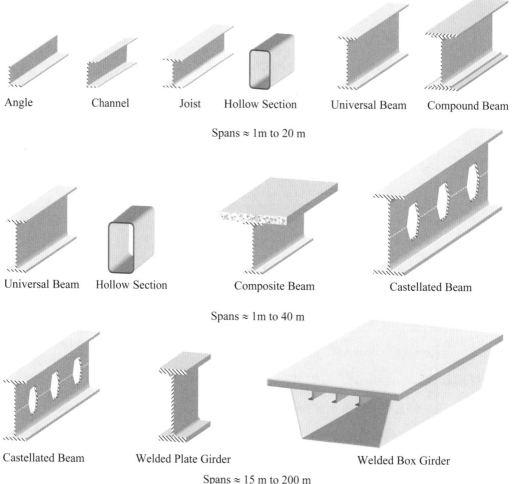

| Angle | Channel | Joist | Hollow Section | Universal Beam | Compound Beam |

Spans ≈ 1m to 20 m

| Universal Beam | Hollow Section | Composite Beam | Castellated Beam |

Spans ≈ 1m to 40 m

| Castellated Beam | Welded Plate Girder | Welded Box Girder |

Spans ≈ 15 m to 200 m

Figure 6.29

For lightly loaded and small spans such as roof purlins and side sheeting rails, the use of hot-rolled angle sections or channel sections is appropriate. Cold-formed sections pressed from thin sheet and galvanised, and provided by proprietary suppliers, are frequently used.

In small to medium spans hot-rolled joists, universal beams (UBs), hollow sections and UBs with additional welded flange plates (compound beams) are often used. If the span and/or magnitude of loading dictates that larger and deeper sections are required, castellated beams formed by welding together profiled cut UB sections, plate girders or box girders can be fabricated in which the webs and flanges are individual plates welded together.

While careful detailing can minimize torsional effects, when they are considered significant, hollow tube sections are more efficient than open sections such as UBs, universal columns (UCs), angles and channels.

The section properties of all hot-rolled sections and cold-formed sections are published by their manufacturers; those for fabricated sections must be calculated by the designer.

The span of a beam is defined in Clause 4.2.1.2 of BS 5950-1:2000 as the distance between points of effective support. In general, unless the supports are wide columns or piers then the span can be considered as the centre-to-centre of the actual supports or columns.

The most widely adopted section to be found in building frames is the Universal Beam. The design of beams to satisfy the requirements of BS 5950-1:2000 includes the consideration of:

♦ section classification,
♦ shear capacity,
♦ moment capacity (including lateral torsional buckling),
♦ deflection,
♦ web bearing,
♦ web buckling,
♦ torsional capacity (not required for the design of most beams).

6.3.4.1   Section Classification (Clause 3.5)

The section classification of structural elements is described in detail in Section 6.2.5 of this text.

6.3.4.2   Shear Capacity (Clause 4.2.3)

Generally in the design of beams for buildings, the effects of shear are negligible and will not significantly reduce the value of the moment capacity. It is evident from the elastic shear stress distribution in an I-beam, as shown in Figure 6.30 that the web of a cross-section is the primary element which carries the shear force.

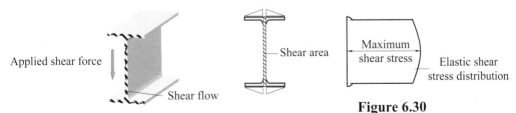

Figure 6.30

In situations such as at internal supports of continuous beams where there is likely to be high coincident shear and moment effects which may induce significant principal stresses (see Figure 6.31), it is sometimes necessary to consider the reduction in moment capacity caused by the effects of the shear.

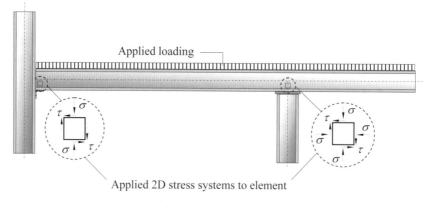

Applied 2D stress systems to element

Figure 6.31

The shear capacity of a beam is defined in the code as:

$$P_v = 0.6 \, p_y \, A_v$$

where:
$0.6 \, p_y$ is approximately equal to the yield stress of steel in shear,
A_v is the shear area as defined in Clause 4.2.3.

When the applied ultimate shear force (F_v) is equal to or greater than 60% of P_v (i.e. $F_v \geq 0.36 \, p_y \, A_v$) the moment capacity of a beam should be reduced as specified in Clause 4.2.5.3.

When the aspect ratio of a web (d/t) is greater than 70ε for rolled sections or 62ε for welded sections, the possibility of shear buckling should be considered in accordance with Clause 4.4.5. In the design of webs of variable thickness and/or which contain large holes (e.g. castellated beams), the code requires that shear stresses be calculated from first principles, assuming elastic behaviour and a maximum shear stress not exceeding $0.7p_y$; the design of such webs is not considered in this text.

6.3.4.3   Example 6.14: Shear Check of a Simply Supported Beam

A simply supported $406 \times 178 \times 74$ UB S 275 is required to span 4.5 m and to carry an ultimate design load of 40 kN/m. Check the suitability of the section with respect to shear.

Solution:
Section properties: $406 \times 178 \times 74$ UB S 275
t $=$ 9.5 mm, D $=$ 412.8 mm, d $=$ 360.4 mm, d/t $=$ 37.9

Design shear force at the end of the beam $F_v = \dfrac{40 \times 4.5}{2} = 90$ kN

Clause 4.2.3 $P_v = 0.6 \, p_y \, A_v$

For a rolled UB section $\qquad A_v = tD$

Shear area $\qquad A_v = (9.5 \times 412.8) = 3.922 \times 10^3 \text{ mm}^2$

Clause 3.1.1 $\qquad$ Web thickness $\qquad t = 9.5 \text{ mm}$

Table 9 $\qquad t \le 16 \text{ mm} \qquad p_y = 275 \text{ N/mm}^2 \text{ and } \varepsilon = 1.0$

Clause 3.5 $\qquad$ Since the beam is subject to pure bending the neutral axis will be at mid-depth.

Section Classification:

Table 11 $\qquad \dfrac{d}{t} = 37.9 < 80\varepsilon \qquad\qquad\qquad$ **Web is plastic**

Clause 4.2.3 $\qquad$ Shear capacity $\quad P_v = \dfrac{0.6 \times 275 \times 3.922 \times 10^3}{10^3} = 647 \text{ kN}$

$$\gg F_v \text{ (90 kN)}$$

$\dfrac{d}{t} < 70\varepsilon \quad$ therefore no need to check shear buckling.

This value indicates the excessive reserve of shear strength in the web.

In cases where bolt holes are required in the web, no allowance is required provided that:

$\qquad A_{v.net} \ge 0.85A_v / K_e$

where:

$A_{v.net}$ is the net shear area after deducting bolt holes,

$K_e \quad$ is the effective net area coefficient from Clause 3.4.3.

When $A_{v.net} < 0.85A_v / K_e$ the net shear capacity is modified such that $P_v = 0.7 p_y K_e A_{v.net}$ as indicated in Clause 6.2.3.

6.3.4.4 Moment Capacity (Clauses 4.2.5 and 4.3)

The moment capacity of a beam is determined by a number of factors such as:

(i)	design strength	Table 9	
(ii)	section classification	Table 12	
(iii)	elastic section modulus	Z	
(iv)	plastic section modulus	S	
(v)	co-existent shear	Clauses (4.2.5.2 and 4.2.5.3)	and
(vi)	lateral restraint to the compression flange	Clause 4.3	

The criteria (i) to (v) are relatively straightforward to evaluate, however criterion (vi) is related to the lateral torsional buckling of beams and is much more complex. The design of beams in this text are considered in two categories:

(a) beams in which the compression flange is fully restrained and lateral torsional buckling cannot occur, and

(b) beams in which either no lateral restraint or only intermittent lateral restraint is provided to the compression flange.

Refer to Chapter 4, Section 4.4 of this text relating to lateral torsional buckling.

6.3.4.4.1. Effective Length

The provision of lateral and torsional restraints to a beam introduces the concept of **effective length**. The effective length of a compression flange is the equivalent length between restraints over which a pin-ended beam would fail by lateral torsional buckling. The values to be used in assessing this are given in Tables 13 and 14 for beams and cantilevers respectively. The values adopted depend on three factors relating to the degree of lateral and torsional restraint at the position of the intermittent restraints. They are:

(a)   the existence of torsional restraints,
(b)   the degree of lateral restraint of the compression flange,
(c)   the type of loading.

In the case of beams (Table 13) factors (a) and (b) give rise to seven possible conditions.

(a)   When nominal torsional restraint exists, as indicated in Clause 4.2.2, and the compression flange is fully restrained:

(i)   both the compression and tension flanges are fully restrained against rotation on plan,
(ii)   the compression flange is fully restrained against rotation on plan,
(iii) both flanges are partially restrained against rotation on plan, or
(iv)   the compression flange is partially restrained against rotation on plan,
(v)   both flanges are free to rotate on plan.

(b)   When both flanges are free to rotate on plan and the compression flange is unrestrained:

(i)   partial torsional restraint against rotation about the longitudinal axis provided by the connection of the bottom flange to the supports,
(ii)   partial torsional restraint against rotation about the longitudinal axis is provided only by pressure of the bottom flange onto supports.

Similar conditions exist in Table 14 for cantilevers.
   Guidance is given in Clause 4.3.3 to assist designers in assessing the degree of torsional restraint which exists.

(c)   Type of loading:
       A beam load is considered **normal** unless both the beam and the load are free to deflect laterally and so induce lateral torsional buckling by virtue of the combined freedom; in this case the load is a **destabilising** load. In an efficiently designed braced structural system, destabilising loads should not normally arise. In some instances the existence of such a load is unavoidable e.g. the sidesway induced in crane-gantry girders by the horizontal surge loads (see Figure 6.32).

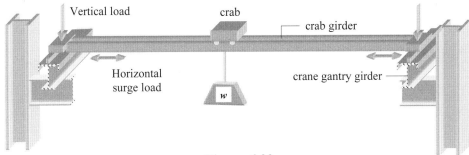

Figure 6.32

In Table 14 it can be seen that destabilising loads are a particular problem for cantilevers since it may be difficult to achieve torsional rigidity at either the free or the fixed end. In addition the bottom flange, which is in compression, may not be as readily restrained as the top flange.

6.3.4.4.2. Moment Capacity (M_c) of Beams With Full Lateral Restraint

The moment capacity (M_c) of beams with the compression flange fully restrained is determined using the following equations which are given in Clauses 4.2.5.2 and 4.2.5.3 for low coincident shear and high coincident shear respectively.

For Plastic and Compact Sections:
Clause 4.2.5.2   **Low Shear** $F_v \le 0.6P_v$:

$$M_c = p_y S$$
$$\le 1.2p_y Z \quad \text{for simply supported beams and cantilevers and}$$
$$\le 1.5p_y Z \quad \text{generally.}$$

The limitations based on the elastic section modulus are to ensure that plasticity does not occur at service loads.

Clause 4.2.6   **High Shear** $F_v > 0.6P_v$

$$M_c = p_y(S - \rho S_v)$$
$$\le 1.2p_y Z \quad \text{for simply supported beams and cantilevers and}$$
$$\le 1.5p_y Z \quad \text{generally.}$$

where:
S is the plastic modulus,
S_v is either
  (i)   the plastic modulus of the shear area for sections defined in Clause 4.2.3 or
  (ii)   equal to $(S - S_f)$ the plastic modulus of the effective cross-section excluding the shear area A_v defined in Clause 4.2.3.
ρ is defined by: $[2(F_v/P_v) - 1]^2$

For Semi-Compact Sections:

Clause 4.2.5.2　　**Low Shear** $F_v \leq 0.6P_v$:

$$M_c = \quad p_y Z \quad \text{or alternatively} \quad M_c = p_y S_{eff}$$
$$\leq \ 1.2p_y Z \quad \text{for simply supported beams and cantilevers and}$$
$$\leq \ 1.5p_y Z \quad \text{generally.}$$

Clause 4.2.6　　**High Shear** $F_v > 0.6P_v$

$$M_c = p_y[Z - (\rho S_v/1.5)] \quad \text{or alternatively } M_c = p_y[S_{eff} - (\rho S_v)]$$
$$\leq \ 1.2p_y Z \quad \text{for simply supported beams and cantilevers and}$$
$$\leq \ 1.5p_y Z \quad \text{generally.}$$

where:
S_{eff} is the effective plastic modulus as indicated in Clause 3.5.6.

For Slender Sections:
Clause 4.2.5.2　　**Low Shear** $F_v \leq 0.6P_v$:

$$M_c = \ p_y Z_{eff}$$

Clause 4.2.6　　**High Shear** $F_v > 0.6P_v$

$$M_c = p_y[Z_{eff} - (\rho S_v/1.5)]$$

where:
Z_{eff} is the effective plastic modulus as indicated in Clause 3.6.2.

6.3.4.5　　Example 6.15: Moment Capacity of Beam with Full Lateral Restraint

A single span beam is simply supported between two columns and carries a reinforced concrete slab in addition to the column and loading shown in Figure 6.33. Using the characteristic loads indicated, select a suitable section considering section classification, shear and bending only. Assume S 275 steel and that dead loads are inclusive of self-weights.

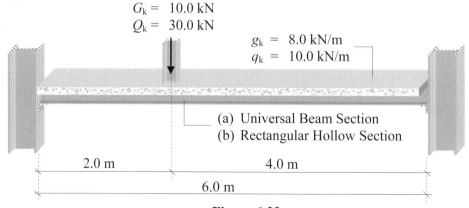

$$G_k = \ 10.0 \text{ kN}$$
$$Q_k = \ 30.0 \text{ kN}$$
$$g_k = \ 8.0 \text{ kN/m}$$
$$q_k = \ 10.0 \text{ kN/m}$$

(a) Universal Beam Section
(b) Rectangular Hollow Section

2.0 m　　　　　　　　　　4.0 m

6.0 m

Figure 6.33

6.3.4.6   Solution to Example 6.15

Contract : Beams Job Ref. No. : Example 6.15 Part of Structure : Fully Restrained Beam Calc. Sheet No. : 1 of 3	Calcs. by : W.McK. Checked by : Date :

References	Calculations	Output
	(a) Design a suitable Universal Beam Section	
	G_k = 10.0 kN g_k = 8.0 kN/m Q_k = 30.0 kN q_k = 10.0 kN/m	
Table 2	Design point load $= (1.4 \times 10) + (1.6 \times 30)$ $=$ 62 kN Design UDL $= (1.4 \times 8) + (1.6 \times 10)$ $=$ 27.2 kN/m	
	By proportion: Vertical reaction at A = V_A = $$\left(27.2 \times 3\right) + \left(\frac{62 \times 4}{6}\right) = 122.9 \text{ kN}$$ Vertical reaction at C = V_C = $$\left(27.2 \times 3\right) + \left(\frac{62 \times 2}{6}\right) = 102.3 \text{ kN}$$	
	Shear Force Diagram	Design shear force = 122.9 kN
	Position of zero shear $x = \dfrac{6.5}{27.2} = 0.24$ m Maximum bending moment occurs at position of zero shear	
	The maximum bending moment = shaded area = $$M_x = \frac{(122.9 + 68.5)2}{2} + \left(\frac{0.24 \times 6.5}{2}\right) = 192.2 \text{ kNm}$$	Design bending moment = 192.2 kNm

References	Calculations	Output
	Contract : Beams Job Ref. No. : Example 6.15 **Part of Structure : Fully Restrained Beam** **Calc. Sheet No. : 2 of 3**	**Calcs. by : W.McK.** **Checked by :** **Date :**

References	Calculations	Output
	The compression flange is fully restrained and assume low shear	
Table 9	(a) Consider a universal beam section: Assume the flange thickness $T < 16$ mm $\therefore p_y = 275$ N/mm^2 $\therefore S_{x, \text{ required}} \geq \dfrac{192.2 \times 10^6}{275} = 698.9 \times 10^3$ mm^3 A trial beam size can be selected from published section tables: try a $305 \times 165 \times 46$ UB S 275	
Section Tables	Section Properties: D = 306.6 mm d = 265.2 mm B = 165.7 mm T = 11.8 mm t = 6.7 mm b/T = 7.02 d/t = 39.6 mm S_x = 720 × 10^3 mm^3 Z_x = 646 × 10^3 mm^3	
Table 11	(i) Section Classification $\varepsilon = \left(\dfrac{275}{p_y}\right)^{\frac{1}{2}} = 1.0$ Flange: Outstand element of compression flange rolled section $b/T = 7.02 < 9.0\varepsilon$	**Flange is plastic**
	Web: Neutral axis at mid-depth (i.e. bending only) $d/t = 39.6 < 80\varepsilon$	**Web is plastic**
Clause 4.2.3	$< 70\varepsilon$ therefore no need to check shear buckling	**Section is plastic**
Clause 4.2.3	(ii) Shear Design shear force = $F_v = 122.9$ kN P_v = $(0.6p_yA_v)$ = $0.6p_ytD$ P_v = $(0.6 \times 275 \times 6.7 \times 306.6)/10^3$ = 338.9 kN $P_v > F_v$	**Section is adequate in shear**
Clause 4.2.5.2	(iii) Bending 60% P_v = (0.6×338.9) = 203.3 kN > 68.5 kN Low shear **Note:** 68.5 kN is the coincident shear at the position of maximum bending moment.	

Clause 4.2.5.1
 	
Clause 4.2.5.2
	$M_c \leq 1.2p_y Z_x$ for simply supported beams
	    = $(1.2 \times 275 \times 646 \times 10^3)/10^6$ = 213.2 kNm
	$M_c = p_y S_x$
	    = $(275 \times 720 \times 10^3)/10^6$ = 198 kNm
	
	Maximum applied moment = M_x = 192.2 kNm
	
	$M_c > M_x$
		Critical value of M_c = 198 kNm
		
		Section is adequate in bending

Contract : **Beams** **Job Ref. No. : Example 6.15** **Part of Structure :** **Fully Restrained Beam** **Calc. Sheet No. : 3 of 3**		**Calcs. by : W.McK.** **Checked by :** **Date :**

References	Calculations	Output
Table 9	(b) Consider a rectangular hollow section: The flange thickness $t < 16$ mm $\therefore p_y = 275$ N/mm^2 $$\therefore S_{x,\ required} \geq \frac{192.2 \times 10^6}{275} = 698.9 \times 10^3 \text{ mm}^3$$ A trial beam size can be selected from published section tables: try a $300 \times 200 \times 8$ RHS S275	
Section Tables	Section Properties: $D = 300$ mm $\quad B = 200$ mm $\quad A = 7680$ mm^2 $Z_x = 648 \times 10^3$ mm^3 $\quad S_x = 779 \times 10^3$ mm^3 $b/t = 22.0 \quad\quad\quad d/t = 34.5$ mm	
Table 12	(i) Section Classification $\quad \varepsilon = \left(\dfrac{275}{p_y}\right)^{\frac{1}{2}} = 1.0$ Flange: Hot finished rectangular hollow section with compression due to bending. $b/T = 22.0 < 28.0\varepsilon$ Web: Neutral axis at mid-depth (i.e. bending only) $d/t = 39.6 < 64\varepsilon$	**Flange is plastic** **Web is plastic**
Clause 4.2.3	$< 70\varepsilon$ therefore no need to check shear buckling	**Section is plastic**
Clause 4.2.3	(ii) Shear Design shear force $= F_v = 122.9$ kN $P_v = (0.6p_yA_v) = 0.6p_yAD/(D+B)$ $P_v = [(0.6 \times 275 \times 7680 \times 300)/(300 + 200)]/10^3 = 760.3$ kN $$P_v \gg F_v$$	**Section is adequate in shear**
Clause 4.2.5.2	(iii)Bending $60\%\ P_v = (0.6 \times 760.3) = 456.2$ kN $\gg 68.5$ kN Low shear **Note:** 68.5 kN is the coincident shear at the position of maximum bending moment.	
Clause 4.2.5.1 Clause 4.2.5.2	$M_c \leq 1.2p_y Z_x$ for simply supported beams $= (1.2 \times 275 \times 648 \times 10^3)/10^6 = 213.8$ kNm $M_c = p_y S_x$ $= (275 \times 779 \times 10^3)/10^6 = 214.2$ kNm Maximum applied moment $= M_x = 192.2$ kNm $$M_c > M_x$$	**Critical value of** $M_c = 213.8$ **kNm** **Section is adequate in bending**

6.3.4.7   Moment Capacity of Beams with Intermittent Lateral Restraint

As indicated in Section 4.4 of Chapter 4, the moment capacity of a beam in which the compression flange is not fully restrained is known as the **Buckling Resistance Moment** (M_b) and is defined in Clause 4.3.6.4 as:

♦ $M_b = p_b S_x$ for plastic and compact cross-sections

♦ $M_b = p_b Z_x$
         or alternatively   for semi-compact sections
      $= p_b S_{x,eff}$

♦ $M_b = p_b Z_{x,eff}$ for slender sections

Tha appropriate values of S_x, $S_{x,eff}$, Z_x, and $Z_{x,eff}$ can be found in Section Property Tables or calculated in accordance with Clauses 3.5.6 and 3.6.7 as illustrated in Section 6.2.6.

The bending strength p_b is dependent on the design strength p_y the **Equivalent Slenderness**–λ_{LT}, and the type of member, i.e. rolled or fabricated by welding. The characteristic shape of the bending moment diagram also influences the value of moment at which lateral torsional buckling will occur.

Consider an unrestrained length of beam which is subject to a maximum bending moment M_1 at one end and M_2 at the other, as shown in Figures 6.34 (a) and (b). In case (a) the beam is subject to single curvature and in case (b) to double curvature. The length of flange in compression in case (a) is greater than that in (b) and consequently it is more likely to buckle. This behaviour is allowed for in the code by introducing a modification factor called the **Equivalent Uniform Moment Factor**–m_{LT} (see Clause 6.3.4.7.2).

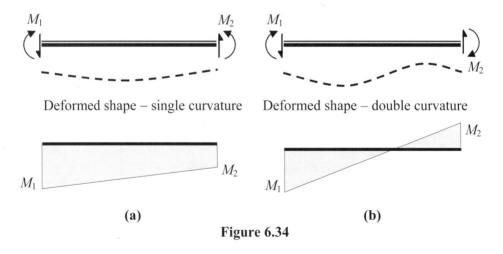

Deformed shape – single curvature   Deformed shape – double curvature

(a)   **(b)**

Figure 6.34

6.3.4.7.1.   Equivalent Slenderness (λ_{LT})

The equivalent slenderness is defined in Clause 4.3.6.7 in terms of:

L_e the effective length for lateral torsional buckling as given in Clause 4.3.5, Tables 13 and 14, *u* a buckling parameter which for rolled **I**-, **H**- or channel sections with equal flanges can be taken as 0.9 and for welded three-plate girders with equal flanges as 1.0,

x a torsional index which may be taken as D/T for rolled sections and welded three-plate girders provided that *u* is taken as 0.9 or 1.0 as appropriate.

(An alternative, more accurate assessment of the buckling parameter and the torsional index can be evaluated using the equations given in Clause B.2.3 of Appendix B.)

v a slenderness factor which may be obtained from Table 19. The values given in Table 19 depend on three factors:

$$\eta = \frac{I_{yc}}{I_{yc} + I_{yt}};$$ where I_{yc} and I_{yt} are the second moments of area of the tension

and compression flanges respectively about the minor axis of the section. In the case of sections with equal flanges $\eta = 0.5$,

$\lambda = L_E/r_y,$

x = torsional index as above.

(As with *u* and *x*, an alternative, more accurate assessment of *v* can be evaluated using the equations given in Clauses 4.3.6.7 and B.2.4.1 of Appendix B.)

β_w a ratio defined in the code in Clause 4.3.6.9 as:

$$= 1.0 \qquad\qquad\qquad\qquad \text{for plastic and compact sections,}$$

$$= Z_x/S_x \qquad \text{when } M_B = p_b Z_x$$
$$\qquad\qquad \text{or alternatively} \left.\vphantom{\begin{array}{c}1\\1\\1\end{array}}\right\} \quad \text{for semi-compact cross-sections}$$
$$= S_{x,eff}/S_x \quad \text{when } M_B = p_b Z_x$$

$$= Z_{x,eff}/S_x \qquad\qquad\qquad \text{for slender cross-sections}$$

Clause 4.3.6.7   The Equivalent Slenderness $= \lambda_{LT} = uv\lambda\sqrt{\beta_w}$

6.3.4.7.2. Equivalent Uniform Moment Factor (m_{LT})
The equivalent uniform moment factor is given in Table 18 of the code for various load cases. This factor is essentially a modification factor to the bending strengths given in Tables 16 and 17 in which it is assumed that the force in the compression flange is constant throughout the unrestrained length of a beam, i.e. the applied moment is assumed to be uniform. If the compression force decreases along the unrestrained length there is less tendency for lateral torsional buckling to occur.

Consider Figures 6.35(a) to (d) in which the unrestrained length of a beam is subject to four different bending moment diagrams, all of which have a *maximum* value at one end equal to *M* and a value at the other end equal to βM.

In cases (b), (c) and (d) an equivalent uniform bending moment diagram, which is assumed to have the same buckling effect as the actual bending moment diagram, is indicated by a broken line.

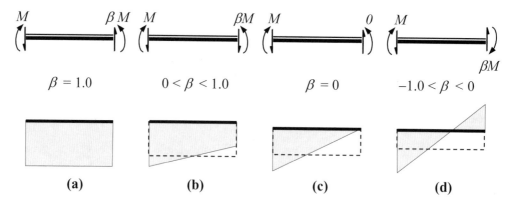

$$\text{Equivalent Uniform Moment} = m_{LT}M$$

Figure 6.35

The values of p_b in the code are based on the assumption of a bending moment diagram as indicated in Figure 6.35(a), i.e. $m_{LT} = 1.0$. In all other cases this *underestimates* the actual bending strength and consequently M_b should be enhanced. This is reflected in Clause 4.3.6.2 as follows:

Clause 4.3.6.2 $M_x \leq M_b/m_{LT}$ (**Note:** $m_{LT} \leq 1.0$)

                                                $\leq M_{cx}$

where:
M_b is the buckling resistance moment,
M_{cx} is the major axis moment capacity of the cross-section,
M_x is the *maximum* major axis moment applied to the unrestrained length being considered,
m_{LT} is the equivalent uniform moment factor for lateral torsional buckling

Note: It is important to check M_{cx} since m_{LT} can be as low as 0.44, leading to the moment capacity being the more critical case.

A general load case and a number of standard load cases to determine m_{LT} are given in the code in Table 18.
The rigorous method for evaluating M_B as given in Clause 4.3.6.4 can be simplified in the case of rolled **I**-, **H**- or channel sections with equal flanges to give a conservative estimate of their buckling resistance moment as indicated in Clause 4.3.7. The evaluation M_b is illustrated using both methods in Examples 6.15 to 6.19.

6.3.4.8   Example 6.16: Beam with No Lateral Restraint

The existing outbuildings in an old stable-block are to be modified to provide facilities for a motor vehicle repair shop. A cross-section of the intended new roof structure is shown in Figure 6.36.

Using the design data given, check the suitability of the proposed new steel beam with respect to shear and bending.

Design Data:

Characteristic dead load including self-weight   (based on the plan area)   2.25 kN/m^2
Characteristic imposed load   (based on the plan area)   1.5 kN/m^2
Span of beam between the centres of bearing   6.55 m
Conditions of restraint at supports:

Compression flange laterally restrained
Beam fully restrained against torsion
Both flanges are free to rotate on plan

Assume that the existing roof trusses and proposed new structure **do not** provide lateral restraint to the compression flange.

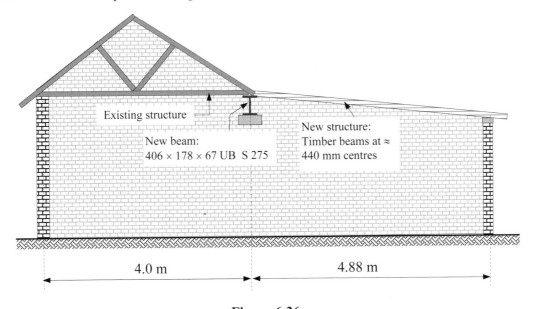

Figure 6.36

6.3.4.9   Solution to Example 6.16

Contract : **Beams** **Job Ref. No. : Example 6.16** Part of Structure : **Beam with No Lateral Restraint** Calc. Sheet No. : **1 of 5**	Calcs. by : **W.McK.** Checked by : Date :	
References	**Calculations**	**Output**
	The existing timber roof trusses and the new timber beams are supported on the top flange of the beam but are not considered to provide any lateral restraint to the compression flange. The plan area supported by the universal beam comprises a width of 2.0 m from the existing structure and 2.44 m from the new one.	

References	Calculations	Output

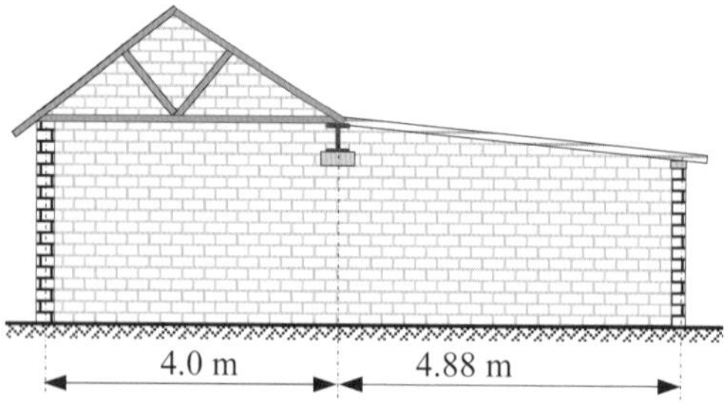

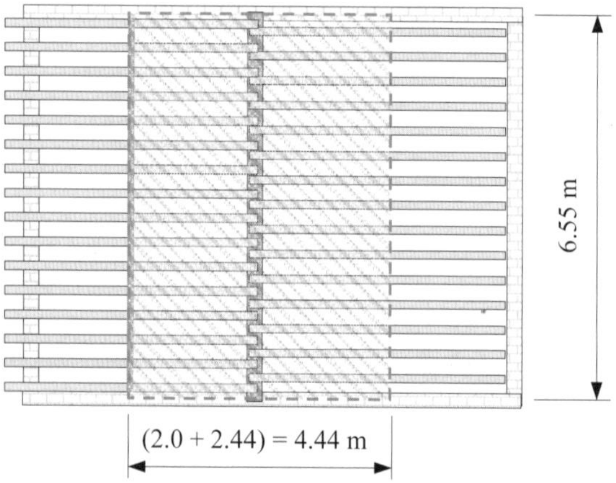

Contract : Beams   Job Ref. No. : Example 6.16
Part of Structure : Beam with No Lateral Restraint
Calc. Sheet No. : 2 of 5

Calcs. by : W.McK.
Checked by :
Date :

4.0 m   4.88 m

6.55 m

$(2.0 + 2.44) = 4.44$ m

Plan area supported $= (4.44 \times 6.55) = 29.1$ m^2

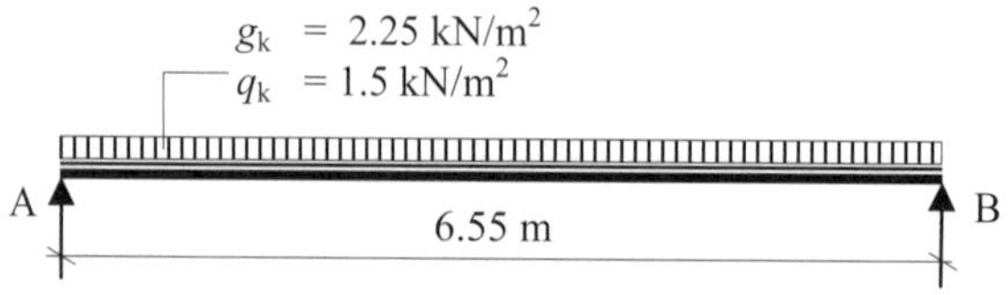

$g_k = 2.25$ kN/m^2
$q_k = 1.5$ kN/m^2

A ▲   6.55 m   ▲ B

Table 2   Design load $= \{[(1.4 \times 2.25) + (1.6 \times 1.50)] \times 29.1\}$
$= 161.5$ kN

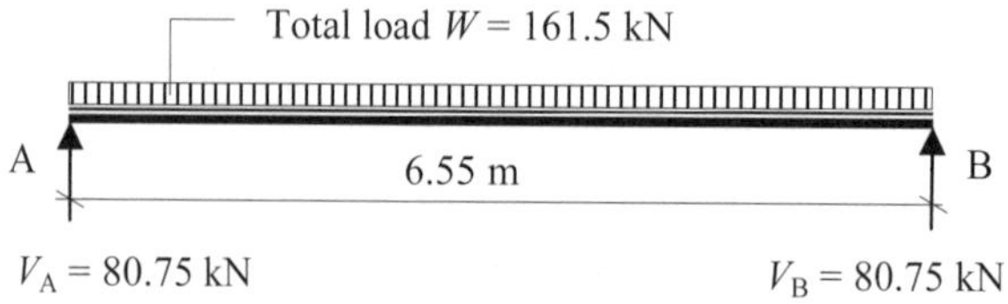

Total load $W = 161.5$ kN

A ▲   6.55 m   ▲ B

$V_A = 80.75$ kN $V_B = 80.75$ kN

References	Calculations	Output

Contract : Beams   Job Ref. No. : Example 6.16 | Calcs. by : W.McK.
Part of Structure : Beam with No Lateral Restraint | Checked by :
Calc. Sheet No. : 3 of 5 | Date :

80.75 kN

Shear Force Diagram   80.75 kN

Design shear force = 80.75 kN

$WL/8$

Bending Moment Diagram

The maximum bending moment $M_x =$ [(161.5 × 6.55)/8.0]

$= 132.2$ kNm

Design bending moment = 132.2 kNm

Check the suitability of a 406 × 178 × 67 UB S 275 section

Section Properties:
$D = 409.4$ mm $d = 360.4$ mm $B = 178.8$ mm
$T = 14.3$ mm $t = 8.8$ mm $b/T = 6.25$
$d/t = 41.0$ $x = 30.5$ $u = 0.88$
$S_x = 1350 \times 10^3$ mm^3 $Z_x = 1190 \times 10^3$ mm^3 $r_y = 39.9$ mm

Table 9

Flange thickness $T < 16$ mm ∴ $p_y = 275$ N/mm^2

Table 11

Section Classification $\varepsilon = \left(\dfrac{275}{p_y}\right)^{\frac{1}{2}} = 1.0$

Flange: Outstand element of compression flange rolled section
$b/T = 6.25 \ < \ 9.0\varepsilon$

Flange is plastic

Web is plastic

Web: Neutral axis at mid-depth (i.e. bending only)
$d/t = 41.0 < 80\varepsilon$

Section is plastic

Clause 4.2.3 $< 70\varepsilon$ therefore no need to check shear buckling

Clause 4.2.3

Shear:
Design shear force $F_v = 80.75$ kN
$P_v = (0.6p_y A_v) = 0.6p_y tD$
$P_v = (0.6 \times 275 \times 8.8 \times 409.4)/10^3 = 594.4$ kN
$P_v \ >> \ F_v$

Section is adequate in shear

Contract : Beams Job Ref. No. : Example 6.16	Calcs. by : W.McK.
Part of Structure : Beam with No Lateral Restraint	Checked by :
Calc. Sheet No. : 4 of 5	Date :

References	Calculations	Output
Clause 4.2.5.1	Bending: $M_c \leq 1.2 p_y Z_x$ for simply supported beams $= (1.2 \times 275 \times 1190 \times 10^3) / 10^6 = 392.7$ kNm The coincident shear at the point of maximum bending moment is equal to zero; use Clause 4.2.5.2 for low shear.	
Clause 4.2.5.2	$M_c = p_y S_x$ $= (275 \times 1350 \times 10^3)/10^6 = 371.3$ kNm Maximum applied moment $= M_x = 132.2$ kNm $\qquad\qquad M_c > M_x$	**Critical value of** $M_c = 371.3$ kNm
Clause 4.3.6.2	Lateral torsional buckling: $M_x \leq M_B / m_{LT}$ $(\leq M_c)$	
Table 18	$m_{LT} = 0.925$ for an unrestrained length with a uniformly distributed load.	
Clause 4.3.6.4	**Rigorous Method:** $M_B = p_b \times S_x$ where p_b is determined from Table 16 for rolled sections using λ_{LT} and p_y	
Clause 4.3.6.7	$\lambda_{LT} = uv\lambda \sqrt{\beta_w}$ where $\lambda = \dfrac{L_E}{r_y}$	
Clause 4.3.5.1	There is no intermediate restraint along the compression flange. Both flanges are free to rotate on plan. There are lateral and torsional restraints at the supports.	
Table 13	The loading condition is normal and $L_E = 1.0 L_{LT} = 1.0L$ $\qquad\qquad\qquad\qquad\qquad\qquad\qquad L_E = 6550$ mm, $\qquad\qquad\qquad\qquad \lambda = \dfrac{6550}{39.9} = 164.2$	
Section Tables	Buckling parameter $u = 0.88$ Torsional index $x = 30.5$	
Clause 4.3.6.7	v is the slenderness factor which is given in Table 19 and depends on λ/x and η $\eta = 0.5$ for UB sections; $\lambda/x = (164.2/30.5) = 5.38$	
Table 19	<table><tr><td>λ/x</td><td>$\eta = 0.5$</td></tr><tr><td>5.0</td><td>0.82</td></tr><tr><td>5.5</td><td>0.79</td></tr></table> $v = 0.82 - \left[(0.82 - 0.79) \times \dfrac{0.38}{0.5}\right]$ $\qquad = 0.8$	

Contract : Beams Job Ref. No. : Example 6.16	Calcs. by : W.McK.
Part of Structure : Beam with No Lateral Restraint	Checked by :
Calc. Sheet No. : 5 of 5	Date :

References	Calculations	Output
Clause 4.3.6.9	$\beta_w = 1.0$ for plastic sections	
	$\lambda_{LT} = uv\lambda \sqrt{\beta_w} = (0.88 \times 0.8 \times 164.2 \times \sqrt{1.0}) = 115.6$	
Table 16	$\lambda_{LO} = 34.3 < \lambda_{LT}$ ∴ check for lateral torsional buckling	
	$\lambda_{LT} = 115.6$ and $p_y = 275$ N/mm^2	
Table 16		

λ_{LT}	S 275
115	102
120	96

$p_b = 102 - \left[(102-96) \times \dfrac{0.6}{5.0}\right]$

$= 101.3$ N/mm^2

Clause 4.3.6.4 — $M_B = (p_b \times S_x) = (101.3 \times 1350 \times 10^3)/10^6 = 136.8$ kNm

Clause 4.3.6.2 — $M_x \le M_B / m_{LT} = (136.8/0.925) = 147.9$ kNm $> M_x$

Since $M_x < M_c$ and $< M_B / m_{LT}$ the section is adequate

Output: Section is adequate in bending

Simplified Method for Equal Flanged Rolled Sections:

Clause 4.3.7 — $M_B = (p_b \times S_x)$ where p_b is determined from Table 20 for rolled sections using $[(\beta_w)^{0.5} L_E/r_y]$ and D/T

Clause 4.3.6.9

$\beta_w = 1.0$ for plastic sections

$L_E/r_y = 164.2$

$[(\beta_w)^{0.5} L_E/r_y] = [(1.0)^{0.5} \times 164.2] = 164.2$

$D/T = (409.4/14.3) = 28.6$

Table 20

$[(\beta_w)^{0.5} L_E/r_y]$	D/T			
	20	25	30	35
155	127	114	105	99
160	124	111	101	95
165	121	107	98	92
170	118	104	95	89

$[(\beta_w)^{0.5} L_E/r_y]$	D/T			
	20	25	30	35
164.2	-	107.6	98.5	-

$p_{b,} = 107.6 - \left[(107.6-98.5) \times \dfrac{3.6}{5.0}\right]$

$= 101$ N/mm^2

Clause 4.3.7 — $M_B = (p_b \times S_x) = (101 \times 1350 \times 10^3)/10^6 = 136.4$ kNm

Clause 4.3.6.2 — $M_x \le M_B / m_{LT} = (136.4/0.925) = 147.5$ kNm $> M_x$

Output: Section is adequate in bending

Since $M_x < M_c$ and $< M_B / m_{LT}$ the section is adequate

6.3.4.10 Example 6.17: Beam 1 with Intermittent Lateral Restraint

A simply supported floor beam in a braced steel industrial frame supports grid flooring and a column as shown in Figure 6.37. Using the design data given, check the suitability of the steel beam with respect to shear and bending.

Design Data:

Characteristic uniformly distributed dead load (including self-weight)	8.0 kN/m
Characteristic uniformly distributed imposed load	10.0 kN/m
Characteristic dead load in column	10.0 kN
Characteristic imposed load in column	30.0 kN

Conditions of restraint at supports:

Compression flange laterally restrained
Beam fully restrained against torsion
Both flanges are partially restrained against rotation on plan

Assume the compression flange to be laterally restrained by the tie beam and that the grid flooring does not provide lateral restraint to the compression flange.

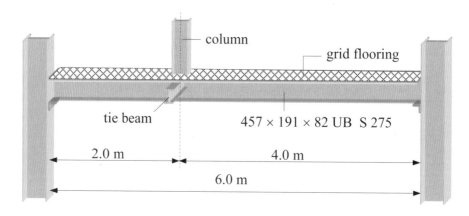

Figure 6.37

6.3.4.11 Solution to Example 6.17

Contract : **Beams** Job Ref. No. : **Example 6.17**	**Calcs. by : W.McK.**
Part of Structure : **Beam 1 – Intermittent Restraint**	**Checked by :**
Calc. Sheet No. : **1 of 5**	**Date :**

References	Calculations	Output
	The industrial grid flooring is supported on the top flange of the beam but not considered to provide any lateral restraint to the compression flange. Lateral restraint is assumed to be provided by the tie beams at the 2.0 m position from the left-hand end. The self-weight of the tie beam is included in the dead load.	

Contract : **Beams** Job Ref. No. : **Example 6.17** Part of Structure : **Beam 1 – Intermittent Restraint** Calc. Sheet No. : **2 of 5**	Calcs. by : **W.McK.** Checked by : Date :

References	Calculations	Output

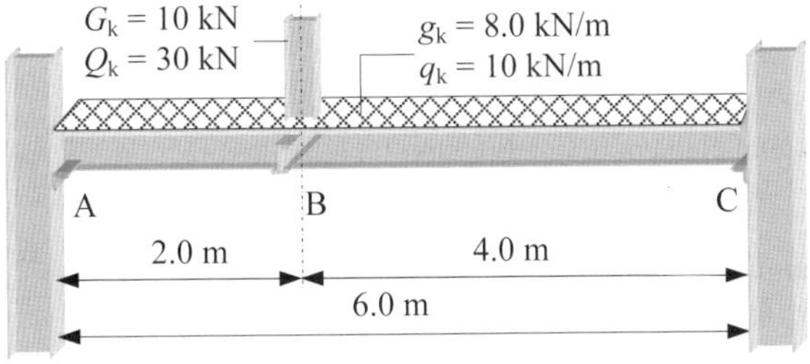

This beam has the same span and the same loading condition as in Example 6.15 and consequently the same design shear force and bending moment diagrams.

Design point load   =   62.0 kN
Design distributed load   =   27.2 kN/m

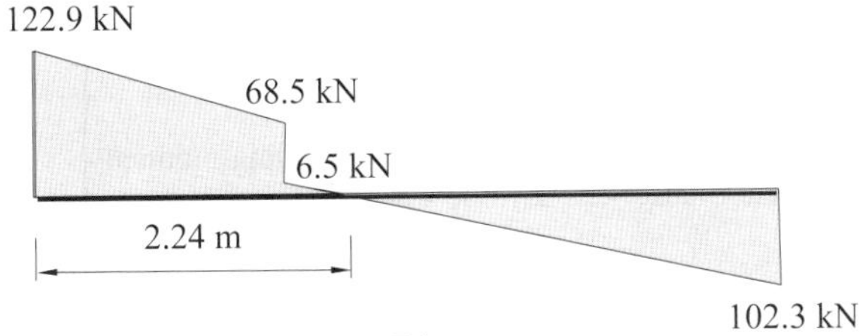

Shear Force Diagram

The design shear force F_v =   122.9 kN

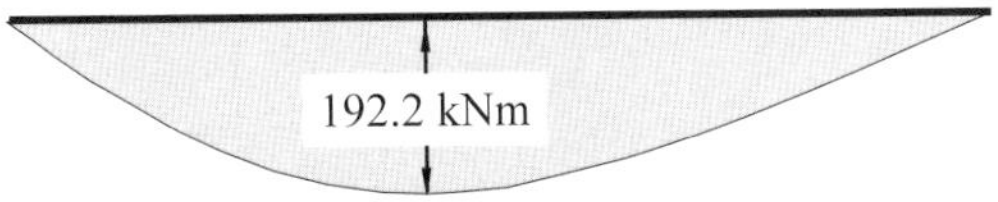

Bending Moment Diagram

The maximum bending moment at B   = M_x = 192.2 kNm

Check the suitability of a 457 × 191 × 82 UB S 275 section

Section Properties:
D =   460.0 mm d =   407.6 mm B = 191.3 mm
T =   16.0 mm t =   9.9 mm b/T = 5.98
d/t =   41.2 x =   30.8 u = 0.879
S_x = 1830 × 10³ mm³ Z_x = 1610 × 10³ mm³ r_y = 42.3 mm

Table 9 | Flange thickness $T \leq 16$ mm $\therefore p_y = 275$ N/mm²

References	Calculations	Output
Table 11	Section Classification $\qquad \varepsilon = \left(\dfrac{275}{p_y}\right)^{\frac{1}{2}} = 1.0$	
	Flange: Outstand element of compression flange rolled section $b/T = 5.98 < 9.0\varepsilon$	**Flange is plastic**
	Web: Neutral axis at mid-depth (i.e. bending only) $d/t = 41.2 < 80\varepsilon$	**Web is plastic**
Clause 4.2.3	$\qquad\qquad < 70\varepsilon$ therefore no need to check shear buckling	**Section is plastic**
Clause 4.2.3	Shear: Design shear force $F_v = 122.9$ kN $P_v = (0.6p_y A_v) \qquad = 0.6p_y tD$ $P_v = (0.6 \times 275 \times 9.9 \times 460.0)/10^3 \quad = \quad 751.4$ kN $\qquad\qquad\qquad\qquad\qquad P_v \gg F_v$	**Section is adequate in shear**
Clause 4.2.5.1	Bending: $M_c \leq 1.2p_y Z_x$ for simply supported beams $\quad = (1.2 \times 275 \times 1610 \times 10^3)/10^6 \quad = \quad 531.3$ kNm	
	The coincident shear at the point of maximum bending moment is equal to zero; use Clause 4.2.5.2 for low shear.	
Clause 4.2.5.2	$M_c = p_y S_x$ $\quad = (275 \times 1830 \times 10^3)/10^6 \qquad = 503.3$ kNm	**Critical value of $M_c = 503.3$ kNm**
	Maximum applied moment $\quad = M_x = 192.2$ kNm $\qquad\qquad\qquad\qquad\qquad M_c > M_x$	
Clause 4.3.6.2	Lateral torsional buckling: $M_x \leq M_B / m_{LT} \qquad (\leq M_c)$	
	The intermediate lateral restraint occurs at B, resulting in two unrestrained lengths AB = 2.0 m and BC = 4.0 m. Clearly in this instance section BC is the more critical unrestrained length. m_{LT} can be determined from Table 18 for the general case as follows:	
Table 18		
	$m_{LT} = 0.2 + \dfrac{(0.15M_2 + 0.5M_3 + 0.15M_4)}{M_{max}} \geq 0.44$	

Contract : **Beams** Job Ref. No. : **Example 6.17**	Calcs. by : **W.McK.**
Part of Structure : **Beam 1 – Intermittent Restraint**	Checked by :
Calc. Sheet No. : **4 of 5**	Date :

References	Calculations	Output

62.0 kN

27.2 kN/m

2.0 m   4.0 m

6.0 m

122.9 kN   102.3 kN

$$M_2 = [(102.3 \times 3.0) - (27.2 \times 3.0 \times 1.5)] = 184.5 \text{ kNm}$$
$$M_3 = [(102.3 \times 2.0) - (27.2 \times 2.0 \times 1.0)] = 150.2 \text{ kNm}$$
$$M_4 = [(102.3 \times 1.0) - (27.2 \times 1.0 \times 0.5)] = 88.7 \text{ kNm}$$

$$m_{LT} = 0.2 + \frac{[(0.15 \times 184.5) + (0.5 \times 150.2) + (0.15 \times 88.7)]}{192.2} \geq 0.44$$

$$= 0.2 + 0.6$$
$$= 0.8$$

Clause 4.3.6.4

Rigorous Method:
$M_B = p_b \times S_x$ where p_b is determined from Table 16 for rolled sections using λ_{LT} and p_y

Clause 4.3.6.7

$\lambda_{LT} = uv\lambda\sqrt{\beta_w}$ where $\lambda = \dfrac{L_E}{r_y}$;

Clause 4.3.5.1

There is no intermediate restraint along the compression flange.
Both flanges are free to rotate on plan.
There are lateral and torsional restraints at A, B and C.

Table 13

The loading condition is normal and $L_E = 1.0L_{LT} = 1.0L$
$L_E = 4000$ mm,

$$\lambda = \frac{4000}{42.3} = 94.6$$

Section Tables

Buckling parameter $u = 0.879$
Torsional index $x = 30.8$

v is the slenderness factor which is given in Table 19 and depends on λ/x and η

Clause 4.3.6.7

$\eta = 0.5$ for UB sections; $\lambda/x = (94.6/30.8) = 3.07$

Table 19

λ/x	$\eta = 0.5$
3.0	0.91
3.5	0.89

$v \approx 0.91$

Contract : **Beams** Job Ref. No. : **Example 6.17** Part of Structure : **Beam 1 – Intermittent Restraint** Calc. Sheet No. : **5 of 5**	Calcs. by : **W.McK.** Checked by : Date :

References	Calculations	Output
Clause 4.3.6.9	β_w = 1.0 for plastic sections λ_{LT} = $uv\lambda\sqrt{\beta_w}$ = $(0.879 \times 0.91 \times 94.6 \times \sqrt{1.0})$ = 75.7	
Table 16	λ_{LO} = 34.3 < λ_{LT} ∴ check for lateral torsional buckling λ_{LT} = 75.7 and p_y = 275 N/mm²	
Table 16	$\begin{array}{c\|c}\lambda_{LT} & \text{S 275}\\\hline 75 & 176\\\hline 80 & 165\end{array}$ p_b = $176 - \left[(176-165)\times\dfrac{0.7}{5.0}\right]$ = 174.5 N/mm²	
Clause 4.3.6.4	M_B = $(p_b \times S_x)$ = $(174.5 \times 1830 \times 10^3)/10^6$ = 319.3 kNm	
Clause 4.3.6.2	$M_x \le$ M_B / m_{LT} = (319.3/ 0.8) = 399.1 kNm >> M_x Since $M_x < M_c$ and < M_B / m_{LT} the section is adequate.	**Section is adequate in bending.**
Clause 4.3.7	**Simplified Method for Equal Flanged Rolled Sections:** M_B = $(p_b \times S_x)$ where p_b is determined from Table 20 for rolled sections using $[(\beta_w)^{0.5} L_E/r_y]$ and D/T	
Clause 4.3.6.9	β_w = 1.0 for plastic sections L_E/r_y = 94.6 $[(\beta_w)^{0.5} L_E/r_y]$ = $[(1.0)^{0.5} \times 94.6]$ = 94.6 D/T = (460/16) = 28.75	
Table 20	$\begin{array}{c\|c\|c\|c\|c}[(\beta_w)^{0.5}L_E/r_y] & \multicolumn{4}{c}{D/T}\\\hline & 20 & 25 & 30 & 35\\\hline 85 & 200 & 193 & 188 & 184\\\hline 90 & 193 & 185 & 179 & 175\\\hline 95 & 186 & 177 & 171 & 167\\\hline 100 & 180 & 170 & 164 & 159\end{array}$ $\begin{array}{c\|c\|c\|c\|c}[(\beta_w)^{0.5}L_E/r_y] & \multicolumn{4}{c}{D/T}\\\hline & 20 & 25 & 30 & 35\\\hline 94.6 & - & 177.6 & 171.6 & -\end{array}$ $p_{b,}$ = $177.6 - \left[(177.6-171.6)\times\dfrac{3.75}{5.0}\right]$ = 173.1 N/mm²	
Clause 4.3.7 Clause 4.3.6.2	M_B = $(p_b \times S_x)$ = $(173.1 \times 1830 \times 10^3)/10^6$ = 316.8 kNm $M_x \le$ M_B / m_{LT} = (316.8/0.8) = 396 kNm >> M_x Since $M_x < M_c$ and < M_B / m_{LT} the section is adequate	**Section is adequate in bending**

6.3.4.12 Example 6.18: Rectangular Hollow Section

A footbridge is required to span a small ravine in a theme park The proposed structure is shown in Figure 6.38. Using the data provided, check the suitability of the rectangular hollow section with respect to shear and bending.

Design Data:

Characteristic uniformly distributed dead load (including self-weight) 2.0 kN/m^2

Characteristic uniformly distributed imposed load 5.0 kN/m^2

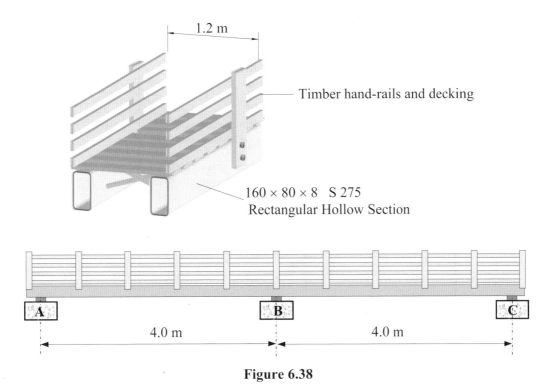

Figure 6.38

6.3.4.13 Solution to Example 6.18

Contract : Beams Job Ref. No. : Example 6.18 Part of Structure : Rectangular Hollow Section Calc. Sheet No. : 1 of 5	Calcs. by : W.McK. Checked by : Date :

References	Calculations	Output
	The timber decking is supported on the top flange of the beams but is not considered to provide any lateral restraint to the compression flange. The loading is assumed to be shared equally between the two rectangular hollow sections.	

Contract : Beams Job Ref. No. : Example 6.18 Part of Structure : Rectangular Hollow Section Calc. Sheet No. : 2 of 5	Calcs. by : W.McK. Checked by : Date :

References	Calculations	Output
Table 2	(calculations below)	

Plan area supported by each beam $= (0.6 \times 1.0) = 0.6 \text{ m}^2/\text{m}$

Maximum design load $= \{[(1.4 \times 2.0) + (1.6 \times 5.0)] \times 0.6\}$
$= 6.48 \text{ kN/m length}$

Minimum design load $= [(1.0 \times 2.0) \times 0.6]$
$= 1.2 \text{ kN/m length}$

There are two load cases to consider:
(i) maximum span moment with the maximum design loading on one span and the minimum design load on the other span, and
(ii) maximum support moment and shear force with the maximum design loading on both spans.

Case (i):

The bending moments and support reactions can be determined using either moment distribution or coefficients from tables of standard loadings.

Moment Distribution:
Fixed-end Moments

$$\text{FEM}_{AB} = 0; \quad \text{FEM}_{AB} = +\frac{wL^2}{8} = +\frac{6.48 \times 4.0^2}{8} = +12.96 \text{ kNm}$$

$$\text{FEM}_{CB} = 0; \quad \text{FEM}_{BC} = +\frac{wL^2}{8} = -\frac{1.2 \times 4.0^2}{8} = -2.40 \text{ kNm}$$

Since the structure and the supports are symmetrical the distribution factors at support B will be equal, i.e.
$\text{DF}_{BA} = \text{DF}_{BC} = 0.5$

	A		B			C
	AB		BA	BC		CB
DF	1.0		0.5	0.5		1.0
FEM	0		+ 12.96	−2.4		0
Balance			− 5.28	− 5.28		
Final Moments	**0**		**+ 7.68**	**−7.68**		**0**

Contract : **Beams** **Job Ref. No. : Example 6.18** **Part of Structure : Rectangular Hollow Section** **Calc. Sheet No. : 3 of 5**	**Calcs. by : W.McK.** **Checked by :** **Date :**

References	Calculations	Output

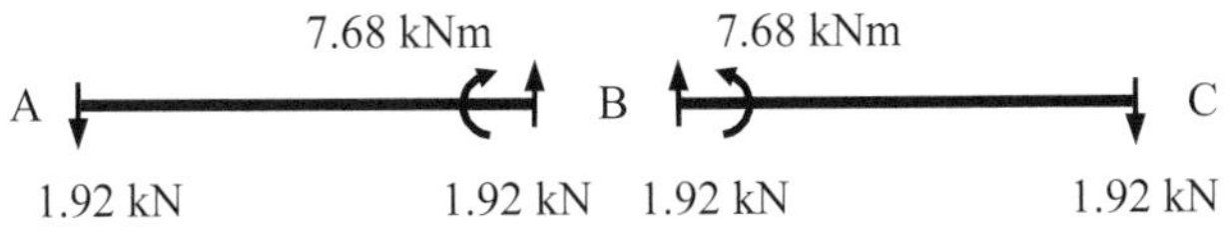

Fixed reactions

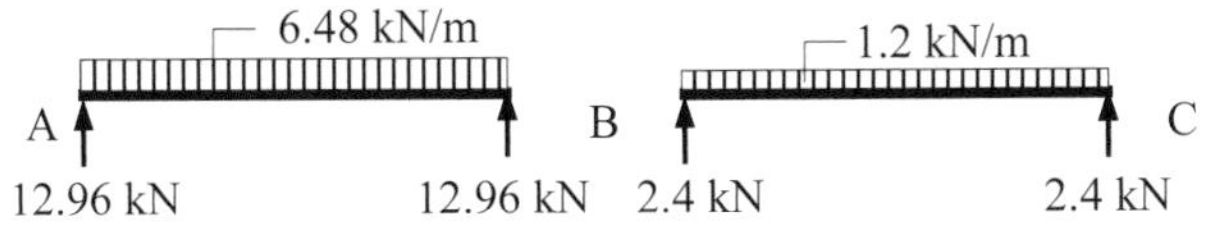

Free reactions

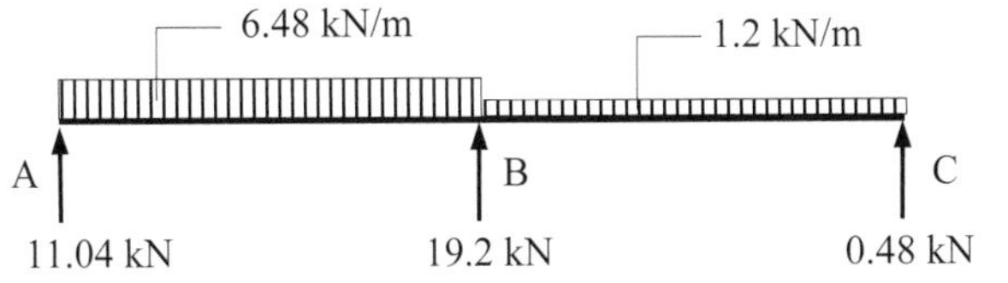

Final reactions

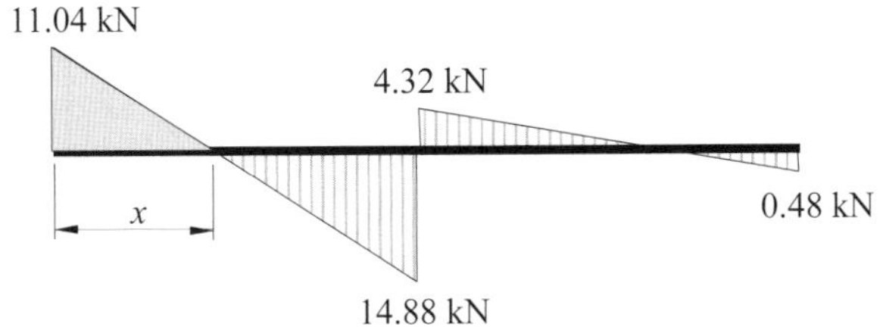

Shear force diagram

The maximum bending moment coincides with the position of

zero shear $x = \dfrac{11.04}{6.48} = 1.7$ m

Maximum bending moment = shaded area
$$= (0.5 \times 1.7 \times 11.04) = 9.38 \text{ kNm}$$

Case (ii):

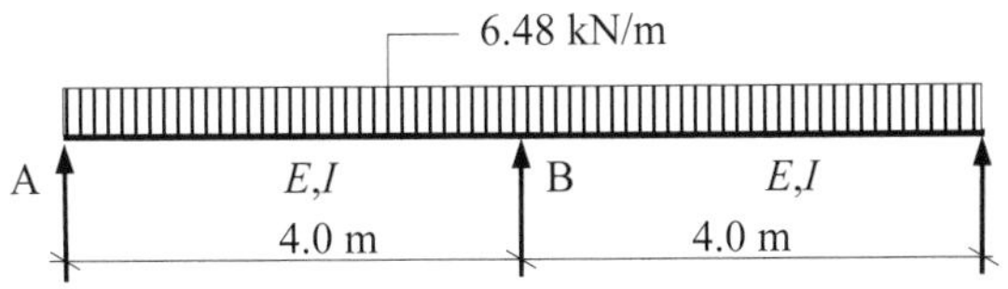

References	Calculations	Output
	$$\text{FEM}_{AB} = 0; \ \text{FEM}_{AB} = + \frac{wL^2}{8} = + \frac{6.48 \times 4.0^2}{8} = +12.96 \text{ kNm}$$ $$\text{FEM}_{CB} = 0; \ \text{FEM}_{BC} = + \frac{wL^2}{8} = - \frac{6.48 \times 4.0^2}{8} = -12.96 \text{ kNm}$$ Since the structure, the supports and the loading are symmetrical there is no out-of-balance moment and no distribution is required. Support moment = 12.96 kNm.	

Contract :   Beams   Job Ref. No. : Example 6.18
Part of Structure : Rectangular Hollow Section
Calc. Sheet No. : 4 of 5

Calcs. by : W.McK.
Checked by :
Date :

12.96 kNm   12.96 kNm

A ⊢————————————————⊣ B ⊢————————————————⊣ C

3.24 kN   3.24 kN   3.24 kN   3.24 kN

Fixed reactions

⌐ 6.48 kN   ⌐ 6.48 kN

A   B   C

12.96 kN   12.96 kN   12.96 kN   12.96 kN

Free reactions

⌐ 6.48 kN/m

A   B   C

9.72 kN   32.4 kN   9.72 kN

Final reactions

9.72 kN   16.2 kN

x

16.2 kN

9.72 kN

Shear force diagram

The maximum bending moment coincides with the position of zero shear $x = \dfrac{9.72}{6.48} = 1.5 \text{ m}$

Span bending moment = shaded area
$$= (0.5 \times 1.5 \times 9.72) = 7.29 \text{ kNm}$$

Design shear force from case (ii) = 16.2 kN

Design bending moment from case (ii) = 12.96 kNm

Contract : Beams Job Ref. No. : Example 6.18	Calcs. by : W.McK.
Part of Structure : Rectangular Hollow Section	Checked by :
Calc. Sheet No. : 5 of 5	Date :

References	Calculations	Output
	Check the suitability of a 160 × 80 × 8 RHS S 275 section	
	Section Properties: D = 160 mm B = 80 mm t = 8.0 mm b/t = 7.0 d/t = 17.0 r_y = 31.8 mm $S_x = 175 \times 10^3$ mm^3 $Z_x = 136 \times 10^3$ mm^3 A = 3520 mm^2	
Table 9	Flange thickness t ≤ 16 mm ∴ p_y = 275 N/mm^2	
Table 12	Section classification $\varepsilon = \left(\dfrac{275}{p_y}\right)^{\frac{1}{2}} = 1.0$	
	Flange: Compression due to bending b/t = 7.0 < 28ε and ≤ (80ε − d/t)	**Flange is plastic**
	Web: Neutral axis at mid-depth (i.e. bending only) d/t = 17.0 < 64ε	**Web is plastic**
Clause 4.2.3	< 70ε therefore no need to check shear buckling	**Section is plastic**
Clause 4.2.3	Shear: Design shear force F_v = 122.9 kN P_v = (0.6$p_y A_v$) = [(0.6$p_y AD$) / ($D + B$)] P_v = [(0.6 × 275 × 3250 × 160)/(160 + 80)]/10^3 = 357.5 kN P_v >> 16.2 kN	**Section is adequate in shear**
Clause 4.2.5.1	Bending: M_c ≤ 1.5$p_y Z_x$ for continuous beams = (1.5 × 275 × 136 × 10^3)/ 10^6 = 56.1 kNm	
Clause 4.2.5.2	M_c = $p_y S_x$ = (275 × 175 × 10^3)/10^6 = 48.13 kNm	**Critical value of M_c = 48.13 kNm**
	Maximum applied moment = M_x = 12.96 kNm M_c >> M_x	
Clause 4.3.6.1	Lateral torsional buckling: In the case of RHS sections, resistance to lateral torsional buckling need not be checked unless L_E/r_y exceeds the limiting value given in Table 15 for the relevant value of D/B.	
Table 15	D/B = (160/80) = 2.0 The limiting value of L_E/r_y = [340 × (275/p_y)] = 340 The limiting length of beam = (340 × r_y) = (340 × 31.8) = 10.81 m Since the limiting value > the length of beam in compression then there is no need to check lateral torsional buckling.	**Section is adequate in bending**

354   *Design of Structural Elements*

6.3.4.14 Deflection (Clause 2.5.2)

In Clause 2.5.1 of the code, one of the serviceability limit states to be considered is deflection. Recommendations for limiting values of deflection under various circumstances are given in Table 8.

Limitations on the deflections of beams are necessary to avoid consequences such as:

♦ damage to finishes, e.g. to brittle plaster, or ceiling tiles,
♦ unnecessary alarm to occupants of a building,
♦ misalignment of door frames causing difficulty in opening doors,
♦ misalignment of crane rails resulting in derailment of crane-gantries.

There are large variations in what are considered by practising engineers to be acceptable deflections for different circumstances. If situations arise in which a designer considers the recommendations given in Table 8 to be too lenient or too severe (e.g. conflicting with the specification of suppliers or manufacturers) then individual engineering judgement must be used.

The values in **Table 8** relating to beams give a $\dfrac{\text{span}}{\text{coefficient}}$[1] ratio calculated using the service imposed loads only. The coefficient varies from 180 for cantilevers to 360 for the deflection of beams supporting brittle finishes. In most circumstances, the dead load deflection will have occurred prior to finishes being fixed and the building being in use, and will not therefore cause any additional problem while the building is in service. Unfactored loads are used since it is under service conditions that deflection may be a problem. Additional values are given for portal frames and crane-gantry girders; they are not considered here.

In a simply supported beam, the maximum deflection induced by the applied loading always approximates the mid-span value if it is not equal to it. A number of standard, frequently used load cases for which the elastic deformation is required are given in Appendix 3 of this text.

In many cases beams support complex load arrangements which do not lend themselves to either an individual load case or a combination of the load cases given in Appendix 3. Since the values in Table 8 are recommendations for maximum values, approximations in calculating deflection are normally acceptable. Provided that deflection is not the governing design criterion, a calculation which gives an approximate answer is usually adequate. The *Steel Designers' Manual* (ref. 70) provides a range of coefficients which can be used either to calculate deflections or to determine the minimum I value (second moment of area), to satisfy any particular $\dfrac{\text{span}}{\text{coefficient}}$ ratio.

An *equivalent uniformly distributed load* technique which can be used for estimating actual deflections or required I values for simply supported spans is given in this text.

---

[1] A ratio is used instead of a fixed value since this limits the curvature of the beam which depends on the span.

6.3.4.14.1. Equivalent UDL Technique

(a) Estimating deflection

Consider a single-span, simply supported beam carrying a *non-uniform* loading which induces a maximum bending moment of $M_{maximum}$ as shown in Figure 6.39.

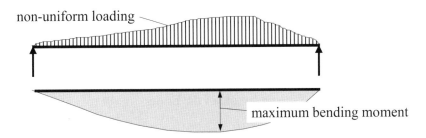

non-uniform loading

maximum bending moment

Bending Moment Diagram

Figure 6.39

The equivalent U.D.L. (w_e) which would induce the same *magnitude* of maximum bending moment (**note:** the position may be different), on a simply supported span carrying a *uniform* loading can be determined from:

$$\text{Maximum bending moment} = M_{maximum} = \frac{w_e L^2}{8}$$

$$\therefore w_e = \frac{8 M_{maximum}}{L^2}$$

where W_e is the equivalent uniform distributed load.

The maximum deflection of the beam carrying the uniform loading will occur at the mid-span and be equal to $\delta_{maximum,\ uniform\ loading} = \dfrac{5 w_e L^4}{384EI}$

Using this expression, the maximum deflection of the beam carrying the non-uniform loading can be estimated by substituting for the w_e term, i.e.

$$\delta_{maximum,\ non\text{-}uniform\ loading} \approx \frac{5 w_e L^4}{384EI} = \frac{5 \times \left(\dfrac{8 M_{maximum}}{L^2}\right) L^4}{384\ EI} = \frac{0.104\ M_{maximum}\ L^2}{EI}$$

(b) Estimating the required second moment of area (*I*) value

Assuming a building in which brittle finishes are to be used, from Table 8 in the code:

$$\delta_{actual} \leq \frac{\text{span}}{360} \qquad \therefore \quad \frac{0.104\ M_{maximum}\ L^2}{EI} \leq \frac{L}{360}$$

where M_{maximum} is the maximum bending moment due to unfactored imposed loads only.

$$\therefore I \geq \frac{37.4 M_{\text{maximum}} L}{E}$$

Note: Care must be taken to ensure that a consistent system of units is used. A similar calculation can be carried out for any other $\dfrac{\text{span}}{\text{coefficient}}$ ratio.

6.3.4.14.2. Example 6.19: Deflection of Simply Supported Beam

A simply supported beam is required to carry characteristic imposed loads, as indicated in Figure 6.40. Check the suitability of a 305 × 165 × 46 UB with respect to deflection, assuming brittle finishes to the underside of the beam.

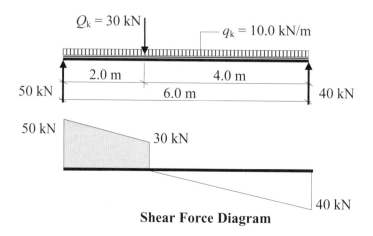

Shear Force Diagram

Figure 6.40

Maximum bending moment due to unfactored imposed loads

$$= \frac{(50+30)2}{2}$$

$$= \quad 80 \text{ kNm}$$

$$\therefore \ \delta_{\text{actual}} \approx \frac{0.104\, M_{\text{maximum}}\, L^2}{EI} \leq \frac{\text{span}}{360}$$

Clause 3.1.3   Modulus of Elasticity $E = 205 \text{ kN/mm}^2$

Section property tables: $I_{\text{xx}} = 9900 \times 10^4 \text{ mm}^4$

$$\therefore \ \delta_{\text{actual}} \approx \frac{0.104 \times 80 \times 10^6 \times 6000^2}{205 \times 10^3 \times 9900 \times 10^4} = 14.8 \text{ mm}$$

Table 8 $\dfrac{\text{span}}{360} = \dfrac{6000}{360} = 16.7 \text{ mm} > \delta_{\text{actual}}$

Section is adequate with respect to deflection

This check could have been carried out more accurately using the values given in Appendix 3 of the text, as shown in Figure 6.41.

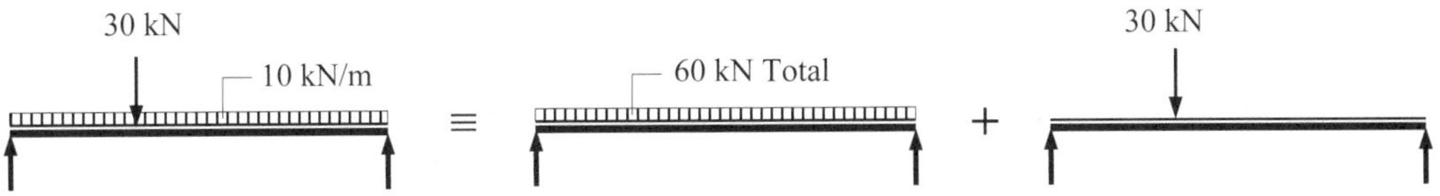

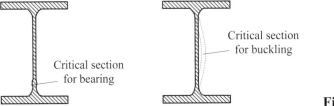

Figure 6.41

$$\delta_{actual} = \left\{ \frac{5WL^3}{384EI} + \frac{PL^3}{48EI} \left[\frac{3a}{L} - 4\left(\frac{a}{L}\right)^3 \right] \right\}$$

$$\delta_{actual} = \left\{ \frac{5 \times 60 \times 6000^3}{384 \times 205 \times 9899 \times 10^4} + \frac{30 \times 6000^3}{48 \times 205 \times 9899 \times 10^4} \left[\frac{3 \times 2}{6} - 4\left(\frac{2}{6}\right)^3 \right] \right\}$$

$$\delta_{actual} = (8.3 + 5.7) = 14 \text{ mm}$$

In this case the approximate technique overestimates the deflection by less than 5%. Provided that the estimated deflection is no more than 95% of the deflection limit from Table 8, the approximate answer should be adequate for design purposes and a more accurate calculation is not required.

6.3.4.15 Web Bearing and Web Buckling

In addition to shear failure of a web as discussed in Section 6.3.4.2, there are two other modes of failure which may occur:

(i)   web bearing   and
(iii) web buckling.

At locations of heavy concentrated loads such as support reactions or where columns are supported on a beam flange, additional stress concentrations occur in the web. This introduces the possibility of the web failing in a buckling mode similar to a vertical strut, or by localised bearing failure at the top of the root fillet, as shown in Figure 6.42.

Critical section
for buckling

Critical section
for bearing

Figure 6.42

The code specifies two local capacities relating to these modes of failure. When either of these is less than the applied concentrated force it will be necessary to provide additional strength to the web. In most cases this requires the design of load bearing stiffeners.

There may be other reasons for utilizing stiffeners, such as enhancing torsional stiffness at supports and points of lateral restraint, as discussed previously. In both cases at the design stage it is usually necessary to make assumptions regarding the provision of bearing plates at supports or cap plates/base plates on columns to provide stiff bearing. In the code, Clause 4.5.1.3 defines the stiff bearing length b_1 as 'that length which cannot deform appreciably in bending'. The value of b_1 is determined as illustrated in Figure 13 of the code. In cases where a bearing plate exists, an additional length may be included assuming a dispersion of load at 45° through the bearing plate.

6.3.4.15.1. Web Bearing (Clause 4.5.2.1)

The bearing capacity (P_{bw}) is calculated on the basis of an effective bearing area over which the design strength of the web is assumed to act. It is determined using:

$$P_{bw} = (b_1 + nk)tp_{yw}$$

where:
b_1 is the stiff bearing length as indicated in Clause 4.5.1.3 and Figure 13,
n $= (2 + 0.6b_e/k)$ but ≤ 5 at the end of a member, and
  $= 5.0$ in all other cases,
k $= (T + r)$ for *rolled* **I**- and **H**- sections,
  $= T$ for *welded* **I**- and **H**- sections,
r is the root radius from Section Tables,
t is the web thickness,
p_{yw} is the design strength of the web.
as shown in Figure 6.43 for rolled sections.

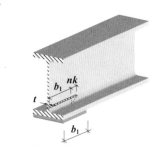

$n = 2$
Bearing ***at the end of*** a member

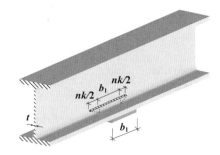

$n = 5$
Bearing ***within the length of*** a member

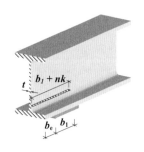

$n = (2 + 0.6b_e/k)$ ≤ 5
Bearing ***near the end of*** a member

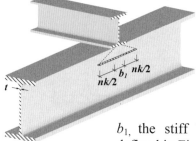

b_1, the stiff bearing length, is defined in Figure 13 of the code

$n = 5$
Bearing ***within the length of*** a member

Figure 6.43

6.3.4.15.2. Web Buckling (Clause 4.5.3.1)

In the buckling check the web is considered to be a fixed-end strut between the flanges. It is assumed that the flange through which the concentrated load is applied is restrained against rotation relative to the web and lateral movement relative to the other flange, as shown in Figure 6.44.

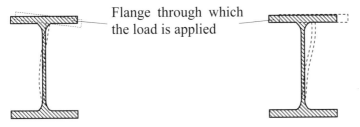

Flange through which the load is applied

Flange rotation relative to the web   Relative lateral movement between the flanges

Figure 6.44

In general, the load carrying capacity of a strut is dependent on its *bearing capacity* of the unstiffened web (P_{bw}) as indicated in Clause 4.5.2.1, and is given by:

$$P_x = \frac{25\,\varepsilon\,t}{\sqrt{(b_1 + nk)d}}\,P_{bw}$$

where:
b_1, n, k, t are as defined previously,
d is the depth of the web

In cases where the distance a_e (see Figure 6.45) is $< 0.7d$, P_x should be modified as follows:

$$P_x = \left(\frac{a_e + 0.7d}{1.4d}\right)\frac{25\,\varepsilon\,t}{\sqrt{(b_1 + nk)d}}\,P_{bw}$$

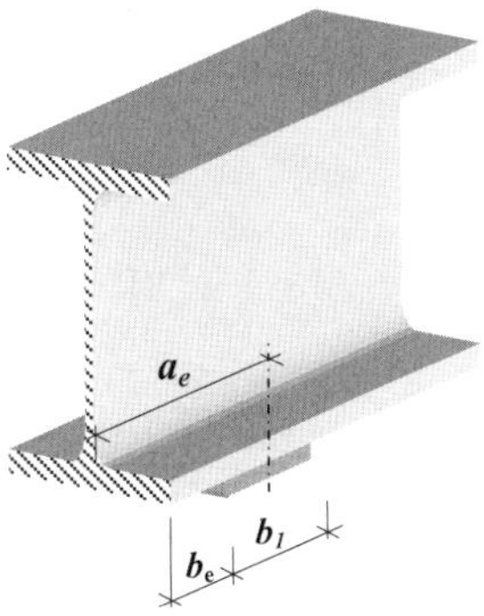

Figure 6.45

In cases where a flange is not restrained against rotation and lateral movement as in Figure 6.44, the value of P_{bw} should be reduced further by using:

$$P_{xr} = \frac{0.7d}{L_E}\,P_{bw}$$ where: L_E is the effective length of the web, acting as a compression

member determined in accordance with Clause 4.7.2 for the appropriate end restraint conditions.

In the case of square and rectangular hollow sections when the flange is not welded to a bearing plate, additional effects of secondary moments which are induced in the web due to eccentricity of loading as shown in Figure 6.46 must be allowed for.

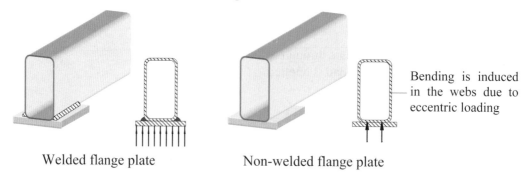

Welded flange plate   Non-welded flange plate

Bending is induced in the webs due to eccentric loading

Figure 6.46

Reference should be made to the *Steelwork Design Guides...* (refs. 68, 69) which contain detailed information relating to the relevant bearing factors in such cases.

6.3.4.15.3. Example 6.20 Web Bearing and Web Buckling

A universal beam section supported on a 90 × 90 × 10 angle section as shown in Figure 6.47 is subjected to a concentrated load of 100 kN transferred through the bottom flange. Check the suitability of the section with respect to web bearing and web buckling.

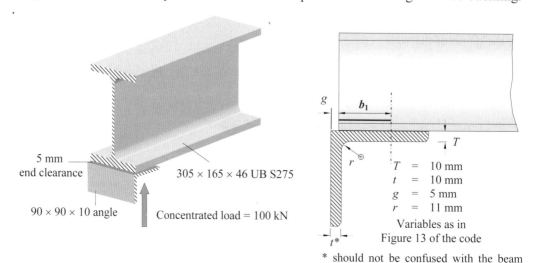

5 mm end clearance

305 × 165 × 46 UB S275

90 × 90 × 10 angle   Concentrated load = 100 kN

T = 10 mm
t = 10 mm
g = 5 mm
r = 11 mm

Variables as in Figure 13 of the code

* should not be confused with the beam web thickness t used in P_{bw} and P_x.

Figure 6.47

Solution:
Web Bearing Capacity:
Clause 4.5.2.1 $P_{bw} = (b_1 + nk)tp_{yw}$
Figure 13 $b_1 = (t + T + 0.8r - g) = [10 + 10 + (0.8 \times 11) - 5] = 23.8$ mm
                        $k = (T + r)$ where T is the flange thickness and r is the root radius of the UB.

Section Tables $T = 11.8$ mm $r = 8.9$ mm $t = 6.7$ mm $d = 265.2$ mm
$$k = (T + r) = (11.8 + 8.9) = 20.7 \text{ mm}$$
$$b_e = 5 \text{ mm}$$
$$n = [2 + (0.6b_e)/k] \leq 5$$
$$= [2 + (0.6 \times 5)/20.7] = 2.14$$

Table 9 $p_{yw} = 275 \text{ N/mm}^2$

$$\therefore \quad P_{bw} = (b_1 + nk)tp_{yw} = \{[23.8 + (2.14 \times 20.7)] \times (6.7 \times 275)\}/10^3$$
$$= 125.5 \text{ kN}$$

$$P_{bw} > 100 \text{ kN} \qquad \qquad \textbf{Section is adequate in web bearing}$$

Web Buckling Capacity:

Clause 4.5.3.1 The web buckling capacity P_x is influenced by a_e, the distance from the load or reaction of the unstiffened web to the nearer end of the member

$$a_e = b_1/2 = 23.8/2 = 11.9 \text{ mm} < 0.7d$$

$$\therefore \quad P_x = \left(\frac{a_e + 0.7d}{1.4d}\right)\frac{25\varepsilon t}{\sqrt{(b_1 + nk)d}}P_{bw}$$

$$= \left[\frac{11.9 + (0.7 \times 265.2)}{(1.4 \times 265.2)}\right]\frac{(25 \times 1.0 \times 6.7)}{\sqrt{[23.8 + (2.14 \times 20.7)] \times 265.2}} \times 125.5$$

$$= 83.2 \text{ kN}$$

$$P_x < 100 \text{ kN}$$
Section is inadequate with respect to web buckling: stiffeners required

6.3.4.16 Example 6.21: Beam 2 with Intermittent Lateral Restraint

It is proposed to use a $457 \times 191 \times 67$ UB S 275 as a main roof beam spanning 12.0 m between two columns as shown in Figure 6.48. The main beam supports three secondary beams, one at mid-span and the other two at the quarter-span points, each of which provides lateral restraint to the compression flange. Using the design data given, check the section with respect to:

♦ shear,
♦ bending,
♦ deflection,
♦ web buckling and
♦ web bearing.

Design Data:
Characteristic uniformly distributed dead load (including self-weight)   5.0 kN/m
Characteristic uniformly distributed imposed load   6.0 kN/m
Characteristic dead load reaction from each secondary beam   10.0 kN
Characteristic imposed load reaction from each secondary beam   20.0 kN

Conditions of restraint at supports:

> *Compression flange laterally restrained*
> *Beam fully restrained against torsion*
> *Both flanges are partially restrained against rotation on plan*

The beam is connected to the columns at each end by flexible end plates.

Assume that $\delta_{max} \approx \dfrac{0.104 ML^2}{EI}$ where M is the maximum bending moment due to the *characteristic imposed load only.*

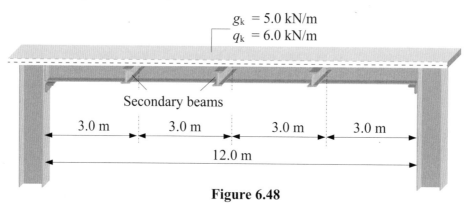

Figure 6.48

6.3.4.17 Solution to Example 6.21

Contract : **Beams** Job Ref. No. : **Example 6.21**	**Calcs. by : W.McK.**
Part of Structure :Beam 2 – Intermittent Restraint	**Checked by :**
Calc. Sheet No. : 1 of 6	**Date :**

References	Calculations	Output
	$G_k = 10$ kN $G_k = 10$ kN $G_k = 10$ kN $Q_k = 20$ kN $Q_k = 20$ kN $Q_k = 20$ kN $g_k = 5$ kN/m $q_k = 6$ kN/m A B C D E 3.0 m 3.0 m 3.0 m 3.0 m 12.0 m Design uniformly distributed load $= [(1.4 \times 5.0) + (1.6 \times 6.0)]$ $\qquad = 16.6$ kN/m Design point loads $\qquad = [(1.4 \times 10) + (1.6 \times 20)]$ $\qquad = 46$ kN $V_A = V_B = [(16.6 \times 12.0) + (3 \times 46)]/2 = 168.6$ kN	

Contract : **Beams** Job Ref. No. : **Example 6.21** **Part of Structure :Beam 2 - Intermittent Restraint** **Calc. Sheet No. : 2 of 6**	**Calcs. by : W.McK.** **Checked by :** **Date :**

References	Calculations	Output
	168.6 kN 118.8 kN 72.8 kN 23.0 kN Shear Force Diagram A B C D E 431.1 kNm 431.1 kNm 574.8 kNm Bending Moment Diagram Check the suitability of a $457 \times 191 \times 67$ UB S 275 section Section Properties: D = 453.4 mm d = 407.6 mm B = 189.9 mm T = 12.7 mm t = 8.5 mm b/T = 7.48 d/t = 48.0 x = 37.9 u = 0.872 $S_x = 1470 \times 10^3$ mm^3 $Z_x = 1300 \times 10^3$ mm^3 r_y = 41.2 mm I_x = 29400×10^4 mm^4	
Table 9	Flange thickness $T \le 16$ mm $\therefore p_y = 275$ N/mm^2	
Table 11	Section Classification $\varepsilon = \left(\dfrac{275}{p_y}\right)^{\frac{1}{2}} = 1.0$	
	Flange: Outstand element of compression flange rolled section $b/T = 7.48 < 9.0\varepsilon$	**Flange is plastic**
	Web: Neutral axis at mid-depth (i.e. bending only) $d/t = 48.0 < 80\varepsilon$	**Web is plastic**
Clause 4.2.3	$< 70\varepsilon$ therefore no need to check shear buckling	**Section is plastic**
Clause 4.2.3	Shear: Design shear force $F_v = 168.6$ kN P_v = $(0.6p_yA_v)$ = $0.6p_ytD$ P_v = $(0.6 \times 275 \times 8.5 \times 453.4)/10^3$ = 635.9 kN $P_v \gg F_v$	**Section is adequate in shear**
Clause 4.2.5.1	Bending: $M_c \le 1.2p_y Z_x$ for simply supported beams = $(1.2 \times 275 \times 1610 \times 10^3)/10^6$ = 531.3 kNm At B, C and D where the maximum bending moments occur the coincident shear force $\ll 0.6P_v$ therefore use the low shear equations.	

Contract : Beams Job Ref. No. : Example 6.21	Calcs. by : W.McK.
Part of Structure :Beam 2 - Intermittent Restraint	Checked by :
Calc. Sheet No. : 3 of 6	Date :

References	Calculations	Output
Clause 4.2.5.2	$M_c = p_y S_x$ $= (275 \times 1470 \times 10^3)/10^6 \qquad = 404.3$ kNm Maximum applied moment $\quad = M_x = 574.8$ kNm $\qquad\qquad\qquad\qquad M_c < M_x$ Try a $533 \times 210 \times 92$ UB S 355 Section Properties: $D = 533.1$ mm $d = 476.5$ mm $B = 209.3$ mm $T = 15.6$ mm $t = 10.1$ mm $b/T = 6.71$ $d/t = 47.2$ $x = 36.4$ $u = 0.873$ $S_x = 2360 \times 10^3$ mm^3 $Z_x = 2070 \times 10^3$ mm^3 $r_y = 45.1$ mm $I_x = 55200 \times 10^4$ mm^4	**Critical value of** $M_c = 404.3$ kNm
Table 9	Flange thickness $T \le 16$ mm $\therefore p_y = 355$ N/mm^2	
Table 11	Section Classification $\qquad \varepsilon = \left(\dfrac{275}{355}\right)^{\frac{1}{2}} = 0.88$	
	Flange: Outstand element of compression flange rolled section $b/T = 6.71 < 9.0\varepsilon$	**Flange is plastic**
	Web: Neutral axis at mid-depth (i.e. bending only) $d/t = 47.2 < 80\varepsilon$	**Web is plastic**
Clause 4.2.3	$\qquad\qquad < 70\varepsilon$ therefore no need to check shear buckling Shear capacity is greater than previous section.	**Section is plastic**
Clause 4.2.5.1	Bending: $M_c \le 1.2 p_y Z_x$ for simply supported beams $= (1.2 \times 355 \times 2070 \times 10^3)/10^6 = 881.8$ kNm	
Clause 4.2.5.2	$M_c = p_y S_x$ $= (355 \times 2360 \times 10^3)/10^6 \qquad = 837.8$ kNm Maximum applied moment $\quad = M_x = 574.8$ kNm $\qquad\qquad\qquad\qquad M_c > M_x$	**Critical value of** $M_c = 837.8$ kNm
Clause 4.3.6.2	Lateral torsional buckling: $M_x \le M_B / m_{LT} \qquad (\le M_c)$ The intermediate lateral restraints occur at B, C and D resulting in four unrestrained lengths AB, BC, CD and DE. Clearly in this instance sections BC and CD are the more critical. m_{LT} can be determined from Table 18 for the general case as follows:	

Contract : **Beams** Job Ref. No. : **Example 6.21** Part of Structure :**Beam 2 - Intermittent Restraint** Calc. Sheet No. : **4 of 6**	**Calcs. by : W.McK.** **Checked by :** **Date :**

References	Calculations	Output

Consider the shear force and bending moment diagrams between B and C.

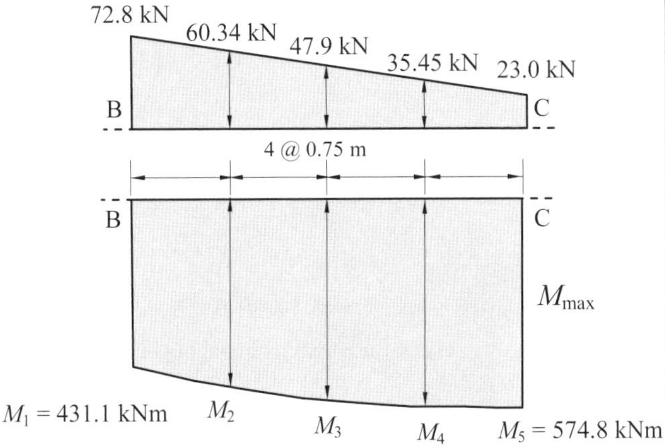

$$m_{LT} = 0.2 + \frac{(0.15M_2 + 0.5M_3 + 0.15M_4)}{M_{max}} \geq 0.44$$

$M_2 = 431.1 + [0.5 \times (72.8 + 60.34) \times 0.75)] = 481.0$ kNm
$M_3 = 431.1 + [0.5 \times (72.8 + 47.9) \times 1.5)] = 521.6$ kNm
$M_4 = 431.1 + [0.5 \times (72.8 + 35.45) \times 2.25)] = 552.9$ kNm

$$m_{LT} = 0.2 + \frac{[(0.15 \times 481.0) + (0.5 \times 521.6) + (0.15 \times 552.9)]}{574.8} \geq 0.44$$

$= 0.2 + 0.72$
$= 0.92$

Table 18 (reference for bending moment diagram)

Clause 4.3.6.4

Rigorous Method:
$M_B = p_b \times S_x$ where p_b is determined from Table 16 for rolled sections using λ_{LT} and p_y

Clause 4.3.6.7
$\lambda_{LT} = uv\lambda \sqrt{\beta_w}$ where $\lambda = \dfrac{L_E}{r_y}$;

Clause 4.3.5.1
Both flanges are free to rotate on plan.
There are lateral and torsional restraints at A, B and C.

Table 13
The loading condition is normal and $L_E = 1.0L_{LT}$
$L_E = 3000$ mm,
$\lambda = \dfrac{3000}{45.1} = 66.5$

Section Tables
Buckling parameter $u = 0.873$
Torsional index $x = 36.4$

Contract : **Beams** Job Ref. No. : **Example 6.21** Part of Structure :**Beam 2 - Intermittent Restraint** Calc. Sheet No. : **5 of 6**	Calcs. by : **W.McK.** Checked by : Date :

References	Calculations	Output
	v is the slenderness factor which is given in Table 19 and depends on λ /x and η	
Clause 4.3.6.7	η = 0.5 for UB sections; λ/x = (66.5/36.4) = 1.83	
Table 19	<table><tr><td>λ/x</td><td>$\eta = 0.5$</td></tr><tr><td>1.5</td><td>0.97</td></tr><tr><td>2.0</td><td>0.96</td></tr></table> $v \approx 0.96$	
Clause 4.3.6.9 Table 16	β_w = 1.0 for plastic sections λ_{LT} = $uv\lambda \sqrt{\beta_w}$ = $(0.873 \times 0.96 \times 66.5 \times \sqrt{1.0})$ = 55.7 λ_{LO} = 30.2 < λ_{LT} ∴ check for lateral torsional buckling λ_{LT} = 55.7 and p_y = 355 N/mm^2	
Table 16	<table><tr><td>λ_{LT}</td><td>S 355</td></tr><tr><td>55</td><td>274</td></tr><tr><td>60</td><td>257</td></tr></table> p_b = $274 - \left[(274-257)\times \dfrac{0.7}{5.0}\right]$ = 271.6 N/mm^2	
Clause 4.3.6.4	M_B = $(p_b \times S_x)$ = $(271.6 \times 2360 \times 10^3)/10^6$ = 640.9 kNm	
Clause 4.3.6.2	$M_x \le M_B / m_{LT}$ = (640.9/ 0.92) = 696.6 kNm $> M_x$ Since $M_x < M_c$ and $< M_B / m_{LT}$ the section is adequate.	**Section is adequate in bending**
Clause 4.3.7	**Simplified Method for Equal Flanged Rolled Sections:** M_B = $(p_b \times S_x)$ where p_b is determined from Table 20 for rolled sections using $[(\beta_w)^{0.5} L_E/r_y]$ and D/T	
Clause 4.3.6.9	β_w = 1.0 for plastic sections L_E/r_y = 66.5 $[(\beta_w)^{0.5} L_E/r_y]$ = $[(1.0)^{0.5} \times 66.5]$ = 66.5 D/T = (533.1/15.6) = 34.17	
Table 20	p_y = 355 N/mm^2	

<table>
<tr><td rowspan="2">$[(\beta_w)^{0.5} L_E/r_y]$</td><td colspan="4">$D/T$</td></tr>
<tr><td>25</td><td>30</td><td>35</td><td>40</td></tr>
<tr><td>60</td><td>-</td><td>286</td><td>284</td><td>-</td></tr>
<tr><td>65</td><td>-</td><td>273</td><td>270</td><td>-</td></tr>
<tr><td>70</td><td>-</td><td>259</td><td>256</td><td>-</td></tr>
<tr><td>75</td><td>-</td><td>246</td><td>242</td><td>-</td></tr>
</table>

Contract : Beams Job Ref. No. : Example 6.21	Calcs. by : W.McK.
Part of Structure :Beam 2 - Intermittent Restraint	Checked by :
Calc. Sheet No. : 6 of 6	Date :

References	Calculations	Output

$[(\beta_w)^{0.5} L_E/r_y]$	D/T			
	25	30	35	40
66.5	-	268.8	265.5	-

$$p_{b,} = 268.8 - \left[(268.8 - 265.5) \times \frac{4.17}{5.0}\right]$$

$$= 266.0 \text{ N/mm}^2$$

Clause 4.3.7
Clause 4.3.6.2

$M_B = (p_b \times S_x) = (266.0 \times 2360 \times 10^3)/10^6 = 627.8 \text{ kNm}$
$M_x \leq M_B / m_{LT} = (627.8/ 0.92) = 682.4 \text{ kNm} > M_x$

Since $M_x < M_c$ and $< M_B / m_{LT}$ the section is adequate.

Output: Section is adequate in bending

Deflection:
Clause 2.5.2
Table 8
Assume that the beam will support a ceiling with a brittle finish.
Deflection due to characteristic imposed loads $\leq$ Span/360

$$\delta_{maximum} \leq (12000/360) = 33.3 \text{ mm}$$

$$\delta_{actual} \approx \frac{0.104 ML^2}{EI}$$

$M = [(66 \times 6.0) - (6.0 \times 6.0 \times 3.0) - (20.0 \times 3.0)]$
$= 228 \text{ kNm}$

$$\therefore \delta_{actual} \approx \frac{\left(0.104 \times 228 \times 10^3 \times 12000^2\right)}{\left(205 \times 55200 \times 10^4\right)} = 30.2 \text{ mm}$$

$$\delta_{actual} < 33.3 \text{ mm}$$

Web Bearing and Web Bearing:
Since flexible end plates are used to connect the beams to the columns and the secondary beams are connected directly to the web of the main beam, there is no need to check web bearing and web buckling.

Output: Web bearing and web buckling checks not required

6.3.4.18 Example 6.22: Beam 3 Beam with Cantilever Span

A beam ABC is connected to a column at A by flexible end plates and simply supported at B on a column cap plate as shown in Figure 6.49. The *top* flange is restrained against lateral torsional buckling for its full length. For the given characteristic loading, check the suitability of the UB section indicated with respect to:

- ♦ shear,
- ♦ bending,
- ♦ deflection,
- ♦ web buckling and
- ♦ web bearing.

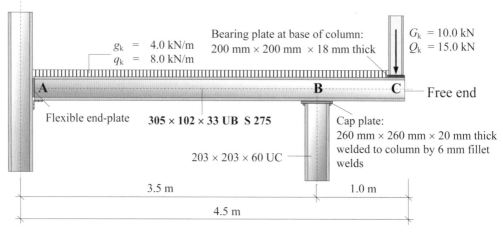

Figure 6.49

6.3.4.19 Solution to Example 6.22

Contract : **Beams** Job Ref. No. : **Example 6.22** Part of Structure : **Beam 3 - with Cantilever Span** Calc. Sheet No. : **1 of 7**	Calcs. by : **W.McK.** Checked by : Date :

References	Calculations	Output

Table 2	Maximum design UDL load $= \;[(1.4 \times 4.0) + (1.6 \times 8.0)]$
	                                     $= \; 18.4$ kN/m length
	Maximum design point load $= \;[(1.4 \times 10.0) + (1.6 \times 15.0)]$
	                                     $= \; 38.0$ kN
	Minimum design UDL load $= \;(1.0 \times 4.0) \;= 4.0$ kN/m length
	Minimum design point load $= \;(1.0 \times 10.0) = 10.0$ kN
	
	There are three load cases to consider:
	(i) maximum span moment with the maximum design loading on the span and the minimum design load on the cantilever,
	(ii) maximum support moment with the maximum design loading on the cantilever,
	(iii) the maximum shear force with maximum loading over the full length.	

Contract : Beams Job Ref. No. : Example 6.22		**Calcs. by : W.McK.**
Part of Structure : Beam 3 - with Cantilever Span		**Checked by :**
Calc. Sheet No. : 2 of 7		**Date :**

References	Calculations	Output

Case (i):

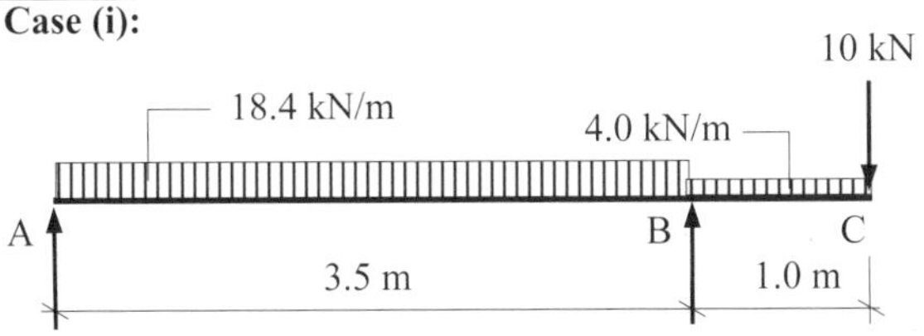

Maximum span moment   =   22.5 kNm   1.56 m from A

Case (ii):

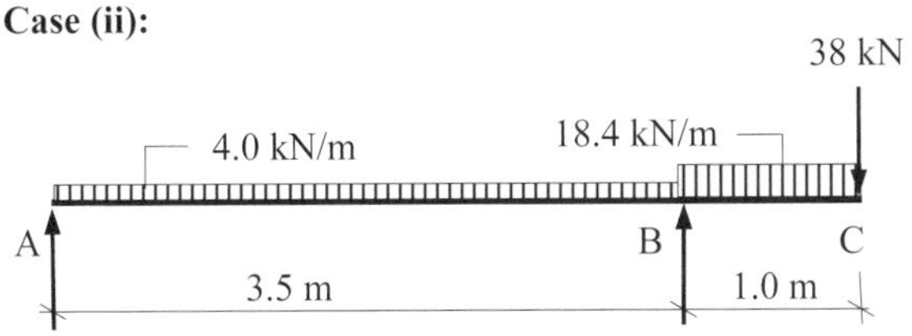

Maximum support moment   =   47.2 kNm

Case (iii):

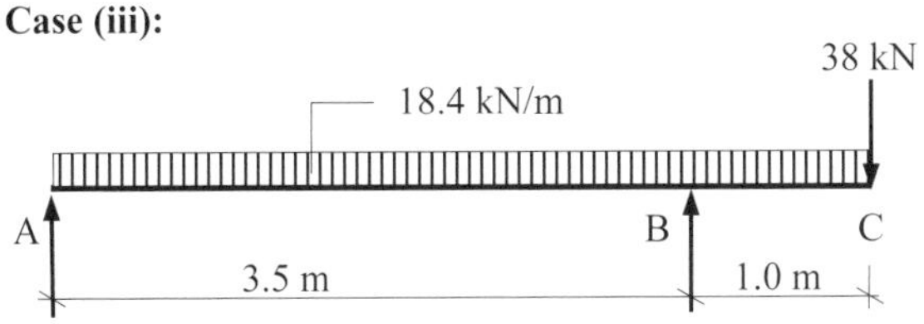

Maximum shear force   =   56.4 kN
Maximum vertical reaction   =   102.1 kN   at support B
**The reader should confirm the values of shear force
bending moments and support reaction indicated above.**

Check the suitability of a 305 × 102 × 33 UB S 275 section
Section Properties:
D =   312.7 mm d =   275.9 mm B = 102.4 mm
T =   10.8 mm t =   6.6 mm b/T = 4.74
d/t =   41.8 x =   31.6 u = 0.867
S_x = 481 × 10^3 mm^3 Z_x = 416 × 10^3 mm^3 r_y = 21.5 mm
I_x =   6500 × 10^4 mm^4 A =   41.8 × 10^2 mm^2
r = 7.6 mm (root radius)

Table 9

Flange thickness $T \le 16$ mm $\therefore p_y = 275$ N/mm^2

Table 11

Section Classification $\varepsilon = \left(\dfrac{275}{p_y}\right)^{\frac{1}{2}} = 1.0$

References	Calculations	Output
	Contract : Beams Job Ref. No. : Example 6.22 **Part of Structure : Beam 3 - with Cantilever Span** **Calc. Sheet No. : 3 of 7**	**Calcs. by : W.McK.** **Checked by :** **Date :**

References	Calculations	Output
	Flange: Outstand element of compression flange rolled section $b/T = 4.74 < 9.0\varepsilon$	**Flange is plastic**
	Web: Neutral axis at mid-depth (i.e. bending only) $d/t = 41.8 < 80\varepsilon$	**Web is plastic**
Clause 4.2.3	$< 70\varepsilon$ therefore no need to check shear buckling	**Section is plastic**
Clause 4.2.3	Shear: Design shear force $F_v = 56.4$ kN $P_v = (0.6p_yA_v) = 0.6p_ytD$ $P_v = (0.6 \times 275 \times 6.6 \times 312.7/10^3 = 340.5$ kN $\qquad\qquad\qquad P_v >> F_v$	**Section is adequate in shear**
Clause 4.2.5.1	Bending: $M_c \leq 1.2p_y Z_x$ for simply supported beams $= (1.2 \times 275 \times 416 \times 10^3)/10^6 = 137.3$ kNm	
	At the position of the maximum bending moments over the support the coincident shear force = 56.5 kN which is $< 0.6P_v$, therefore use the low shear equations.	
Clause 4.2.5.2	$M_c = p_y S_x$ $= (275 \times 481 \times 10^3)/10^6 = 132.3$ kNm	**Critical value of $M_c = 132.3$ kNm**
	Maximum applied moment $= M_x = 47.2$ kNm $\qquad\qquad\qquad M_c > M_x$	
Clause 4.3.6.2	Consider the cantilever: Lateral torsional buckling: $M_x \leq M_B / m_{LT}$ $(\leq M_c)$	
Table 18	Note at the end of Table 18: For cantilevers without lateral restraint $m_{LT} = 1.0$	
Clause 4.3.6.4	**Rigorous Method:** $M_B = p_b \times S_x$ where p_b is determined from Table 16 for rolled sections using λ_{LT} and p_y	
Clause 4.3.6.7	$\lambda_{LT} = uv\lambda \sqrt{\beta_w}$ where $\lambda = \dfrac{L_E}{r_y}$;	
Table 14	At the support the cantilever is continuous with restraint to the top flange. At the tip the cantilever is free.	
Table 13	The loading condition is normal and $L_E = 3.0L_{LT}$ $\qquad\qquad\qquad\qquad L_E = 3000$ mm $\qquad\qquad\qquad\qquad \lambda = \dfrac{3000}{21.5} = 139.5$	

Contract : Beams Job Ref. No. : Example 6.22 Part of Structure : Beam 3 - with Cantilever Span Calc. Sheet No. : 4 of 7	Calcs. by : W.McK. Checked by : Date :

References	Calculations	Output
Section Tables	Buckling parameter u = 0.867 Torsional index x = 31.6 v is the slenderness factor which is given in Table 19 and depends on λ/x and η	
Clause 4.3.6.7	η = 0.5 for UB sections; λ/x = (139.5/31.6) = 4.41	
Table 19	<table><tr><td>λ/x</td><td>$\eta = 0.5$</td></tr><tr><td>4.0</td><td>0.86</td></tr><tr><td>4.5</td><td>0.84</td></tr></table> $v = 0.86 - \left[(0.86 - 0.84) \times \dfrac{0.41}{5.0}\right] = 0.844$	
Clause 4.3.6.9	β_w = 1.0 for plastic sections $\lambda_{LT} = uv\lambda\sqrt{\beta_w}$ = $(0.867 \times 0.844 \times 139.5 \times \sqrt{1.0})$ = 102.1	
Table 16	λ_{LO}= 34.0 $<$ λ_{LT} $\therefore$ check for lateral torsional buckling λ_{LT} = 102.1 and p_y = 275 N/mm^2	
Table 16	<table><tr><td>λ_{LT}</td><td>S 355</td></tr><tr><td>100</td><td>125</td></tr><tr><td>105</td><td>117</td></tr></table> p_b = $125 - \left[(125 - 117) \times \dfrac{2.1}{5.0}\right]$ = 121.6 N/mm^2	
Clause 4.3.6.4	M_B = $(p_b \times S_x)$ = $(121.6 \times 481 \times 10^3)/10^6$ = 58.5 kNm	**Section is adequate in bending**
Clause 4.3.6.2	$M_x \le$ M_B / m_{LT} = (58.5/ 1.0) = 58.5 kNm $> M_x$ Since $M_x < M_c$ and $< M_B / m_{LT}$ the section is adequate.	
Clause 4.3.7	**Simplified Method for Equal Flanged Rolled Sections:** M_B = $(p_b \times S_x)$ where p_b is determined from Table 20 for rolled sections using $[(\beta_w)^{0.5} L_E/r_y]$ and D/T	
Clause 4.3.6.9	β_w = 1.0 for plastic sections L_E/r_y = 139.5 $[(\beta_w)^{0.5} L_E/r_y]$ = $[(1.0)^{0.5} \times 139.5]$ = 139.5	
Table 20	D/T = (312.7/10.8) = 29 p_y = 275 N/mm^2	

$[(\beta_w)^{0.5} L_E/r_y]$	D/T			
	20	25	30	35
130	-	-	-	-
135	-	130	121	-
140	-	126	117	-
145	-	-	-	-

Contract : **Beams**　　Job Ref. No. : **Example 6.22**	Calcs. by : **W.McK.**
Part of Structure : **Beam 3 - with Cantilever Span**	**Checked by :**
Calc. Sheet No. : **5 of 7**	**Date :**

References	Calculations	Output
	(table and calculations below)	

$[(\beta_w)^{0.5} L_E/r_y]$	D/T			
	20	25	30	35
139.5	-	126.4	117.4	-

$$p_b = 126.4 - \left[(126.4-117.4)\times\frac{4.5}{5.0}\right] = 118.3 \text{ N/mm}^2$$

Clause 4.3.7
Clause 4.3.6.2

$M_B = (p_b \times S_x) = (118.3 \times 481 \times 10^3)/10^6 = 56.9 \text{ kNm}$

$M_x \le M_B / m_{LT} = (56.9/1.0) = 56.9 \text{ kNm} > M_x$

Since $M_x < M_c$ and $< M_B/m_{LT}$ the section is adequate.

Output: Section is adequate in bending

Deflection: Cantilever

Clause 2.5.2
Table 8

Assume that the beam will support a ceiling with a brittle finish.
Deflection due to characteristic imposed loads $\le$ Length/180
$\delta_{maximum} \le (1000/180) = 5.55 \text{ mm}$
For a simply supported span 'a' with a cantilever end = 'b',
total load 'W' along the cantilever and point load P at the end:

$$\delta_{actual} \approx \left[\frac{Wb^3}{8EI}+\frac{Wab^2}{6EI}\right]+\left[\frac{Pb^3}{3EI}+\frac{Pab^2}{3EI}\right]$$

$$= \frac{W}{EI}\left[\frac{b^3}{8}+\frac{ab^2}{6}\right]+\frac{P}{3EI}\left[b^3+ab^2\right]$$

Clause 3.1.3

$EI = (205 \times 6500 \times 10^4) = 13325 \times 10^6 \text{ kNmm}^2$
$a = 3500 \text{ mm};$　　　　　$b = 1000 \text{ mm}$
$W = (8.0 \times 1.0) = 8.0 \text{ kN}$　　$P = 15.0 \text{ kN}$
$\delta_{actual} \approx$

$$\frac{8.0}{EI}\left[\frac{1000^3}{8}+\frac{3500\times1000^2}{6}\right]+\frac{15}{3EI}\left[1000^3+3500\times1000^2\right]$$

$$= \frac{28166.7\times10^6}{13325\times10^6} = 2.11 \text{ mm}$$

$$\delta_{actual} < 5.55 \text{ mm}$$

Deflection: Span
$\delta_{maximum} \le (3500/200) = 17.5 \text{ mm}$

Appendix 3

$$\delta_{actual} \approx \left[\frac{5WL^3}{384EI}\right] = \frac{5\times(8.0\times3.5)\times3500^3}{384\times13325\times10^6} = 1.17 \text{ mm}$$

$$\delta_{actual} < 17.5 \text{ mm}$$

The actual deflections will be slightly less than these values due to the minimum dead loads which have been ignored.

Output: Section is adequate with respect to deflection

References	Calculations	Output
	Contract : Beams Job Ref. No. : Example 6.22 **Part of Structure : Beam 3 - with Cantilever Span** **Calc. Sheet No. : 6 of 7** **Calcs. by : W.McK.** **Checked by :** **Date :**	

References	Calculations	Output
	Consider support B:	
	Web Bearing Capacity:	
Clause 4.5.2.1	$P_{bw} = (b_1 + nk)tp_{yw}$	
Figure 13	$203 \times 203 \times 60$ UC	
	Note: $g = 0$	
	$D_c = 209.6$ mm	
	$t = 20$ mm	
	$s = 6$ mm	
	b_1	
	$\begin{aligned} b_1 &= D_C + \{2 \times [t + (0.8 \times s) - g]\} \\ &= 209.6 + \{2 \times [20 + (0.8 \times 6)]\} \\ &= 259.2 \text{ mm} \end{aligned}$	
	$k = (T + r)$ where T is the flange thickness and r is the root radius of the UC.	
Section Tables	$T = 10.8$ mm $\quad r = 7.6$ mm $\quad t = 6.6$ mm $d = 275.9$ mm	
	$k = (T + r) = (10.8 + 7.6) = 18.4$ mm $n = 5$	
Table 9	$p_{yw} = 275$ N/mm^2	
	$\begin{aligned} P_{bw} &= (b_1 + nk)tp_{yw} \\ &= \{[259.2 + (5 \times 18.4)] \times (6.6 \times 275)\}/10^3 = 637.4 \text{ kN} \end{aligned}$	**Section is adequate with respect to web bearing at B**
	Maximum concentrated load $= F_x = 102.1$ kN $\ll P_{bw}$	
	Web Buckling Capacity:	
Clause 4.5.3.1	$a_e > 0.7d$ $P_x = \dfrac{25\varepsilon t}{\sqrt{(b_1 + nk)d}} P_{bw}$	
	$= \dfrac{(25 \times 1.0 \times 6.6)}{\sqrt{[259.2 + (5 \times 18.4)] \times 275.9}} \times 637.4 = 337.9 \text{ kN}$	**Section is adequate with respect to web buckling at B**
	$F_x = 102.1$ kN $\ll P_x$	

References	Calculations	Output
	Contract : Beams Job Ref. No. : Example 6.22 **Part of Structure : Beam 3 - with Cantilever Span** **Calc. Sheet No. : 7 of 7**	**Calcs. by : W.McK.** **Checked by :** **Date :**

References	Calculations	Output
	Consider the free end of the cantilever: Web Bearing Capacity:	
Clause 4.5.2.1	$P_{bw} = (b_1 + nk)tp_{yw}$ 200 mm wide × 20 mm thick base plate	
Table 9	$b_1 = 200$ mm $k = (T + r) = (10.8 + 7.6) = 18.4$ mm $n = [2 + (0.6b_e)/k] \le 5$ $b_e = 0$ $\quad = [2 + (0)] = 2.0 \le 5$ $p_{yw} = 275$ N/mm^2 $P_{bw} = (b_1 + nk)tp_{yw}$ $\quad = \{[200 + (2.0 \times 18.4)] \times (6.6 \times 275)\}/10^3 = 430.8$ kN Maximum concentrated load $= F_x = 38.0$ kN $\ll P_{bw}$ Web Buckling Capacity:	**Section is adequate with respect to web bearing at C**
Clause 4.5.3.1	$a_e = 100$ mm $< 0.7d$ $P_x = \left(\dfrac{a_e + 0.7d}{1.4d}\right)\dfrac{25\varepsilon t}{\sqrt{(b_1 + nk)d}} P_{bw}$ $\quad = \left[\dfrac{100 + (0.7 \times 275.9)}{(1.4 \times 275.9)}\right]\dfrac{(25 \times 1.0 \times 6.6)}{\sqrt{[200 + (2.03 \times 18.4)] \times 275.9}} \times 430.8$ $\quad = 210.8$ kN $\quad\quad\quad\quad\quad\quad F_x = 38.0$ kN $\ll P_x$	**Section is adequate with respect to web buckling at C**

6.3.5 *Elements Subject to Combined Axial and Flexural Loads*

6.3.5.1 Introduction

While many structural members have a single dominant effect, such as axial loading or bending, there are numerous elements which are subjected to both types of loading at the same time. The behaviour of such elements is dependent on the interaction characteristics of the individual components of load. Generally, members resisting combined tension and bending are less complex to design than those resisting combined compression and bending, since the latter are more susceptible to associated buckling effects. The combined effects can occur for several reasons such as eccentric loading or rigid frame action, as illustrated in Figures 6.50 (a) and (b) respectively, or concentrated loads between the nodes in continuous lattice girders members, as indicated in (c).

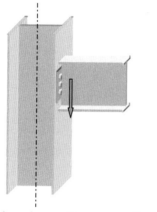

Simple 'pinned end' connection inducing an axial load and a **nominal** secondary moment in the column section (see Clause 4.7.7).

(a)

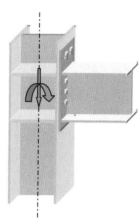

Rigid 'moment' connection inducing an axial load and a **primary** moment in the column section (see Clause 4.8.3).

(b)

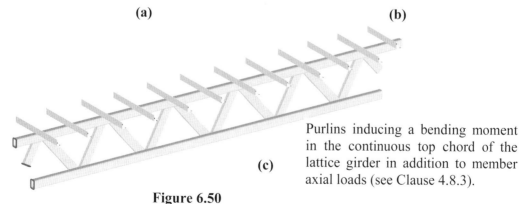

(c) Purlins inducing a bending moment in the continuous top chord of the lattice girder in addition to member axial loads (see Clause 4.8.3).

Figure 6.50

6.3.5.2 Combined Tension and Bending

Consider a structural member subjected to concentric axial loading, as shown in Figure 6.51(a).

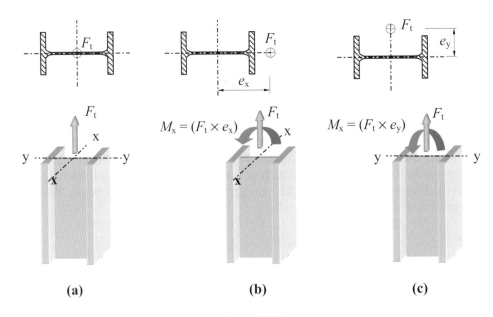

Figure 6.51

The limiting value of applied axial load F_t can be determined by using the equation:

$$F_t = p_y A_e = P_t \quad \therefore \quad \frac{F_t}{P_t} = 1.0$$

Similarly, if the applied load is eccentric to the X–X axis, as shown in 6.51(b):

$$M_x = M_{cx} \quad \therefore \quad \frac{M_x}{M_{cx}} = 1.0$$

In the case of eccentricity about the Y–Y axis, as shown in Figure 6.51(c):

$$M_y = M_{cy} \quad \therefore \quad \frac{M_y}{M_{cy}} = 1.0$$

If these limits are plotted on three-dimensional orthogonal axes, then they represent the members' capacity under each form of loading acting singly. Figure 6.52 is a linear inter-action diagram. Any point located within the boundaries of the axes and the interaction surface represents a combination of applied loading F, M_x and M_y for which

$$\frac{F_t}{P_t} + \frac{M_x}{M_{cx}} + \frac{M_y}{M_{cy}} \leq 1.0 \qquad \text{(Equation 1)}$$

and which can be safely carried by the section.

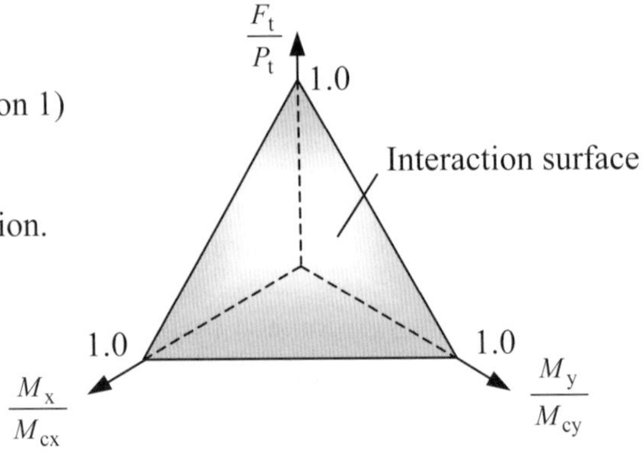

Figure 6.52

Equation (1), which represents a linear approximation of member behaviour, is used in BS 5950 in Clause 4.8.2.2 in the simplified method of design for tension members with moments.

A more rigorous analysis allowing for plastic behaviour of plastic and compact sections results in an interaction surface as shown in Figure 6.53. The precise shape of the surface is dependent on the cross-section for which the diagram is constructed. This non-linear surface is represented by:

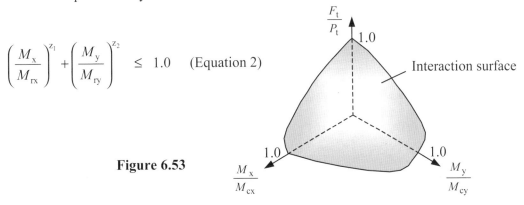

$$\left(\frac{M_x}{M_{rx}}\right)^{z_1} + \left(\frac{M_y}{M_{ry}}\right)^{z_2} \leq 1.0 \quad \text{(Equation 2)}$$

Figure 6.53

where:
M_{rx} is the reduced plastic moment capacity about the major axis due to axial loading,
M_{ry} the reduced moment capacity about the minor axis due to axial loading
        (see Section 6.3.5.3 regarding reduced plastic moment capacity),
z_1 $= 2.0$ for I and H sections with equal flanges and solid and hollow circular sections,
   $= 5/3$ for solid and hollow rectangular sections,
   $= 1.0$ for all other cases,
z_2 $= 2.0$ for solid and hollow circular sections,
   $= 5/3$ for solid and hollow rectangular sections,
   $= 1.0$ for all other cases.

The values of M_{rx} and M_{ry} are available from published Section Properties Tables. *BS 5950-1:2000* gives Equation (2) as a more economic alternative to using Equation (1) when designing members subject to both axial tension and bending.

Note: *It is also necessary to check the resistance of such members to lateral torsional buckling in accordance with Clause 4.3, assuming the value of the axial load to be zero.*

6.3.5.3 Reduced Moment Capacity

In Section 6.2.5.2 and Figure 6.4 the plastic stress distribution is shown for a cross-section subject to pure bending. When an axial load is applied at the same time, this stress diagram should be amended. Consider an I-section subjected to an eccentric compressive load as shown in Figure 6.51(b).

The load on the cross-section comprises two components: an axial load F and a bending moment ($F \times e$). The stress diagram can be considered to be the superposition of two components, as shown in Figure 6.54. When the axial load is relatively low then sufficient material is contained within the web to resist the pure axial effects, as shown in

Figure 6.54(a). There is a small reduction in the material available to resist bending and hence the reduced bending stress diagram is as shown in Figure 6.54(b). The stress diagram relating to combined axial and bending effects is shown in Figure 6.54(c), indicating the displaced plastic neutral axis.

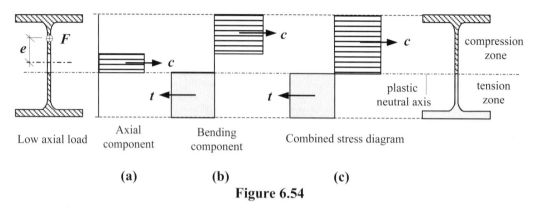

Low axial load | Axial component | Bending component | Combined stress diagram

(a)   (b)   (c)

Figure 6.54

When the axial load is relatively high then in addition to the web material, some of the flange material is required to resist the axial load, as shown in Figure 6.55(a), and hence a larger reduction in the bending moment capacity occurs, as shown in Figure 6.55(b). As before the plastic neutral axis is displaced, as shown in Figure 6.55(c).

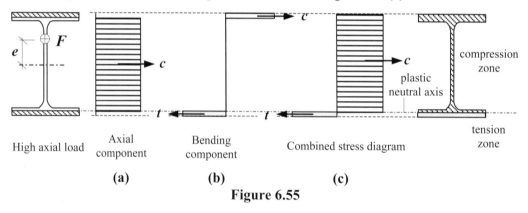

High axial load | Axial component | Bending component | Combined stress diagram

(a)   (b)   (c)

Figure 6.55

The published Section Property Tables provide a *change formula* for each section, to enable a reduced plastic section modulus to be evaluated depending on the amount of material required to resist the axial component of the applied loading.

6.3.5.4 Combined Compression and Bending

The behaviour of members subjected to combined compression and bending is much more complex than those with combined tension and bending. A comprehensive explanation of this is beyond the scope of this text and can be found elsewhere (refs. 68, 70). Essentially, there are three possible modes of failure to consider:

  (i)   a combination of column buckling and simple uniaxial bending;
  (ii)   a combination of column buckling and lateral torsional beam buckling;
  (iii) a combination of column buckling and biaxial beam bending.

As with combined tension and bending, an interaction diagram can be constructed to illustrate the behaviour of sections subjected to an axial load F_c, and bending moments M_x and M_y about the major and minor axes respectively; this is shown in Figure 6.56.

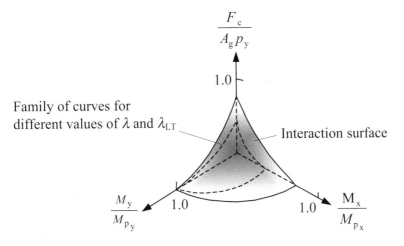

Figure 6.56

Although similar, this interaction diagram differs from that shown in Figure 6.53 in that it relates to *slender* members. The precise intercept on the F_c and M_x axes will depend on the slenderness of the member. Clearly, short, stocky members will intercept the $\dfrac{F_c}{A_g p_y}$ axis at a value of 1.0; as the slenderness λ increases, the intercept decreases. Similarly, in sections which are fully restrained against lateral torsional buckling the intercept on the $\dfrac{M_x}{M_{p_x}}$ axis will equal 1.0, decreasing as the equivalent slenderness λ_{LT} increases. *The minor axis strength is not affected by overall buckling of the member and the intercept is always equal to 1.0.*

Members, except those with slender cross-sections, which are subject to combined compression and bending should be checked for **cross-section capacity** using the simplified approach, as given in Clause 4.8.3.2:

$$\frac{F_c}{A_g p_y} + \frac{M_x}{M_{cx}} + \frac{M_y}{M_{cy}} \leq 1.0 \qquad \text{(Equation 3)}$$

which differs slightly from the equation given in Clause 4.8.2.2 for combined tension and bending, or alternatively using the more exact method:

$$\left(\frac{M_x}{M_{rx}}\right)^{z_1} + \left(\frac{M_y}{M_{ry}}\right)^{z_2} \leq 1.0 \qquad \text{(Equation 4)}$$

as before for plastic and compact cross-sections.

For members with slender cross-sections the following equation should be used:

$$\frac{F_c}{A_{eff}\,p_y} + \frac{M_x}{M_{cx}} + \frac{M_y}{M_{cy}} \quad \leq \quad 1.0 \qquad \text{(Equation 5)}$$

The difference between equations (3) and (5) is the use of the effective cross-sectional area A_{eff} as defined in Clause 3.6

In addition to checking the cross-section capacity of a member, **buckling resistance** should be checked in accordance with Clause 4.8.3.3. A simplified conservative approach for this check, which is suitable in most circumstances, is given in Clause 4.8.3.3.1. The following two relationships should be satisfied:

$$\frac{F_c}{P_c} + \frac{m_x M_x}{p_y Z_x} + \frac{m_y M_y}{p_y Z_y} \quad \leq \quad 1.0 \qquad \text{(Equation 6)}$$

$$\frac{F_c}{P_{cy}} + \frac{m_{LT} M_{LT}}{M_b} + \frac{m_y M_y}{p_y Z_y} \quad \leq \quad 1.0 \qquad \text{(Equation 7)}$$

where:

F_c is the axial compression

M_b is the buckling resistance moment, generally from Clause 4.3,

M_{LT} is the maximum major axis moment in the segment length L governing M_b,

m_{LT} is the equivalent uniform moment factor for lateral torsional buckling given in Table 18 (see Clause 4.8.8.3.4),

M_x is the maximum major axis moment in the segment length L_x governing P_{cx},

m_x is the equivalent uniform moment factor for major axis flexural buckling given in Table 26 (see Clause 4.8.8.3.4),

M_y is the maximum minor axis moment in the segment length L_y governing P_{cy},

m_y is the equivalent uniform moment factor for minor axis flexural buckling given in Table 26 (see Clause 4.8.8.3.4),

P_c is the smaller of P_{cx} and P_{cy},

P_{cx} is the compression resistance from Clause 4.7.4, considering buckling about the major axis only,

P_{cy} is the compression resistance from Clause 4.7.4, considering buckling about the minor axis only,

Z_x is the elastic section modulus about the major axis, ⎫ In the case of slender

Z_y is the elastic section modulus about the minor axis. ⎬ cross-sections the value
                                                  ⎭ of Z_{eff} should be used.

In Clause 4.8.3.3.2 a more rigorous analysis is indicated which utilizes alternative equations which will produce a more economic design, but in most cases this is unnecessary. The method is not dealt with in this text.

6.3.5.5 Example 6.23: Lattice Top Chord with Axial and Bending Loads

A lattice girder supports a series of beams at the mid-span points between the nodes in the top chord, as shown in Figure 6.57. Using the data given, check the suitability of the top chord to resist the combined axial load and bending moment.

Data:

Characteristic imposed load due to each beam	50 kN
Ultimate limit state design load due to each beam	100 kN
Ultimate limit state design axial load in the top chord members AB and BC	1000 kN

254 × 146 × 31 Castellated UB S 275 bearing on welded flange plate

Top and bottom chords: 300 × 200 × 10 RHS S 275

5.0 m	5.0 m	5.0 m	5.0 m

Figure 6.57

6.3.5.6 Solution to Example 6.23

Contract : Lattice Girder Job Ref. No. : Example 6.23	**Calcs. by : W.McK.**
Part of Structure: Chord with Axial and Bending Load	**Checked by :**
Calc. Sheet No. : 1 of 6	**Date :**

References	Calculations	Output
	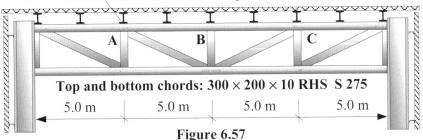 100 kN 100 kN 100 kN 100 kN 100 kN 100 kN 100 kN A B C 1000 kN 1000 kN Consider members AB or BC: Design axial load $F_x = 1000$ kN Design bending moment $M_x \approx (WL/8)$ (This allows for the continuity of the top girder) $\quad\quad\quad\quad\quad = [(100 \times 5.0)/8.0] = 62.5$ kNm Section Properties: **300 × 200 × 10 RHS S 275** $D = 300$ mm $\quad B = 200$ mm $\quad t = 10.0$ mm $b/t = 17.0 \quad\quad d/t = 27.0 \quad\quad r_x = 112$ mm $r_y = 81.3$ mm $\quad S_x = 956 \times 10^3$ mm^3 $\quad Z_x = 788 \times 10^3$ mm^3 $A_g = 94.9$ cm^2 $\quad I_{xx} = 11800$ cm^4	
Table 9	Flange thickness $t \leq 16$ mm $\therefore p_y = 275$ N/mm^2	
Table 12	Section Classification $\quad \varepsilon = \left(\dfrac{275}{p_y}\right)^{\frac{1}{2}} = 1.0$	

Contract : Lattice Girder Job Ref. No. : Example 6.23 Part of Structure: Chord with Axial and Bending Load Calc. Sheet No. : 2 of 6	Calcs. by : W.McK. Checked by : Date :

References	Calculations	Output
Table 12	Flange: Compression due to bending $b/t = 17.0\ <\ 28.0\varepsilon$	**Flange is plastic**
	Web: Generally $d/t = 27.0 \le 40\varepsilon$	**Web is plastic**
Clause 4.2.3	$< 70\varepsilon$ therefore no need to check shear buckling	**Section is plastic**
Clause 4.7.3(c) Clause 4.10	Effective length in the plane of the truss: 	
Table 22	$L_{Ex} \approx (0.85 \times L) = (0.85 \times 5000) = 4250$ mm	
	Effective length perpendicular to the plane of the truss: 	
Table 22	$L_{Ey} \approx (1.0 \times L) = (1.0 \times 2500) = 2500$ mm	
Clause 4.7.2	$\lambda_x = \dfrac{L_{Ex}}{r_x} = \dfrac{4250}{112} = 37.95$	
	$\lambda_y = \dfrac{L_{Ey}}{r_y} = \dfrac{2500}{81.3} = 30.75$	
Clause 4.7.4	$P_c = A_g p_c$	
Table 23	x–x axis Use strut curve a y–y axis Use strut curve a	
Table 24 (1)	$\lambda_x = 37.95$; $p_y = 275$ N/mm²; $\therefore$ $p_c = 261.6$ N/mm² $\lambda_y = 30.75$; $p_y = 275$ N/mm²; $\therefore$ $p_c = 266.6$ N/mm²	
	$P_x = (9490 \times 261.6)/10^3 = 2482$ kN $P_y = (9490 \times 266.6)/10^3 = 2530$ kN <div align="right">$P_c \gg F_c$</div>	**Compression resistance is adequate**
Clause 4.2.3	Shear: $P_v = (0.6\,p_y A_v) = 0.6\,p_y AD / (D + B)$ $P_v = [(0.6 \times 275 \times 9490 \times 300)/(300 + 200)]/10^3$ $= 939.5$ kN $F_v \approx (0.5 \times 100) = 50$ kN $P_v \gg F_v$	**Section is adequate in shear**

Contract : Lattice Girder Job Ref. No. : Example 6.23	Calcs. by : W.McK.
Part of Structure: Chord with Axial and Bending Load	Checked by :
Calc. Sheet No. : 3 of 6	Date :

References	Calculations	Output
Clause 4.2.5.1	Bending: $M_c \leq 1.5 p_y Z_x$ for beams generally $= (1.5 \times 275 \times 788 \times 10^3)/10^6 = 325$ kNm	
Clause 4.2.5.2	Use the low shear equations. $M_{cx} = p_y S_x$ $= (275 \times 956 \times 10^3)/10^6 = 262.9$ kNm	**Critical value of** $M_c = 262.9$ **kNm**
	Maximum applied moment $= M_x = 62.5$ kNm $$M_c \gg M_x$$	
Clause 4.3.6 Clause 4.3.6.1 Table 15	Lateral torsional buckling: $M_B = M_c$ provided that $L_E/r_y \leq$ Table 15 value $D/B = (300/200) = 1.5$ Limiting value of $L_E/r_y = [515 \times (275/p_y)] = 515$ $\gg \lambda_x$ and λ_y $M_B = p_y S_x = 262.9$ kNm	
Clause 4.8.3 Clause 4.8.3.2	**Compression members with moments:** Cross-section capacity $$\frac{F_c}{A_g p_y} + \frac{M_x}{M_{cx}} + \frac{M_y}{M_{cy}} \leq 1.0$$ Since there is no bending about the y–y axis $\dfrac{M_y}{M_{cy}} = 0$ $A_g p_y = (9490 \times 275)/10^3 = 2609$ kN $\therefore \left[\dfrac{F_c}{A_g p_y} + \dfrac{M_x}{M_{cx}}\right] = \left[\dfrac{1000}{2609} + \dfrac{62.5}{262.9}\right] = 0.62 \quad \leq 1.0$	
Clause 4.8.3.3.1	Buckling resistance: Both relationships must be satisfied. $$\frac{F_c}{P_c} + \frac{m_x M_x}{p_y Z_x} + \frac{m_y M_y}{p_y Z_y} \leq 1.0$$ $$\frac{F_c}{P_{cy}} + \frac{m_{LT} M_{LT}}{M_b} + \frac{m_y M_y}{p_y Z_y} \leq 1.0$$ $F_c = 1000$ kN, $M_x = 62.5$ kNm, $M_{LT} = 62.5$ kNm $p_y Z_x = (275 \times 788 \times 10^3)/10^6 = 216.7$ kNm $P_c = 2482$ kN (**Note:** smaller of P_{cx} and P_{cy}) $P_{cy} = 2530$ kN $M_B = 262.9$ kNm	

Contract : Lattice Girder Job Ref. No. : Example 6.23	Calcs. by : W.McK.
Part of Structure: Chord with Axial and Bending Load	Checked by :
Calc. Sheet No. : 4 of 6	Date :

References	Calculations	Output
Clause 4.8.3.3.4	m_{LT} is for lateral torsional buckling and relates to the restraints on the y–y axis. Consider the bending moment diagram due to the point load in the top chord between A and B. 62.5 kNm 62.5 kNm A B 62.5 kNm Unrestrained length $L_{LT} = 2.5$ m	
Table 18	$M = -62.5$ kNm, $\beta M = +62.5$ kNm $\therefore$ $\beta = -1.0$ $m_{LT} = 0.44$	
Clause 4.8.3.3.4	m_x relates to major axis flexural buckling and can be determined from Table 26 for the pattern of major axis moments over the segment length L_x governing P_{cx}, i.e. the length between the nodes A and B.	
Table 26	M_1 M_5 A M_2 M_4 B M_3 $M_1 = M_3 = M_5 = 62.5$ kNm $M_2 = M_4 = 0$ $m = 0.2 + \left[\dfrac{0.1M_2 + 0.6M_3 + 0.1M_4}{M_{max}}\right] \geq \dfrac{0.8M_{24}}{M_{max}}$ where M_{24} is the maximum moment in the central half of the segment. $m = 0.2 + \left[\dfrac{0 + (0.6 \times 62.5) + 0}{62.5}\right] = 0.8$ $\geq \left[\dfrac{(0.8 \times 62.5)}{62.5}\right] = 0.8$ $\left[\dfrac{1000}{2482} + \dfrac{(0.8 \times 62.5)}{216.7}\right] = (0.4 + 0.23) = 0.63 \leq 1.0$ $\left[\dfrac{1000}{2530} + \dfrac{(0.44 \times 62.5)}{262.9}\right] = (0.4 + 0.1) = 0.5 \leq 1.0$	**The section is adequate with respect to combined axial load and bending**

Contract : Lattice Girder Job Ref. No. : Example 6.23 Part of Structure: **Chord with Axial and Bending Load** Calc. Sheet No. : **5 of 6**	Calcs. by : **W.McK.** Checked by : Date :

References	Calculations	Output
	Web Bearing Capacity: 254 × 146 × 31 Castellated UB 200 mm × 200 mm flange plate welded to RHS section 	
Clause 4.5.1.5	Refer to publication: *Steelwork Design Guide to BS 5950-1 : 2000 Volume 1 Section Properties and Member Capacities 6th Edition* Published jointly by The Steel Construction Institute and The British Constructuinal Steelwork Association Limited. **Note 9.3** Square and rectangular hollow sections: bearing, buckling and shear capacities for unstiffened webs. b_1 = $(w + 5t)$ w = B (castellated beam flange width) = 146.1 mm t = web thickness of RHS = 10 mm b_1 = $[156.1 + (5 \times 10)]$ = 206.1 mm P_{bw} = $(b_1 + nk)2\,t\,p_{yw}$ where n = 5 for continuity over bearing, k = t for hollow sections, P_{bw} = $(b_1 + nk)2\,t\,p_{yw}$ = $\{[206.1 + (5 \times 10)] \times 2 \times 10 \times 275\}/\,10^3$ = 1409 kN Maximum concentrated load $F_x = 100$ kN $P_{bw} \gg F_x$	**Section is adequate with respect to web bearing**

Contract : Lattice Girder Job Ref. No. : Example 6.23	Calcs. by : W.McK.
Part of Structure: Chord with Axial and Bending Load	Checked by :
Calc. Sheet No. : 6 of 6	Date :

References	Calculations	Output
	Web Buckling Capacity:	

Web Buckling Capacity:

$P_x = (b_1 + n_1)\, 2\, t\, p_c$

where

n_1 is the length obtained by dispersion at 45° through half the depth of the section,

t is the thickness of the web,

p_c is the compressive strength based on:

$\lambda = 1.5 \left[\dfrac{D-2t}{t} \right] \sqrt{3}$ using strut curve c

$n_1 = (0.5 \times 300) = 150 \text{ mm}$

$\lambda = 1.5 \left[\dfrac{300 - (2 \times 10)}{10} \right] \sqrt{3} = 72.7$

Table 24(5)

$\lambda = 72.7, \quad p_y = 275 \text{ N/mm}^2, \quad \therefore \ p_c = 175.6 \text{ N/mm}^2$

$P_x = [(206.1 + 150) \times 2 \times 10 \times 175.6]/ 10^3 = 1250 \text{ kN}$

$F_x = 100 \text{ kN} \ \ll P_x$

Output: Section is adequate with respect to web buckling

It is evident from the calculations that the web bearing and web buckling resistances are considerably greater than the applied concentrated load. In situations where a concentrated load is not applied through a *welded* flange plate, the additional secondary moments induced in the web must be taken into account as indicated in the above publication.

Deflection:

Clause 2.5.2
Table 8

Deflection due to characteristic imposed loads ≤ span/200

$\delta_{\text{maximum}} \leq (5000/200) = 25 \text{ mm}$

Appendix 3

$\delta_{\text{actual}} < \left[\dfrac{PL^3}{48EI} \right]$ which is the value for a simply supported

span with a point load at the centre. If this deflection is less than the Table 8 value then clearly the actual deflection will be adequate.

Clause 3.1.3

$EI = (205 \times 11800 \times 10^4) = 24190 \times 10^6 \text{ kNmm}^2$

$L = 5000 \text{ mm} \qquad P = 50.0 \text{ kN}$

$\delta_{\text{actual}} < \left[\dfrac{50 \times 5000^3}{48 \times 24190 \times 10^6} \right] = 5.4 \text{ mm}$

Output: Section is adequate with respect to deflection

$\delta_{\text{actual}} < 25 \text{ mm}$

6.3.5.7 Simple Construction

The design of members subject to axial loads combined with significant moments (e.g. due to rigid-frame action or applied loads as in Example 6.23), is carried out as in the previous section.

In 'simple' structures (i.e. structures in which an alternative structural form such as bracing, shear walls etc. exists to resist lateral loads) the beam-to-column connections are assumed to be pinned. A simplified interaction equation is included in the code to allow for the nominal moments which develop due to the eccentricity of member end reactions. This expression, which has not been justified by experimental evidence, has appeared in a similar style in previous codes and experience in use over many years has demonstrated its suitability.

The method demonstrated in the preceding section is **not** intended for use with members subject to nominal moments only as in simple construction. In Clause 4.7.7 guidelines are given to evaluate the nominal moments as follows.

Condition 1:
When a beam is supported on a cap plate, as shown in Figure 6.58, the load should be considered to be acting at the face of the column, or edge of packing if used, towards the span of the beam.

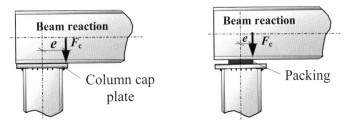

Figure 6.58

Condition 2:
When a roof truss is supported using simple connections (see Figure 6.59) which cannot develop significant moments, the eccentricity may be neglected and a concentric axial load may be assumed at this point.

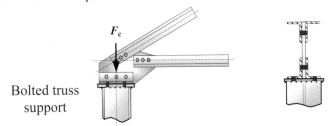

Figure 6.59

Condition 3:
In all other cases, such as the beam connections shown in Figure 6.60, the eccentricity of loading should be taken as 100 mm from the face of the column or at the centre of the length of stiff bearing, whichever gives the greater eccentricity.

Note: The stiff length of bearing of a supporting element is defined in Clause 4.5.1.3 and Figure 13 of the code. In most cases when designing a column the designer does not know the precise details of, for example a seating angle, and the assumption of 100 mm eccentricity will be adopted.

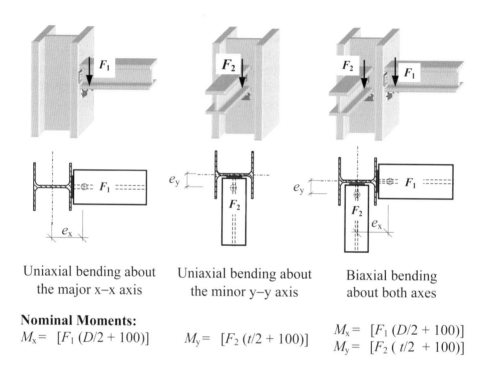

| Uniaxial bending about the major x–x axis | Uniaxial bending about the minor y–y axis | Biaxial bending about both axes |

Nominal Moments:

$M_x = [F_1 (D/2 + 100)]$ $\qquad$ $M_y = [F_2 (t/2 + 100)]$ $\qquad$ $M_x = [F_1 (D/2 + 100)]$
$M_y = [F_2 (t/2 + 100)]$

Figure 6.60

The interaction equation adopted for simplified structures as given in Clause 4.7.7 of the code is:

$$\frac{F_c}{P_c} + \frac{M_x}{M_{bs}} + \frac{M_y}{p_y Z_y}$$

where:
F_c $\quad$ is the axial compression
M_{bs} $\quad$ is the buckling resistance moment for simple columns,
M_x $\quad$ is the nominal moment about the major axis,
M_y $\quad$ is the nominal moment about the minor axis,
P_c $\quad$ is the compression resistance from Clause 4.7.4,
p_y $\quad$ is the design strength,
Z_y $\quad$ is the elastic section modulus about the minor axis.

In general when using this equation the value of M_{bs} should be taken as M_B determined as described in Clause 4.3.6.3 (except for circular, square and rectangular hollow sections) using the equivalent slenderness of columns $\lambda_{LT} = \mathbf{0.5L/r_y}$, where L is the distance between levels at which the column is laterally restrained in both directions.

In the case of circular or square hollow sections and rectangular hollow sections within the limiting values of L_E/r_y, given in Table 15, the buckling resistance moment M_{bs} should be taken as equal to M_c calculated as in Clause 4.2.5.

In the design of multi-storey columns in simple construction cost penalties are normally incurred when purchasing and/or transporting sections longer than approximately 18.0 m. Many multi-storey columns are spliced during erection to avoid this or to produce a more economic design by reducing the section sizes used in upper levels. It is normal to consider such columns to be effectively continuous at their splice locations.

In simple construction the beams supported by a column are considered to be simply supported with their end reactions assumed at an eccentricity as indicated in Figure 6.60. The net nominal moments induced at any one level by the eccentricities of the beams should be divided between the column lengths above and below that level in proportion to the stiffness, I/L, of each length. When the ratio of the stiffnesses of each length is less than or equal to 1.5 then the net moment may be divided equally. The effects of the nominal moments can be assumed to be limited to the level at which they are applied and not to affect any other level.

When considering loading it may be assumed that all beams supported by a column at any one level are fully loaded. This alleviates the need to consider various load combinations to determine the most critical axial load and moment effects.

Despite this guidance given in Clause 4.7.7 it may be prudent for a designer to consider load cases which maximize the bending moment in cases where there can be a significant difference in loading between two adjacent floors.

6.3.5.7.1. Example 6.24: Industrial Unit

An industrial unit comprises a series of braced rectangular frames as shown in Figure 6.61. A travelling crane is supported on a runner beam attached to the underside, at the mid-span point of the rafters.

Using the design data provided, check the suitability of a $203 \times 133 \times 25$ UB S 275 for the columns of a typical internal frame.

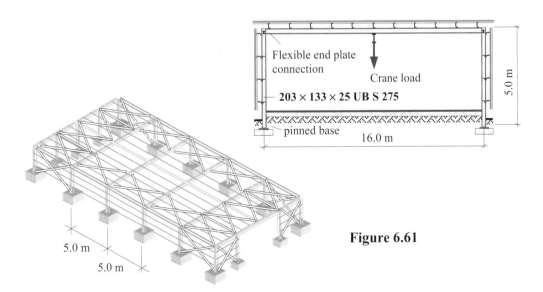

Figure 6.61

Design Data:

Characteristic dead load due to the sheeting, purlins and services	0.5 kN/m^2
Characteristic imposed load	0.75 kN/m^2
Characteristic dead load due to side sheeting	0.3 kN/m^2
Characteristic dead load due to crane	5.0 kN
Characteristic imposed load	40.0 kN

Ignore wind loading

6.3.5.7.2. Solution to Example 6.24

Contract : Industrial Unit Job Ref. No. : Example 6.24 **Part of Structure : Side Column** **Calc. Sheet No. : 1 of 3**	**Calcs. by : W.McK.** **Checked by :** **Date :**

References	**Calculations**	**Output**
	Area of roof supported by typical internal frame: $= (5.0 \times 16.0)$ $= 80 \text{ m}^2$ Design dead load due to roof sheeting etc.: $= 1.4 \times (0.5 \times 80)$ $= 56 \text{ kN}$ Design imposed load on roof: $= 1.6 \times (0.75 \times 80)$ $= 96 \text{ kN}$ Total distributed design load on rafter $= (56 + 96)$ $= 152 \text{ kN}$ Design load on rafter due to crane: $= (1.4 \times 5.0) + (1.6 \times 40)$ $= 71 \text{ kN}$ Area of side sheeting supported by one column: $\approx (5.0 \times 5.0)$ $= 25 \text{ m}^2$ Design dead load due to side sheeting $= (1.4 \times 25 \times 0.3)$ $= 10.5 \text{ kN}$ Design loads applied to typical internal frame. 152 kN 10.5 kN 71 kN 10.5 kN Column Loads: Load imposed on column from end of rafter $= 0.5 \times (152 + 71)$ $= 111.5 \text{ kN}$ Total axial load on column $= (111.5 + 10.5)$ $= 122 \text{ kN}$	

Contract : Industrial Unit Job Ref. No. : Example 6.24 **Part of Structure : Side Column** **Calc. Sheet No. : 2 of 3**	**Calcs. by : W.McK.** **Checked by :** **Date :**	

References	Calculations	Output
	Check the suitability of a $203 \times 133 \times 25$ UB S 275 section Section Properties: $D = 203.2$ mm $\quad d = 172.4$ mm $\quad B = 133.2$ mm $T = 7.8$ mm $\quad t = 5.7$ mm $\quad b/T = 8.54$ $d/t = 30.2$ $\quad x = 25.6$ $\quad u = 0.877$ $r_y = 31.0$ mm $\quad r_x = 85.6$ mm $\quad A = 32.0 \times 10^2$ mm^2 $S_x = 258 \times 10^3$ mm^3 $\quad Z_y = 46.2 \times 10^3$ mm^3	
Table 9	Flange thickness $T \le 16$ mm $\therefore p_y = 275$ N/mm^2	
Table 11	Section Classification $\qquad \varepsilon = \left(\dfrac{275}{p_y}\right)^{\frac{1}{2}} = 1.0$	
	Flange: Outstand element of compression flange rolled section $b/T = 8.54 < 9.0\varepsilon$	**Flange is compact**
	In this case the web should be checked for 'web generally', since both bending and axial loads are present; if the section is plastic or compact when considering web to be in compression throughout, no precise calculation is required.	
	Web: Generally $d/t = 30.2 < 40\varepsilon$	**Web is plastic**
Clause 4.7.7	An eccentricity equal to $(D/2 + 100)$ should be assumed to evaluate the bending moment since the precise details of the seating angles are not known at this stage. Eccentricity $\quad e = \left(\dfrac{203.2}{2}+100\right) = 201.6$ mm Design bending moment $= (111.5 \times 0.2016) = 22.5$ kNm	**Section is plastic**
	Interaction equation: $\dfrac{F_c}{P_c}+\dfrac{M_x}{M_{bs}}+\dfrac{M_y}{p_y Z_y}$ Since there is no moment about the y–y axis $\dfrac{M_y}{p_y Z_y} = 0$	
Table 22	Effective length of column $L_E = 1.0L = 5000$ mm	
Clause 4.7.2	$\lambda_x = \dfrac{L_{Ex}}{r_x} = \dfrac{5000}{85.6} = 58.4$ $\lambda_y = \dfrac{L_{Ey}}{r_y} = \dfrac{5000}{31.0} = 161.3$	

References	Calculations	Output
Contract : Industrial Unit Job Ref. No. : Example 6.24 Part of Structure : Side Column Calc. Sheet No. : 3 of 3		Calcs. by : W.McK. Checked by : Date :

References	Calculations	Output
Clause 4.7.4	$P_c = A_g p_c$	
Table 23	For a rolled I-section $T \le 40$ mm Buckling about the x–x axis Use strut curve a Buckling about the y–y axis Use strut curve b	
Table 24 (1) Table 24 (4)	$\lambda_x = 58.4;$ $p_y = 275$ N/mm²; $\therefore$ $p_c = 241.4$ N/mm² $\lambda_y = 161.3;$ $p_y = 275$ N/mm²; $\therefore$ $p_c = 65.2$ N/mm² The critical value of $p_c = 65.2$ N/mm²	
	$P_c = (3200 \times 65.2)/10^3 = 208.6$ kN $> F_c$	
Clause 4.3.6.4	Bending: $M_B = p_b S_x$	
Clause 4.7.7	$\lambda_{LT} = 0.5L/r_y = (0.5 \times 5000)/31.0 = 80.65$ **Note:** L in this equation equals the actual length	
Table 16	$\lambda_{LT} = 80.65$ $\therefore$ $p_b = 163.6$ N/mm²	
	$M_{bs} = M_B = (163.6 \times 258 \times 10^3)/10^6 = 42.2$ kNm	
Clause 4.7.7	$\dfrac{F_c}{P_c} + \dfrac{M_x}{M_{bs}} = \dfrac{122}{208.6} + \dfrac{22.5}{42.2} = 1.12 > 1.0$	**The section is 12% overstressed**

6.3.5.7.3. Example 6.25: Multi-storey Column in Simple Construction

The floor plan and longitudinal cross-section of a two-storey, two-bay braced steelwork frame is shown in Figures 6.62 (a) and (b). Beams on grid-line 1 at the roof and first floor level are small tie members, all others are substantial members which are connected to the column flanges by web and seating cleats.

Check the suitability of a $203 \times 203 \times 52$ UC S 275 for the column B2 at section x–x indicated.

Design Data:

Characteristic dead and imposed loads on the roof and floor level are as indicated in Figure 6.62(a)

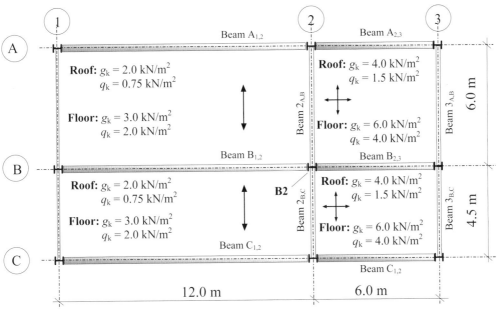

Figure 6.62(a)

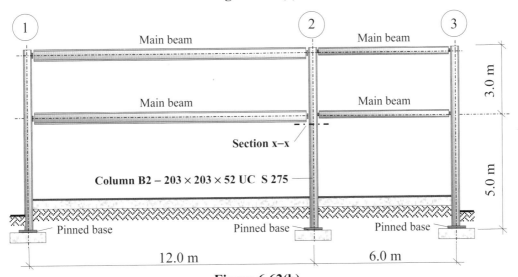

Figure 6.62(b)

Column B2 supports four beams at each level. The slabs between grid-lines 1 and 2 are one-way spanning and supported by beams $A_{1,2}$, $B_{1,2}$ and $C_{1,2}$. The slabs between grid-lines 2 and 3 are two-way spanning and supported by beams $A_{2,3}$, $B_{2,3}$ $C_{2,3}$ and $2_{A,B}$, $2_{B,C}$, $3_{A,B}$, and $3_{B,C}$. Beams $B_{1,2}$, $B_{2,3}$, $2_{A,B}$ and $2_{B,C}$ impose loads on column B2 at both the roof and first floor levels.

Since the column is the same section throughout, the critical section for design will be section x–x, as indicated in Figure 6.62(b). At this location all loading originating from the roof beams is considered to be axial, while the first floor beams will induce nominal moments due to the eccentricity of the connections in addition to their axial effects.

6.3.5.7.4. Solution to Example 6.25

References	Calculations	Output
	Contract : Simple Frame Job Ref. No. : Example 6.25 **Part of Structure: Multi-storey Column** **Calc. Sheet No. : 1 of 5**	**Calcs. by : W.McK.** **Checked by :** **Date :**

The load distribution from the one-way and two-way spanning slabs will be as follows:

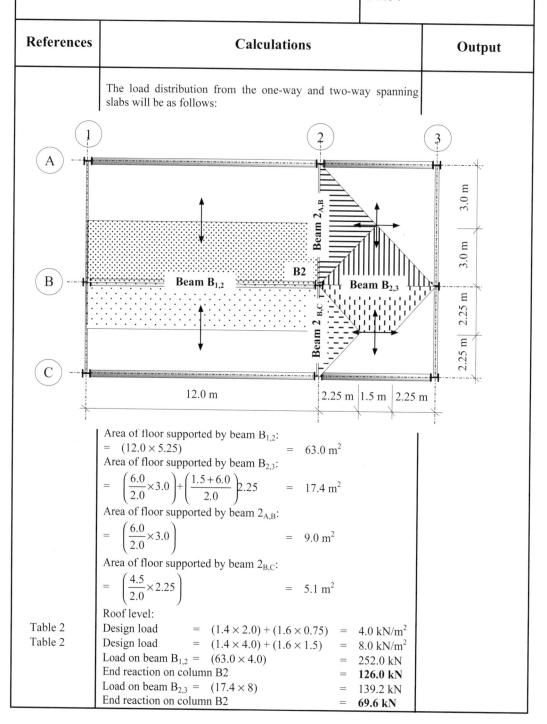

Area of floor supported by beam $B_{1,2}$:
$= (12.0 \times 5.25)$ $= 63.0 \text{ m}^2$

Area of floor supported by beam $B_{2,3}$:
$= \left(\dfrac{6.0}{2.0} \times 3.0\right) + \left(\dfrac{1.5+6.0}{2.0}\right)2.25 = 17.4 \text{ m}^2$

Area of floor supported by beam $2_{A,B}$:
$= \left(\dfrac{6.0}{2.0} \times 3.0\right)$ $= 9.0 \text{ m}^2$

Area of floor supported by beam $2_{B,C}$:
$= \left(\dfrac{4.5}{2.0} \times 2.25\right)$ $= 5.1 \text{ m}^2$

Roof level:

Table 2   Design load $= (1.4 \times 2.0) + (1.6 \times 0.75) = 4.0 \text{ kN/m}^2$

Table 2   Design load $= (1.4 \times 4.0) + (1.6 \times 1.5) = 8.0 \text{ kN/m}^2$

Load on beam $B_{1,2} = (63.0 \times 4.0)$ $= 252.0 \text{ kN}$

End reaction on column B2 $= \mathbf{126.0 \text{ kN}}$

Load on beam $B_{2,3} = (17.4 \times 8)$ $= 139.2 \text{ kN}$

End reaction on column B2 $= \mathbf{69.6 \text{ kN}}$

Contract : Simple Frame Job Ref. No. : Example 6.25 Part of Structure: Multi-storey Column Calc. Sheet No. : 2 of 5	Calcs. by : W.McK. Checked by : Date :

References	Calculations	Output
	Load on beam $2_{A,B} = (9.0 \times 8)$ $= 72.0$ kN	
	End reaction on column B2 $= \mathbf{36.0}$ kN	
	Load on beam $2_{B,C} = (5.1 \times 8)$ $= 40.8$ kN	
	End reaction on column B2 $= \mathbf{20.4}$ kN	
	Axial load transmitted from the roof to column B2:	
	$= (126.0 + 69.6 + 36.0 + 20.4)$ $= \mathbf{252.0}$ kN	
	First Floor level:	
	Despite the indication in Clause 4.7.7 that pattern loading need not be considered, there can be a significant difference between the reactions from beams $B_{1,2}$ and $B_{2,3}$. In this example two load cases will be considered for floor loadings;	
	Case (i): maximum loading on all beams,	
	Case (ii): maximum loading on beam $B_{1,2}$ and $2_{A,B}$ and minimum loading on beam $B_{2,3}$ and $2_{B,C}$.	
Table 2	Maximum design load $= (1.4 \times 3.0) + (1.6 \times 2.0) = 7.4$ kN/m^2	
	Minimum design load $= (1.0 \times 3.0)$ $= 3.0$ kN/m^2	
	Maximum design load $= (1.4 \times 6.0) + (1.6 \times 4.0) = 14.8$ kN/m^2	
	Minimum design load $= (1.0 \times 6.0)$ $= 6.0$ kN/m^2	
	Check the suitability of a $203 \times 203 \times 52$ UB S 275 section	
	Section Properties:	
	$D = 206.2$ mm $\quad d = 160.8$ mm $\quad B = 204.3$ mm	
	$T = 12.5$ mm $\quad t = 7.9$ mm $\quad b/T = 8.17$	
	$d/t = 20.4$ $\quad x = 15.8$ $\quad u = 0.848$	
	$r_y = 51.8$ mm $\quad r_x = 89.1$ mm $\quad A = 66.3 \times 10^2$ mm^2	
	$S_x = 567 \times 10^3$ mm^3 $\quad Z_y = 174 \times 10^3$ mm^3	
	Case (i):	
	Load on beam $B_{1,2} = (63.0 \times 7.4)$ $= 466.2$ kN	
	End reaction on column B2 $= \mathbf{233.1}$ kN	
	Load on beam $B_{2,3} = (17.4 \times 14.8)$ $= 257.5$ kN	
	End reaction on column B2 $= \mathbf{128.8}$ kN	
	Load on beam $2_{A,B} = (9.0 \times 14.8)$ $= 133.2$ kN	
	End reaction on column B2 $= \mathbf{66.6}$ kN	
	Load on beam $2_{B,C} = (5.1 \times 14.8)$ $= 75.5$ kN	
	End reaction on column B2 $= \mathbf{37.8}$ kN	
	Axial load transmitted from the first floor to column B2:	
	$= (233.1 + 128.8 + 66.6 + 37.8)$ $= \mathbf{466.3}$ kN	
	Self-weight of 3.0 m length of column:	
	$= (1.4 \times 0.52 \times 3.0)$ $= \mathbf{2.2}$ kN	
	Total axial load at section x–x:	
	$= (252.0 + 466.3 + 2.2)$ $= \mathbf{720.5}$ kN	

References	Calculations	Output

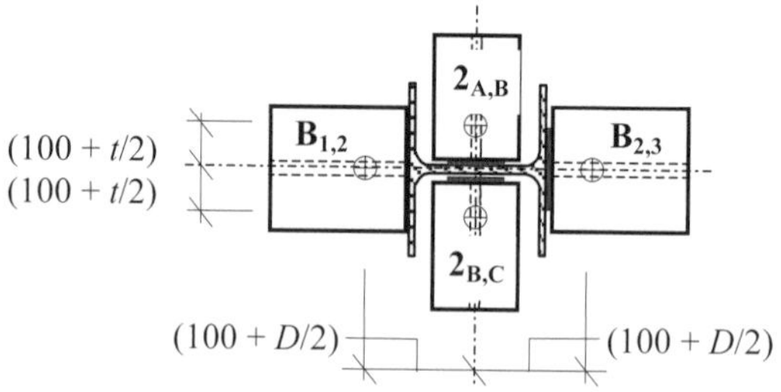

Contract : Simple Frame Job Ref. No. : Example 6.25
Part of Structure: Multi-storey Column
Calc. Sheet No. : 3 of 5

Calcs. by : W.McK.
Checked by :
Date :

Clause 4.7.7

Net nominal moment about the x–x axis:
$= [(B_{1,2} - B_{2,3}) \times (100 + D/2)]$
$B_{1,2} = 233.1$ kN, $B_{2,3} = 128.8$ kN, $D = 206.2$ mm

$= [(233.1 - 128.8) \times (100 + 206.2/2)]/10^3$ $= 21.18$ kNm

Net nominal moment about the y–y axis:
$= [(2_{A,B} - 2_{B,c}) \times (100 + t/2)]$
$2_{A,B} = 66.6$ kN, $2_{B,C} = 37.8$ kN, $t = 7.9$ mm

$= [(66.6 - 37.8) \times (100 + 7.9/2)] /10^3$ $= 3.0$ kNm

Since the column is continuous and the same section above and below section x–x :

$$\frac{I_{upper}}{L_{upper}} = \frac{I}{3} = 0.3\,I ; \qquad \frac{I_{lower}}{L_{lower}} = \frac{I}{5} = 0.2\,I$$

$$\frac{I_u/L_u}{I_L/L_L} = \frac{0.33}{0.2} = 1.65 > 1.5$$

The net nominal moments applied at the first floor level should be divided in proportion to the stiffnesses of each length of column.

Lower length at section x–x:

$$M_x = 21.18 \times \left(\frac{0.2}{0.2 + 0.33} \right) = \mathbf{8.0\ kNm}$$

$$M_y = 3.0 \times \left(\frac{0.2}{0.2 + 0.33} \right) = \mathbf{1.13\ kNm}$$

Table 9

Flange thickness $T \le 16$ mm $\therefore p_y = 275$ N/mm^2

Table 11

Section Classification $\varepsilon = \left(\dfrac{275}{p_y} \right)^{\frac{1}{2}} = 1.0$

Contract : Simple Frame Job Ref. No. : Example 6.25 Part of Structure: Multi-storey Column Calc. Sheet No. : 4 of 5	Calcs. by : W.McK. Checked by : Date :

References	Calculations	Output
	Flange: Outstand element of compression flange rolled section $b/T = 8.17 < 9.0\varepsilon$	**Flange is compact**
	In this case the web should be checked for 'web generally' since both bending and axial loads are present, if the section is plastic or compact when considering web to be in compression throughout, no precise calculation is required.	
	Web: Generally $d/t = 20.4 < 40\varepsilon$	**Web is plastic**
	Interaction equation: $\dfrac{F_c}{P_c} + \dfrac{M_x}{M_{bs}} + \dfrac{M_y}{p_y Z_y}$	**Section is plastic**
Table 22	Effective length of column Assuming that the beams provide directional restraint to one end of the column about both axes: $L_E = 0.85L = (0.85 \times 5000) = 4250$ mm	
Clause 4.7.2	$\lambda_x = \dfrac{L_{Ex}}{r_x} = \dfrac{4250}{89.1} = 47.7$	
	$\lambda_y = \dfrac{L_{Ey}}{r_y} = \dfrac{4250}{51.8} = 82.0$	
Clause 4.7.4	$P_c = A_g p_c$	
Table 23	For a rolled **H**-section $T \le 40$ mm Buckling about the x–x axis Use strut curve b Buckling about the y–y axis Use strut curve c	
Table 24 (3) Table 24 (5)	$\lambda_x = 47.7$; $p_y = 275$ N/mm²; ∴ $p_c = 239.5$ N/mm² $\lambda_y = 82.0$; $p_y = 275$ N/mm²; ∴ $p_c = 157.0$ N/mm² **The critical value of $p_c = 157.0$ N/mm²**	
	$P_c = (6630 \times 157.0)/10^3 = 1041$ kN $> F_c$	
Clause 4.3.6.4	Bending: $M_B = p_b S_x$	
Clause 4.7.4	$\lambda_{LT} = 0.5L/r_y = (0.5 \times 5000)/51.8 = 48.3$ **Note:** L in this equation equals the actual length	

Contract : Simple Frame Job Ref. No. : Example 6.25 Part of Structure: Multi-storey Column Calc. Sheet No. : 5 of 5	Calcs. by : W.McK. Checked by : Date :

References	Calculations	Output
Table 16 Clause 4.7.7	$\lambda_{LT} = 48.3, \quad \therefore \quad p_b = 242.0 \text{ N/mm}^2$ $M_{bs} = M_B = (242.0 \times 567 \times 10^3)/10^6 = 137.2 \text{ kNm}$ $p_y Z_y = (275 \times 174 \times 10^3)/10^6 = 47.9 \text{ kNm}$ $\dfrac{F_c}{P_c} + \dfrac{M_x}{M_{bs}} + \dfrac{M_y}{p_y Z_y} = \dfrac{720.5}{1041} + \dfrac{8.0}{137.2} + \dfrac{1.13}{47.9} = 0.77 \quad < \quad 1.0$	
	Case (ii): Load on beam $B_{1,2} = (63.0 \times 7.4) = 466.2 \text{ kN}$ End reaction on column B2 = **233.1 kN** Load on beam $B_{2,3} = (17.4 \times 6.0) = 104.4 \text{ kN}$ End reaction on column B2 = **52.2 kN** Load on beam $2_{A,B} = (9.0 \times 14.8) = 133.2 \text{ kN}$ End reaction on column B2 = **66.6 kN** Load on beam $2_{B,C} = (5.1 \times 6.0) = 30.6 \text{ kN}$ End reaction on column B2 = **15.3 kN** Axial load transmitted from the first floor to column B2: $= (233.1 + 52.2 + 66.6 + 15.3) = $ **367.2 kN** Self-weight of 3.0 m length of column: $= (1.4 \times 0.52 \times 3.0) = $ **2.2 kN** Total axial load at section x–x: $= (252.0 + 367.2 + 2.2) = $ **621.4 kN**	
Clause 4.7.7	Net nominal moment about the x–x axis: $= [(B_{1,2} - B_{2,3}) \times (100 + D/2)]$ $B_{1,2} = 233.1 \text{ kN}, \quad B_{2,3} = 52.2 \text{ kN}, \quad D = 206.2 \text{ mm}$ $= [(233.1 - 52.2) \times (100 + 206.2/2)]/10^3 = 36.7 \text{ kNm}$ Net nominal moment about the y–y axis: $= [(2_{A,B} - 2_{B,c}) \times (100 + t/2)]$ $2_{A,B} = 66.6 \text{ kN}, \quad 2_{B,C} = 15.3 \text{ kN}, \quad t = 7.9 \text{ mm}$ $= [(66.6 - 15.3) \times (100 + 7.9/2)]/10^3 = 5.3 \text{ kNm}$ The net nominal moments applied at the first floor level should be divided in proportion to the stiffnesses of each length of column. Lower length at section x–x: $M_x = 36.7 \times \left(\dfrac{0.2}{0.2 + 0.33} \right) = $ **13.85 kNm**	
Clause 4.7.7	$M_y = 5.3 \times \left(\dfrac{0.2}{0.2 + 0.33} \right) = $ **2.0 kNm** $\dfrac{F_c}{P_c} + \dfrac{M_x}{M_{bs}} + \dfrac{M_y}{p_y Z_y} = \dfrac{621.4}{1041} + \dfrac{13.85}{137.2} + \dfrac{2.0}{47.9} = 0.74 \quad < \quad 1.0$	The section is adequate with respect to combined axial loading and bending

6.3.6 *Connections*

6.3.6.1 Introduction

Traditionally, the consulting engineer has been responsible for the design and detailing of structural frames and individual members, while in many instances the fabricator has been responsible for the design of connections and consideration of local effects. Codes of Practice tend to give detailed specific advice relating to members and relatively little guidance on connection design. This has resulted in a wide variety of acceptable methods of design and details to transfer shear, axial and bending forces from one structural member to another.

Current techniques include the use of black bolts, high-strength friction-grip (H.S.F.G.) bolts, fillet welds, butt welds and, more recently, the use of flow-drill techniques for rolled hollow sections. In addition there are numerous proprietary types of fastener available.

Since fabrication and erection costs are a significant proportion of the overall cost of a steel framework, the specification and detailing of connections is also an important element in the design process.

The basis of the design of connections must reflect the identified load paths throughout a framework, assuming a realistic distribution of internal forces, and must have regard to local effects on flanges and webs. If necessary, localised stiffening must be provided to assist load transfer.

All buildings behave as complex three-dimensional systems exhibiting interaction between principal elements such as beams, columns, roof and wall cladding, floors and connections. *BS 5950-1:2000* Clause 2.1.2 specifies four methods of design which may be used in the design of steel frames:

(i) **Simple Design** (Clause 2.1.2.2)
In simple design it is assumed that lateral stability of the framework is provided by separate identified elements such as shear walls, portal action, or bracing. The beams are designed assuming them to be pinned at the ends, and any moments due to eccentricities of connections are considered as nominal moments when designing the columns.

It is important that sufficient flexibility exists in the connection detail (i.e. the flexible end-plates, fin-plates, web cleats etc.,) to permit rotation at the joint as assumed in the design.

(ii) **Continuous Design** (Clause 2.1.2.3)
In continuous design, full continuity is assumed at connections transferring shear, axial and moment forces between members. In addition it is assumed that adequate stiffness exists at the joints to ensure minimum relative deformation of members and hence maintaining the integrity of the angles between them.

(iii) **Semi-Continuous Design** (Clause 2.1.2.4)
In this technique partial continuity is assumed between members. The moment -rotation characteristics of the connection details are used both in the analysis of the framework and the design of the connections. The complexity and lack of readily available data renders this method impractical at the present time.

(iv) **Experimental Verification**   (Clause 2.1.2.5)
Loading tests may be carried out to determine the suitability of a structure with respect to strength, stability and stiffness if any of the methods (i) to (iii) are deemed inappropriate.

The uses of bolts and welds as used in simple and rigid connections are illustrated in Figure 6.63.

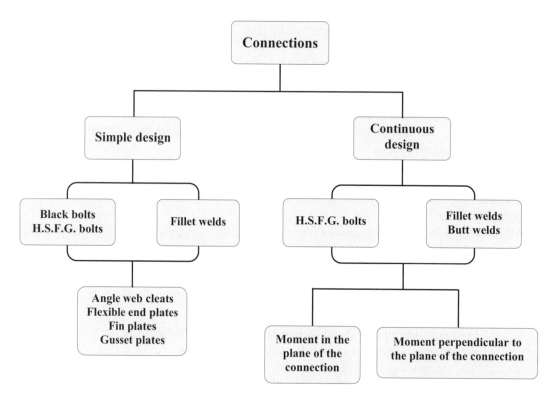

Figure 6.63

The design of connections requires analysis to determine the magnitude and nature of the forces which are to be transmitted between members. In both bolted and welded connections this generally requires the evaluation of a resultant shear force and, in the case of moment connections, may include combined tension and shear forces.

6.3.6.2 Simple Connections

These are most frequently used in pin-jointed frames and braced structures in which lateral stability is provided by diagonal bracing or other alternative structural elements. Typical examples of the use of simple connections, such as in single or multi-storey braced frames, or the flange cover plates in beam splices, are shown in Figure 6.64.

In each case, the shear force to be transmitted is shared by the number of bolts or area of weld used and details such as end/edge distances and fastener spacings are specified to satisfy the code requirements.

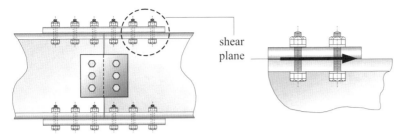

Cover plate to beam flange in beam splice

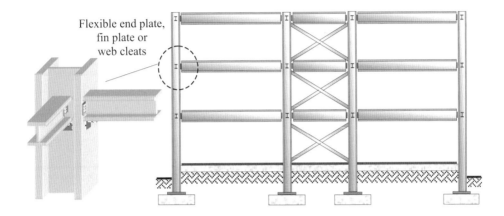

Beam to column connection in braced multi-storey frame

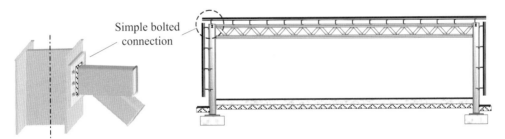

Roof truss to column connection in single-storey braced frame

Figure 6.64

6.3.6.3 Moment Connections

These are used in locations where, in addition to shear and axial forces, moment forces must be transferred between members to ensure continuity of the structure. Typical examples of this occur in web cover plates in beam splices, unbraced single or multi-storey continuous frames, support brackets with the moment either in the plane of or perpendicular to the plane of the connection, and as shown in Figures 6.65 and 6.66.

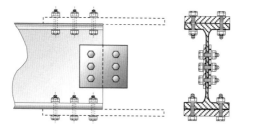

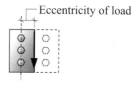

Eccentricity of load

Web cover plate to
beam web connection

Beam web splice

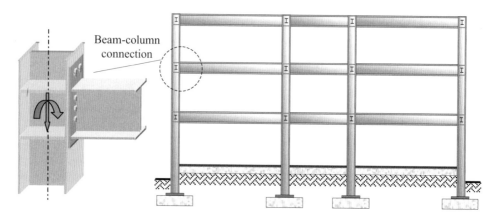

Beam-column
connection

Unbraced multi-storey frame

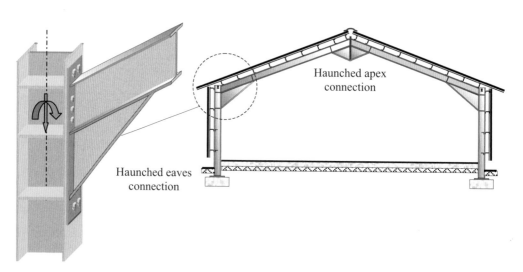

Haunched apex
connection

Haunched eaves
connection

Unbraced pitched-roof portal frame

Figure 6.65

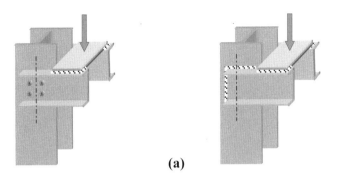

(a)

Bolted and welded brackets with moment in the plane of the connection

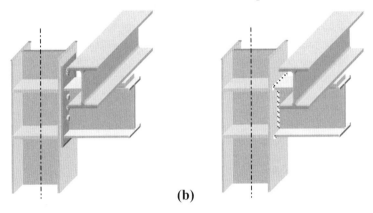

(b)

Bolted and welded brackets with moment perpendicular to the plane of the connection

Figure 6.66

A rigorous approach to the design of moment connections is given in the publication *Joints in Steel Construction:Moment connections* produced by The Steel Construction Institute (ref. 67).

A simplified, more traditional approach is indicated in this text. The following analysis techniques are frequently adopted when evaluating the maximum bolt/weld forces in both types of moment connection.

6.3.6.3.1. Applied Moment in the Plane of the Connection

(i) Bolted connection
The bolts with the maximum force induced in them are those most distant from the centre of rotation of the bolt group and with the greatest resultant shear force when combined with the vertical shear force, i.e. the bolts in the top and bottom right-hand corners.

Consider a group of six bolts which are subjected to an eccentric axial load as illustrated in Figure 6.67.

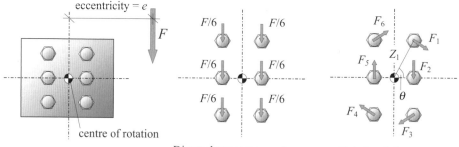

Figure 6.67

Each of the bolts has the same vertical shear due to the force F. The bolts with the maximum rotational shear force due to the moment induced by the eccentric force (i.e. $F \times e$) are numbers 1, 3, 4 and 6 which are most distant from the centre-of-rotation.

In general terms, considering n number of bolts, the resultant maximum shear force on a bolt is given by the vector summation of the direct shear component and the rotational shear component. Bolts F_1 and F_3 have the maximum resultant as follows:

Total vertical component $= F_V = \dfrac{F}{n} + \dfrac{(Fe \times z_1)}{\displaystyle\sum_{n=1}^{\text{number of bolts}} z^2} \times \mathrm{Cos}\,\theta$

Total horizontal component $= F_H = \dfrac{(Fe \times z_1)}{\displaystyle\sum_{n=1}^{\text{number of bolts}} z^2} \times \mathrm{Sin}\,\theta$

Resultant maximum shear force $= F_R = \sqrt{\left(F_V^2 + F_H^2\right)} \;\leq\;$ Bolt shear strength

where
F is the applied vertical force,
e is the eccentricity of the applied force about the centre-of-rotation,
Z is the perpendicular distance from the centre-of-rotation to the line of action of the rotational shear force in a bolt,
Z_1 is the maximum value of Z,
θ is the angle between the horizontal axis and Z_1,
n is the number of bolts.

(ii) Welded connection
In a welded connection the extreme fibres of the weld most distant from the centre-of-gravity of the fillet weld group are subjected to the maximum stress as indicated in Figure 6.68.
The resultant maximum shear force/mm length of fillet weld is found in a similar manner as for bolts.

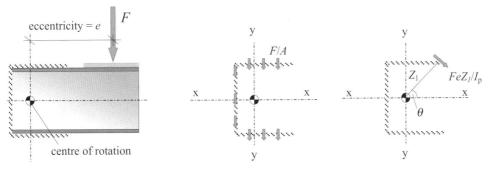

Figure 6.68

Total vertical component/mm $\quad = F_{\mathrm{v}} \quad = \dfrac{F}{A} + \dfrac{(Fe \times Z_1)}{I_{\mathrm{p}}} \times \mathrm{Cos}\theta$

Total horizontal component/mm $\quad = F_{\mathrm{h}} \quad = \dfrac{(Fe \times Z_1)}{I_{\mathrm{p}}} \times \mathrm{Sin}\theta$

Resultant maximum shear force/mm $\ = F_{\mathrm{R}} \quad = \sqrt{(F_{\mathrm{v}}^2 + F_{\mathrm{h}}^2)} \quad \leq \ \text{weld strength}$

where
F, e, Z_1 and θ are as before,
A is the area of the weld group,
I_{p} is the polar moment of area of the weld $= (I_{\mathrm{xx}} + I_{\mathrm{yy}})$

6.3.6.3.2. Applied Moment Perpendicular to the Plane of the Connection

(i) Bolted connection
In this type of connection it is often assumed that the bracket will rotate about the bottom row of bolts. While this assumption is not necessarily true, it is adequate for design purposes. In this case the bolts will be subjected to combined tension and shear forces, as shown in Figure 6.69.

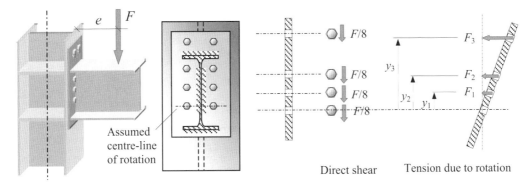

Figure 6.69

In general terms, considering n the number of bolts and m the number of vertical columns of bolts, the maximum tensile force on a bolt is found by the consideration of the assumed rotation of the bracket about the bottom line of bolts, as indicated in Figure 6.69. Bolts which are furthest from the line of rotation, i.e. distance y_3, have the maximum tension force, which can be determined from:

Maximum shear force $\quad = F_v \quad = \dfrac{F}{n}$

Maximum tensile force $\quad = F_{maximum} \quad = \dfrac{\left(Fe \times y_{maximum}\right)}{m \sum y^2}$

The bolts must be designed for the combined effects of the vertical shear and horizontal tension.

(ii) Welded connection
In welded connections the maximum stress in the weld is determined by vectorial summation of the shear and bending forces/mm, as shown in Figure 6.70.

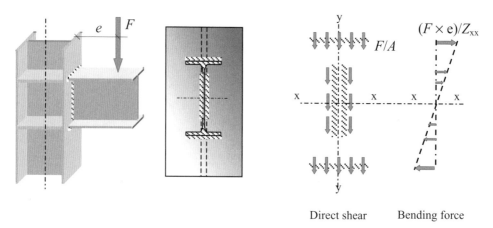

Direct shear   Bending force

Figure 6.70

Maximum shear force/mm $\quad = F_v \quad = \dfrac{F}{A}$

Maximum bending force/mm $\quad = F_{bending} \quad = \dfrac{\left(F \times e\right)}{Z_{xx}}$

where
A is the area of the weld group,
Z_{xx} is the elastic section modulus of the weld group.

Resultant maximum shear force/mm $\quad = F_R = \sqrt{\left(F_v^2 + F_{bending}^2\right)} \qquad \leq$ weld strength

6.3.6.4 Bolted Connections

The dimensions and strength characteristics of bolts commonly used in the U.K. are specified in BS 4190 (black hexagon bolts) and BS 4395 (High-Strength-Friction-Grip bolts). Washer details are specified in BS 4320. BS 5950 mentions several strength grades: e.g. *Grade 4.6* which is mild steel, *Grade 8.8* and *Grade 10.9* which are high-strength-steel.

The most commonly used bolt diameters are 16, 20, 24 and 30 mm; 22 mm and 27 mm diameter are also available, but are not preferred. Combinations of bolts, washers and nuts should match those specified in Table 2 of *BS 5950-2:2001*.

The usual method of forming site connections is to use bolts in clearance holes which are 2 mm larger than the bolt diameter for bolts less than or equal to 24 mm dia., and 3 mm larger for bolts of greater diameter. Such bolts are untensioned and called *non-preloaded bolts*; referred to in this text as *black bolts*. In circumstances where slip is not permissible, such as when full continuity is assumed (e.g. rigid-design), vibration, impact or fatigue is likely, or connections are subject to stress reversal (other than that due to wind loading), *preloaded bolts*, referred to in this text as *high strength friction grip* bolts, should be used.

Traditionally bolts in which only a short length of the shank has been threaded have been used in connections. The current trend is to specify fully threaded bolts (i.e. the full length of the shank) in some instances longer than is required, resulting in considerably fewer bolt lengths being required for any particular construction.

6.3.6.4.1. Black Bolts

Black bolts transfer shear at the connection by bolt shear at the interface and bearing on the bolts and plates as shown in Figure 6.71.

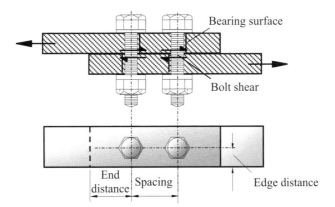

Figure 6.71

6.3.6.4.2. Bolt Spacing and Edge Distances

In Clauses 6.2.1, 6.2.2 and Table 29, *BS 5950-1:2000* specifies minimum and maximum distances for the spacing of bolts, in addition to end and edge distances from the centre-line of the holes to the plate edges. The requirement for minimum spacing is to ensure that local crushing in the wake of a bolt does not affect any adjacent bolts. The maximum spacing requirement is to ensure that the section of plate between bolts does not buckle when it is in compression.

The requirement for minimum end and edge distances is to ensure that no end or edge splitting or tearing occurs and that a smooth flow of stresses is possible. Lifting of the edges between the bolts is prevented by specifying a maximum edge distance. The values specified are illustrated in Figure 6.72.

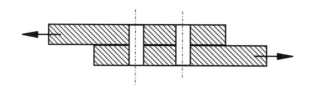

End distance – same as for edge distance

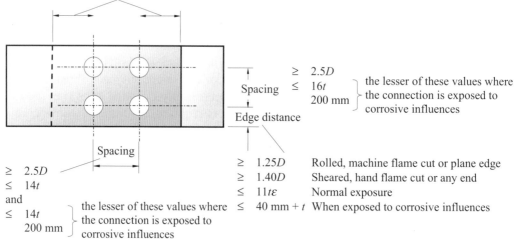

$\geq$ 2.5D
Spacing $\leq$ 16t   } the lesser of these values where
             200 mm   } the connection is exposed to
                         corrosive influences
Edge distance

$\geq$ 1.25D   Rolled, machine flame cut or plane edge
$\geq$ 1.40D   Sheared, hand flame cut or any end
$\leq$ 11tε   Normal exposure
$\leq$ 40 mm + t   When exposed to corrosive influences

$\geq$ 2.5D
$\leq$ 14t
and
$\leq$ 14t   } the lesser of these values where
   200 mm   } the connection is exposed to
               corrosive influences

where
t   is the thickness of the thinner part,
D   is the hole diameter,
ε   is as before.

Figure 6.72

6.3.6.4.3.   Bolt Shear Capacity

The shear capacity of a black bolt is given in Clause 6.3.2.1 by:

$$P_s = p_s A_s$$

where:

A_s is the cross-section resisting shear: normally this is based on the root of the thread (tensile area A_t); if the shear surface coincides with the full bolt shank then the shank area (A) based on the nominal bolt diameter can be used[*]

p_s is given in Table 30 as 160 N/mm² for Grade 4.6 bolts and 375 N/mm² for Grade 8.8 bolts.

---

[*] The adoption of fully threaded bolts will result in A_s being equal to the tensile area.

The shear capacity P_s is reduced in the following circumstances:

- ♦ when using steel packing (**note:** total thickness of packing $\leq 4d/3$),
- ♦ when requiring large grip lengths (i.e. $> 5d$),
- ♦ when using long joints such as in splices (with > 2 rows of bolts),
- ♦ when using kidney-shaped holes,

to allow for effects such as possible bolt bending and/or an unequal distribution of force within the bolts. In each case P_s is equal to the minimum value determined from the following equations as appropriate:

$$P_s \leq p_s A_s \left(\frac{9d}{8d + 3t_{pa}} \right) \qquad \text{When the thickness of steel packing} \quad t_{pa} > d/3$$

$$\leq p_s A_s \left(\frac{8d}{3d + T_g} \right) \qquad \text{When the total thickness of plies} \quad T_g > 5d$$

$$\leq p_s A_s \left(\frac{5500 - L_j}{5000} \right) \qquad \text{The lap length of splice or connection } L_j > 500 \text{ mm}$$

In the case of connections which have two bolts, one in a standard clearance hole and the other in a kidney-shaped slot, the shear capacity is reduced by 20% such that:

$P_s = 0.8 p_s A_s$ as indicated in Clause 6.3.2.4.

6.3.6.4.4. Plate Shear Capacity
The effect of bolt holes on the shear capacity of a plate may be ignored provided that:

$$A_{v,net} \geq 0.85 A_v / K_e$$

where:
$A_{v,net}$ is the net shear area after deducting bolt holes,
K_e, A_v are the effective net area coefficient and the shear area as defined in Clause 3.4.3.

6.3.6.4.5. Block Shear Capacity
The possibility of failure by block shear as shown in Figure 6.73 (i.e. by combined shearing of the plate through the bolt holes parallel to the applied load and tensile failure in a plane perpendicular to the applied load) is considered by ensuring that the applied force F_r is less than the block shear capacity given by:

$$P_r = 0.6 p_y t [L_v + K_e(L_t - kD_t)]$$

where
D_t is the hole size for the tension face – generally the hole diameter, but for slotted holes the dimension perpendicular to the direction of the load transfer should be

used,

k is a coefficient with values as follows:

         0.5   for a single line of bolts
         2.5   for two lines of bolts,

L_t is the length of the tension face,
L_v is the length of the shear face,
t is the thickness,
K_e is as before.

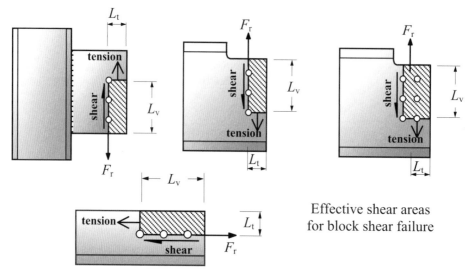

Figure 6.73

6.3.6.4.6. Bolt Bearing Capacity

The bearing capacity of a black bolt is given in Clause 6.3.3.2 as:

$$P_{bb} = dt_p p_{bb}$$

where:

t_p is the thickness of the plate, or if the bolts are countersunk, the thickness of the part minus half the depth of the countersinking,

d is the nominal bolt diameter, and

p_{bb} is given in Table 31 as 460 N/mm² for Grade 4.6 bolts and 1000 N/mm² for Grade 8.8 bolts. In the case of H.S.F.G. bolts:

         General grade to BS 4395-1   Bolt diameter ≤ 24 mm = 1000 N/mm²
                                                      ≥ 27 mm = 900 N/mm²
         Higher grade to BS 4395-2   = 1300 N/mm²

6.3.6.4.7. Plate Bearing Capacity

In Clause 6.3.3.3 the bearing capacity of a connected plate is:

$$P_{bs} = k_{bs} dt_p p_{bs}$$
$$\leq 0.5 k_{bs} e t_p p_{bs}$$

where:

d is the nominal diameter of the bolt,
t_p is the thickness of the plate,
e is the end distance,
p_{bs} is given in Table 32 as:

$$460 \text{ N/mm}^2 \text{ for S 275 steel}$$
$$550 \text{ N/mm}^2 \text{ for S 355 steel}$$
$$670 \text{ N/mm}^2 \text{ for S 460 steel.}$$

k_{bs} is a coefficient for black bolts which satisfy the standard dimensions given in Table 33 for a range of bolt sizes and hole types as follows:

k_{bs} = 1.0   for bolts in standard clearance holes

     = 0.7   ⎫ for bolts in oversized holes,
            ⎬ for bolts in short slotted holes

     = 0.5   ⎫ for bolts in long slotted holes,
            ⎬ for bolts in kidney-shaped holes.

In most cases plate bearing will be more critical than bolt bearing: compare the value of p_{bb} and $(k_{bs} \times p_{bs})$. Generally in structural connections Grade 8.8 bolts are used for main connections, Grade 4.6 frequently being adopted for holding down bolts.

6.3.6.4.8. Bolt Tension Capacity

Connections in which the bolts are subjected to tension as shown in Figure 6.74 induce secondary tensile forces in the bolts due to *prying action* as indicated. The magnitude of the additional force is dependent on the stiffness and the details of the parts being connected.

Provided that the flanges are relatively thick, that the spacing of the bolts is not excessive and that the edge distance is sufficiently large, the prying force Q will be small in relation to F_t and can be neglected. This is the situation in most cases, however further information about evaluating prying forces can be found in the *Steel Designers Manual* (ref. 70).

If the prying forces are accounted for in the total applied axial load (F_{tot}), then as indicated in Clause 6.3.4.3 of the code using '...*the more exact method...*', the tension capacity of the bolt can be determined from:

$$P_t = p_t A_t \quad [\geq F_{tot} = (F_t + Q)]$$

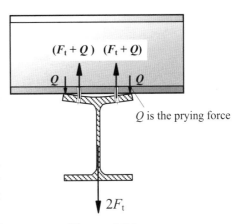

Figure 6.74

where:
p_t is the tension strength of bolts as given in Table 34,
A_t is the tensile area of the bolt (normally based on the root thread diameter),
F_t is the applied axial load in the bolt.

It is necessary in this case to ensure that both of the connected parts satisfy one or more of the conditions specified in Clause 6.3.4.3.

A simplified method is given in Clause 6.3.4.2 in which the prying force is allowed for by ensuring that the applied tension force, F_t not including the prying force, does not exceed the 'nominal' tension capacity given by:

$$P_{nom} = 0.8 p_t \, A_t \qquad\qquad [\geq F_t]$$

where p_t, A_t and F_t are as before.

It is necessary in this case to ensure that the connected parts satisfy both of the following conditions:

- the cross-centre spacing of the bolt lines should not exceed 55% of the flange width or end-plate width as shown in Figure 24 of the code,
- if a connected part is designed assuming double curvature bending (as in Figure 25b of the code) its moment capacity per unit width should be taken as $(p_y t_p^2)/6$, where t_p is the thickness of the connected part.

6.3.6.4.9. Bolts Subject to Combined Shear and Tension

When black bolts are subject to both shear and tension simultaneously, then in addition to satisfying the individual requirements for shear and tension as discussed previously they should also satisfy the following relationship:

$$\frac{F_s}{P_s} + \frac{F_t}{P_{nom}} \leq 1.4 \qquad \begin{array}{l}\text{in cases where the simplified method has been} \\ \text{used to allow for prying action under tension}\end{array}$$

$$\frac{F_s}{P_s} + \frac{F_{tot}}{P_t} \leq 1.4 \qquad \begin{array}{l}\text{in cases where the more exact method has been} \\ \text{used to allow for prying action under tension}\end{array}$$

where:
F_s is the applied shear force,
F_t is the applied tensile force,
F_{tot} is the applied tensile force including Q for prying action,
P_s is the shear capacity,
P_{nom} is the nominal tension capacity $= 0.8 p_y A_t$,
P_t is the bolt tension capacity $= p_y A_t$.

6.3.6.4.10. High Strength Friction Grip Bolts (H.S.F.G.)

H.S.F.G. bolts are manufactured from high strength steel so that they can be tightened to give a high shank tension. The shear force at the connection is considered to be transmitted by friction between two plates as shown in Figure 6.75.

The bolts must be used with hardened steel washers to prevent damage to the connected parts. The surfaces in contact must be free of mill scale, rust, paint grease etc., since these would reduce the coefficient of friction (**slip factor**) between the surfaces.

It is essential to ensure that bolts are tightened up to the required tension, otherwise slip will occur at service loads and the joint will behave as an ordinary bolted joint. There are several techniques which are used to achieve the correct shank tension:

♦ **Torque Wrench**
   A power or hand-operated tool which is used to induce a specified torque to the nut.

♦ **Load-indicating Washers and Bolts**
   These have projections which squash down as the bolt is tightened. A feeler gauge is used to measure when the gap has reached the required size.

♦ **Part-turning**
   The nut is tightened up and then forced a further half to three-quarters of a turn, depending on the bolt length and diameter.

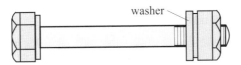

Parallel shank High Strength Friction Grip bolt

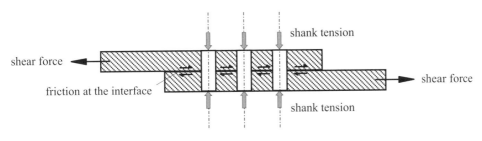

Figure 6.75

H.S.F.G. bolts are generally used in clearance holes. The clearances are the same as for ordinary bolts. The design of H.S.F.G. bolts when used in shear, tension, and combined shear and tension is set out in Clause 6.4 of BS 5950-1:2000 and must be undertaken assuming one of three following conditions:

(i) *neglecting pretensioning* of the bolts, i.e as for black bolts in Clause 6.3. – note: it is not necessary to impose the restrictions relating to packing, large grip lengths or long joints when using H.S.F.G. bolts as black bolts,

(ii) *non-slip under service loads* – note: additional checks for shear and bearing after slip has occurred and before the factored loads have been reached must be carried out as indicated in Clause 6.4.4,

(iii) *non-slip under factored loads*.

6.3.6.4.11. H.S.F.G. Bolt Shear Capacity

The shear capacity, known as the *slip resistance*, is given in Clauses 6.4.2 and 6.4.4 as follows:

$$P_{sL} \quad = \quad 1.1 \, K_S \, \mu \, P_o$$

$\leq P_s$ as given in Clause 6.3.2,

$\leq P_{bg}$ where: For connections designed to be

            $P_{bg} = 1.5 d t_p p_{bs}$ non-slip under *service* loading

               $\leq 0.5 e t_p p_{bs}$

        e, t_p and p_{bs} are as before.

$$= \quad 0.9 \, K_S \, \mu \, P_o$$

                                    For connections designed to be non-slip under *factored* loading

where:

K_S is a coefficient allowing for the type of hole as follows:

     $= 1.0$ for bolts in standard clearance holes

     $= 0.85$ for bolts in oversized holes, short slotted holes and long slotted holes loaded perpendicular to the slot

     $= 0.7$ long slotted holes loaded parallel to the slot *

P_o is the minimum shank tension specified in BS 4604,

μ is the slip factor given in Table 35 of the code and is dependent on the preparation of the surfaces to be connected, classified as follows;

     Class A   Blasted with shot or grit:

                     loose rust removed, no pitting,

                     spray metallized with aluminium,

                     spray metallized with a zinc based coating that has been $\mu = 0.5$

                     demonstrated to provide a slip factor of at least 0.5

     Class B   Blasted with shot or grit:

                     spray metallized with zinc $\mu = 0.4$

     Class C   Wire brushed or flame cleaned:

                     loose rust removed tight mill scale $\mu = 0.3$

     Class D   Untreated or galvanised:

                     untreated. $\mu = 0.2$

*Note: *Connections with H.S.F.G. bolts in long slotted holes, loaded parallel to the slot and connections with waisted shank H.S.F.G. bolts, should always be designed such that $P_{sL} = 0.9 \, K_S \, \mu \, P_o$*

6.3.6.4.12. H.S.F.G. Combined Shear and Tension

When H.S.F.G. bolts are subjected to an external tensile force, the clamping action, and hence the friction force available to resist shear, is reduced. The interaction equation given in Clause 6.4.5 for this situation is:

$$\frac{F_s}{P_{sL}} + \frac{F_{tot}}{1.1 P_o} \leq 1.0 \quad \text{and} \quad F_{tot} \leq A_t p_t$$

with $P_{sL} = 1.1 \, K_S \, \mu \, P_o$

For connections designed to be non-slip under *service* loading.

$$\frac{F_s}{P_{sL}} + \frac{F_{tot}}{0.9 P_o} \leq 1.0$$

with $P_{sL} = 0.9 \, K_S \, \mu \, P_o$

For connections designed to be non-slip under *factored* loading.

where F_s, F_{tot}, K_s, μ and P_o are as before.

Holes for H.S.F.G. bolts should satisfy the standard dimensions given in Table 36 (which is similar to those given for black bolts in Table 33) as indicated in Clause 6.4.6.

6.3.6.5 Example 6.26: Single Lap Joint

A lap joint is shown in Figure 6.76 in which a single Grade 8.8, 16 mm diameter black bolt connects two 10 mm thick S 275 plates. There is one shear interface and it is assumed that the bolt is fully threaded.

(a)   Check the minimum and maximum edge and end distances
(b)   Check the load capacity of the connection with respect to:
    (i)   bolt shear,
    (ii)   bolt bearing,
    (iii)   plate bearing,
    (iv)   block shear, and
    (v)   plate tension capacity.

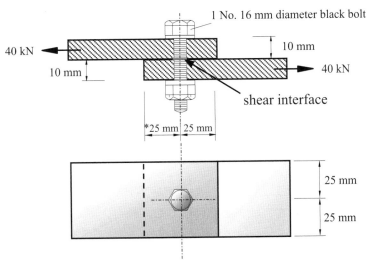

* It is desirable to adopt the minimum edge distance +5 mm to
accommodate any enlargement which may be necessary on site.

Figure 6.76

6.3.6.5.1. Solution to Example 6.26

References	Calculations	Output
	Contract : Connections Job Ref. No. : Example 6.26 **Part of Structure : Bolted Single Lap Joint** **Calc. Sheet No. : 1 of 2**	**Calcs. by : W.McK.** **Checked by :** **Date :**

References	Calculations	Output

Clause 6.2.2.4
Table 29
 	
Clause 6.2.2.5
	Assuming that all edges are either rolled or machine flame cut.
	(a)
	Since 16 mm diameter bolts are being used:
	Clearance holes are: $(16 + 2)$ $=$ 18 mm diameter
	Minimum edge/end distance $\geq 1.25D$ $=$ $(1.25 \times 18) =$ 22.5 mm
	Maximum edge/end distance $\leq 11t\varepsilon$ $=$ (11×10) $=$ 110 mm
	        Actual edge/end distance $=$ 25 mm $\geq$ $1.25D$
	                        $\leq$ $11t\varepsilon$
		Minimum and maximum end and edge distances are acceptable

Clause 6.3.2.1 Table 30 Appendix 7	(b) (i) Bolt shear capacity $P_s = p_s A_s$ Grade 8.8 bolts $p_s = $ 375 N/mm^2 Tensile area of 16 mm dia. bolt $A_t = $ 157 mm^2 $P_s = (375 \times 157)/10^3$ $=$ 58.9 kN $\geq$ 40 kN	**Bolts are adequate in shear**
Clause 6.3.3.2 Table 31	(ii) Bolt bearing capacity $P_{bb} = d t_p p_{bb}$ $d = 16$ mm, $t = 10$ mm $p_{bb} = $ 1000 N/mm^2 $P_{bb} = (16 \times 10 \times 1000)/10^3$ $=$ 160 kN $\geq$ 40 kN	**Bolts are adequate in bearing**
Clause 6.3.3.3 Table 32	(iii) Plate bearing capacity $P_{bs} = k_{bs} d t_p p_{bs}$ $\leq$ $0.5 k_{bs} e t_p p_{bs}$ $e = 25$ mm, $t = 10$ mm, $k_{bs} = 1.0$ and $p_{bs} = 460$ N/mm^2 In this case $e < 2d$ and $0.5 k_{bs} e t_p p_{bs}$ will govern. $P_{bs} = (0.5 \times 1.0 \times 25 \times 10 \times 460)/10^3$ $=$ 57.5 kN $\geq$ 40 kN	**Plate is adequate in bearing**
Clause 6.2.4	(iv) Block shear capacity $P_r = 0.6 p_y t[L_v + K_e(L_t - kD_t)]$ $t = 10$ mm, $D_t = 18$ mm, $L_v = 25$ mm, $L_t = 25$ mm, $k = 0.5$ $P_r = (0.6 \times 275 \times 10) \times \{25.0 + 1.2[25.0 - (0.5 \times 18)]\}/10^3$ $=$ 73 kN $\geq$ 40 kN	**Plate is adequate with respect to block shear**

25 mm

25 mm

Contract : Connections Job Ref. No. : Example 6.26 Part of Structure : Bolted Single Lap Joint Calc. Sheet No. : 2 of 2	Calcs. by : W.McK. Checked by : Date :

References	Calculations	Output
Clause 4.6.1 Clause 3.4.3 Table 9 Clause 3.4.3	(v) Plate tension capacity $\quad P_t = p_y A_e$ $A_e \quad = \quad \Sigma a_e$ $a_e \quad = \quad K_e a_n < a_g$ $p_y \quad = \quad 275 \text{ N/mm}^2$ $K_e \quad = \quad 1.2,$ $a_n \quad = \quad [(50 - 18) \times 10] \quad = \quad 320 \text{ mm}^2$ $a_g \quad = \quad (50 \times 10) \quad\quad\quad = \quad 500 \text{ mm}^2$ $A_e \quad = \quad (1.2 \times 320) \quad\quad = \quad 384 \text{ mm}^2$ $P_t \quad = \quad (275 \times 384)/10^3 \quad = \quad 105.6 \text{ kN} \quad\quad \geq \quad 40 \text{ kN}$ From (i) to (v) the capacity of the connection is governed by the plate bearing strength, i.e. Maximum force which can be transmitted $= 57.5$ kN $> \quad 40$ kN	Plate is adequate with respect to tension The load capacity of the connection is suitable

6.3.6.5.2.   Example 6.27:   Double Lap Point

A lap joint similar to Example 6.26 is shown in Figure 6.77; in this case there are two shear interfaces and four 20 mm diameter Grade 8.8 black bolts. The outer plates are 8 mm thick, whilst the inner plate is 12 mm thick; all are S 275 steel. Determine (a) and (b) as in the previous example.

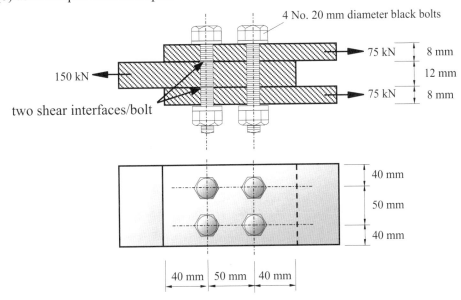

Figure 6.77

6.3.6.5.3. Solution to Example 6.27

References	Calculations	Output
	Contract : Connections Job Ref. No. : Example 6.27 **Part of Structure : Bolted Double Lap Joint** **Calc. Sheet No. : 1 of 2**	**Calcs. by : W.McK.** **Checked by :** **Date :**

References	Calculations	Output
	Assuming that all edges are either rolled or machine flame cut.	
Clause 6.2.2.1	(a) Minimum bolt spacing = $2.5d$ = (2.5×20) = 50 mm $\leq$ 50 mm	
Clause 6.2.1.2	Maximum bolt spacing = $14t$ = (14×8) = 112 mm $\geq$ 50 mm	
Clause 6.2.2.4 Table 29	20 mm diameter bolts are being used. Clearance holes are: $(20 + 2)$ = 22 mm diameter	
Clause 6.2.2.5	Minimum edge/end distance $\geq 1.25D$ = (1.25×22) = 27.5 mm Maximum edge/end distance $\leq 11t\varepsilon$ = (11×8) = 88 mm Actual edge/end distance = 40 mm $\geq 1.25D$ $\leq 11t\varepsilon$	**Minimum and maximum end and edge distances are acceptable**
Clause 6.3.2.1 Table 30 Appendix 7	(b) (i) Bolt shear capacity $P_s = p_s A_s$ Grade 8.8 bolts p_s = 375 N/mm^2 Tensile area of 20 mm dia. bolt A_t = 245 mm^2 $P_s = (375 \times 245)/10^3$ = 91.9 kN	
	Since the bolts have two shear planes, i.e. double shear: Maximum shear force/bolt = (2×91.9) = 183.8 kN/bolt Maximum applied shear force /bolt = 37.5 kN	**Bolts are adequate in shear**
Clause 6.3.3.2 Table 31	(ii) Bolt bearing capacity $P_{bb} = d\,t_p\,p_{bb}$ $d = 20$ mm, $t = 12$ mm p_{bb} = 1000 N/mm^2 **Note:** The aggregate thickness of the outer plates $t_{outer} > t_{inner}$	
	$P_{bb} = (20 \times 12 \times 1000)/10^3$ = 240 kN $\geq$ 37.5 kN	**Bolts are adequate in bearing**
Clause 6.3.3.3 Table 32	(iii) Plate bearing capacity $P_{bs} = k_{bs}\,d\,t_p\,p_{bs}$ $\leq$ $0.5\,k_{bs}\,e\,t_p\,p_{bs}$ $e = 40$ mm, $t = 12$ mm, $k_{bs} = 1.0$ and $p_{bs} = 460$ N/mm^2 Since $e = 2d$ $k_{bs}\,d\,t_p\,p_{bs}$ = $0.5\,k_{bs}\,e\,t_p\,p_{bs}$	
	$P_{bs} = (1.0 \times 20 \times 12 \times 460)/10^3$ = 110.4 kN $\geq$ 37.5 kN	**Plate is adequate in bearing**
Clause 6.2.4	(iv) Block shear capacity $P_r = 0.6\,p_y\,t[L_v + K_e(L_t - kD_t)]$ $t = 12$ mm, $D_t = 22$ mm, $L_v = 90$ mm, $L_t = 90$ mm, $k = 2.5$	

References	Calculations	Output
	$P_r = (0.6 \times 275 \times 12) \times \{90.0 + 1.2[90.0 - (2.5 \times 22)]\}/10^3$ $= 261.3 \text{ kN} \qquad \geq \quad 150 \text{ kN}$	**Plate is adequate with respect to block shear**
Clause 4.6.1 Clause 3.4.3 Table 9 Clause 3.4.3	(v) Plate tension capacity $P_t = p_y A_e$ $A_e = \Sigma a_e \qquad a_e = K_e a_n < a_g$ $p_y = 275 \text{ N/mm}^2$ $K_e = 1.2$ $a_n = [(130 - 44) \times 12] = 1032 \text{ mm}^2$ $a_g = (130 \times 12) = 1560 \text{ mm}^2$ $A_e = (1.2 \times 1032) = 1238.4 \text{ mm}^2$ $P_t = (275 \times 1238.4)/10^3 \qquad = 340.6 \text{ kN}$ $\qquad \geq \quad 150 \text{ kN}$	**Plate is adequate with respect to tension**
	From (i) to (v) the capacity of the connection is governed by the block shear capacity, i.e. Maximum force which can be transmitted $= 261.3 \text{ kN}$ $\qquad > \quad 150 \text{ kN}$	**The load capacity of the connection is suitable**

6.3.6.6 Welded Connections

The most common processes of welding used in connections are methods of *fusion* (arc) welding. There are several techniques of fusion welding which are adopted, both manual and automatic/semi-automatic, such as *manual metal arc* (MMA) and *metal inert gas* (MIG). The MMA process is usually used for short runs in workshops or on site and the MIG process for short runs or long runs in the workshop. The two most widely used types of weld are *fillet* and *butt* welds.

6.3.6.6.1. Fillet Welds

Fillet welds, as illustrated in Figure 6.78, transmit forces by shear through the throat thickness.

The design strength (p_w) is given in BS 5950:Part 1, Table 37, and the effective throat thickness as defined in Figure 29 is normally taken as 0.7 times the leg length for equal leg welds. It is assumed when using this value that the angle between the fusion faces lies between 60° and 90°. When the fusion faces are inclined at an angle between 91° and 120° then the 0.7 coefficient should be modified, for example as indicated in Figure 6.79.

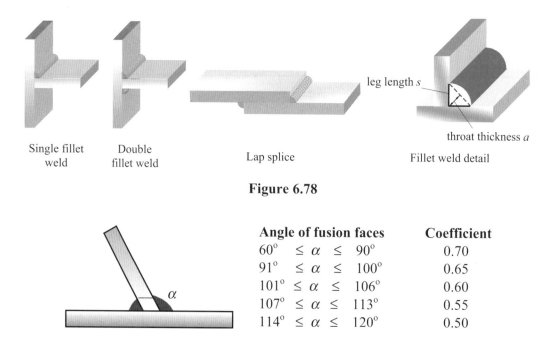

Single fillet weld   Double fillet weld   Lap splice   leg length s   throat thickness *a*   Fillet weld detail

Figure 6.78

Angle of fusion faces	Coefficient
60° ≤ α ≤ 90°	0.70
91° ≤ α ≤ 100°	0.65
101° ≤ α ≤ 106°	0.60
107° ≤ α ≤ 113°	0.55
114° ≤ α ≤ 120°	0.50

Figure 6.79

In situations when $120° \leq \alpha \leq 60°$ poor access to the acute fillet weld and a small throat thickness on the obtuse fillet weld can create problems. In these situations a different type of weld, e.g. a single-sided butt weld, may be more appropriate. BS 5950 indicates in Clause 6.8.1 that the strength of such acute and obtuse fillet welds should be demonstrated by testing in accordance with Section 7 of the code.

Where possible a run of fillet weld should be returned around corners for a distance of not less than twice the leg length. If this is not possible, then the length of weld considered effective for strength purposes should be taken as the overall length less one leg length for each length which does not continue round a corner (Clause 6.8.2). The effective length L of a fillet weld should be: $\geq (4 \times s)$ and ≥ 40 mm

In many connections welds are subject to a complex stress condition induced by multi-directional loading. The strengths of transverse fillet welds and longitudinal fillet welds differ, transverse fillet welds being the stronger. In addition, in side fillet welds, large longitudinal forces are concentrated locally on the member cross-section.

A significant variation in tensile stress occurs across the width of the tensile members when the lateral spacing between weld runs is considerably larger than the length. These effects are limited in BS 5950 by ensuring that the length of weld L is at least equal to T_w, as indicated in Clause 6.7.2.4 and shown in Figure 6.80. In addition in lap joints the minimum lap should not be less than ($4 \times t$ – the thickness of the thinner part joined), as indicated in Clause 6.7.2.3.

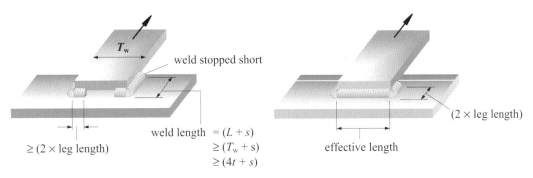

weld stopped short

T_w

weld length $= (L + s)$
$\geq (T_w + s)$
$\geq (4t + s)$

$\geq (2 \times \text{leg length})$

$(2 \times \text{leg length})$

effective length

Side or longitudinal fillet weld

End or transverse fillet weld

Figure 6.80

The code permits two methods of calculating the capacity of fillet welds, the Simple Method (Clause 6.8.7.2) and the Directional Method (Clause 6.8.7.3). The Simple Method involves the vectorial summation of stresses to determine the stress for which fillet welds should be designed; this is illustrated in this text. The Directional Method recognises the different strengths of longitudinal and transverse welds by satisfying an interaction equation.

6.3.6.6.2. Butt Welds

A butt–weld is formed when the cross-section of a member is fully or partially joined by preparation of one or both faces of the connection to provide a suitable angle for welding. A selection of typical end preparations of butt welds is illustrated in Figure 6.81.

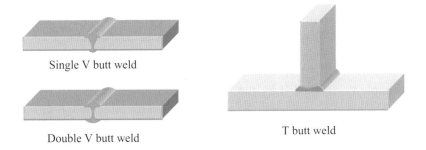

Single V butt weld

Double V butt weld

T butt weld

Figure 6.81

The strength of a full penetration butt weld is defined in Clause 6.9.1 and is generally taken as the capacity of the weaker element joined, provided that appropriate electrodes and consumables are used. In most situations when designing structural frames, full penetration welds are unnecessary and the additional cost involved in preparing such connections is rarely justified unless dynamic or fatigue loadings are being considered. Butt welds are used extensively in the fabrication of marine structures such as ships, submarines, and offshore oil installations, and in pipe-work.

6.3.6.6.3. Example 6.28: Fillet Weld Lap Joint

A flat-plate tie is connected to another structural member as indicated in Figure 6.82. Design a suitable 6 mm side fillet weld to transmit a 75 kN axial force. Assume the electrode classification to be 35.

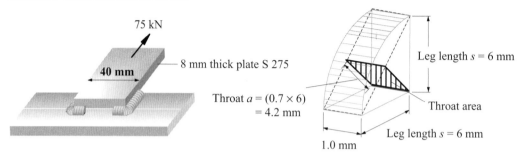

Figure 6.82

Solution:

Clause 6.8.5
Table 37   Design strength of fillet weld p_w = 220 N/mm^2
Clause 6.8.7.2   Simple Method:
Clause 6.8.7.1   Strength of fillet weld/mm length   = $p_w \times$ throat area

$$\text{throat size } a = 0.7 \times \text{leg length } s$$
$$a = (0.7 \times 6.0) = 4.2 \text{ mm}$$
$$\therefore \text{ Strength} = (220 \times 4.2 \times 1.0)/10^3$$
$$= 0.924 \text{ kN/mm length}$$

Figure 27 T_w = 40 mm
$$4s = (4 \times 6) = 24 \text{ mm}$$
Clause 6.8.2   Effective length of weld required, L = 37.5/0.924 = 41 mm/side
$$\geq T_w \text{ and } 4s$$
          Overall length of weld   = (41+ 6) = 47 mm/side
Clause 6.7.2.2   Weld should be returned round each corner at least equal to 2 s = 12 mm

6.3.6.6.4. Example 6.29: Fillet Weld Eccentric Joint

A 65 × 50 × 8 angle tie is welded by the long leg to a gusset plate and is required to transmit characteristic loads as indicated in Figure 6.83. Design suitable 6 mm fillet welds *X* and *Y* as shown.

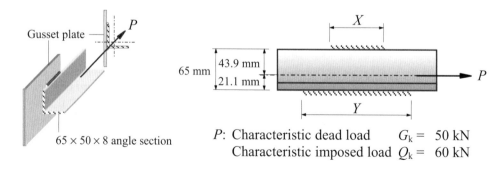

P: Characteristic dead load G_k = 50 kN
Characteristic imposed load Q_k = 60 kN

Figure 6.83

The force is assumed to be transmitted through the centroid of the section. Since the force is applied nearer to weld Y than weld X, weld Y will transmit a proportionately larger share of the total force.

Solution:

Table 2	Design load $= [(1.4 \times 50) + 1.6 \times 60)]$	$= 166$ kN
Clause 6.8.5		
Table 37	Design strength of fillet weld p_w	$= 220$ N/mm^2

Clause 6.8.7.1 Strength of fillet weld/mm length $= p_w \times$ throat area

$$\text{throat size } a = 0.7 \times \text{leg length } s$$
$$a = (0.7 \times 6.0) = 4.2 \text{ mm}$$
$$\therefore \text{Strength} = (220 \times 4.2 \times 1.0)/10^3$$
$$= 0.924 \text{ kN/mm length}$$

Clause 6.8.2 Effective length of weld required, $L = 166/0.924 = 180$ mm

$$\text{Weld } X = \left[180 \times \frac{21.1}{65}\right] = 58.4 \text{ mm} \qquad \geq 4s = 24 \text{ mm}$$
$$\geq 40 \text{ mm}$$
$$\geq T_w = \mathbf{65 \text{ mm}}$$

$$\text{Weld } Y = 180 - 65 = \mathbf{115 \text{ mm}} \qquad \geq 4s = 24 \text{ mm}$$
$$\geq 40 \text{ mm}$$
$$\geq T_w = 65 \text{ mm}$$

$$\text{Overall length of weld } X = (L + 2s) = (65 + 12) = 77 \text{ mm}$$
$$\text{Overall length of weld } Y = (L + 2s) = (115 + 12) = 127 \text{ mm}$$

6.3.6.7   Beam End Connections

There are three types of beam end connections which are commonly used in the fabrication of steelwork:

♦ double angle web cleats,
♦ flexible end plates, and
♦ fin-plates.

The method adopted by any particular fabricator will depend on a number of factors such as the joint geometry and the equipment available. Generally the following characteristics are found:

♦ All three types of connection are capable of transmitting at least 75% of the shear capacity of the beam being connected, depending upon the depth of the plates and/or the number of vertical rows of bolts used.
♦ Fin plates are the most suitable for the connection of beams which are eccentric to columns or connections which are skewed.

◆ End plates are the most suitable when connecting to column webs.
◆ Fabrication and treatment do not present any significant problems for any of the three types of connection.

It is important in simple design to detail beam end connections which will permit end rotation of the connecting beam, thus allowing it to displace in a simply supported profile whilst still maintaining the integrity of the shear capacity. This rotational capacity is provided by the slip of the bolts and the deformation of the connection component parts.

6.3.6.7.1. Double-Angle Web Cleats

Typical angle web cleats comprise $2 / 90 \times 90 \times 10$ angle sections bolted or welded to the web of the beam to be supported, as shown in Figure 6.84.

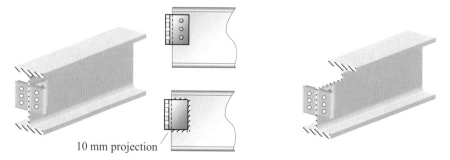

10 mm projection

Beam-to-column flange connection   Beam-to-beam web connection

Figure 6.84

The 10 mm projection of the web cleats beyond the end of the beam is to ensure that when the beam rotates, the bottom flange does not bear on the supporting member. The positioning of the cleats near the top of the beam provides directional restraint to the compression flange. When this positioning is used in addition to a length of cleat equal to approximately 60% of the beam depth, the end of the beam can be assumed to have torsional restraint.

This detail enables an effective length of compression flange of $1.0L$ to be used when designing a beam which is not fully restrained.

6.3.6.7.2. Flexible End Plates

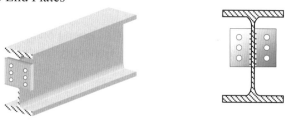

Flexible end plate – beam-to-column or beam-to-beam web connections

Figure 6.85

Connections using flexible end plates are generally fabricated from an 8 mm or 10 mm thick plate which is fillet welded to the web of the beam as shown in Figure 6.85.

As with double angle web cleats, the end plates should be welded near the top flange and be of sufficient length to provide torsional restraint. If full depth plates are used which are relatively thick and which may be welded to both flanges in addition to the web, then the basic assumption of a simple connection in which end rotation occurs will be invalid, – which could lead to overstressing of the other elements.

6.3.6.7.3. Fin Plates

A fin-plate connection comprises a length of plate which is welded to the supporting member, such as a column or other beam web, to enable the supported beam to be bolted on-site as in Figure 6.86. As with angle cleats and flexible end plates, appropriate detailing should ensure that there is sufficient rotational capability to assume a simple connection.

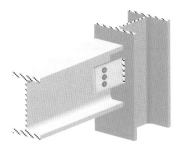 Fin plate connection

Figure 6.86

Experimental data has indicated that torsions induced by the plate being connected on only one side of the web of a beam are negligible and may be ignored.

Recommendations for the detailing of fin plates as given in *Joints in Simple Construction* (ref. 67) are:

♦ the thickness of the fin plate or beam web should be:
    (i) $0.42d$ for grade 50 (S 275) steel,   or
    (ii) $0.5d$ for grade 43 (S 355) steel (where d = bolt diameter)
♦ grade 8.8 bolts are used, un-torqued and in clearance holes,
♦ all end and edge distances on the plate and the beam web are at least $2d$,
♦ the fillet weld leg length is at least 0.8 times the fin plate thickness,
♦ the fin plate is positioned reasonably close to the top flange of the beam,
♦ the fin plate depth is at least 0.6 times the beam depth.

6.3.6.7.4. Example 6.30: Web Cleat, End Plate and Fin Plate Connections

A braced rectangular frame in which simple connections are assumed between the columns and rafter beam is shown in Figure 6.87. Using the data provided, design a suitable connection considering;

    (i)   double angle web cleats,
    (ii)   a flexible end plate,
    (iii)   a fin plate.

Factored design load = 50 kN/m

457 × 191 × 98 UB S 275

(i) double angle web cleats,
(ii) a flexible end plate,
(iii) a fin plate

7.0 m span

Figure 6.87

6.3.6.7.5. Solution to Example 6.30

Contract : **Braced Frame** Job Ref. No. : **Example 6.30** Part of Structure : **Beam–Column Connection** Calc. Sheet No. : **1 of 8**	**Calcs. by : W.McK.** **Checked by :** **Date :**

References	Calculations	Output

	Assume simple supports at the ends of the beam:
	Total design load on the beam = (50 × 7.0) = 350 kN
	Design shear force at the connection = 175 kN
	The use of the following components is assumed where appropriate in each connection:
	(a) M20 grade 8.8 bolts in 22 mm dia. holes,
	(b) 6 mm or 8 mm fillet welds with electrode classification 35,
	(c) grade S 275 steel adopted throughout.
	
	(i) Double Angle Web Cleats:
	Use 2 / 90 × 90 × 10 angle sections
	Minimum length ≈ 60% of beam depth
	  = 0.6 × 467 = 280 mm
	Assume end distances = 40 mm
	Spacing of bolts = 70 mm
	Length of angle required = (2 × 40) + (3 × 70) = 290 mm
	  > 280 o.k.
	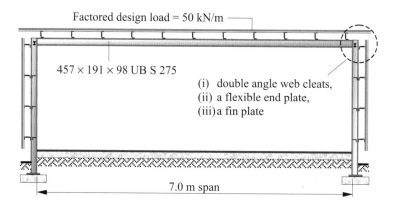	

References	Calculations	Output

Contract : Braced Frame Job Ref. No. : Example 6.30
Part of Structure : Beam–Column Connection
Calc. Sheet No. : 2 of 8

Calcs. by : W.McK.
Checked by :
Date :

Consider the bolt group connecting the beam to the web cleats:
The equivalent loading on the bolt group results in a 'vertical'
shear force acting through the centroid of the group in addition to
a 'rotational' shear force induced by the eccentricity of the load.

Total vertical shear force	F	=	**175 kN**
Vertical shear force/bolt	F_v	=	(175/4)
(Double shear)		=	**43.75 kN/bolt**

Maximum rotational shear force F_h =
$$\dfrac{(F \times e) \times z_{maximum}}{\text{number of bolts} \displaystyle\sum_{n=1} z^2}$$

$$= \dfrac{(175 \times 0.05) \times 0.105}{2\left(0.035^2 + 0.105^2\right)}$$

$$= \quad 37.5 \text{ kN}$$

Resultant maximum shear force F_r = $\sqrt{\left(F_v^2 + F_h^2\right)}$

$$= \quad \sqrt{\left(43.75^2 + 37.5^2\right)}$$

$$= \quad \textbf{57.6 kN/bolt}$$

(i) Web cleats - the following checks should be carried out:

1 bolts: F_r $\leq$ $2P_s$ shear
2 cleats: $F/2$ $\leq$ P_v shear
3 cleats: $F_r/2$ $\leq$ P_{bs} bearing
4 beam web: F_r $\leq$ P_r block shear
5 beam web: F_r $\leq$ P_{bs} bearing
6 bolts: F $\leq$ ΣP_s shear
7 cleats: $F/2$ $\leq$ P_v shear
8 cleats: F $\leq$ P_{bs} bearing

the web cleat
to beam web
connection

the web cleat to
column flange
connection

If the column flange thickness is less than the cleat thickness, it
will be necessary to check the flange for bearing.

References	Calculations	Output
	Contract : Braced Frame Job Ref. No. : Example 6.30	**Calcs. by : W.McK.**
	Part of Structure : Beam–Column Connection	**Checked by :**
	Calc. Sheet No. : 3 of 8	**Date :**

References	Calculations	Output
Clause 6.3.2.1 Table 30 Appendix 7	**1. Bolt shear capacity** $\qquad P_s = p_s A_s$ Grade 8.8 bolts $\qquad\qquad\qquad p_s = $ 375 N/mm^2 Tensile area of 20 mm dia. bolt $\quad A_t = $ 245 mm^2 $P_s = (375 \times 245)/10^3 \qquad\qquad = $ 91.9 kN Since the bolts have two shear planes, i.e. double shear: Maximum shear force/bolt $= \;(2 \times 91.9) = $ 183.8 kN/bolt $\qquad\qquad\qquad\qquad\qquad\qquad \geq \;$ 57.6 kN	Beam bolts are adequate in shear
Clause 4.2.3	**2. Cleat shear capacity:** $P_v = 0.6 p_y A_v$ where $A_v = 0.9 A_o = \;(0.9 \times 290 \times 10) \qquad = $ 2610 mm^2	
Clause 6.2.3 Clause 3.4.3	Bolt holes need not be allowed for if $A_{v,net} \geq \; 0.85 A_v/K_e$ $K_e = \;$ 1.2 for S 275 steel $A_{v,net} \qquad = \;[2610 - (4 \times 22 \times 10)] \qquad = $ 1730 mm^2 $0.85 A_v/K_e = \;(0.85 \times 2610)/1.2 \qquad = $ 1849 mm^2 $\qquad\qquad\qquad\qquad\qquad\qquad\qquad > \; A_{v,net}$	
	$P_v = 0.7 p_y K_e A_{v,net} = \;(0.7 \times 275 \times 1.2 \times 1730)/10^3 \; = $ 400 kN $\qquad\qquad\qquad\quad \geq \;$ Applied shear force/cleat $\quad = $ 87.5 kN	Cleats are adequate in shear
Clause 6.3.3.3 Table 32	**3. Cleat bearing capacity:** $P_{bs} = \;k_{bs} d t_p p_{bs} \;\leq \;0.5 k_{bs} e t_p p_{bs}$ $\quad e = \;$ 40 mm, $\quad t = $ 10 mm, $\quad k_{bs} = 1.0$, $\;p_{bs} = 460$ N/mm^2 Since $e = 2d \quad k_{bs} d t_p p_{bs} \;= \;0.5 k_{bs} e t_p p_{bs}$	
	$P_{bs} = \;(1.0 \times 20 \times 10 \times 460)/10^3 \; = $ 92.0 kN/cleat $\qquad\qquad\qquad \geq \;$ Applied bearing force/cleat $= $ 28.8 kN	Cleats are adequate in bearing
	4. Beam web – Block shear capacity: 40 mm 70 mm 70 mm 70 mm	
Clause 6.2.4	$P_r = \;0.6 p_y \, t[L_v + K_e(L_t - k D_t)]$ $t = 11.4$ mm, $\;D_t = 22$ mm, $\;L_v = 210$ mm, $\;L_t = 40$ mm, $\;k = 0.5$ $P_r = \;(0.6 \times 275 \times 11.4) \times \{210 + 1.2[40 - (0.5 \times 22)]\}/10^3$ $\quad = \;$ 460.5 kN $\qquad\qquad\qquad\qquad \geq \;$ 175 kN	Beam is adequate with respect to block shear

References	Calculations	Output
	Contract : Braced Frame Job Ref. No. : Example 6.30 **Part of Structure : Beam–Column Connection** **Calc. Sheet No. : 4 of 8**	**Calcs. by : W.McK.** **Checked by :** **Date :**

References	Calculations	Output
Clause 6.3.3.3 Table 32	**5. Beam web bearing capacity:** $P_{bs} = k_{bs} d\, t_{web} p_{bs}$ $\quad \leq 0.5\, k_{bs}\, e\, t_{web}\, p_{bs}$ $e = 40$ mm, $\quad t = 11.4$ mm, $\quad k_{bs} = 1.0$ $\quad$ and $p_{bs} = 460$ N/mm^2 Since $e = 2d$ $\quad k_{bs} d\, t_p\, p_{bs} = 0.5\, k_{bs}\, e\, t_p\, p_{bs}$ $P_{bs} = (1.0 \times 20 \times 11.4 \times 460)/10^3$ $\qquad\qquad = 104.9$ kN $\qquad\qquad\qquad\qquad\qquad\qquad\qquad\qquad\qquad \geq 57.6$ kN The bolts are subjected to a concentric shear load, i.e. there is no rotational component, and they are in single shear. 175 kN **6. Bolt shear capacity:** As before $\qquad P_s = 91.9$ kN single shear For eight bolts $\quad \Sigma P_s = (8 \times 91.9)$ $\qquad = 735.2$ kN $\qquad\qquad\qquad\qquad\qquad\qquad\qquad\qquad \geq 175$ kN **7. Cleat shear capacity:** As before $\qquad P_v = 400$ kN $\quad \geq 175/2 = 87.5$ kN **8. Cleat bearing capacity:** As before $\qquad P_{bs} = 92.0$ kN/bolt For eight bolts $\quad \Sigma P_{bs} = (8 \times 92.0)$ $\qquad = 736$ kN $\qquad\qquad\qquad\qquad\qquad\qquad\qquad\qquad \geq 175$ kN **(ii) Flexible end plate-the following checks should be carried out:** 290 mm × 8 mm thick flexible end plate 50 mm / 40 mm / 70 mm / 70 mm / 70 mm / 40 mm 175 kN 90 mm 1 bolts: $\quad F \leq \Sigma P_s \quad$ shear 2 end plate: $\quad F \leq P_v \quad$ shear 3 end plate: $\quad F/2 \leq \Sigma P_{bs} \quad$ bearing 4 beam web: $\quad F \leq P_v \quad$ block shear 5 fillet weld: $\quad F \leq P_{weld} \quad$ shear If the column flange thickness is less than the cleat thickness, it will be necessary to check the flange for bearing.	**Beam web is adequate with respect to bearing** **Column bolts are adequate in shear** **Column cleats are adequate in shear** **Column cleats are adequate in bearing**

References	Calculations	Output
	Contract : Braced Frame Job Ref. No. : Example 6.30 **Part of Structure : Beam–Column Connection** **Calc. Sheet No. : 5 of 8**	**Calcs. by : W.McK.** **Checked by :** **Date :**

References	Calculations	Output
	1. Bolt shear capacity: As before P_s = 91.9 kN single shear For eight bolts ΣP_s = (8×91.9) = 735.2 kN $\geq$ 175 kN	**Column bolts are adequate in shear**
Clause 4.2.3	**2. End plate shear capacity:** P_v = $0.6 p_y A_v$ where $A_v = 0.9A$ = $(0.9 \times 290 \times 8)$ = 2088 mm^2	
Clause 6.2.3 Clause 3.4.3	Bolt holes need not be allowed for if $A_{v,net} \geq$ $0.85 A_v/K_e$ K_e = 1.2 for S 275 steel $A_{v,net}$ = $[2088 - (4 \times 22 \times 8)]$ = 1384 mm^2 $0.85 A_v/K_e$ = $(0.85 \times 2610)/1.2$ = 1849 mm^2 > $A_{v,net}$	
	P_v = $0.7 p_y K_e A_{v,net}$ = $(0.7 \times 275 \times 1.2 \times 1384)/10^3$ = 319 kN $\geq$ Applied shear force/cleat = 87.5 kN	**End plate is adequate in shear**
Clause 6.3.3.3 Table 32	**3. End plate bearing capacity:** P_{bs} = $k_{bs} d t_p p_{bs}$ $\leq$ $0.5 k_{bs} e t_p p_{bs}$ e = 40 mm, $t = 8$ mm, $k_{bs} = 1.0$, $p_{bs} = 460$ N/mm^2 Since $e = 2d$ $k_{bs} d t_p p_{bs}$ = $0.5 k_{bs} e t_p p_{bs}$	
	P_{bs} = $(1.0 \times 20 \times 8 \times 460)/10^3$ = 73.6 kN/bolt For eight bolts ΣP_{bs} = (8×73.6) = 588.8 kN $\geq$ 175 kN	**End plate is adequate in bearing**
Clause 4.2.3	**4. Beam web shear capacity:** P_v = $0.6 p_y A_v$ where $A_v = 0.9A$ = $(0.9 \times 290 \times 11.4)$ = 2975 mm^2 P_v = $(0.6 \times 275 \times 2975)/10^3$ = 490 kN $\geq$ 175 kN	**Beam web is adequate in shear**
Clause 6.8.7.2 Table 37	**5. End plate/beam weld fillet weld:** P_{weld} = $(p_w \times$ effective length of weld $\times$ throat thickness) Assume Electrode classification 35 and 6 mm fillet welds: p_w = 220 N/mm^2	
Clause 6.8.2	Effective length = [total length − (2 × leg length)] = $[290 - (2 \times 6)]$ = 278 mm/side Total effective length = (2×278) = 556 mm	
Clause 6.8.3	Throat size a = $(0.7 \times s) = (0.7 \times 6)$ = 4.2 mm	
	P_{weld} = $(220 \times 556 \times 4.2)/10^3$ = 513.7 kN $\geq$ 175 kN	**6 mm fillet weld both sides is adequate**

References	Calculations	Output

Contract : Braced Frame Job Ref. No. : Example 6.30
Part of Structure : Beam–Column Connection
Calc. Sheet No. : 6 of 8

Calcs. by : W.McK.
Checked by :
Date :

(iii) Fin plate – the following checks should be carried out:

175 kN | $e = 60$ mm

6 mm fillet welds

40 mm
70 mm
70 mm
70 mm
40 mm

290 mm × 100 mm × 10 mm thick fin plate

60 mm 40 mm

1	bolts:	F_r	$\leq$ P_s	shear
2	beam web:	F_r	$\leq$ P_{bs}	bearing
3	beam web:	F	$\leq$ P_r	block shear
4	fin plate:	F	$\leq$ P_v	shear
5	fin plate:	F_r	$\leq$ P_{bs}	bearing
6	fin plate:	$(F \times e)$	$\leq$ $p_y S_x$	bending
7	fillet weld:	$F_{combined}$	$\leq$ P_{weld}	(bending + shear)

As in the case of web cleats, the bolt group is subjected to the combined effects of direct shear and rotational shear. In this case:

Total vertical shear force
Vertical shear force/bolt
(Single shear)

$F = $ **175 kN**
$F_v = (175/4)$
$= $ **43.75 kN/bolt**

Maximum rotational shear force F_h

$$= \frac{(F \times e) \times z_{maximum}}{\displaystyle\sum_{n=1}^{number\ of\ bolts} z^2}$$

$$= \frac{(175 \times 0.06) \times 0.105}{2(0.035^2 + 0.105^2)}$$

$$= 45.0 \text{ kN}$$

Resultant maximum shear force F_r

$$= \sqrt{\left(F_v^2 + F_h^2\right)}$$

$$= \sqrt{\left(43.75^2 + 45.0^2\right)}$$

$$= 62.8 \text{ kN/bolt}$$

1. Bolt shear capacity:
As before
(single shear)

$P_s = $ 91.9 kN
$\geq$ 62.8 kN

Bolts are adequate with respect to shear

Contract : Braced Frame Job Ref. No. : Example 6.30	Calcs. by : W.McK.
Part of Structure : Beam–Column Connection	Checked by :
Calc. Sheet No. : 7 of 8	Date :

References	Calculations	Output
Clause 6.3.3.3	**2. Beam web bearing capacity:** As before P_{bs} = 104.9 kN $\geq$ 62.8 kN	Beam web is adequate with respect to bearing
	3. Beam web – Block shear capacity: As before P_r = 460.5 kN $\geq$ 175 kN	Beam web is adequate with respect to block shear
Clause 4.2.3	**4. Fin plate shear capacity:** P_v = $0.6 p_y A_v$ where $A_v = 0.9A$ = $(0.9 \times 290 \times 10)$ = 2610 mm^2	
Clause 6.2.3 Clause 3.4.3	Bolt holes need not be allowed for if $A_{v,net} \geq$ $0.85A_v/K_e$ K_e = 1.2 for S 275 steel $A_{v,net}$ = $[2610 - (4 \times 22 \times 10)]$ = 1730 mm^2 $0.85A_v/K_e$ = $(0.85 \times 2610)/1.2$ = 1849 mm^2 $>$ $A_{v,net}$	
	P_v = $0.7 p_y K_e A_{v,net}$ $= (0.7 \times 275 \times 1.2 \times 1730)/10^3$ $= 400$ kN $\geq$ 175 kN	Fin plate is adequate with respect to shear
Clause 6.3.3.3 Table 32	**5. Fin plate bearing capacity:** P_{bs} = $k_{bs} d t_p p_{bs}$ $\leq$ $0.5 k_{bs} e t_p p_{bs}$ e = 40 mm, $t = 10$ mm, $k_{bs} = 1.0$, $p_{bs} = 460$ N/mm^2 Since $e = 2d$ $k_{bs} d t_p p_{bs}$ = $0.5 k_{bs} e t_p p_{bs}$	
	P_{bs} = $(1.0 \times 20 \times 10 \times 460)/10^3$ = 92.0 kN/bolt $\geq$ 62.8 kN	Fin plate is adequate with respect to bearing
Clause 4.3.6.2	**6. Fin plate bending capacity:** $M_x \leq M_b/m_{LT}$ $M_x = (175 \times 0.06)$ = 10.5 kNm $M_b = S_x p_b$ $S_x = \dfrac{tl^2}{4} = \dfrac{10 \times 290^2}{4} = 210.25 \times 10^3$ mm^3	
Appendix B	$\lambda_{LT} = 2.8 \times \left(\dfrac{\beta_w L_E d}{t^2} \right)^{\frac{1}{2}}$	
Clause 4.3.6.9	$m_{LT} = 1.0$, assume $\beta = 1.0$, $L_E = 60$ mm $\lambda_{LT} = 2.8 \times \left(\dfrac{60 \times 290}{10^2} \right)^{\frac{1}{2}} = 36.9$	
Table 12	p_y = 275 N/mm^2 $\lambda_{LT} = 36.9$ $\lambda_{L0} = 34.3 < \lambda_{LT}$ p_b = 263 N/mm^2 $M_b = (210.25 \times 10^3 \times 263) / 10^6$ = 55.3 kNm $\geq$ 10.5 kNm	Fin plate is adequate with respect to bending

Contract : Braced Frame Job Ref. No. : Example 6.30	Calcs. by : W.McK.
Part of Structure : Beam–Column Connection	Checked by :
Calc. Sheet No. : 8 of 8	Date :

References	Calculations	Output
	7. Weld between fin plate and column	
	  290 mm 	
	Consider a pair of welds each of unit throat width × 290 mm long	
	Applied moment $\quad= 10.5$ kNm	
	Applied shear force $\quad= 175$ kN	
	Total length of weld $\quad= (2 \times 290) \qquad = 580$ mm	
	Direct vertical shear force/mm $= \quad 175/580 = 0.3$ kN/mm	
	$$I_x = 2 \times \left(\frac{1 \times 290^3}{12}\right) = 4.06 \times 10^6 \text{ mm}^4$$	
	Distance to the extreme fibre $\quad= 145$ mm	
	Maximum bending force/mm $\quad= \dfrac{\left(10.5 \times 10^3\right) \times 145}{4.06 \times 10^6} = 0.38$ kN/mm	
	Resultant Force/mm $\quad= \sqrt{0.3^2 + 0.375^2} = 0.48$ kN	
Reference (67)	Leg length $s \quad \geq (0.8 \times \text{plate thickness}) \quad= (0.8 \times 10) = 8$ mm	**Adopt 8 mm fillet welds between**
Table 37	Strength of 8 mm fillet weld $\quad= \quad (0.7 \times 8 \times 220)/10^3$	**the fin plate and the column flange**
	$\qquad\qquad\qquad\qquad\qquad = \quad 1.23$ kN/mm	
	$\qquad\qquad\qquad\qquad\qquad \geq \quad 0.48$ kN/mm	

6.3.6.8 Design of Moment Connections

The principal difference between 'simple design' and 'rigid design' of structural frames occurs in the design of the connections between the elements. In the former the connections are assumed to transmit direct and shear forces, in the latter it is necessary to transmit moments in addition to these forces. The moments can be considered to be either:

(i)   in the plane of the connection as shown in Figure 6.66(a) or
(ii)   perpendicular to the plane of the connection as shown in Figure 6.66(b).

In both cases the connections are generally designed using either H.S.F.G. bolts or welding. There are a number of approaches to designing connections in moment-resisting frames. In most connections the problem is to identify the distribution of forces, moments and stresses in the component parts. Other factors which must be considered are the overall stiffness of the connection, and the practical aspects of fabrication, erection and inspection.

In rigid frame design each component of the frame can be designed individually to sustain the bending moments, shear forces and axial loads; the connections must then be designed to transfer these forces. In many instances the connections occur where members change direction, such as at the eaves and ridge of pitched roof portal frames. Such frames are usually transported 'piece-small' from the fabrication shop to the site and provision must be made for site-joints; High Strength Friction Grip bolts are ideally suited for this purpose.

In welded connections, such as the knee joint shown in Figure 6.88, because the compressive forces X and Y are not collinear, an induced compressive force Z exists to maintain the equilibrium of the forces. This force Z acts across the web plate at the corner, and in order to obviate the likelihood of the web plate buckling, corner stiffeners are required as shown, to carry the force Z.

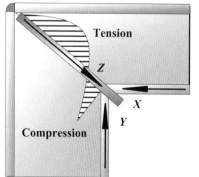

The bending stress lines flowing between column and beam are not able to follow a sudden change of direction exactly and become more concentrated towards the inside corner. This leads to an appreciable shift of the neutral axis towards the inside corner and a redistributed stress diagram, as shown. As the total compression must equal the total tension, an enhanced value of the maximum compressive stress results. This distribution is further modified by the direct compressive stresses in the beam and column.

Figure 6.88

6.3.6.9 Typical Site Connection using H.S.F.G. Bolts

Typical site connections using H.S.F.G. bolts are shown in Figure 6.89. The cap plate transmits the force in the tension flange by means of the H.S.F.G bolts whilst the force in the compression flange is transmitted in direct bearing. Where the depth of the connection in Figure 6.89(a) leads to forces of too large a magnitude to be transmitted reasonably, the depth of the connection can be increased by the introduction of a haunch such as is shown in Figure 6.89(b) for a pitched roof portal frame.

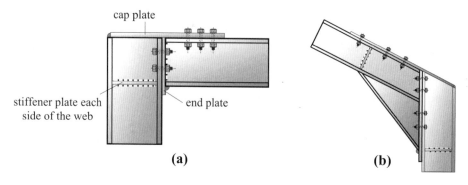

(a)   **(b)**

Figure 6.89

Similar details occur at the ridge connection of pitched portal frames, as shown in Figure 6.90.

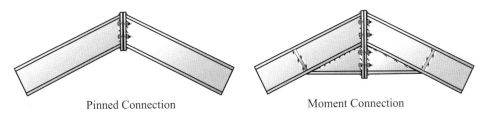

Pinned Connection   Moment Connection

Figure 6.90

6.3.6.9.1. Example 6.31: Moment Connection in Rectangular Portal Frame

The uniform rectangular portal frame shown in Figure 6.91(a) and (b) is subjected to loading which induces moments and shear forces at the knee joint as given in the data below. Using this data, determine a suitable size of H.S.F.G. bolt for the connection between the column and the roof beam. Assume non-slip in service conditions.

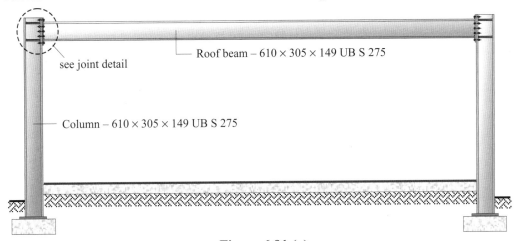

Roof beam – 610 × 305 × 149 UB S 275

see joint detail

Column – 610 × 305 × 149 UB S 275

Figure 6.91 (a)

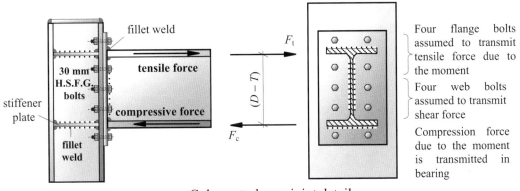

fillet weld

F_t

Four flange bolts assumed to transmit tensile force due to the moment

30 mm H.S.F.G. bolts

tensile force

$(D – T)$

Four web bolts assumed to transmit shear force

stiffener plate

compressive force

F_c

Compression force due to the moment is transmitted in bearing

fillet weld

Column-to-beam joint detail

Figure 6.91 (b)

Design Data:

Column / Beam Section $610 \times 305 \times 149$ UB
Ultimate Design Moment at the knee joint   520 kNm
Ultimate Design Shear Force at the knee joint   350 kN

Section Data: $610 \times 305 \times 149$ **UB**

$D = 609.6$ mm	$d = 537.2$ mm	$I_{xx} = 124.7 \times 10^3$ cm^4	
$B = 304.8$ mm	$b/T = 7.74$	$I_{yy} = 9.308 \times 10^3$ cm^4	
$t = 11.9$ mm	$d/t = 45.1$	$r_{xx} = 25.6$ cm	$x = 32.5$
$T = 19.7$ mm	$r_{yy} = 6.99$ cm	$u = 0.886$	$A = 190$ cm^2

6.3.6.9.2. Solution to Example 6.31

Contract : Portal Frame Job Ref. No. : Example 6.31 Part of Structure : Beam–Column Moment Connection Calc. Sheet No. : 1 of 2	Calcs. by : W.McK. Checked by : Date :

References	Calculations	Output
	$M_x = 520$ kNm $F_s = 350$ kN Bolts to accommodate stress reversal	
	Assume that the moment is transferred by the flange bolts whilst the shear is transferred by the web bolts.	
	Flange force induced by the moment $= F_t = \dfrac{\text{Moment}}{\text{lever} - \text{arm}}$	
	$F_t = \dfrac{520 \times 10^3}{(609.6 - 19.7)} = 881.5$ kN	
	Moment (4 flange bolts):	
Clause 6.4.5	Tension capacity $P_t = 1.1 P_o \geq \dfrac{881.5}{4} = 220.4$ kN	
	Minimum. P_o required $= \dfrac{220.4}{1.1} = 200.4$ kN	**24 mm diameter H.S.F.G. bolts adequate for tension**
	For 24 mm H.S.F.G Bolts $P_o = 207$ kN	
Clause 6.4.2	**Shear** (4 web bolts): Slip resistance $P_{SL} = 4 \times (1.1 \times K_s \mu P_o) \qquad \geq 380$ kN	
Table 6.4.3	where $K_s = 1.0$ and $\mu = 0.4$ (assuming Class B as the condition of the faying surfaces)	

$P_{SL\,provided} = (4 \times 1.1 \times 0.4 \times 207) = 364.3$ kN ≥ 350 kN

References	Calculations	Output
Contract : Portal Frame Job Ref. No. : Example 6.31 **Part of Structure : Beam–Column Moment Connection** **Calc. Sheet No. : 2 of 2**		**Calcs. by : W.McK.** **Checked by :** **Date :**

References	Calculations	Output
Clause 6.4.4	For friction grip connection designed to be non-slip in service: Slip resistance $P_{SL} \leq P_{bg}$ where: $P_{bg} = 1.5 d t_p p_{bs} \leq 0.5 e t_p p_{bs}$ Assume: the end distances for the bolt holes $\geq 3d$ and end plate thickness = 10 mm	**Adopt 24 mm H.S.F.G. bolts.** **The minimum end-distances should be** $\geq$ **72 mm and ensure that at least category B preparation of the faying surfaces is specified.**
Table 32	$P_{bg} = (4 \times 1.5 \times 24 \times 10 \times 460) \leq 662$ kN The welds, end-plates and stiffeners must also be designed to transfer the appropriate loads and in addition the column flange and web should be checked. If necessary additional strengthening such as stiffeners may be required to enhance the buckling or bearing capacity of the column web. The detailed design required for this is beyond the scope of this publication. A rigorous, detailed analysis and design procedure for connections and their component parts can be found in *Joints In Steel Construction: Moment connections*, published by The Steel Construction Institute (ref. 67).	

6.3.6.9.3. Example 6.32: Crane Bracket Moment Connection

An industrial frame building supports a light electric overhead travelling crane on brackets bolted to the main columns, as shown in Figures 6.92(a) and (b) below. Using the design data given, determine a suitable size of H.S.F.G. bolt to connect the brackets to the columns. Assume non-slip in service conditions, end distances and surface conditions and end plate thickness as before.

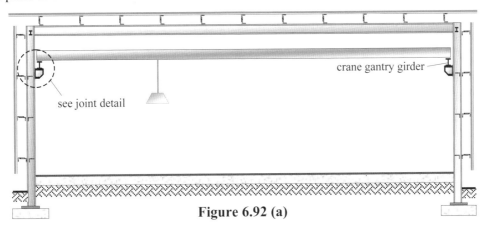

Figure 6.92 (a)

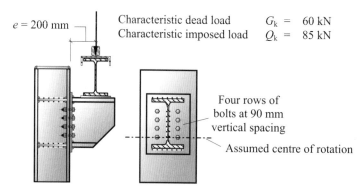

Figure 6.92 (b)

6.3.6.9.4. Solution to Example 6.31

Design Load = $(60 \times 1.4) + (85 \times 1.6) = $ 220 kN

Assuming the bracket rotates at the level of the bottom bolts (this is conservative) then the maximum forces induced in the bolts are:

Maximum shear force/bolt $= F_s = \dfrac{F}{n} = \dfrac{220}{8} = $ 27.5 kN

Maximum tensile force $= F_{maximum} = \dfrac{(Fe \times y_{maximum})}{m \sum y^2}$

where:

n = number of bolts
m = number of vertical columns of bolts
e = eccentricity of applied loads
F = applied load
$y_{maximum}$ = distance from the centre of rotation of the bracket to the most distant bolt
  $\sum y^2 = (90^2 + 180^2 + 270^2) = $ 113.4×10^3 mm^2
  Design Bending Moment $= (220 \times 200)$ $= (44 \times 10^3)$ kNmm

$$F_{maximum} = \frac{44 \times 10^3 \times 270}{2 \times 113.4 \times 10^3} = 52.38 \text{ kN}$$

Assuming 20 mm dia. General Grade H.S.F.G. bolts: $P_o = $ 144 kN; $A_t = $ 245 mm^2

Clause 6.4.5 $\dfrac{F_s}{P_{SL}} + \dfrac{F_{tot}}{1.1 P_o} \leq 1$ but

Table 35 $F_{tot} \leq A_t p_t = (245 \times 590)/10^3 = 145$ kN

Clause 6.4.2 Slip resistance $P_{SL} = (1.1 \times K_s \mu P_o) = (1.1 \times 1.0 \times 0.4 \times 144) = $ 63.4 kN
  $\leq P_{bg} = (1.5 \times 20 \times 10 \times 460)/10^3$ $= 138$ kN

$$\frac{F_s}{P_{SL}} + \frac{F_{tot}}{1.1 P_o} = \left[\frac{27.5}{63.4} + \frac{52.4}{(1.1 \times 144)}\right] = 0.76 \leq 1$$

M20 H.S.F.G. bolts are adequate

7. Design of Structural Timber Elements (BS 5268)

> **Objective:** *to introduce the inherent botanical and structural characteristics of timber and to illustrate the process of design for structural timber elements.*

7.1 Introduction

The inherent variability of a material such as timber, which is unique in its structure and mode of growth, results in characteristics and properties which are distinct and more complex than those of other common structural materials such as concrete, steel and brickwork. Some of the characteristics which influence design and are specific to timber are:

- the moisture content,
- the difference in strength when loads are applied parallel and perpendicular to the grain direction,
- the duration of the application of the load,
- the method adopted for strength grading of the timber.

As a live growing material, every identified tree has a name based on botanical distinction, for example *Pinus sylvestris* is commonly known as Scots pine. The botanical names have a Latin origin, the first part indicating the genus, e.g. *Pinus*, and the second part indicating the species, e.g. *sylvestris*.

A classification such as this is of little value to a structural designer and consequently design codes adopt a classification based on stress grading. Stress grading is discussed in Section 7.2. The growth of a tree depends on the ability of the cells to perform a number of functions, primarily conduction, storage and mechanical support. The stem (or trunk) conducts essential mineral salts and moisture from the roots to the leaves, stores food materials and provides rigidity to enable the tree to compete with surrounding vegetation for air and sunlight. Chemical processes, which are essential for growth, occur in the branches, twigs and leaves in the crown of the tree.

7.1.1 Moisture Content

Unlike most structural materials, the behaviour of timber is significantly influenced by the existence and variation of its moisture content. The moisture content, as determined by oven drying of a test piece, is defined in *Annex H* of BS 5268 as:

$$w = 100(m_1 - m_2)/m_2$$

where:
m_1 is the mass of the test piece before drying (in g)
m_2 is the mass of the test piece after drying (in g)

Moisture contained in 'green' timber is held both within the cells (free water) and within the cell walls (bound water). The condition in which all free water has been removed but the cell walls are still saturated is known as the **fibre saturation point** (FSP). At levels of moisture above the FSP, most physical and mechanical properties remain constant. Variations in moisture content below the FSP cause considerable changes to properties such as weight, strength, elasticity, shrinkage and durability. The controlled drying of timber is known as **seasoning**. There are two methods generally used:

♦ **Air seasoning**, in which the timber is stacked and layered with air-space in open-sided sheds to promote natural drying. This method is relatively inexpensive with very little loss in the quality of timber if carried out correctly. It has the disadvantage that the timber and the space which it occupies are unavailable for long periods. In addition, only a limited control is possible by varying the spaces between the layers and/or by using mobile slatted sides to the sheds.

♦ **Kiln drying**, in which timber is dried out in a heated, ventilated and humidified oven. This requires specialist equipment and is more expensive in terms of energy input. The technique does offer a more controlled environment in which to achieve the required reduction in moisture content and is much quicker.

The anisotropic nature of timber and differential drying out caused by uneven exposure to drying agents such as wind, sun or applied heat can result in a number of defects such as *twisting*, *cupping*, *bowing* and *cracking*, as shown in Figure 7.1.

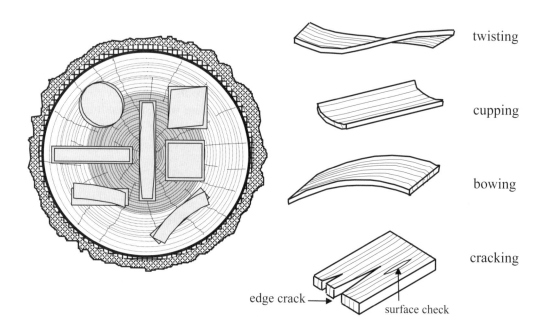

Figure 7.1   Distortions due to differential directional shrinkage

7.1.2 Defects in Timber

In addition to the defects indicated in Figure 7.1 there are a number of naturally occurring defects in timber. The most common and familiar of such defects is a **knot** (see Figure 7.2). Normal branch growth originates near the pith of a tree and consequently its base develops new layers of wood each season which develop with the trunk. The cells of the new wood grow into the lower parts of the branches, maintaining a flow of moisture to the leaves. The portion of a branch which is enclosed within the main trunk constitutes a *live* or *intergrown* knot and has a firm connection with surrounding wood.

When lower branches in forest trees die and drop off as a result of being deprived of sunlight, the dead stubs become overgrown with new wood but have no connection to it. This results in dead or enclosed knots which are often loose and, when cut, fall out.

The presence of knots is often accompanied by a decrease in the physical properties of timber such as the tensile and compressive strength. The reduction in strength is primarily due to the distortion of the grain passing around the knots and the large angle between the grain of the knot and the piece of timber in which it is present. During the seasoning of timber, **checks** often develop around the location of knots.

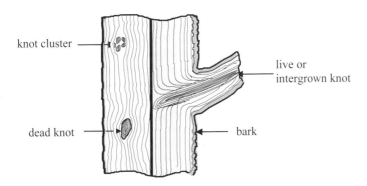

Figure 7.2   Defects due to knots

In a mill when timber is converted from a trunk into suitable commercial sizes (see Figure 7.3) a *wane* can occur when part of the bark or rounded periphery of the trunk is present in a cut length, as shown in Figure 7.4. The effect of a wane is to reduce the cross-sectional area with a resultant reduction in strength.

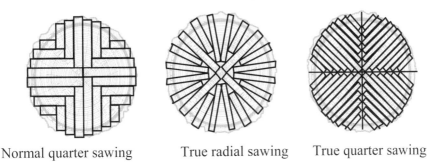

Normal quarter sawing   True radial sawing   True quarter sawing

Figure 7.3   Typical sawing patterns

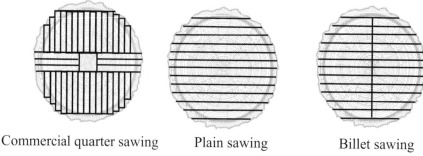

Commercial quarter sawing   Plain sawing   Billet sawing

Figure 7.3 (continued) Typical sawing patterns

A *shake* is produced when fibres separate along the grain: this normally occurs between the growth rings, as shown in Figure 7.4. The effect of a shake in the cross-section is to reduce the shear strength of beams; it does not significantly affect the strength of axially loaded members.

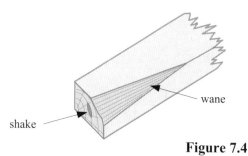

Figure 7.4

7.2   Classification of Timber

The efficient and economic use of any structural material requires a knowledge of its physical characteristics and properties such that a representative mathematical model can be adopted to predict behaviour in qualitative terms. Manufactured materials such as steel and concrete, in which quality can be tightly controlled and monitored during production, readily satisfy this. As a natural product, timber is subject to a wide range of variation in quality which cannot be controlled.

Uniformity and reliability in the quality of timber as a structural material is achieved by a process of selection based on established grading systems, i.e. appearance grading and strength grading:

♦ **Appearance grading** is frequently used by architects to reflect the warm, attractive features of the material such as the surface grain pattern, the presence of knots, colour, etc. In such circumstances the timber is left exposed and remains visible after completion; it may be either structural or non-structural.

♦ All structural (load-bearing) timber must be **strength-graded** according to criteria which reflect its strength and stiffness. In some cases timber may be graded according to both appearance and strength. Strength grading is normally carried out either visually or mechanically, using purpose-built grading machines.

Every piece of strength-graded timber should be marked clearly and indelibly with the following information:

♦ grade/strength class (e.g. GS, SS, C16, C24),
♦ species or specific combination (e.g. ER – *redwood*, B/D – *Douglas fir (British)*),
♦ number of the British Standard used (e.g. BS 4978),
♦ company and grader/machine used,
♦ company logo/mark of the certification body,
♦ timber condition (e.g. KD – *kiln dried*, WET).

A typical grading mark/stamp is illustrated in Figure 7.5.

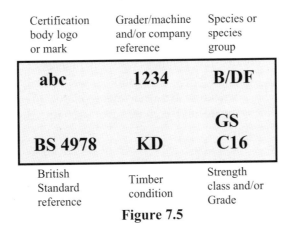

Figure 7.5

7.2.1 *Visual Strength Grading*

As implied by the name, this method of grading is based on the physical observation of strength-reducing defects such as knots, rate of growth, cracks, wane, bowing, etc. Since the technique is based on the experience and judgement of the grader it is inherently subjective. In addition, important properties such as density, which has a significant influence on stiffness, and strength are not considered.

Numerous grading rules and specifications have been developed throughout Europe, Canada and the USA during the last fifty years. In the UK, visual grading is governed by the requirements of *BS 4978:1996 Specification for softwood grades for structural use* and the Eurocode *BS EN 518:1995 Structural timber – Grading – Requirements for visual grading standards.*

Visual defects considered when assessing timber strength include: location and extent of knots, slope of grain, rate of growth, fissures, wane, distortions such as bowing, springing, twisting, cupping, resin and bark pockets, and insect damage. Two strength grades are specified: **General Structural** (GS) and **Special Structural** (SS), the latter being the higher quality material. Timber which contains abnormal defects such as compression wood, insect damage such as worm holes, or fungal decay (not sapstain), or which is likely to impair the serviceability of the pieces, is excluded from the grades.

7.2.2   *Machine Strength Grading*

The requirements for machine strength grading are specified in *BS EN 519:1995 Structural Timber – Grading – Requirements for machine strength graded timber and grading machines.* Timber is classified into:

- nine classes of poplar and coniferous species ranging from the weakest grade C14 to the highest grade C40,
- six classes for deciduous species ranging from the weakest grade D30 to the highest grade D70.

In each case the number following either the 'C' or the 'D' represents the characteristic bending strength of the timber. In BS 5268-2:2002 two additional strength classes, TR20 and TR26, are also given; this is intended for use in the design of trussed rafters.

The inherently subjective nature of visual strength grading results in a lower yield of higher strength classes than would otherwise be achieved. Machine strength grading is generally carried out by conducting bending tests on planks of timber which are fed continuously through a grading machine. The results of such tests produce a value for the modulus of elasticity. The correlation between the modulus of elasticity and strength properties such as bending, tensile and compressive strength can be used to define a particular grade/class of timber.

Visual grading enables a rapid check on a piece of timber to confirm, or otherwise, an assigned grade; this is not possible with machine grading. The control and reliability of machine grading is carried out either by destructive testing of output samples (output controlled systems), or regular and strict control/adjustments of the grading machine settings. In Europe the latter technique is adopted since this is more economic than the former when using a wide variety of species and relatively low volumes of production.

7.3   Material Properties

The strength of timber is due to certain types of cells (called tracheids in softwoods and fibres in hardwoods) which make up the many minute hollow cells of which timber is composed. These cells are roughly polygonal in cross-section and the dimension along the grain is many times larger than across it.

The principal constituents of the cells are cellulose and lignin. Individual cell walls comprise four layers, one of which is more significant with respect to strength than the others. This layer contains chains of cellulose which run nearly parallel to the main axis of the cell. The structure of the cell enhances the strength of the timber in the grain direction.

Density, which is expressed as mass per unit volume, is one of the principal properties affecting strength. The heaviest species, i.e. those with most wood substance, have thick cell walls and small cell cavities. They also have the highest densities and consequently are the strongest species. Numerous properties in addition to strength, e.g. shrinkage, stiffness and hardness, increase with increasing density.

When timber is seasoned, the cell contents dry out leaving only cell walls. Shrinkage occurs during the drying process as absorbed moisture begins to leave the cell walls. The cell walls become thinner as they draw closer together. However the length of the cell layers is only marginally affected. A consequence of this is that as shrinkage occurs the width and thickness change but the length remains the same. The degree to which shrinkage occurs is dependent upon its initial moisture content value and the value at which it stabilises in service. A number of defects such as bowing, cupping, twisting and surface checks are a direct result of shrinkage.

Since timber is hygroscopic, and can absorb moisture whilst in service, it can also swell until it reaches an equilibrium moisture content.

Anisotropy is a characteristic of timber because of the long fibrous nature of the cells and their common orientation, the variation from early to late wood, and the differences between sapwood and hardwood. The elastic modulus of a fibre in a direction *along* its axis is considerably greater than that *across* it, resulting in the strength and elasticity of timber parallel to the grain being much higher than in the radial and tangential directions.

The slope of the grain can have an important effect on the strength of a timber member. Typically a reduction of 4% in strength can result from a slope of 1 in 25, increasing to an 11% loss for slopes of 1 in 15.

The strength of timber is also affected by the rate of growth as indicated by the width of the annual growth rings. For most timbers the number of growth rings to produce the optimum strength is approximately in the range of 6–15 per 25 mm measured radially. Timber which has grown either much more quickly or much more slowly than that required for the optimum growth rate is likely to be weaker.

In timber from a tree which has grown with a pronounced lean, wood from the compression side (compression wood) is characterised by much greater shrinkage than normal. In softwood planks containing compression wood, bowing is likely to develop in the course of seasoning and the bending strength will be low. In hardwoods, the tension wood has abnormally high longitudinal shrinkage and although stronger in tension is much weaker in longitudinal compression than normal wood.

Like many materials, e.g. concrete, the stress–strain relationship demonstrated by timber under load is linear for low stress values. For all species the strains for a given load increase with moisture content. A consequence of this is that the strain in a beam under constant load will increase in a damp environment and decrease as it dries out again.

Timber demonstrates viscoelastic behaviour (creep) as high stress levels induce increasing strains with increasing time. The magnitude of long-term strains increases with higher moisture content. In structures where deflection is important, the duration of the loading must be considered. This is reflected in BS 5268-2:2002 by the use of modifying factors applied to admissible stresses depending on the type of loading, e.g. long-term, medium-term, short-term and very short-term.

The cellular structure of timber results in a material which is a poor conductor of heat. The air trapped within its cells greatly improves its insulating properties. Heavier timbers having smaller cell cavities are better conductors of heat than lighter timbers. Timber does expand when heated, but this effect is more than compensated for by the shrinkage caused by loss of moisture.

The fire resistance of timber generally compares favourably with other structural materials and is often better than most. Steel is subject to loss of strength, distortion, expansion and collapse, whilst concrete may spall and crack.

Whilst small timber sections may ignite easily and support combustion until reduced to ash, this is not the case with large structural sections. At temperatures above 250°C material at the exposed surface decomposes, producing flammable gases and charcoal. These gases, when ignited, heat the timber to a greater depth and the fire continues. The charcoal produced during the fire is a poor conductor and will eventually provide an insulating layer between the flame and the unburned timber.

If there is sufficient heat, charcoal will continue to char and smoulder at a very slow rate, particularly in large timber sections with a low surface:mass ratio. Fire authorities usually consider that a normal timber door will prevent the spread of fire to an adjoining room for about 30 minutes. The spread of fire is then often due to flames and hot gases permeating between the door and its frame or through cracks between door panels and styles produced by shrinkage (see *BS 5268:Part 4 – Fire resistance of timber structures*).

The durability of timbers in resisting the effects of weathering, chemical or fungal attack varies considerably from one species to another. In general the heartwood is more durable to fungal decay than the sapwood. This is due to the presence of organic compounds within the cell walls and cavities which are toxic to fungi and insects.

Provided timber is kept dry, or is continuously immersed in fresh water, decay will generally not be a problem. Where timber is used in seawater, particularly in harbours, there is always a risk of severe damage due to attack by molluscs. The pressure impregnation of timber with suitable preservatives will normally be sufficient to prevent damage due to fungal, insect or mollusc attack.

7.3.1 *Permissible Stress Design*

When using permissible stress design, the margin of safety is introduced by considering structural behaviour under working/service load conditions and comparing the stresses thereby induced with permissible values. The permissible values are obtained by dividing the failure stresses by an appropriate factor of safety. The applied stresses are determined using elastic analysis techniques, i.e.

$$Stress\ induced\ by\ working\ loads \leq \frac{failure\ stress}{factor\ of\ safety}$$

This is the philosophy adopted in the current timber design code BS 5268-2:2002 as used in this text. Since BS 5268 is a permissible stress design code, mathematical modelling of the behaviour of timber elements and structures is based on assumed elastic behaviour.

The laws of structural mechanics referred to are those well established in recognised 'elastic theory', as follows.

♦ The material is *homogeneous,* which implies that its constituent parts have the same physical properties throughout its entire volume. This assumption is clearly violated in the case of timber. The constituent fibres are large in relation to the mass of which they form a part when compared with a material such as steel in which there are millions of very small crystals per square centimetre, randomly distributed and of similar quality creating an amorphous mass. In addition, the presence of defects such as knots, shakes etc. as described in Sections 7.1.1 and 7.1.2 represent the inclusion of elements with differing physical properties.

◆ The material is *isotropic,* which implies that the elastic properties are the same in all directions. The main constituent of timber is cellulose, which occurs as long chain molecules. The chemical/electrical forces binding the molecules together in these chains are much stronger than those which hold the chains to each other. A consequence of this is that considerable differences in elastic properties occur according to the grain orientation. Timber exhibits a marked degree of anisotropic behaviour. The design of timber is generally based on an assumption of orthotropic behaviour with three principal axes of symmetry: the longitudinal axis (parallel to the grain), the radial axis and the tangential axis, as shown in Figure 7.6. The properties relating to the tangential and radial directions are often treated together and regarded as properties perpendicular to the grain.

◆ The material obeys *Hooke's Law,* i.e. when subjected to an external force system the deformations induced will be directly proportional to the magnitude of the applied force. A typical stress–strain curve for a small wood specimen (with as few variations or defects as possible and loaded for a short-term duration) exhibits linearity prior to failure when loaded in tension or compression, as indicated in Figures 7.7(a) and (b).

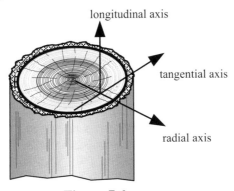

Figure 7.6

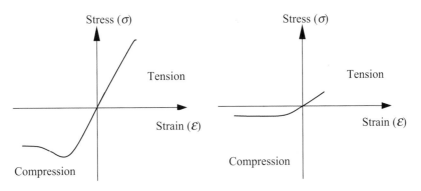

(a) load parallel to the grain   (b) load perpendicular to the grain

Figure 7.7

In Figure 7.7(a) the value of tensile strength is greater than that of the compressive strength. In both compression and tension linear behaviour occurs. In the case of compression ductility is present before failure occurs, whilst in tension a brittle, sudden failure occurs. These characteristics are reflected in the interaction behaviour of timber elements designed to resist combined bending and axial stresses.

♦   The material is *elastic*, which implies that it will recover completely from any deformation after the removal of load. Elastic behaviour is generally observed in timber subject to compression up to the limit of proportionality. The elastic properties of timber in tension are more sensitive to the level of moisture content. Permanent strain occurs at very low stress levels in timber which contains a high percentage of moisture.

♦   The *modulus of elasticity* is the same in tension and compression. This assumption is reasonable for both compression and tension. The value is much lower when the load is applied perpendicular to the grain than when it is applied parallel to the grain, as shown in Figure 7.7. Two values of modulus of elasticity are given in the code for each timber grade; $E_{minimum}$ and E_{mean}. The value to be used in any given circumstance is given in the code in the pertinent clauses.

♦   *Plane sections remain plane* during deformation. During bending this assumption is violated and is reflected in a non-linear bending stress diagram throughout cross-sections subject to a moment.

The behaviour and properties of timber do not satisfy the basic assumptions used in simple elastic theory. These deficiencies are accommodated in the design processes by the introduction of numerous modification factors and a factor of safety, which are applied to produce an admissible stress which is then compared with the applied stress induced by the applied load system calculated using elastic theory. Extensive research and development during the latter half of the twentieth century has resulted in more representative mathematical models of timber behaviour and is reflected in Eurocode 5, which adopts a limit state approach to design. This code will eventually be utilised for timber design in the UK.

7.4   Modification Factors

The inherently variable nature of timber and its effects on structural material properties such as stress–strain characteristics, elasticity and creep has resulted in more than eighty different modification factors which are used in converting grade stresses (see Section 7.2) to permissible stresses for design purposes. In general, when designing to satisfy strength requirements (e.g. axial, bending or shear strength) the following relationship must be satisfied:

applied stress   ≤   permissible stress

The applied stresses are calculated using elastic theory, and the permissible stresses are determined from the code using the appropriate values relating to the strength classification multiplied by the modification factors which are relevant to the stress condition being considered. Symbols are defined relating to stresses and other variables in Clause 1.4 of *BS 5268-2:2002* as follows:

a distance;
A area;
b breadth of beam, thickness of web, or lesser transverse dimension of a tension or compression member;
d diameter;
E modulus of elasticity;
F force or load;
h depth of member, greater transverse dimension of a tension or compression member;
i radius of gyration;
K modification factor (always with a subscript);
L length; span;
m mass;
M bending moment;
n number;
r radius of curvature;
t thickness; thickness of laminations;
u fastener slip;
α angle between the direction of the load and the direction of the grain;
η eccentricity factor;
θ angle between the longitudinal axis of a member and a connector axis;
λ slenderness ratio;
σ stress;
τ shear stress;
ω moisture content.

In many instances subscripts are also used to identify various types of force, stress or geometry; these are as follows:

a) type of force, stress etc.:
    c compression;
    m bending;
    t tension;
b) significance:
    a applied;
    adm permissible;
    e effective;
    mean arithmetic mean;

c) geometry:

       apex   apex;
       r   radial;
       tang   tangential;
       $\parallel$ parallel (to the grain);
       $\perp$ perpendicular (to the grain);
       α angle.

The following examples illustrate the use of these symbols and subscripts:

$\sigma_{m,a,\perp}$ $\equiv$ applied bending stress perpendicular to the grain
$\sigma_{m,adm,\perp}$ $\equiv$ permissible bending stress perpendicular to the grain
$\sigma_{c,a,\parallel}$ $\equiv$ applied compressive stress parallel to the grain
$\sigma_{c,adm,\parallel}$ $\equiv$ permissible compressive stress parallel to the grain
τ_a $\equiv$ applied shear stress.

Whilst not given in this Clause the subscript 'g' is often used to identify grade stresses. As mentioned previously, the permissible stress is evaluated by multiplying the grade stress for a particular strength class by the appropriate modification factors, e.g.

$$\sigma_{m,adm,\parallel} \;=\; \sigma_{m,g,\parallel} \times K_2 \times K_3 \times K_6 \times K_7 \times K_8$$

where:

K_2 relates to the *moisture content* of the timber;
K_3 relates to the *duration of the load* on the timber;
K_6 relates to the *shape of the cross-section* of the element being considered;
K_7 relates to the *depth* of the section being considered;
K_8 relates to the existence of structural elements enabling *load sharing*.

In each case a definition and the method of evaluating a coefficient is given in the code. In this text a table is given in each section which identifies each coefficient and reference clause numbers for the coefficients which are pertinent to the structural elements being considered.

Three of the most frequently used coefficients are K_2, K_3 and K_8:

K_2:   The value of K_2 is governed by the average moisture content likely to be attained in service conditions. This is allowed for in the code by identifying a *service class* for the particular element being designed, as given in Clause 1.6.4. The service classes are:

Service Class 1:   This is characterized by a moisture content in the materials corresponding to a temperature of 20°C and the relative humidity of the surrounding air only exceeding 65% for a few weeks per year. In such moisture conditions most timber will attain an average moisture content not exceeding 12%.

Service Class 2:   This is characterized by a moisture content in the materials corresponding to a temperature of 20°C and the relative humidity of the surrounding air only exceeding 85% for a few weeks per year. In such moisture conditions most timber will attain an average moisture content not exceeding 20%.

Service Class 3: This is characterized, due to climatic conditions, by moisture contents higher than service class 2.

In Table 16 of the code values of K_2 are given varying from 0.6 in the case of compression parallel and perpendicular to the grain to 0.9 for shear parallel to the grain. These values reflect the non-uniform influence of moisture content on the mechanical properties of timber.

K_3: The grade stresses given in the code relate to the strength of timber subject to long-term permanent loads. Extensive research and testing has established that the short-term strength of timber is considerably higher than can be expected in the long-term. The value of K_3 used is therefore dependent on the duration of loading being considered, i.e. long-term, medium-term, short-term or very short-term as defined in Table 17 of the code, e.g.

Long-term: dead + permanent imposed
Medium-term: dead + snow, dead + temporary imposed
Short-term: dead + imposed + wind,   dead + imposed + snow + wind
Very short-term: dead + imposed + wind

Each of these examples is qualified in the notes relating to Table 17 in the code.

K_8: When designing structures in which four or more members, which are no greater than 610 mm apart, are connected by structural elements which provide lateral distribution of load (i.e. load-sharing) the grade stresses can be enhanced by multiplying by K_8 as indicated in Clause 2.9. Typical elements, which provide lateral distribution of load, are purlins, binders, boarding, battens, etc.

    In addition, the mean modulus of elasticity[*] can be used to calculate the displacements induced by both dead and imposed loads. *This does not apply to flooring systems which support mechanical plant and equipment or storage of systems which are subject to vibrations.*

    Provisions for built-up beams, trimmer joists, lintels and laminated beams are given separately in Clause 2.10.10, 2.10.11 and Section 3 of the code.
[*]*The use of K_8 does not extend to the calculation of factor K_{12} in which the E value is used when designing load-sharing columns.*

7.5 Flexural Members

Beams are the most commonly used structural elements, for example as floor joists, and as trimmer joists around openings, rafters, etc. The cross-section of a timber beam may be one of a number of frequently used sections such as those indicated in Figure 7.8.

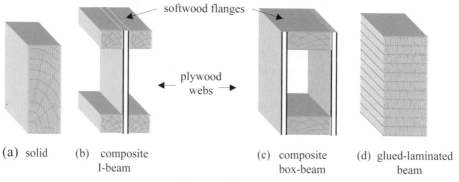

(a) solid   (b) composite   (c) composite   (d) glued-laminated
                        I-beam   box-beam   beam

Figure 7.8

The principal considerations in the design of all beams are:

- ◆ shear,
- ◆ bending,
- ◆ deflection,
- ◆ bearing, and
- ◆ lateral stability.

In the case of ply-web beams such as the composite I and box beams shown in Figures 7.8(b) and (c), the additional phenomenon of rolling shear, which is specific to plywood, must also be considered.

The size of timber beams may be governed by the requirements of:

- ◆ the elastic section modulus (Z), to limit the bending stresses and ensure that neither lateral torsional buckling of the compression flange nor fracture of the tension flange induces failure,
- ◆ the cross-section, to ensure that the vertical and/or horizontal shear stresses do not induce failure,
- ◆ the second moment of area, to limit the deflection induced by bending and/or shear action to acceptable limits.

Generally, the bearing area actually provided at the ends of a beam is much larger than is necessary to satisfy the permissible bearing stress requirement. Whilst lateral stability should be checked, it is frequently provided to the compression flange of a beam by nailing of floor boards, roof decking etc. (see Section 7.7.5). Similarly the proportions of solid timber beams are usually such that lateral instability is unlikely.

The detailed design of solid, ply-web and glued laminated beams is explained and illustrated in the examples given in Sections 7.7, 7.8 and 7.9 respectively. In each case the relevant modification factors (K values), their application and value/location are summarised.

7.6   Effective Span

Most timber beams are designed as simply supported and the effective span which should be used is defined in Clause 2.10.3 of *BS 5268-2:2002*, as illustrated in Figure 7.9.

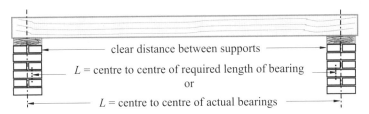

Figure 7.9

Since the required bearing length on most beams is relatively small when compared with the actual span it is common practice to assume an effective span equal to:

♦ the clear distance between the supports + 50 mm for solid beams, and
♦ the clear distance between the supports + 100 mm for ply-web beams.

In the case of long span beams (e.g. in excess of 10.0 m), or heavily loaded beams with consequently larger end reactions, the validity of this assumption should be checked.

7.7   Solid Rectangular Beams

The modification factors, which are pertinent when designing solid timber beams, are summarized in Table 7.1.

7.7.1   Shear (Clause 2.10.4)

The grade and hence permissible stresses given in the BS relate to the maximum shear stress parallel to the grain for a particular species or strength class. In solid beams of rectangular cross-section the maximum horizontal shear stress occurs at the level of the neutral axis, and is equal to $1.5 \times$ the average value:

$$\tau_{a,||} = \frac{1.5V}{A}$$

where:
$\tau_{a,||}$ maximum applied horizontal shear stress,
V maximum applied vertical shear force,
A cross-sectional area.
The magnitude of $\tau_{a,||}$ must not exceed $\tau_{adm,||}$ given by:

$$\tau_{adm,||} = \tau_{g,||} \times K_2 \times K_3 \times K_5 \times K_8$$

where:
$\tau_{g,||}$ grade stress parallel to the grain
K_2, K_3, K_5 and K_8 are modification factors used when appropriate (see Section 7.7.6 for notched beams).

$$\tau_{a,||} \leq \tau_{adm,||}$$

Factors	Application	Clause Number	Value/Location
K_2	Service class 3 sections (wet exposure): all stresses	2.6.2	Table 16
K_3	Load duration: all stresses (does not apply to E or G)	2.8	Table 17
K_4	Bearing stress	2.10.2	Table 18
K_5	Shear at notched ends: shear stress	2.10.4	Equations given
K_6	Cross-section shape: bending stress	2.10.5	1.18 for ○ 1.41 for ◆
K_7	Depth of section: bending stress	2.10.6	Equations given
K_8	Load-sharing: all stresses	2.9	1.1
K_9	Load-sharing: modulus of elasticity of trimmer joists and lintels	2.10.11	Table 20

Table 7.1   Modification Factors – solid beams

7.7.2   Bending (Clause 2.10)

As indicated in Section 1.9 of Chapter 1, the applied bending stress is determined using simple elastic bending theory:

$$\sigma_{m,a,||} = \frac{M_a}{Z}$$

where:

$\sigma_{m,a,||}$ maximum applied bending stress parallel to the grain,
M_a maximum applied bending moment,
Z elastic section modulus about the axis of bending (usually the x–x axis).

The permissible bending stress is given by:

$$\sigma_{m,adm,||} = \sigma_{m,g,||} \times K_2 \times K_3 \times K_6 \times K_7 \times K_8$$

where:

$\sigma_{m,g,||}$ grade bending stress parallel to the grain,

K_2 , K_3, K_6 , K_7 and K_8 are modification factors used when appropriate (see Table 7.1 and Section 7.4). **Note:** $K_6 = 1.0$ for rectangular cross-sections.

$$\sigma_{m,a,||} \leq \sigma_{m,adm,||}$$

7.7.3   Deflection (Clause 2.10.7)

In the absence of any special requirements for deflection in buildings, it is customary to adopt an arbitrary limiting value based on experience and good practice. The recommended value adopted in BS 5268 : Part 2 is $(0.003 \times \text{span})$ when fully loaded. In the case of domestic floor joists there is an additional recommendation of limiting deflection to less than or equal to 14 mm.

These limitations are intended to minimize the risk of cracking/damage to brittle finishes (e.g. plastered ceilings), unsightly sagging or undesirable vibration under dynamic loading. The magnitude of the actual deflection induced by the applied loading can be estimated using the coefficients given in Appendix 1 or the equivalent uniform load technique described in Section 1.7 of Chapter 1.

The calculated deflection for solid beams is usually based on the bending action of the beam ignoring the effects of shear deflection (this is considered when designing ply-web beams in Section 7.8).

$$\delta_{actual} \leq 0.003 \times span \quad and$$
$$\leq 14 \text{ mm for domestic floor joists}$$

7.7.4   Bearing (Clause 2.10.2)

The behaviour of timber under the action of concentrated loads, e.g. at positions of support, is complex and influenced by both the length and location of the bearing, as shown in Figures 7.10 (a) and (b). The grade stress for compression perpendicular to the grain is used to determine the permissible bearing stress.

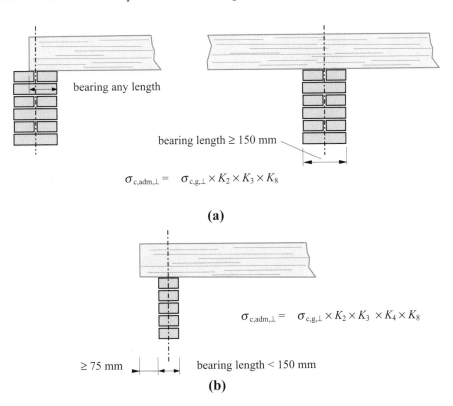

bearing any length

bearing length ≥ 150 mm

$$\sigma_{c,adm,\perp} = \sigma_{c,g,\perp} \times K_2 \times K_3 \times K_8$$

(a)

$$\sigma_{c,adm,\perp} = \sigma_{c,g,\perp} \times K_2 \times K_3 \times K_4 \times K_8$$

≥ 75 mm   bearing length < 150 mm

(b)

Figure 7.10

Note: In case (b), an additional modification factor K_4 for bearing stress has been included.

The actual bearing stress is determined from:

$$\sigma_{c,a,\perp} = \frac{P}{A_b}$$

where:
P   applied concentrated load,
A_b actual bearing area provided.

$$\sigma_{c,a,\perp} \leq \sigma_{c,adm,\perp}$$

The actual bearing area is the net area of the contact surface and allowance must be made for any reduction in the width of bearing due to wane, as shown in Figure 7.11. In timber engineering, pieces of wood with wane are frequently not used and consequently this can often be ignored.

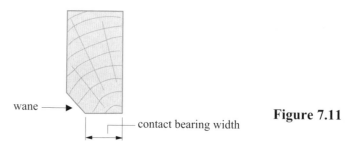

wane ⟶

contact bearing width

Figure 7.11

7.7.5   *Lateral Stability (Clause 2.10.8)*
A beam in which the depth and length are large in comparison to the width (i.e. a slender cross-section) may fail at a lower bending stress value due to lateral torsional buckling, as shown in Figure 7.12.

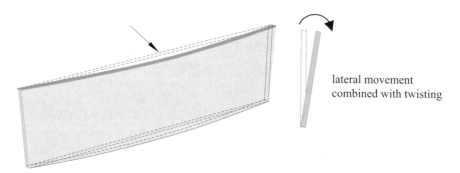

lateral movement
combined with twisting

Figure 7.12

The critical value of bending moment which induces this type of failure is dependent on several parameters, such as: the relative cross-section dimensions (i.e. aspect ratio), shape, modulus of elasticity (E), shear modulus (G), span, degree of lateral restraint to the compression flange, and the type of loading.

This problem is accommodated in *BS 5628-Part 2:2001* by using a simplified approach based on practical experience, in which limiting ratios of **maximum depth to maximum breadth** are given relating to differing restraint conditions. In 'Table 19' of BS 5268, values of limiting ratios are given varying from '2', when no restraint is provided to a beam, to a maximum of '7', for beams in which the top and bottom edges are fully laterally restrained.

Provision is made in the BS for designers to undertake a rigorous analysis, if desired, to determine the critical moment which will induce lateral torsional buckling of a beam. The vast majority of beams which are designed are of such proportions and have such restraint conditions that this analysis is unnecessary. The calculations relating to the critical moment are outwith the scope of this text: further information can be found in the *Timber Designers' Manual* (ref. 71), or *Step 2: Structural Timber Education Programme* (ref. 75).

7.7.6 *Notched Beams (Clause 2.10.9)*

It is often necessary to create notches or holes in beams to accommodate fixing details such as gutters, reduced fascias and connections with other members. In such circumstances high stress concentrations occur at the locations of the notches or holes. Whilst notches and holes should be kept to a minimum, when they are necessary cuts with square re-entrant corners should be avoided. This can be achieved by providing a fillet or taper or cutting the notch to a pre-drilled hole, typically of 8 mm diameter.

7.7.6.1 *Effect on Shear Strength (Clause 2.10.4)*

The projection of a notch beyond the inside edge of the bearing line at the point of support reduces the shear capacity of a beam. There are two situations to consider, as shown in Figure 7.13.

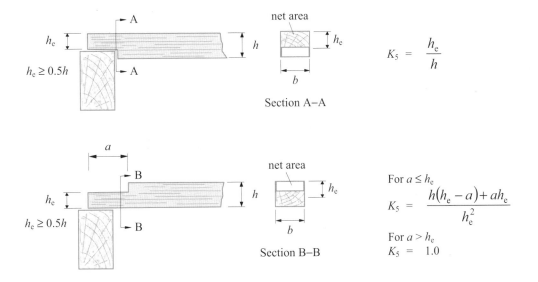

Figure 7.13

The reduction in shear capacity is reflected in the use of the net area and a reduction factor K_5, as indicated.

$$\text{shear capacity} \ = \ \tau_{\text{adm, } ||} \times h_e \times b$$

where:

$\tau_{\text{adm, } ||}$ permissible shear stress   (see Section 7.7.1),
h_e effective depth of the beam,
b breadth of the beam.

7.7.6.2 Effect on Bending Strength *(Clause 2.10.9)*

The calculated bending strength of notched beams is based on the net cross-section, as shown in Figure 7.14.

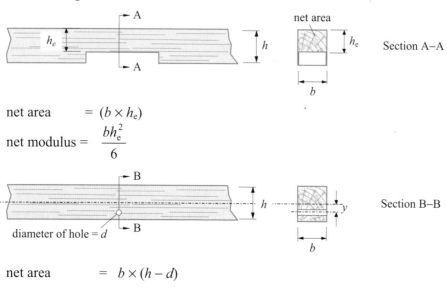

net area $= (b \times h_e)$

net modulus $= \dfrac{bh_e^2}{6}$

net area $= \ b \times (h - d)$

net modulus $= \dfrac{b}{6h}\left(h^3 - d^3 - 12dy^2\right)$

Figure 7.14

    When considering simply supported floor and roof joists which are not more than 250 mm deep and which satisfy the restrictions indicated in Figures 7.15(a) and (b), the effects of notches and holes can be neglected.

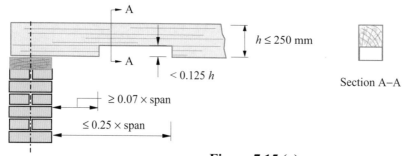

Figure 7.15 (a)

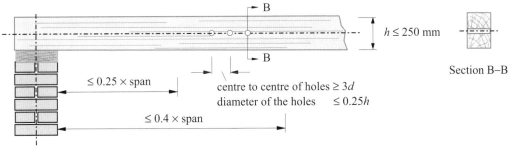

Figure 7.15 (b)

7.7.7 *Example 7.1: Suspended Timber Floor System*

Consider the design of a suspended timber floor system in a domestic building in which the joists at 500 mm centres are simply supported by timber beams on load-bearing brickwork, as shown in Figure 7.16 (a). The support beams are notched at the location of the wall, as shown in Figure 7.16(b).

♦ Determine a suitable section size for the tongue and groove floor boards.
♦ Determine a suitable section size for the joists.
♦ Check the suitability of the main support beams.

Design data:

Centre of timber joists	500 mm
Distance between the centre-lines of the brickwork wall	4.5 m
Strength class of timber for joists and tongue and groove boarding and beams	C22
Imposed loading (long-term)	3.0 kN/m²
Exposure condition	Service Class 1

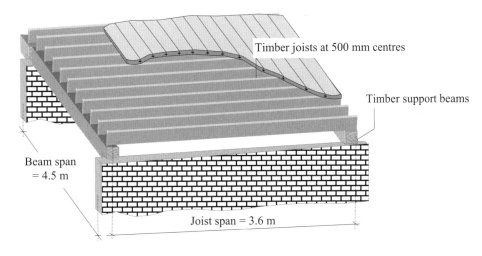

Figure 7.16 (a)

100 mm wide × 475 mm deep beam

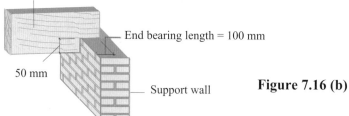

End bearing length = 100 mm

50 mm

Support wall

Figure 7.16 (b)

7.7.7.1 Solution to Example 7.1

Contract : **Solid Beams Job Ref. No. : Example 7.1**	**Calcs. by : W.McK.**
Part of Structure : **Suspended floor system**	**Checked by :**
Calc. Sheet No. : **1 of 8**	**Date :**

References	Calculations	Output
Clause 2.6 Table 8	Bending parallel to grain $\sigma_{m,g,\parallel} = 6.8\ \text{N/mm}^2$ Compression perpendicular to grain (assume wane is permitted) $\sigma_{c,g,\perp} = 1.7\ \text{N/mm}^2$ Shear parallel to grain $\tau_{m,g,\parallel} = 0.71\ \text{N/mm}^2$ Modulus of elasticity $E_{mean} = 9700\ \text{N/mm}^2$ Modulus of elasticity $E_{min} = 6500\ \text{N/mm}^2$ Average density $\rho_k = 410\ \text{kg/m}^3$ **Note:** A value of characteristic density is also given for use when designing joints. **Tongue and groove floor boarding:** Consider 1.0 m width of flooring and 19 mm thick boarding:  ◄────────── 1000 mm width ──────────► Self-weight $= \dfrac{(0.019 \times 410) \times 9.81}{10^3}$ = 0.08 kN/m² Imposed loading $=$ 3.0 kN/m² Total load $= (3.0 + 0.08)$ = 3.08 kN/m²	
Clause 2.10	**Bending:** *Permissible stress* $\sigma_{m,adm,\parallel} = \sigma_{m,g,\parallel} \times K_2 \times K_3 \times K_6 \times K_7 \times K_8$	
Clause 2.6.2 Clause 2.8 Table 17 (note a)	K_2 – wet exposure does not apply in this case K_3 – load duration for uniformly distributed imposed floor loads	$K_3 = 1.0$

References	Calculations	Output
Contract : Solid Beams Job Ref. No. : Example 7.1 Part of Structure : Suspended floor system Calc. Sheet No. : 2 of 8		Calcs. by : W.McK. Checked by : Date :

References	Calculations	Output
Clause 2.10.5 K_6 Clause 2.10.6 K_7 Clause 2.9	– shape factor does not apply in this case – depth of section < 72 mm K_8 – load sharing stresses tongue and groove boarding has adequate provision for lateral load distribution $\sigma_{m,adm,\|\|} = (6.8 \times 1.0 \times 1.17 \times 1.1) = 8.75 \text{ N/mm}^2$ span of boards = joist spacing = 500 mm Allowing for the continuity of the boards over the supports reduces the bending moment $M_{x,\text{ maximum}} \approx \dfrac{wL^2}{10} = \dfrac{3.08 \times 0.5^2}{10} = 0.077 \text{ kNm}$ Minimum section modulus required: $Z_{min} \geq \dfrac{\text{maximum bending moment}}{\text{permissible stress}} = \dfrac{0.077 \times 10^6}{8.75}$ $\approx 8.8 \times 10^3 \text{ mm}^3/\text{metre width}$ $Z = \dfrac{bh^2}{6} \quad \therefore h \geq \sqrt{\dfrac{6Z}{b}} \quad \therefore h \geq \sqrt{\dfrac{6 \times 8.8 \times 10^3}{1000}}$ $h \geq 7.3 \text{ mm};$ assume an additional 3 mm for wear $h \geq 7.3 + 3 = 10.3 \text{ mm}$ Try 16 mm thick	$K_7 = 1.17$ $K_8 = 1.1$
Clause 2.10.7 Clause 2.9	**Deflection:** Since load-sharing exists use E_{mean} to calculate deflection. Since the boards are continuous, assume the end span deflection (i.e. a propped cantilever) is approximately equal to 50% of a simply supported span: $\delta_{max} \approx (0.5 \times \delta_{\text{simply supported span}}) = \dfrac{1}{2}\left(\dfrac{5W_{total}L^3}{384EI} \right)$ $\delta_{max} = 0.5 \times \dfrac{5 \times 3.08 \times 0.5 \times 10^3 \times 500^3}{384 \times 9700 \times \left(\dfrac{1000 \times 16^3}{12} \right)} \approx 0.38 \text{ mm}$ $\delta_{\text{permissible}} \leq 0.003 \times 500 = 1.5 \text{ mm}$ $\delta_{max} \ll \delta_{\text{permissible}}$	**Adopt a minimum thickness of 16 mm for the tongue and groove boarding**

References	Calculations	Output
	Contract : Solid Beams Job Ref. No. : Example 7.1 **Part of Structure : Suspended floor system** **Calc. Sheet No. : 3 of 8**	**Calcs. by : W.McK.** **Checked by :** **Date :**

References	Calculations	Output
	Joists at 500 centres: Dead load due to self-weight of joist: assume $=$ 0.1 kN/m Dead load due to t & g boarding $= (0.08 \times 0.5) =$ 0.04 kN/m Imposed loading $= (3.0 \times 0.5) =$ 1.5 kN/m Total load $= (0.1 + 0.04 + 1.5) =$ 1.64 kN/m	
Clause 2.10	**Bending:** *Permissible stress* $\sigma_{m,adm,\|\|} = \sigma_{m,g,\|\|} \times K_2 \times K_3 \times K_6 \times K_7 \times K_8$	
Clause 2.6.2 Clause 2.8 Table 17 (note a)	K_2 – wet exposure $\quad$ does not apply in this case K_3 – load duration $\quad$ for uniformly distributed imposed (long-term) $\quad\quad\quad$ floor loads	$K_3 = 1.0$
Clause 2.10.5 Clause 2.10.6	K_6 – shape factor; $\quad$ does not apply in this case K_7 – depth of section; $\quad$ assume $h \leq 300$ mm This assumption should be checked at a later stage and modified if necessary.	$K_7 \approx 1.0$
Clause 2.9	K_8 – load sharing stresses;this applies since tongue and groove $\quad$ boarding provides adequate lateral distribution of loading $\quad\quad$ and the spacing of the joists ≤ 610 mm $\sigma_{m,adm,\|\|} = (6.8 \times 1.0 \times 1.0 \times 1.1) = 7.48$ N/mm^2 Span of joists: assume centre-to-centre of bearings $L = 3.6$ m $M_{x,\,maximum} = \dfrac{wL^2}{8} = \dfrac{1.64 \times 3.6^2}{8} = 2.66$ kNm $Z_{min} \geq \dfrac{2.66 \times 10^6}{7.48} = 0.36 \times 10^6$ mm^3	$K_8 = 1.1$
Clause 2.10.7	**Deflection:** $\delta_{permissible} \leq 0.003 \times \text{span} = (0.003 \times 3600) = 10.8$ mm $\quad\quad\quad\quad \leq 14.0$ mm	
Clause 2.9	Since load-sharing exists and assuming floor is *not* intended for mechanical plant and equipment, storage or subject to vibration (e.g. a gymnasium), use E_{mean} to calculate the deflection.	

Contract : Solid Beams Job Ref. No. : Example 7.1	Calcs. by : W.McK.
Part of Structure : Suspended floor system	Checked by :
Calc. Sheet No. : 4 of 8	Date :

References	Calculations	Output
	$$\delta_{maximum} \approx \frac{5W_{total}L^3}{384E_{mean}I}, \quad E_{mean} = 9700 \text{ N/mm}^2$$ $$= \frac{5 \times \left(3.6 \times 1.64 \times 10^3\right) \times 3600^3}{384 \times 9700 \times I_{xx}} = \frac{369.8 \times 10^6}{I_{xx}}$$ since $\delta_{maximum} \leq 10.8,$ $\quad \dfrac{369.8 \times 10^6}{I_{xx}} \leq 10.8$ $$I_{xx} \geq \frac{369.8 \times 10^6}{10.8} = 34.2 \times 10^6 \text{ mm}^4$$ Try a 75 mm × 200 mm joist: Cross-sectional area $A = (75 \times 200) = 15.0 \times 10^3 \text{ mm}^2$ Section modulus $\quad Z_{xx} = \dfrac{75 \times 200^2}{6} = 0.5 \times 10^6 \text{ mm}^3$ $> 0.36 \times 10^6 \text{ mm}^3$ Second moment of area $\quad I_{xx} = \dfrac{75 \times 200^3}{12} = 50 \times 10^6 \text{ mm}^4$ $> 34.2 \times 10^6 \text{ mm}^4$	
Clause 2.10.8 Table 19	$\dfrac{h}{b} = \dfrac{225}{75} = 3.0 \quad < \quad 5.0$	Lateral support is adequate
Clause 2.10.2	**Bearing:** 100 mm End reaction $\quad = (1.64 \times 3.6)/2.0 \quad = 2.95 \text{ kN}$ Bearing area $\quad = (100 \times 75) \quad = 7.5 \times 10^3 \text{ mm}^2$ $$\sigma_{c,a,\perp} = \frac{2.95 \times 10^3}{7.5 \times 10^3} = 0.39 \text{ N/mm}^2$$ $$\sigma_{c,adm,\perp} = \sigma_{c,g,\perp} \times K_2 \times K_3 \times K_4 \times K_8$$	

Contract : Solid Beams Job Ref. No. : Example 7.1 Part of Structure : Suspended floor system Calc. Sheet No. : 5 of 8	Calcs. by : W.McK. Checked by : Date :

References	Calculations	Output
Clause 2.6.2 Clause 2.8 Table 17 (note a) Clause 2.10.2	As before K_2 does not apply $K_3 = 1.0$ bearing length = 100 mm For any length at the end of a member K_4 does not apply.	
Clause 2.9	Since load-sharing applies $K_8 = 1.1$ $\sigma_{c,adm,\perp} = (1.7 \times 1.0 \times 1.1) = 1.87 \text{ N/mm}^2$ $\geq 0.39 \text{ N/mm}^2$	**Joist is adequate with respect to bearing**
Section 7.6.1 of this text	**Shear:** Maximum shear stress on rectangular section $\tau = \dfrac{1.5\,V}{A}$ where V = design value of shear force = 2.95 kN $\tau_{a,\parallel} = \dfrac{1.5 \times 2.95 \times 10^3}{15 \times 10^3} = 0.295 \text{ N/mm}^2$ $\tau_{adm,\parallel} = \tau_{g,\parallel} \times K_2 \times K_3 \times K_5 \times K_8$ As before K_2 does not apply and $K_3 = 1.0$ Since the end of the beam is not notched, K_5 does not apply	
Clause 2.9	Load-sharing applies $K_8 = 1.1$ $\tau_{adm,\parallel} = (0.71 \times 1.1) = 0.78 \text{ N/mm}^2$ $> 0.39 \text{ N/mm}^2$	**Joist is adequate with respect to shear** **Adopt 75 × 200 Grade C22 Timber Joists at 500 mm centres**
	Main beams: 100 mm × 475 mm deep Joists at 500 mm centres Span $L = 4.5$ m Cross-sectional area $A = (100 \times 475) = 47.5 \times 10^3 \text{ mm}^2$ Section modulus $Z_{xx} = \dfrac{100 \times 475^2}{6} = 3.76 \times 10^6 \text{ mm}^3$ Second moment of area $I_{xx} = \dfrac{100 \times 475^3}{12} = 893 \times 10^6 \text{ mm}^4$	

References	Calculations	Output
	Contract : Solid Beams Job Ref. No. : Example 7.1 **Part of Structure : Suspended floor system** **Calc. Sheet No. : 6 of 8**	**Calcs. by : W.McK.** **Checked by :** **Date :**

References	Calculations	Output
	Self-weight $= \dfrac{(0.0475 \times 410) \times 9.81}{10^3} = 0.19$ kN/m	
	Point loads at 500 mm centres $= 2.95$ kN	
	Vertical reaction at the supports $= (10 \times 2.95)/2 + (0.19 \times 2.25)$ $= 15.18$ kN (for bearing stress)	
	Shear force at the supports $= (8 \times 2.95)/2 + (0.19 \times 2.25)$ $= 12.23$ kN	
Clause 2.10	**Bending:** *Permissible stress* $\sigma_{m,adm,||} = \sigma_{m,g,||} \times K_2 \times K_3 \times K_6 \times K_7 \times K_8$	
Clause 2.6.2 Clause 2.8 Table 17 Clause 2.10.5	K_2 – wet exposure does not apply in this case K_3 – load duration loading can be assumed to be (long-term) uniformly distributed K_6 – shape factor does not apply in this case	$K_3 = 1.0$
Clause 2.10.6	K_7 – depth of section; $h > 300$ mm $K_7 = 0.81\dfrac{\left(h^2 + 92300\right)}{\left(h^2 + 56800\right)} = 0.81\dfrac{\left(475^2 + 92300\right)}{\left(475^2 + 56800\right)} = 0.91$	$K_7 = 0.91$
Clause 2.9	K_8 – load sharing does not apply in this case $\sigma_{m,adm,||} = (6.8 \times 1.0 \times 0.91) = 6.19$ N/mm^2 Self-weight bending moment $= \dfrac{wL^2}{8} = \dfrac{0.19 \times 4.5^2}{8}$ $= 0.48$ kNm $M_{x,\,maximum} = \{(12.23 \times 2.25) - [2.95 \times (0.25 + 0.75 + 1.25 + 1.75)]\}$ $+ 0.48$ $= 16.2$ kNm $Z_{min} \geq \dfrac{16.2 \times 10^6}{6.19} = 2.62 \times 10^6$ mm^3 $< 3.76 \times 10^6$ mm^3	**Section is adequate with respect to bending**
Clause 2.10.7	**Deflection:** $\delta_{permissible} \leq 0.003 \times \text{span} = (0.003 \times 4500) = 13.5$ mm ≤ 14.0 mm Since load-sharing does not exist use $E_{minimum}$ to calculate the deflection.	

Contract : Solid Beams Job Ref. No. : Example 7.1 Part of Structure : Suspended floor system Calc. Sheet No. : 7 of 8	Calcs. by : W.McK. Checked by : Date :

References	Calculations	Output
(See Chapter 1 Section 1.7)	Use the equivalent uniform distributed load method to estimate the deflection. $\delta_{maximum} \approx \dfrac{0.104 M_{maximum} L^2}{E_{minimum} I}$, $E_{min} = 6500$ N/mm^2 $= \dfrac{0.104 \times 16.2 \times 10^6 \times 4500^2}{6500 \times 893 \times 10^6}$ $= 5.9$ mm $< \quad 13.5$ mm	Section is adequate with respect to deflection
Clause 2.10.8 Table 19 Clause 2.10.2	$\dfrac{h}{b} = \dfrac{475}{100} = 4.75 \qquad < \quad 5.0$ **Bearing:** 100 mm wide $\times$ 475 mm deep beam 100 mm 50 mm Support wall	Lateral support is adequate
	End reaction $= 15.18$ kN Bearing area $= (100 \times 100) \quad = 10.0 \times 10^3$ mm^2 $\sigma_{c,a,\perp} = \dfrac{15.18 \times 10^3}{10.0 \times 10^3} = 1.52$ N/mm^2 $\sigma_{c,adm,\perp} = \sigma_{c,g,\perp} \times K_2 \times K_3 \times K_4 \times K_8$	
Clause 2.6.2 Clause 2.8 Table 17 (note a) Clause 2.10.2	As before K_2 does not apply $K_3 = 1.0$ bearing length $= 100$ mm	
Clause 2.9	For any length at the end of a member K_4 does not apply. Load-sharing does not apply (i.e. Ignore K_8)	

Contract : Solid Beams Job Ref. No. : Example 7.1 Part of Structure : Suspended floor system Calc. Sheet No. : 8 of 8	Calcs. by : W.McK. Checked by : Date :

References	Calculations	Output
	$\sigma_{c,adm,\perp}$ $= 1.7 \times 1.0$ $= 1.7 \text{ N/mm}^2$ $> 1.58 \text{ N/mm}^2$	**Adopt bearing length $\geq$ 100 mm**
	Shear:	
Section 7.6.1 of	Maximum shear stress on rectangular section $\quad \tau = \dfrac{1.5\,V}{A}$	
this text	where V = design value of shear force = 12.23 kN	
Clause 2.10.4	Effective depth $h_e = (475 - 50) = 425 \text{ mm}$	
	$\tau_{a,||} = \dfrac{1.5 \times 12.23 \times 10^3}{(425 \times 100)} = 0.43 \text{ N/mm}^2$	
	$\tau_{adm,||} = \tau_{g,||} \times K_2 \times K_3 \times K_5 \times K_8$	
	As before K_2 does not apply and $K_3 = 1.0$ Since the end of the beam is notched, K_5 does apply	**Beam is adequate with respect to shear**
	$K_5 = \dfrac{h_e}{h} = \dfrac{425}{475} = 0.89$	
	$\tau_{adm,||} = (0.71 \times 0.89) = 0.63 \text{ N/mm}^2$ $> 0.43 \text{ N/mm}^2$	**Adopt 100 × 475 Grade C22 Timber Beams with $\geq$ 100 mm length of bearing on the brick-wall support.**

7.8 Ply-web Beams

In situations where heavy loads and/or long spans require beams of strength and stiffness which are not available as solid sections, ply-web construction of I or Box-sections are frequently used (see Figure 7.8). The increased size of ply-web beams (e.g. 500 mm deep) and consequent strength/weight characteristics permit larger spacings (typically 1.2 m to 4.0 m) than solid beams, but can still be sufficiently close to enable the use of standard cladding and ceiling systems. In addition, they are frequently able to accommodate services, and insulation materials. The expansion of timber-framed housing in the U.K. has resulted in the use of smaller ply-web beams, typically 200 mm to 400 mm deep for floor joists and roof framing. A considerable saving in weight can be achieved over solid timber joists, and problems often associated with warping, cupping, bowing, twisting and splitting of sawn timber joists can be significantly reduced.

In most cases, since ply-web beams are hidden, the surface finishes including features such as nail heads, holes and glue marks need not be disguised. If desired, surface treatment can be carried out to enhance the appearance, but this will incur additional cost.

The construction of ply-web beams comprises four principal components:

♦ web,
♦ stiffeners,
♦ flanges,
♦ joints between flanges and the web.

The manufacture should comply with the requirements of *BS 6446:1984 Manufacture of glued structural components of timber and wood based panel products.*

The modification factors appropriate to plywood beams are given in Table 7.2.

Factors	Application	Clause Number	Value/Location
K_2	Service class 3 sections (wet exposure): all stresses. Does not apply to plywood	2.6.2	Table 16
K_3	Load duration: all stresses. Does not apply to plywood	2.8	Table 17
K_4	Bearing stress	2.10.2	Table 18
K_5	Shear at notched ends: shear stress	2.10.4	Equations given
K_8	Load-sharing: all stresses	2.9	1.1
K_{36}	Load-duration and service classes for plywood	4.5	Table 39
K_{37}	Stress concentration factor – rolling shear	4.7	0.5

Table 7.2   Modification factors for ply-web beams

7.8.1   Web

The primary purpose of the **web** is to resist stresses induced by shear forces. In the majority of cases the material used for the web is plywood; other wood-based panel materials such as particle board and fibreboard are also suitable. The most commonly used material is Finnish birch-faced plywood (also known as Combi), in which the outer veneers are always birch whilst the inner plies alternate between birch and softwood. This type of plywood is more readily available than, for example, Finnish all-birch plywood which may be more appropriate when the web is highly stressed. There are a number of alternative timbers such as Canadian Douglas fir, Swedish softwood, and American construction and industrial plywoods which can also be used. The appropriate grade stresses for the materials are given in Tables 40 to 56 of BS 5268-Part 2:2002.

The construction of the webs is normally carried out such that butt end joints do not occur at the mid-span location and full 2440 mm panels are used where possible. In Finnish birch-faced plywood the face grain is normally perpendicular to the span, whilst in cases where Douglas fir is used the face grain is normally parallel to the span.

There are two types of shear stress which must be resisted by the web of a ply-web beam:

- panel shear,   and
- rolling shear.

7.8.1.1 Panel Shear

The maximum horizontal shear stress induced in a beam subjected to bending and vertical shear forces occurs at the level of the neutral axis, as shown in Figure 7.17, and can be determined as shown in Chapter 1, Section 1.8, using the following equation:

$$\tau_{a,\parallel} = \frac{VA\bar{y}}{bI_{\text{N.A.}}}$$

where:
$\tau_{a,\parallel}$ maximum applied horizontal shear stress,
V maximum applied vertical shear force,
$A\bar{y}$ first moment of area of the material above the neutral axis,
$I_{N.A.}$ second moment of area of the cross-section,
b thickness of the web at the position of the section being considered.

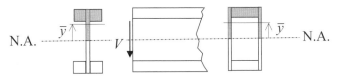

Figure 7.17

Note: If a section is designed on the basis of a transformed section in terms of the flange material as discussed in Chapter 1, Section 1.10, then $\tau_{a,\|}$ must be modified using the modular ratio to obtain equivalent plywood stresses, i.e.

$$\tau_{a,\| \text{ plywood}} = \tau_{a,\| \text{ transformed}} \times \alpha$$

where:

α the modular ratio equal to $\dfrac{E_w}{E_f}$

This is illustrated in Example 7.2.

The magnitude of $\tau_{a,\| \text{ plywood}}$ must not exceed $\tau_{adm,\| \text{ plywood}}$, given by:

$$\tau_{adm,\|} = \tau_{g,\|} \times K_8 \times K_{36}$$

where:

$\tau_{g,\|}$ grade stress of plywood given in Tables 34 to 37 of the code
K_8 modification factor to allow for load sharing,
K_{36} modification factor to allow for differing load-duration and/or different service classes.

7.8.1.2 Rolling Shear

The physical construction of plywood, in which alternate veneers have grain directions which are mutually perpendicular, enables a mode of failure called *rolling shear* to occur (see Figure 7.18); there is a tendency for the material fibres to roll across each other creating a horizontal shear failure plane. This phenomenon can occur at locations where plywood is joined to other members/materials, either at the interface with the plywood or between adjacent veneers of the plywood. In ply-web beams the rolling shear must be checked at the connection of the web to the flanges and sufficient thickness (*T*) of flange must be available to transfer the horizontal shear force at this location.

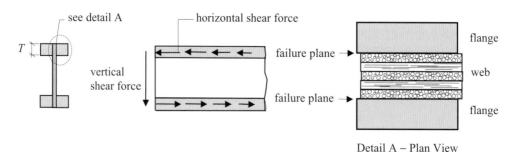

Figure 7.18

The magnitude of the rolling shear stress can be determined using the same equation as for panel shear with the critical section considered being section *x–x* at the interface between the web and the flange, as shown in Figure 7.19:

$$\tau_{a,||} = \frac{VA\bar{y}}{bI_{N.A.}}$$

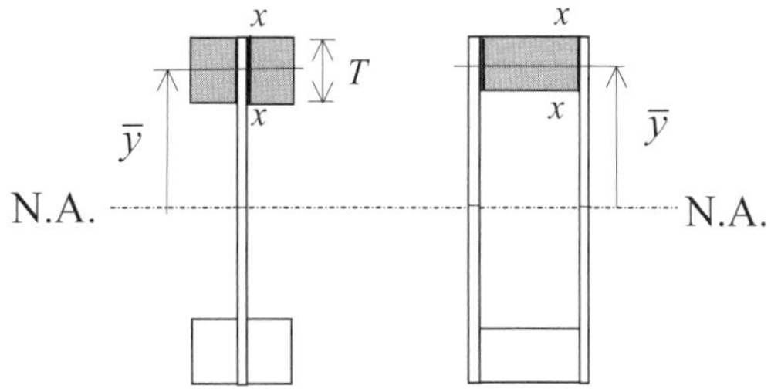

Figure 7.19

where V, $A\bar{y}$ and $I_{N.A.}$ are as before. The value of b is equal to (flange thickness × no. of interfaces between the flange and the web). In the beam shown above $b = 2T$.

As for panel shear, when using a transformed section the calculated value of τ should be modified using the modular ratio to obtain equivalent plywood stresses.

$$\tau_{a,rolling} = \tau_{a,rolling,transformed} \times \alpha$$
$$\tau_{adm,rolling} = \tau_g \times K_8 \times K_{36} \times K_{37}$$

As before the actual stress must be less than or equal to the permissible value:

$$\tau_{a,rolling} \leq \tau_{adm,rolling}$$

where: α, K_8 and K_{36} are as previously defined, and K_{37} is a stress concentration factor given in Clause 4.7 and is equal to 0.5.

7.8.1.3 Web Stiffeners

Where webs are slender and at locations such as supports and points of application of concentrated load, there is the possibility of failure caused by buckling of the web. BS 5268-2:2002 does not give any guidance on the design of either non-loadbearing (intermediate) or loadbearing web stiffeners other than to indicate that they are required where appropriate. Most proprietary suppliers of ply-web beams advise web stiffener details based on the results of full scale tests of their product. Design methods are illustrated in various publications, notably by the Council of Forest Industries of British Columbia (COFI) publication *Fir Plywood Web Beam Design* (ref. 73) and *Timber Designers' Manual* (ref. 71). Reference should be made to these publications for further information regarding stiffeners.

7.8.2 Flanges

The primary purpose of the **flanges** is to resist tensile and compressive stresses induced by bending effects and/or axial loads. Their construction is normally carried out using continuous or finger-jointed structural timber such as European whitewood, Douglas fir, larch or redwoods, the first of these being the most commonly used. Alternatively plywood or glued-laminated components can be used.

7.8.2.1 Bending

There are a number of techniques which can be used to determine the bending moment capacity of a ply-web beam. The method adopted in this text assumes that the full cross-section, i.e. the web and the flanges, contribute to the bending resistance. Analysis to determine bending stresses is carried out assuming a transformed section, as indicated in Chapter 1, Section 1.10, where:

$$\sigma_{m,a} \;=\; \frac{bending\ moment}{transformed\ elastic\ section\ modulus} = \frac{M}{Z}$$

When using this method it is necessary to ensure that:

♦ the calculated stresses in the extreme fibres of the flanges do not exceed the permissible bending stress parallel to the grain as indicated in Section 7.7.2 for solid beams:
$$\sigma_{m,a} \;\leq\; \sigma_{m,adm,||}$$
where:
$$\sigma_{m,adm,||} \;=\; \sigma_{m,g,||} \times K_2 \times K_3 \times K_8$$

♦ the tension and compressive stresses induced by the bending moment in the plywood web do not exceed the appropriate values for the face grain orientation as indicated in Tables 40 to 56 and Clause 4.7 of the Code:

i.e.
$$\sigma_{t,a} \;=\; \frac{M}{Z} \times \alpha \;\leq\; \sigma_{t,adm} \quad \text{and}$$

$$\sigma_{c,a} \;=\; \frac{M}{Z} \times \alpha \;\leq\; \sigma_{c,adm}$$

where:
$\sigma_{t,adm} = \sigma_{t,g} \times K_8 \times K_{36}$,
$\sigma_{c,adm} = \sigma_{c,g} \times K_8 \times K_{36}$,
Z is the value of section modulus for the transformed section,
α is the modular ratio of the web and flange materials.

The above equations relating to the plywood apply to face grain in either the parallel or the perpendicular directions.

In many cases load-sharing will not occur and the K_8 value of 1.1 will not apply. This is a conservative interpretation of Clause 2.9 relating to load-sharing systems. Some designers interpret this Clause more widely and include the K_8 value when lateral load distribution does exist and the number of individual pieces of timber within a cross-section is greater than four.

Note: The *bending* stresses given in Tables 40 to 56 apply to stresses induced when bending is about either axis *in the plane* of the board, as indicated in Figure 7.20.

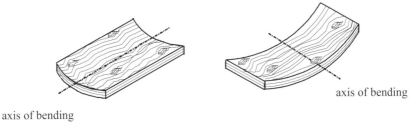

axis of bending

axis of bending

Figure 7.20   In-plane axes of bending

In the case of ply-web beams, the axis of bending is perpendicular to the plane of the board as shown in Figure 7.21, and consequently the *tensile* and *compressive* stresses are used.

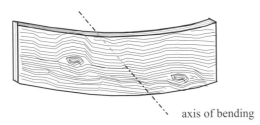

axis of bending

Figure 7.21   Out-of-plane axis of bending

In most cases the governing criteria will relate to the flange material.

7.8.2.2   *Deflection (Clauses 2.10.7 and 4.7)*

Traditionally when calculating the deflection of beams, only the component due to the bending action is considered. This is due to the fact that in other materials, for example steel, the shear modulus is considerably higher as a percentage of the true elastic modulus than is the case in timber. A consequence of this is that when considering the deflection of timber beams the effect of shearing deformation may be significant. In Clause 4.7 the code indicates that shear deflection should be taken into account when designing ply-web beams such that:

$$\delta_{\text{actual}} \approx (\delta_{\text{bending}} + \delta_{\text{shear}}) \le 0.003 \times \text{span}$$

7.8.2.2.1 Bending deflection (δ_{bending})

The calculation of the bending component of the overall deflection is based on elastic deformation using the standard deflection formulae given in Appendix 3, or the equivalent UDL technique discussed in Section 1.7 of Chapter 1. The bending rigidity (*EI*) of the section must be determined using the modular ratio to account for the different elastic moduli of the flange and web materials. This can be achieved using the following expression:

$$EI = (EI)_{\text{flange}} + (EI)_{\text{web transformed}}$$
$$EI = E_{\text{flange}}(I_{\text{flange}} + \alpha I_{\text{web}})$$

474   *Design of Structural Elements*

7.8.2.2.2 Shear deflection (δ_{shear})

A number of factors, such as the cross-sectional dimensions, the shear modulus of the web (G_{web}), and the position and intensity of the loads, influence the shear deflection of a beam. A number of complex analytical expressions have been developed to determine the magnitude of the shear deflection – see Roark (ref. 45), COFI (ref. 73) – but a simplified equation may be used to give an acceptable, approximate value:

$$\delta_{shear} \approx \frac{M}{G_{web} A_w}$$

where:
M bending moment at mid-span,
A_w area of the web,
G_{web} modulus of rigidity of the web given in Tables 40 to 56 of the Code.

7.8.3   *Lateral stability (Clauses 2.10.8 and 2.10.10)*

The lateral stability of a built-up beam can be assessed by calculation assuming the compression flange to be a column subject to sideways buckling between points of lateral restraint. An alternative is also given in the Code in which differing lateral restraint conditions are required, depending on the ratio of the second moment of area in the x–x direction to that in the y–y direction. These requirements are given in paragraphs a) to f) of Clause 2.10.10 of the Code and are summarised here in Table 7.3. This is similar to those given in Table 19 for solid and laminated beams which are dependent on the depth/breadth ratio.

Clause in BS 5268	Ratio	Requirement
2.10.10 (a)	$\dfrac{I_{xx}}{I_{yy}} \leq 5$	no lateral support required
2.10.10 (b)	$5 < \dfrac{I_{xx}}{I_{yy}} \leq 10$	ends of beam to be held in position at the bottom flange at supports
2.10.10 (c)	$10 < \dfrac{I_{xx}}{I_{yy}} \leq 20$	beam to be held in line at the ends
2.10.10 (d)	$20 < \dfrac{I_{xx}}{I_{yy}} \leq 30$	one edge to be held in line
2.10.10 (e)	$30 < \dfrac{I_{xx}}{I_{yy}} \leq 40$	beam to be restrained by bridging or other bracing at intervals of not more than 2.4 m
2.10.10 (f)	$40 < \dfrac{I_{xx}}{I_{yy}}$	compression flange should be fully restrained

Table 7.3   Lateral Restraint Requirements

7.8.4 Example 7.2: Ply-web Beam Design

A proposed temporary platform is to be constructed using decking supported by the timber ply-web I-beam sections indicated in Figure 7.22. Using the design data given, check the suitability of the proposed beam section with respect to:

 i) bending,
 ii) panel shear,
 iii) rolling shear   and
 iv) deflection.

Design data:

Characteristic dead load (including self-weight)	0.3 kN/m
Characteristic imposed load (medium term)	3.0 kN/m
Span of beam	5.4 m
Exposure condition	Service Class 2

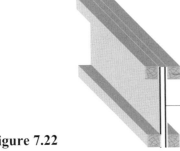

Flanges: 69 mm wide × 44 mm deep
British grown Douglas fir Grade SS

Web: 400 mm deep × 21 mm thick
Finnish combi-plywood: thick veneer, with 13 plies

Figure 7.22

7.8.4.1 Solution to Example 7.2

<table>
<tr>
<td colspan="2">Contract : Ply-web Beam Job Ref. No : Example 7.2
Part of Structure : Temporary Platform
Calc. Sheet No. : 1 of 4</td>
<td>Calcs. by : W.McK.
Checked by :
Date :</td>
</tr>
<tr>
<td>References</td>
<td>Calculations</td>
<td>Output</td>
</tr>
<tr>
<td>

Table 17
Table 2

Table 8</td>
<td>

Medium-term design load = (0.3 + 3.0)   = 3.3 kN/m
Design shear force   = (3.3 × 5.4)/2   = 8.91 kN

Design bending moment $= \dfrac{wL^2}{8} = \dfrac{3.3 \times 5.4^2}{8} = 12.02$ kNm

Load duration factor $K_3 = 1.25$ for medium-term
Grade stresses:
British grown Douglas fir Grade SS – Strength Class C18

$\sigma_{m,g,||} = 5.8$ N/mm^2 $\qquad \sigma_{t,g,||} = 3.5$ N/mm^2

$\sigma_{c,g,||} = 7.1$ N/mm^2 $\qquad \sigma_{c,g,\perp} = 2.2$ N/mm^2 (no wane)

$\tau_{g,||} = 0.67$ N/mm^2

</td>
<td></td>
</tr>
</table>

Contract : Ply-web Beam Job Ref. No : Example 7.2	Calcs. by : W.McK.
Part of Structure : Temporary Platform	Checked by :
Calc. Sheet No. : 2 of 4	Date :

References	Calculations	Output

References column / **Calculations** / **Output**

E_{mean} = 9,100 N/mm^2 $\qquad$ E_{min} = 6000 N/mm^2

Table 53

Finnish combi-plywood (21 mm thick, 13 plies)

$\sigma_{t,g,\perp}$ = 14.82 N/mm^2 $\qquad$ $\sigma_{c,g,\perp}$ = 7.73 N/mm^2

$\sigma_{b,g}$ = 3.0 N/mm^2 $\qquad$ τ_p = 4.43 N/mm^2

$\tau_{roll,g}$ = 0.79 N/mm^2 $\qquad$ $E_{b,\perp}$ = 3200 N/mm^2

$E_{c,\perp}$ = $E_{t,\perp}$ = 3800 N/mm^2 $\qquad$ $G_\perp$ = 285 N/mm^2

Table 8
Table 53

E_{flange} = 9,100 N/mm^2 $\qquad$ E_{web} = 3200 N/mm^2

Using a transformed section, the equivalent thickness of the transformed web is given by

$t_{transfromed}$ = t_{actual} × modular ratio (α)

$$= 21 \times \frac{E_{web}}{E_{flange}} \quad \therefore t^* = 21 \times \frac{3200}{9100} = 7.38 \text{ mm}$$

Transformed section properties:

69 mm 69 mm

44 mm

7.38 mm

312 mm

44 mm

A^* = (400 × 145.38) − (138 × 312)

$\quad$ = 15.1 × 10^3 mm^2

$$I^*_{xx} = \frac{145.38 \times 400^3}{12} - \frac{138 \times 312^3}{12}$$

$\quad$ = 426.1 × 10^6 mm^4

$$I^*_{yy} = \frac{2 \times 44 \times 145.38^3}{12} + \frac{312 \times 7.38^3}{12}$$

$\quad$ = 22.54 × 10^6 mm^4

$$Z^*_{xx} = \frac{426.1 \times 10^6}{200}$$

$\quad$ = 2.13 × 10^6 mm^3

$$\frac{I_{xx}}{I_{yy}} = \frac{426.1}{22.54} = 18.9$$

Clause 2.10.10

The beams should be held in line at the ends. The decking which is attached to the beams should achieve this.

Bending:

$$\sigma_{m,a,\parallel} = \frac{12.02 \times 10^6}{2.13 \times 10^6} = 5.64 \text{ N/mm}^2$$

$\sigma_{m,adm,\parallel}$ = $\sigma_{m,g,\parallel} \times K_2 \times K_3 \times K_8$

Contract : Ply-web Beam Job Ref. No : Example 7.2 Part of Structure : Temporary Platform Calc. Sheet No. : 3 of 4	Calcs. by : W.McK. Checked by : Date :

References	Calculations	Output
	Assume no load-sharing $\therefore K_8 = 1.0$ Medium term loading $\therefore K_2 = 1.25$ $\sigma_{m,adm,\|\|} = 5.8 \times 1.25 = 7.25$ N/mm > 5.64 N/mm^2	**Flanges are adequate in bending**
Clause 4.7	*Compression flange of plywood:* $\sigma_{c,a} = \sigma_{t,a} = \dfrac{M}{Z} \times \alpha = 5.64 \times \dfrac{3200}{9100} = 1.98$ N/mm^2 $\sigma_{c,adm,\perp} = \sigma_{c,g,\perp} \times K_8 \times K_{36}$ as before $K_8 = 1.0$	
Clause 4.5 Table 39	$K_{36} = 1.33 \therefore \sigma_{c,adm,\perp} = 7.73 \times 1.33 = 10.28$ N/mm^2 $\gg 1.98$ N/mm^2	
Clause 4.7	*Tension flange of plywood:* $\sigma_{t,a} = 1.98$ N/mm^2 $\sigma_{t,adm,\perp} = \sigma_{t,g,\perp} \times K_8 \times K_{36}$ $\sigma_{t,adm,\perp} = 14.82 \times 1.33 = 19.71$ N/mm^2 $\gg 1.98$ N/mm^2	**Plywood is adequate in bending**
See Section 7.8.1.1	*Panel shear:* The panel shear stress is given by: $$\tau_{p,a} = \dfrac{VA\bar{y}}{bI_{NA}}$$ V = maximum shear force = 8.91 kN $A\bar{y}$ = 1st moment of area above the neutral axis $= (138 \times 44 \times 178) + (200 \times 7.38 \times 100)$ $= 1.228 \times 10^6$ mm^3 $\tau_{p,a,transf} = \dfrac{8.91 \times 10^3 \times 1.228 \times 10^6}{7.38 \times 426.1 \times 10^6} = 3.48$ N/mm^2 $\tau_{p,a,plywood} = \tau_{p,a,transf} \times \alpha = 3.48 \times \dfrac{3200}{9100} = 1.22$ N/mm^2 $\tau_{p,adm,plywood} = \tau_{p,g} \times K_{36} = 4.43 \times 1.33 = 5.89$ N/mm^2 > 1.22 N/mm^2 *Rolling shear:* N.A.	**Plywood is adequate in panel shear**

Contract : Ply-web Beam Job Ref. No : Example 7.2 Part of Structure : Temporary Platform Calc. Sheet No. : 4 of 4	Calcs. by : W.McK. Checked by : Date :

References	Calculations	Output
	The rolling shear stress is given by $\tau_{p,a} = \dfrac{VA\bar{y}}{bI_{NA}}$	
	V = maximum shear force = 8.91 kN	
	$A\bar{y}$ = 1st moment of area of the flanges above the neutral axis	
	$\quad = (2 \times 69 \times 44 \times 178)$	
	$\quad = 1.081 \times 10^6 \text{ mm}^3$	
	$\tau_{p,a,transf} \quad = \dfrac{8.91 \times 10^3 \times 1.081 \times 10^6}{2 \times 44 \times 426.1 \times 10^6} = 0.26 \text{ N/mm}^2$	
	$\tau_{roll,a,plywood} \quad = \tau_{roll,a,transf} \times \alpha = 0.26 \times \dfrac{3200}{9100} = 0.09 \text{ N/mm}^2$	
Clause 4.7 Table 39	$\tau_{roll,adm,plywood} = \tau_{roll,g} \times K_{36} \times K_{37} = (0.79 \times 1.33 \times 0.5)$ (Note the 0.5 factor in Clause 4.7) $= 0.53 \text{ N/mm}^2$ $\gg 0.09 \text{ N/mm}^2$ (**Note:** The K_3 load duration factor does not apply to plywood.)	**Plywood is adequate in rolling shear**
	Deflection:	
Clause 2.10.7	$\delta_{adm} = 0.003 \times 5400 = 16.2 \text{ mm}$	
Clause 4.7	$\delta_{actual} \approx \delta_{bending} + \delta_{shear}$	
Section 7.8.2.2.1	$\delta_{bending} \approx \dfrac{5WL^3}{384EI}$	
	$EI = E_{flange}(I_{flange} + \alpha I_{web})$	
	$I_{flange} = 4\left\{\dfrac{69 \times 44^3}{12} + \left(69 \times 44 \times 178^2\right)\right\} = 386.73 \times 10^6 \text{ mm}^4$	
	$\alpha I_{web} = \left\{\dfrac{3200}{9100} \times \dfrac{21 \times 400^3}{12}\right\} \qquad = 39.38 \times 10^6 \text{ mm}^4$	
	$EI \quad = 9100 \, (386.73 + 39.38) \times 10^6 = 3877 \times 10^9 \text{ Nmm}^2$	
	$\delta_{bending} = \dfrac{5 \times (3.3 \times 5.4) \times 10^3 \times 5400^3}{384 \times 3877 \times 10^9} = 9.42 \text{ mm}$	
Section 7.8.2.2.2	$\delta_{shear} \approx \dfrac{M}{G_{web}A_w} \; ; \quad G = 285 \text{ N/mm}^2$	
Table 43	$\delta_{shear} \approx \dfrac{12.02 \times 10^6}{285 \times (21 \times 400)} = 5.02 \text{ mm}$	
	$\delta_{actual} \approx (9.42 + 5.02) = 14.44 \text{ mm} \quad < 19.2 \text{ mm}$	
	Note: The shear deflection makes a significant contribution to the total deflection of the beam.	**Ply-web beam is adequate in deflection**

7.9   Glued Laminated Beams (Glulam)

One of the fastest growing and most successful structural material industries in the UK is that related to the use of **glulam**. Traditionally until the 1970's and with the exception of specialist uses such as in aircraft and marine components, glulam was purpose-made for a limited number of types of structure such as swimming pools, churches or footbridges. The availability of standard glulam components such as straight, curved or cambered members has been made possible by the introduction of improved, modern high-volume production plants. This has resulted in an ever-expanding range of uses, e.g. timber lintel beams in domestic housing, large portal frames in conference and leisure centres, and even structures such as the 162 m diameter dome of the Tacoma Sports and Convention Centre in Washington State, USA. Glulam has many advantages, such as:

♦ members can be straight or curved in profile and uniform or variable in cross-section,
♦ the strength:weight ratio is high, enabling the dead load due to the superstructure to be kept to a minimum with a consequent saving in foundation construction,
♦ factory production allows a high standard of material quality to be achieved,
♦ timbers of large cross-section have a superior performance in fire than alternatives such as concrete and steel,
♦ when treated with appropriate preservatives, softwood laminated timber is very durable in wet exposure situations; in addition it also has a high resistance to chemical attack and aggressive/polluted environments,
♦ there is no need for expansion joints because of the low coefficient of thermal expansion,
♦ defects such as knots are restricted to the thickness of one lamination and their effects on overall structural behaviour are significantly reduced,
♦ large spans are possible within the constraints of transportation to site.

An indication of the range of structures for which glulam is suitable is given in Figures 7.23 (a) and (b).

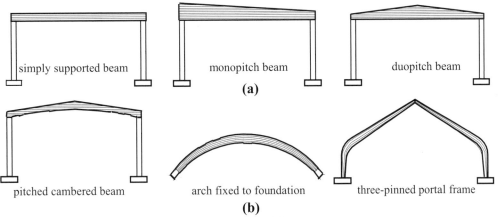

simply supported beam   monopitch beam   duopitch beam

(a)

pitched cambered beam   arch fixed to foundation   three-pinned portal frame

(b)

Figure 7.23

7.9.1   *Manufacture of Glulam*

Glulam members are fabricated by gluing together accurately prepared timber laminations in which the grain is in the longitudinal direction. *BS 5628:Part 2*, Clause 3.2 states that *'Members may be horizontally laminated from two strength classes, provided that the strength classes are not more than three classes apart in table 7 (i.e. C24 and C16 may be horizontally laminated, but C24 and C14 may not), and the members are fabricated so that not less than 25% of the depth of both the top and the bottom of the members is of the superior strength class.'*

This is illustrated in the examples shown in Figure 7.24:

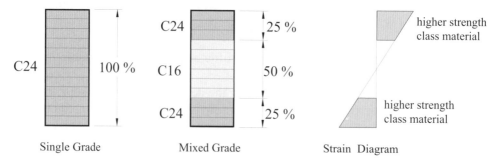

Figure 7.24

When using mixed grade laminations it is evident that the higher strength laminations are used to resist the larger bending stresses in the outer fibres of the cross-section.

7.9.1.1 *Finger-Joints*

The length of glulam members frequently exceeds the length of commercially available solid timber, resulting in the need to finger-joint together individual planks to make laminations of the required length.

A typical finger-joint is shown in Figure 7.25:

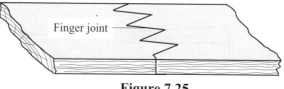

Figure 7.25

The finger-joint is cut into the end-grain of each plank and the planks are pressed together after applying adhesive.

In Clause 6.10.2 of the BS it is stated that Finger joints should be manufactured in accordance with BS EN 385 1995.

7.9.1.2 *Adhesives*

The most widely used adhesives are the phenol-resorcinol-formaldehyde (PRF) type. The adhesive, which is a liquid, is used with a 'hardener' containing formaldehyde and various inert fillers. Pure resorcinal, which is expensive, is usually replaced with alternative, cheaper chemicals (phenols), which also react with formaldehyde. The chemical bond formed during this reaction is a carbon-to-carbon type which is very strong and durable.

The resulting adhesives are fully water-, boil- and weather-resistant. In addition, PRF adhesives do not decompose or ignite in a fire, and delamination will not occur under normal circumstances. The chemical 'pH' value is approximately 7 (i.e. neutral), and hence alkaline or acidic damage/corrosion will not occur in either the timber or any metal components.

7.9.1.3 Surface finishes
Finished glulam beams are normally planed on their sides to remove any residual adhesive which has been squeezed out of their joints during manufacture. Modern planing machinery usually provides a finish of sufficient quality that subsequent sanding is unnecessary.

7.9.1.4 Preservation treatment
The preservative treatment to glulam members tends to be applied to the finished product rather than individual laminations prior to gluing. The preservatives discussed in Chapter 1, Section 1.5 apply to glulam members and should comply with the requirements of *BS 5268-5* as indicated in Clause 7.7.1 of *BS 5268-2*.

7.9.2 Vertically Laminated Beams
Vertically laminated beams in which the applied load is parallel to the laminate joints usually occur in elements such as laminated columns subjected to both axial and flexural loads.

7.9.3 Horizontally Laminated Beams
Horizontally laminated beams are beams in which the laminations are parallel to the neutral plane. The loading is applied in a direction perpendicular to the plane of the laminations. The admissible stresses are determined by modifying the grade stresses. There are two possibilities to consider:

♦ where the strength class is specified, and
♦ where a specific grade and species are specified.

Strength Class specified
The grade stresses are determined from Table 10 and modified using factors K_{15} to K_{20} given in Table 24 for individual strength classes and number of laminations. This is in addition to the requirements of Clauses 2.6.2, 2.8, 2.9 and 2.10.

Grade and Species specified
The grade stresses are determined from Tables 11 to 15 for the appropriate grade and species. These values are modified using the factors K_{15} to K_{20} in Table 24 assuming a strength class appropriate to the grade and species as given in Tables 2 to 7. This is in addition to the requirements of Clauses 2.62, 2.8, 2.9 and 2.10.

In addition to the modification factors indicated in Table 7.3, where members are laminated from two strength classes the grade stresses for the superior lamination are used, and those relating to bending, tension and compression parallel to the grain should be multiplied by 0.95 as indicated in Clause 3.2.

The design of glulam beams follows the same pattern as solid rectangular beams, where shear, bending, bearing and deflection are the main criteria to be considered. The following stress criteria should be satisfied:

Shear stress:

$$\tau_{a,||} \leq \tau_{g,||} \times K_2 \times K_3 \times K_5 \times K_8 \times K_{19}$$

where:

$\tau_{a,||}$ = maximum applied horizontal shear stress
$\tau_{g,||}$ = grade stress parallel to the grain
K_2, K_3, K_5, K_8, K_{19} are modification factors used where appropriate

Bending stress:

$$\sigma_{m,a,||} \leq \sigma_{m,g,||} \times K_2 \times K_3 \times K_6 \times K_7 \times K_8 \times K_{15} \times K_{33}$$

where:

$\sigma_{m,a,||}$ = maximum applied bending stress
$\sigma_{m,g,||}$ = grade stress parallel to the grain
K_2, K_3, K_6, K_7, K_8, K_{15}, K_{33} are modification factors used where appropriate

Note: In the case of curved glulam members, equations are given to determine the maximum radial and bending stresses.

Radial stress:

$$\sigma_{r,a} \leq \sigma_{t,g,\perp} \times K_2 \times K_3 \times K_8$$

where:

$\sigma_{r,a}$ =   maximum radial stress, calculated as given in Clauses 3.5.3.3 and 3.5.4.2
$\sigma_{t,g,\perp}$ =   grade tension stress perpendicular to the grain, calculated in accordance with Clause 2.7, i.e. equal to $0.33 \times \tau_{g,||}$
K_2, K_3, K_8, are modification factors used where appropriate

Bearing stress:

$$\sigma_{c,a,\perp} \leq \sigma_{c,g,\perp} \times K_2 \times K_3 \times K_8 \times K_{18}$$

where:

$\sigma_{c,a,\perp}$ = maximum applied bearing stress
$\sigma_{c,g,\perp}$ = grade stress perpendicular to the grain
K_2, K_3, K_8, K_{18}, are modification factors used where appropriate.

In addition to satisfying stress criteria, the deflection due to bending and shear should be considered, i.e.

$$\delta_{actual} \leq 0.003 \times span$$

The manufacture of glulam beams lends itself to introducing an initial pre-camber to offset the deflection due to dead and permanent loads. Where advantage is taken of this, the calculated deflection due to live or intermittent imposed load only should not exceed $0.003 \times$ span.

7.9.3.1 Modification factors

The modification factors which apply to glued laminated beams are summarised in Table 7.4.

Factors	Application	Clause Number	Value/Location
K_2	service class 3 sections (wet exposure): all stresses. – does not apply to plywood	2.6.2	Table 16
K_3	load duration: all stresses	2.8	Table 17
K_4	bearing stress	2.10.2	Table 18
K_5	shear at notched ends: shear stress	2.10.4	Equations given
K_6	bending stresses: form factor	2.10.5	Equations given
K_7	bending stresses: depth factor	2.10.6	Equations given
K_8	load-sharing all stresses	2.9	1.1
K_{15}	bending parallel to grain	3.2	Table 24
K_{16}	tension parallel to grain	3.2	Table 24
K_{17}	compression parallel to grain	3.2	Table 24
K_{18}	compression perpendicular to grain	3.2	Table 24
K_{19}	shear parallel to grain	3.2	Table 24
K_{20}	modulus of elasticity	3.2	Table 24
K_{27}	bending, tension, & shear parallel to grain	3.3	Table 25
K_{28}	modulus of elasticity & compression parallel to grain	3.3	Table 25
K_{29}	compression perpendicular to grain	3.3	Table 25
K_{30}	end joints in horizontal glulam	3.4	Table 26
K_{31}	end joints in horizontal glulam	3.4	Table 26
K_{32}	end joints in horizontal glulam	3.4	Table 26
K_{33}	curved glulam members: bending, tension & compression all parallel to grain	3.5.3.2	Equations given
K_{34}	curved glulam members: calculated bending stresses	3.5.3.2	Equations given
K_{35}	calculated radial stresses in pitched cambered glulam beams	3.5.4.2	Equations given

Note: K_{15} to K_{20} apply to horizontally laminated beams
K_{27} to K_{29} apply to vertically laminated beams
K_{30} to K_{32} apply to individually designed glued end joints in horizontally glued laminated members

Table 7.4   Modification factors for glulam beams

7.9.4   Example 7.3: Glulam Roof Beam Design

The glulam timber beam shown in Figure 7.26 comprises 8/45 mm finished laminations (machined from 50 mm thick timber) each of which is imported redwood, visually classified as SS grade. It is intended to use the beam internally in a building which is continually heated. Assuming an 8.0 m simply supported span, determine the maximum long-term uniformly distributed load (including self-weight) which can be carried considering:

- ◆ bending,
- ◆ shear and
- ◆ deflection.

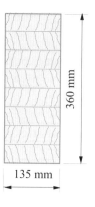

360 mm

135 mm

Figure 7.26

7.9.4.1 *Solution to Example 7.3*

References	Calculations	Output
Contract : Glulam Beam Job Ref. No : Example 7.3 **Part of Structure : Simply Supported Span** **Calc. Sheet No. : 1 of 3**		**Calcs. by : W.McK.** **Checked by :** **Date :**

References	Calculations	Output
	The beam is to be manufactured from imported redwood of SS grade with 8 horizontal laminations each 45 mm thick. (Assume that wane is not permitted)	
Table 2	Imported redwood of grade SS is classified as strength class C24	
Table 8	**Grade stresses:** $\sigma_{m,g,\|\|}$ = 7.5 N/mm^2 $\sigma_{t,g,\|\|}$ = 4.5 N/mm^2 $\sigma_{c,g,\|\|}$ = 7.9 N/mm^2 $\sigma_{c,g,\perp}$ = 2.4 N/mm^2 $\tau_{g,\|\|}$ = 0.71 N/mm^2 E_{mean} = 10,800 N/mm^2 E_{min} = 7,200 N/mm^2 ρ_{mean} = 420 kg/m^3 **Section properties:** Area = 135×360 = 48.6×10^3 mm^2 I_{xx} = $\dfrac{135 \times 360^3}{12}$ = 524.9×10^6 mm^4 Z_{xx} = $\dfrac{135 \times 360^2}{6}$ = 2.92×10^6 mm^3	

References	Calculations	Output

Contract : Glulam Beam Job Ref. No : Example 7.3
Part of Structure :   Simply Supported Span
Calc. Sheet No. : 2 of 3

Calcs. by : W.McK.
Checked by :
Date :

Table 19

$$D/B = \frac{360}{135} = 2.7$$

$$2 < D/B < 3$$

Consider Bending:

Maximum applied bending moment $\quad = \dfrac{w \times 8^2}{8}$

$$= 8.0w \text{ kNm}$$

where w is the total distributed load including self-weight

$$\sigma_{m,a,||} = \frac{8.0w \times 10^6}{2.92 \times 10^6} = 2.74w \text{ N/mm}^2$$

$$\sigma_{m,adm,||} = \sigma_{m,g,||} \times K_2 \times K_3 \times K_6 \times K_7 \times K_8 \times K_{15}$$

Clause 1.6.4
Table 16
Table 17
Clause 2.10.5

Moisture content $\leq 12°C$ $\therefore$ Service class 1
K_2 does not apply (i.e. =1.0)
Long-term loading $K_3 = 1.0$
Form factor $\qquad K_6 = 1.0$

Clause 2.10.6

Depth factor $\quad K_7 = 0.81 \times \dfrac{\left(h^2 + 92,300\right)}{\left(h^2 + 56,800\right)}$

$$K_7 = 0.81 \times \frac{\left(360^2 + 92,300\right)}{\left(360^2 + 56,800\right)} = 0.964$$

Clause 2.9

Assume no load-sharing, so K_8 does not apply

Table 24

Strength class C24, number of laminations = 8

$$K_{15} = 1.39 + (1.43 - 1.39) \times \frac{1}{3} = 1.40$$

Note: Interpolation is permitted for an intermediate number of laminations

$$\sigma_{m,adm,||} = 7.5 \times 1.0 \times \times 1.0 \times 0.964 \times 1.40$$
$$= 10.12 \text{ N/mm}^2$$

Clause 3.2

Note: If more than one strength class is used within a beam
cross-section then $\sigma_{m,g,||}$, $\sigma_{t,g,||}$, $\sigma_{c,g,||}$ value should be
multiplied by 0.95.

Maximum applied stress $\quad = 2.74w \qquad \leq 10.12 \text{ N/mm}^2$

Maximum value of $w \qquad = (10.12/2.74) \quad = \textbf{3.69 kN/m}$

Output:

The ends should be held in position to satisfy the lateral torsional stability requirements.

References	Calculations	Output
	Contract : Glulam Beam Job Ref. No : Example 7.3 **Part of Structure : Simply Supported Span** **Calc. Sheet No. : 3 of 3**	**Calcs. by : W.McK.** **Checked by :** **Date :**

References	Calculations	Output
	Consider Shear: Maximum applied shear force $= (w \times 8.0)/2 = 4.0w$ kN $\tau_{a,\|\|} = 1.5 \times$ average $\tau_{a,\|\|}$ $\qquad = \dfrac{1.5 \times 4.0w \times 10^3}{48.6 \times 10^3} = 0.123w$ N/mm^2 $\tau_{adm,\|\|} = \tau_{g,\|\|} \times K_2 \times K_3 \times K_5 \times K_8 \times K_{19}$ As before $K_2 = 1.0,\qquad K_3 = 1.0$ Assume K_5 for notched ends and K_8 do not apply in this case.	
Table 24	$K_{19} = 2.34$ $\tau_{adm,\|\|} = 0.71 \times 1.0 \times 2.34 = 1.66$ N/mm^2 $\tau_{a.\|\|}$ Maximum applied stress $= 0.123w \leq 1.66$ N/mm^2 Maximum value of $w = (1.66/1.23) = $ **13.5 kN/m**	
Table 24	**Consider Deflection:** $\delta_{actual} = \delta_{bending} + \delta_{shear}$ The mean modulus should be multiplied by $K_{20} = 1.07$ $E = E_{mean} \times K_{20} = 10800 \times 1.07 = 11556$ N/mm^2 $\delta_{actual} = \left[\dfrac{5wL^4}{384EI} + \dfrac{M}{AG}\right]$ where $M = 8.0w$ kNm The value of the shear modulus is normally assumed to be equal to $E/16$ $G = \dfrac{11556}{16} = 722$ N/mm^2 $\delta_{actual} = \dfrac{5 \times w \times (8000)^4}{384 \times 11556 \times 524.9 \times 10^6} + \dfrac{8.0w \times 10^6}{48.6 \times 10^3 \times 722}$ $\qquad = 8.79w + 0.23w = 9.02w$ mm	
Clause 2.10.7	$\delta_{adm} = 0.003 \times$ span $= 0.003 \times 8000 = 24.0$ mm Maximum actual deflection $= 9.02w \leq 24.0$ N/mm^2 Maximum value of $w = (24.0/9.02) = $ **2.66 kN/m**	**The maximum udl which can be carried is governed by the deflection criteria, i.e. 2.66 kN/m**
Clause 3.5.2	A camber can be provided to offset the dead load and/or permanent load deflection. This would permit an increased load to be carried since in this case w is limited by deflection.	

7.10 Axially Loaded Members

7.10.1 Introduction

The design of axially loaded members considers any member where the applied loading induces either axial tension or axial compression. Members subject to axial forces frequently occur in bracing systems, roof trusses or lattice girders.

Frequently, in structural frames sections are subjected to combined axial and bending effects, which may be caused by eccentric connections, wind loading or rigid-frame action. The design of such members is discussed and illustrated in Section 7.11. In *BS 5268-2:2002*, the design of compression members is considered in Clause 2.11 and the design of tension members in Clause 2.12. In both cases the service class, duration of loading and load sharing modification factors apply. In addition, for compression members allowance must be made for the slenderness of the section (see section 7.10.4.1), and in the case of tension members a width factor is also considered. The relevant modification factors which apply to axially loaded members are summarized in Table 7.5.

Factors	Application	Clause Number	Value/Location
K_2	service class 3 sections (wet exposure): all stresses	2.6.2	Table 16
K_3	load duration: all stresses	2.8	Table 17
K_8	load-sharing: all stresses	2.9	1.1
K_9	modulus of elasticity	2.10.11 / 2.11.5	Table 20
K_{12}	slenderness of columns	2.11.5	Table 22
K_{13}	spaced columns: effective length	2.11.10	Table 23
K_{14}	width factor: tensile stresses	2.12.2	Equations given
K_{28}	minimum modulus of elasticity: **compression parallel** to the grain in **vertically laminated** glulam members	2.11.5	Table 25
K_{29}	minimum modulus of elasticity: **compression perpendicular** to the grain in **vertically laminated** glulam members	2.11.5	Table 25

Table 7.5 Modification factors – axially loaded members

7.10.2 Design of Tension Members (Clause 2.12)

The design of tension members is based on the effective area of the cross-section allowing for a reduction due to notches, bolts, dowels, screw holes or any mechanical fastener inserted in the member. Since the grade stresses given in Table 10 of the code apply to material assigned to a strength class and having a width of 300 mm (the *greater* transverse dimension *h*), a modification factor K_{14} must be used. The value to be used for K_{14} is given in Clause 2.12.2 as:

Solid timber members   where $h \leq 72$ mm $K_4 = 1.17$
Solid and glulam members   where $h \geq 72$ mm $K_4 = (300/h)^{0.11}$

The tension capacity of a section is given by:

$$P_t = [\sigma_{t,g,||} \times K_2 \times K_3 \times K_8 \times K_{14}] \times [\text{effective cross-sectional area}]$$

where:
$\sigma_{t,g,||}$ is the grade stress
K_2, K_3, K_8, and K_{14} are modification factors
effective cross-sectional area is the cross-sectional area after allowances have been made as described above.

7.10.3 Example 7.4: Truss Tie Member

A close coupled roof construction comprises a tie (175 mm × 63 mm) connected to two rafters by M8 steel bolts, as shown in Figure 7.27. Check the suitability of the tie to transmit a long-term tensile load of 3.0 kN and a medium-term tensile load of 5.0 kN.

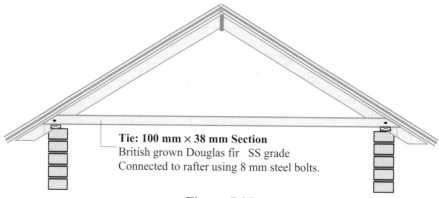

Tie: 100 mm × 38 mm Section
British grown Douglas fir   SS grade
Connected to rafter using 8 mm steel bolts.

Figure 7.27

7.10.3.1 Solution to Example 7.4

Contract : Truss Job Ref. No : Example 7.4	Calcs. by : W.McK.
Part of Structure : Tie	Checked by :
Calc. Sheet No. : 1 of 2	Date :

References	Calculations	Output
Table 1 Table 2 Table 8	Assume Service Class 2 British grown Douglas fir grade SS is classified as C18 **Grade stresses:** Tension parallel to grain $\sigma_{t,g,||} = 3.5$ N/mm^2 $\sigma_{t,adm,||} = \sigma_{t,g,||} \times K_2 \times K_3 \times K_8 \times K_{14}$	

Contract : Truss Job Ref. No : Example 7.4	Calcs. by : W.McK.
Part of Structure : Tie	Checked by :
Calc. Sheet No. : 2 of 2	Date :

References	Calculations	Output
Clause 2.6.2	K_2 – wet exposure does not apply in this case	
	$\dfrac{\text{medium term loading}}{1.25} = \dfrac{5.0}{1.25} = 4.0\text{ kN} > \text{long-term load}$	
	$\therefore$ Check for medium-term loading	
	$\sigma_{t,a,\parallel} = \dfrac{F_t}{\text{effective area}}$	
	effective area = gross area – projected area of bolt hole	
	$= (100 \times 38) - (10 \times 38) = 3420\text{ mm}^2$	
Clause 6.6.1	**Note:** The bolt hole is 2 mm larger than the bolt diameter.	
	$\sigma_{t,a,\parallel} = \dfrac{5.0 \times 10^3}{3420} = 1.46\text{ N/mm}^2$	
Clause 2.8	K_3 – load duration Medium-term $K_3 = 1.25$	
Clause 2.9	K_8 – load sharing stresses does not apply in this case	
Clause 2.12.2	K_{14} – width of section $h > 72$ mm	
	$K_{14} = \left(\dfrac{300}{100}\right)^{0.11} = 1.128$	
	$\sigma_{t,adm,\parallel} = 3.5 \times 1.25 \times 1.128 = 4.9\text{ N/mm}^2$	**Tie is adequate**
	$\qquad\qquad\qquad > 1.46\text{ N/mm}^2$	
	There is considerable reserve of strength in the tie.	

7.10.4 Design of Compression Members (Clause 2.11)

The design of compression members is more complex than that of tension members, and encompasses the design of structural elements referred to as columns, stanchions or struts. The term 'struts' is usually used when referring to members in lattice/truss frameworks, whilst the other two generally refer to vertical or inclined members supporting floors and/or roofs in structural frames.

As with tension members, in many cases they are subjected to both axial and bending effects. This Section deals with those members which are subjected to concentric axial loading.

The dominant mode of failure to be considered when designing struts is axial buckling.

Buckling failure is caused by secondary bending effects induced by factors such as these:

♦ The inherent eccentricity of applied loads due to asymmetric connection details.
♦ Imperfections present in the cross-section and/or profile of a member throughout its length. The allowable deviation from straightness when using either visual or machine grading (i.e. bow not greater than 20 mm lateral displacement over a length of 2.0 m) is inadequate when considering compression members. More severe restrictions, such as $L/300$ for structural timber and $L/500$ for glulam sections, should be considered.
♦ Non-uniformity of material properties throughout a member.

The effects of these characteristics are to introduce initial curvature, secondary bending and consequently premature failure by buckling before the stress in the material reaches the failure value. The stress at which failure will occur is influenced by several variables, e.g.

♦ the cross-sectional shape of the member,
♦ the slenderness of the member,
♦ the permissible stress of the material.

A practical and realistic assessment of the critical slenderness of a strut is the most important criterion in determining the compressive strength.

7.10.4.1 Slenderness

Slenderness is evaluated using **either**:

(i) $\lambda = \dfrac{L_e}{i}$

where:
λ is the slenderness ratio,
L_e is the effective length with respect to the axis of buckling being considered,
i is the radius of gyration with respect to the axis of buckling being considered.

or for a rectangular section the alternative given in (ii) can be used:

(ii) $\lambda = \dfrac{L_e}{b}$

where:
λ and L_e are as above and
b is the lesser transverse dimension of the member being considered.
Note: For a rectangular section of breadth b and depth d

$$i = \sqrt{\frac{I}{A}} = \sqrt{\frac{db^3}{12}} = \frac{b}{2\sqrt{3}}$$

e.g.   when $\lambda = 20$ $\therefore \dfrac{L_e}{i} = 20$ $\therefore \dfrac{L_e \times 2\sqrt{3}}{b} = 20 \therefore \dfrac{L_e}{b} = 5.77$

In BS 5268-2 values for both $\dfrac{L_e}{i}$ and $\dfrac{L_e}{b}$ are given in Table 22. In most cases rectangular

sections are used in timber design, therefore the slenderness can be evaluated using $\dfrac{L_e}{b}$.

Limiting values of slenderness are given in the code to reduce the possibility of premature failure in long struts, and to ensure an acceptable degree of robustness in a member. These limits, which are given in Clause 2.11.4, are shown in Table 7.6.

7.10.4.1.1 Effective Length

The effective length is considered to be the actual length of the member between points of restraint multiplied by a coefficient to allow for effects such as stiffening due to end connections of the frame of which the member is a part. Appropriate values for the coefficients are given in Table 21 of the code and illustrated in Figure 7.28. Alternatively the effective length can be considered to be '*the distance between adjacent points of zero bending between which the member is in single curvature*' as indicated in Clause 2.11.3.

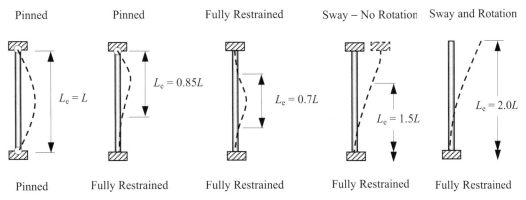

Figure 7.28 Effective Lengths (L_e)

Type of member	$\lambda_{maximum}$	L_e/b
(a) any compression member carrying dead and imposed loads other than loads resulting from wind; (b) any compression member, however loaded, which by its deformation will adversely affect the stress in another member carrying dead and imposed loads other than wind	180	52
(c) any member normally subject to tension or combined tension and bending arising from dead and imposed loads, but subject to a reversal of axial stress solely from the effect of wind; (d) any compression member carrying self-weight and wind loads only (e.g. wind bracing)	250	72.3

Table 7.6

When considering members in triangulated frameworks (other than trussed rafters which are considered in BS 5268:Part 3:1985), with **continuous compression members**, e.g. the top chord of a lattice girder, the effective length may be taken as:

- between $0.85 \times$ distance *l* between the node points   and
           $1.0 \times$ distance *l* between the node points
  for buckling in the plane of the framework *and*
- between $0.85 \times$ actual distance between effective lateral restraints   and
           $1.0 \times$ actual distance between effective lateral restraints
  for buckling perpendicular to the plane of the framework.

The value used depends on the degree of fixity and the distribution of load between the node points.

When considering **non-continuous compression members** such as web elements within a framework (i.e. uprights and diagonals in compression), the effective length is influenced by the type of end connection. A single bolt or connector which permits rotation at the ends of the member is assumed to be a pin, and the effective length should be taken as $1.0 \times$ length between the bolts or connectors. Where glued gusset plates are used, then partial restraint exists. The effective length both in the plane of the frame and perpendicular to the plane of the frame should be taken as $0.9 \times$ the actual distance between the points of intersection of the lines passing through the centroids of the members connected. These requirements are given in Clause 2.11.11 of the Code.

All other cases should be assessed and the appropriate factor from Table 21 used to determine the effective length.

The compression capacity of a section is given by:

$$P_c = [\sigma_{c,g,||} \times K_2 \times K_3 \times K_8 \times K_{12}] \times [K_{17} \text{ or } K_{18}] \times [\text{effective cross-sectional area}]$$

where:

$\sigma_{c,g,||}$ is the grade stress

K_2, K_3, K_8, K_{12}, K_{17} and K_{18} are modification factors

effective cross-sectional area is the cross-sectional area after allowances have been made for open holes, notches etc. – if a bolt is inserted into a hole with only nominal clearance, then no deduction is required.

The K_{12} factor applies when the slenderness of a member is greater than 5. The values of K_{12} are given in Table 22 and require the value of slenderness (L_e/i or L_e/b), and $E/\sigma_{c,||}$.

The value of E to be used is the minimum value. If appropriate, the E_{min} value can be modified by K_9 from Table 20 for members comprising two or more pieces connected together in parallel and acting together to support the loads. In the case of vertically laminated members the E_{min} value can be multiplied by K_{28} from Table 25. For horizontally laminated members the mean modulus modified by K_{20} from Table 24 should be used. The value of $\sigma_{c,g,||}$ should be modified using only K_2 for moisture content and K_3 for load duration. The size factor for compression members does not apply, since the grade compression stresses given in Tables 10 to 15 apply to all solid timber members and laminations graded in accordance with BS 4978, BS 5756 or BS EN 519.

A consequence of the inclusion of the K_3 factor in determining K_{12} is that all relevant loading conditions should be considered separately, since K_{12} and subsequently $\sigma_{c,adm,||}$ is different in each case.

7.10.5 Example 7.5: Concentrically Loaded Column

A symmetrically loaded internal column is required to support four beams as shown in Figure 7.29. The top can be considered to be held in position but not in direction, and the bottom to be restrained in both direction and position about both axes.

A lateral restraint is provided 2.2 m from the base as indicated. Assuming Service Class 2, check the suitability of a 75 mm × 150 mm, strength class C18 section to resist a long-term load of 12.0 kN and a medium-term load of 20.0 kN.

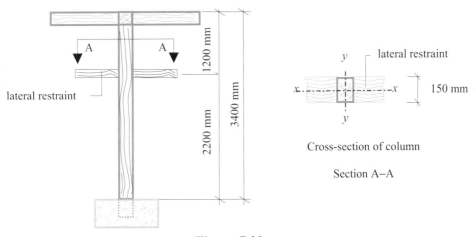

Figure 7.29

7.10.5.1 Solution to Example 7.5

Contract : Struts Job Ref. No. : Example 7.5	Calcs. by : W.McK.
Part of Structure : Axially Loaded Column	Checked by :
Calc. Sheet No. : 1 of 3	Date :

References	Calculations	Output
Table 8	Strength Class C18 $\sigma_{c,g,||} = 7.1 \text{ N/mm}^2$, $E_{min} = 6000 \text{ N/mm}^2$ K_2 (wet exposure), and K_8 (load sharing), do not apply	
Table 17	long-term loading $K_3 = 1.0$ medium-term loading $K_3 = 1.25$	

References	Calculations	Output

Contract : Struts   Job Ref. No. : Example 7.5
Part of Structure :   Axially Loaded Column
Calc. Sheet No. : 2 of 3

Calcs. by : W.McK.
Checked by :
Date :

Clause 2.11.3
Table 21

Clause 2.11.4

Slenderness:

x–x axis L_{exx} = (0.85×3400) =   2890 mm

$\dfrac{L_{exx}}{h}$ = $\dfrac{2890}{150}$ =   22 < 52   (i.e. $L_e/i < 180$)

y–y axis L_{eyy} ≥ (1.0×1200) =   1200 mm
            or   ≥ (0.85×2200) =   1870 mm   *critical*

Clause 2.11.4

$\dfrac{L_{eyy}}{b}$ = $\dfrac{1870}{75}$ =24.93 <   52 (i.e. $L_e/i < 180$)

critical slenderness = 24.93

Clause 2.11.5
Table 22

Long-term:

$E/\sigma_{c,||}$ is determined using E_{min} and $\sigma_{c,||} = \sigma_{c,g,||} \times K_3$

E_{min} =   6000 N/mm^2

$\sigma_{c,||}$ = 7.1×1.0 =   7.1 N/mm^2

$E/\sigma_{c,||} =$ $\dfrac{6000}{7.1}$ =   845; $\lambda = 24.93$

The value of K_{12} can be determined by interpolation
from Table 22:

Extract from Table 22

L_e/b $E/\sigma_{c,||}$	23.1	26.0
800	0.497	0.43
900	0.522	0.456

A conservative, approximate answer can be found
using the lowest value corresponding with the lowest $E/\sigma_{c,||}$
and highest λ value, i.e. in this case 0.43. If the
section is adequate with this value then it will also be
adequate with a more accurate determination of K_{12}

$\sigma_{c,adm,||}$ = $\sigma_{c,g,||} \times K_3 \times K_{12}$

$\sigma_{c,adm,||}$ ≈ $7.1 \times 1.0 \times 0.43 =$ 3.05 N/mm^2

$\sigma_{c,a,||}$ = $\dfrac{F_c}{\text{Effective Area}}$

Contract : Struts Job Ref. No. : Example 7.5 Part of Structure : Axially Loaded Column Calc. Sheet No. : 3 of 3	Calcs. by : W.McK. Checked by : Date :

References	Calculations	Output
	Effective area = gross area = (150×75) = 11250 mm^2 $\sigma_{c,a,\|\|}$ = $\dfrac{12 \times 10^3}{11250}$ = 1.1 N/mm^2 < 3.05 N/mm^2	**Long-term capacity is adequate**
Table 22	**Medium-term:** $E/\sigma_{c,\|\|}$ is determined using E_{min} and $\sigma_{c,\|\|} = \sigma_{c,g,\|\|} \times K_3$ E_{min} = 6000 N/mm^2 $\sigma_{c,\|\|}$ = 7.1×1.25 = 8.88 N/mm^2 $E/\sigma_{c,\|\|}$ = $\dfrac{6000}{8.88}$ = 675.7; $\lambda = 24.93$ The value of K_{12} can be determined by interpolation from Table 22 or as above using an approximate value. If necessary a more precise value can be determined.	

Extract from Table 19

L_e/b $E/\sigma_{c,\|\|}$	23.1	26.0
600	0.43	0.363
700	0.467	0.399

References	Calculations	Output
	$K_{12} \approx$ 0.363 $\sigma_{c,adm,\|\|}$ = $\sigma_{c,g,\|\|} \times K_3 \times K_{12}$ $\sigma_{c,adm,\|\|}$ $\approx$ $7.1 \times 1.25 \times 0.363 = 3.22$ N/mm^2 $\sigma_{c,a,\|\|}$ = $\dfrac{F_c}{\text{Effective Area}}$ Effective area = gross area = (150×75) = 11250 mm^2 $\sigma_{c,a,\|\|}$ = $\dfrac{20 \times 10^3}{11250}$ = 1.78 N/mm^2 < 3.22 N/mm^2	**Medium-term capacity is adequate**

7.11 Members Subject to Combined Axial and Flexural Loads

7.11.1 Introduction

Many structural elements such as beams and truss members are subjected to a single dominant effect, i.e. applied bending or axial stresses. Secondary effects which also occur are often insignificant and can be neglected. There are, however, numerous elements in which the combined effects of bending and axial stresses must be considered, e.g. rigid-jointed frames such as portals, chords in lattice girders with applied loading between the node points, and columns with eccentrically applied loading.

The relevant modification factors which apply to members subject to combined loading are those which apply to the individual types, i.e. axially loaded and flexural members, and are summarized in Table 7.7.

Factors	Application	Clause Number	Value/Location
K_2	service class 3 sections (wet exposure): all stresses	2.6.2	Table 16
K_3	load duration: all stresses	2.8	Table 17
K_8	load-sharing: all stresses	2.9	1.1
K_9	modulus of elasticity	$2.10.11 - 2.11.5$	Table 20
K_{12}	slenderness of columns	2.11.5	Table 22
K_{13}	spaced columns: effective length	2.11.10	Table 23
K_{14}	width factor: tensile stresses	2.12.2	Equations given
$K_{15} - K_{20}$	horizontally laminated members	3.2	Table 24
$K_{28} - K_{29}$	vertically laminated members	3.3	Table 25
K_{33} & K_{34}	curved laminated members	3.5.3	Equations given
K_{35}	pitched cambered laminated members	3.5.4	Equations given

Table 7.7   Modification factors

7.11.2 Combined Bending and Axial Tension (Clause 2.12.3)

The interaction equation for members subject to combined bending and axial tension is given in Clause 2.12.3 as:

$$\frac{\sigma_{m,a,||}}{\sigma_{m,adm,||}} + \frac{\sigma_{t,a,||}}{\sigma_{t,adm,||}} \leq 1.0$$

where:

$\sigma_{m,a,||}$ is the applied bending stress,

$\sigma_{m,adm,||}$ is the permissible bending stress,

$\sigma_{t,a,||}$ is the applied tension stress,

$\sigma_{t,adm,||}$ is the permissible tension stress.

The values of $\sigma_{m,adm,||}$ and $\sigma_{t,adm,||}$ are evaluated using the modification factors where appropriate. The value of $\sigma_{m,a,||}$ should represent the maximum value of the bending stress in the cross-section. In some instances bending will occur about both the x–x and y–y axes simultaneously, in which case the applied bending stress is equal to:

$$\sigma_{m,a,||} = (\sigma_{m,ax,||} + \sigma_{m,ay,||})$$

where:

$\sigma_{m,ax,||}$ is the maximum bending stress due to bending about the x–x axis of the section,

$\sigma_{m,ay,||}$ is the maximum bending stress due to bending about the y–y axis of the section.

The expression given in the code is a linear interaction formula and could have been written using separate terms for bending about the x–x and y–y axes as:

$$\frac{\sigma_{m,ax,||}}{\sigma_{m,adm,||}} + \frac{\sigma_{m,ay,||}}{\sigma_{m,adm,||}} + \frac{\sigma_{t,a,||}}{\sigma_{t,adm,||}} \le 1.0$$

7.11.3 Example 7.6: Ceiling Tie

Using the data given, check the suitability of the tie in the truss indicated in Example 7.4 when it is subjected to a combined medium-term bending moment of 0.5 kNm and a medium-term axial load of 5.0 kN.

Data:
Tie: 100 mm × 38 mm Section
Strength Class TR26
The tie is connected to the rafters using 8 mm steel bolts

7.11.3.1 Solution to Example 7.6

Contract : Truss Tie Job Ref. No : Example 7.6 Part of Structure : Combined Tension and Bending Calc. Sheet No. : 1 of 2		Calcs. by : W.McK. Checked by : Date :
References	**Calculations**	**Output**
Table 1	Assume Service Class 2 Strength Class T26	
Table 9	**Grade stresses:** Tension parallel to grain $\sigma_{t,g,||} = $ 6.0 N/mm^2 $\sigma_{t,adm,||} = \sigma_{t,g,||} \times K_2 \times K_3 \times K_8 \times K_{14}$ Bending parallel to the grain $\sigma_{m,g,||} = $ 10.0 N/mm^2 $\sigma_{m,adm,||} = \sigma_{m,g,||} \times K_2 \times K_3 \times K_6 \times K_7 \times K_8$	
Clause 2.6.2 Clause 2.8 Clause 2.10.5	K_2 – wet exposure does not apply in this case K_3 – load duration Medium-term K_3 = 1.25 K_6 – shape factor Rectangular K_6 = 1.0	

Contract : Truss Tie Job Ref. No : Example 7.6 Part of Structure : Combined Tension and Bending Calc. Sheet No. : 2 of 2	Calcs. by : W.McK. Checked by : Date :

References	Calculations	Output
Clause 2.10.6	K_7 – depth of section 72 mm $< h <$ 300 mm $K_7 = \left(\dfrac{300}{h}\right)^{0.11} = \left(\dfrac{300}{100}\right)^{0.11} = 1.128$	
Clause 2.9 Clause 2.12.2	K_8 – load sharing stresses does not apply in this case K_{14} – width of section for tension members $K_{14} = \left(\dfrac{300}{h}\right)^{0.11} = \left(\dfrac{300}{100}\right)^{0.11} = 1.128$ $\sigma_{t,adm,||} = \sigma_{t,g,||} \times K_2 \times K_3 \times K_8 \times K_{14}$ $= (6.0 \times 1.25 \times 1.128) = 8.46 \text{ N/mm}^2$ $\sigma_{m,adm,||} = \sigma_{m,g,||} \times K_2 \times K_3 \times K_6 \times K_7 \times K_8$ $= (10.0 \times 1.25 \times 1.0 \times 1.128) = 14.1 \text{ N/mm}^2$ **Section properties:** Effective area = gross area – projected area of bolt hole $= (100 \times 38) - (10 \times 38) = 3420 \text{ mm}^2$	
Clause 6.6.1	**Note:** The bolt hole is 2 mm larger than the bolt diameter. Effective section modulus $= \dfrac{38 \times 100^2}{6} = 63.33 \times 10^3 \text{ mm}^3$ **Axial stress:** $\sigma_{t,a,||} = \dfrac{F_t}{\text{effective area}} = \dfrac{5.0 \times 10^3}{3420} = 1.46 \text{ N/mm}^2$ **Bending stress:** $\sigma_{m,a,||} = \dfrac{\text{bending moment}}{\text{section modulus}} = \dfrac{0.5 \times 10^6}{63.33 \times 10^3} = 7.9 \text{ N/mm}^2$ **Combined tension and bending:**	
Clause 2.12.3	$\dfrac{\sigma_{m,a,||}}{\sigma_{m,adm,||}} + \dfrac{\sigma_{t,a,||}}{\sigma_{t,adm,||}} \leq 1.0$ $\dfrac{7.9}{14.1} + \dfrac{1.46}{8.46} = 0.71 \leq 1.0$ Only one load case has been considered here, and in addition the design of the connections may require a larger section size than is necessary for the combined axial and bending stresses.	The section is adequate for combined tension and bending

7.11.4 Combined Bending and Axial Compression (Clause 2.11.6)

As indicated previously, in slender members subjected to axial compressive loads there is a tendency for lateral instability to occur. This type of failure is called **buckling** and is reflected in the modification factor K_{12} which is used to reduce the permissible compressive stress in a member.

The interaction diagram for combined bending and compression is more complex than that for combined tension and bending. For stocky members (i.e. slenderness ratio values λ lower than approximately 30), a non-linear relationship exists between axial stresses and bending stresses. For members with a high slenderness ratio, a linear approximation is more realistic as shown in Figure 7.30.

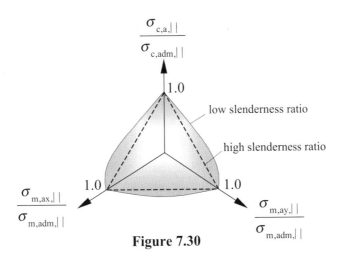

Figure 7.30

The non-linear behaviour in members with low slenderness ratios is a consequence of the higher values of compressive stress required to induce failure. Unlike timber subjected to tension, which exhibits a linear stress-strain curve until failure is reached, timber subjected to compressive stresses exhibits considerable plasticity (i.e. non-linear behaviour) after the initial elastic linear deformation. At high values of slenderness the stresses are relatively low when buckling occurs and hence a linear response is more appropriate, whilst at low values of slenderness relatively high values of stress occur at failure and hence a non-linear approximation is more realistic.

The interaction formula given in Clause 2.11.6 of the code for combined bending and axial compression is:

$$\frac{\sigma_{m,a,||}}{\sigma_{m,adm,||}\left(1-\frac{1.5\sigma_{c,a,||}}{\sigma_e}\times K_{12}\right)}+\frac{\sigma_{c,a,||}}{\sigma_{c,adm,||}} \leq 1.0$$

where:
$\sigma_{m,a,||}$ is the applied bending stress,
$\sigma_{m,adm,||}$ is the permissible bending stress,
$\sigma_{c,a,||}$ is the applied tension stress,

$\sigma_{c,adm,||}$ is the permissible tension stress,

σ_e is the Euler critical stress $= \dfrac{\pi^2 EI}{L^2}$

The values of $\sigma_{m,adm,||}$ and $\sigma_{c,adm,||}$ are determined using the modification factors where appropriate. The value of $\sigma_{m,a,||}$ should represent the maximum value of the bending stress in the cross-section. In some instances bending will occur about both the x–x and y–y axes simultaneously, in which case:

$$\sigma_{m,a,||} = (\sigma_{m,ax,||} + \sigma_{m,ay,||})$$

where:

$\sigma_{m,ax,||}$ is the maximum bending stress due to bending about the x–x axis of the section,
$\sigma_{m,ay,||}$ is the maximum bending stress due to bending about the y–y axis of the section.

Note: In glulam members the value of $\sigma_{m,adm,||}$ about the x–x axis and $\sigma_{m,adm,||}$ about the y–y axis may differ.

In the first term, the inclusion of the expression $\dfrac{1.5\sigma_{c,a,||}}{\sigma_e} \times K_{12}$ is to allow for the

additional applied bending moment $(P \times \delta)$, induced by the eccentricity of the axial load after deformation of the member. If no axial load existed this term would equal zero and the interaction equation would reduce to:

$$\frac{\sigma_{m,a,||}}{\sigma_{m,adm,||}} \leq 1.0 \qquad \text{i.e.} \quad \sigma_{m,a,||} \leq \sigma_{m,adm,||}$$

which is the same as that required for a member subject to bending only.

7.11.5 Example 7.7: Stud Wall

The rear wall of a small shelter is constructed from a stud wall as shown in Figure 7.31. The vertical studs are built into concrete at the base, support a roof structure at the top, and are positioned at 500 mm centres with noggings at mid-height.

Assume that the cladding material does not contribute to the structural stability of the framed structure.

Using the design data given, show that the studs indicated are suitable.

Design Data:
Characteristic axial load (medium term)   1.75 kN/stud
Characteristic wind load   0.6 kN/m²
Exposure condition   Service Class 3
Strength Class   C24
Section size   47 mm × 97 mm

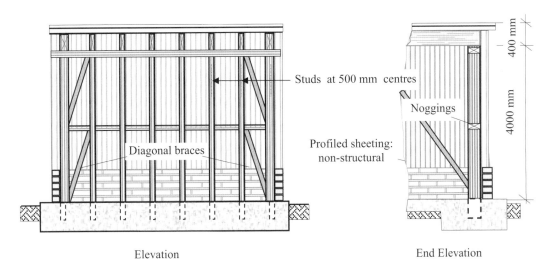

Elevation   End Elevation

Figure 7.31

7.11.5.1 *Solution to Example 7.7*

References	Calculations	Output
	Contract : Shelter **Job Ref. No : Example 7.7** **Calcs. by : W.McK.** **Part of Structure : Stud Wall** **Checked by :** **Calc. Sheet No. : 1 of 4** **Date :**	
Table 8	Strength Class 24 **Grade stresses:** Compression parallel to grain $\sigma_{c,g,\|\|}$ = 7.9 N/mm^2 $\sigma_{c,adm,\|\|} = \sigma_{c,g,\|\|} \times K_2 \times K_3 \times K_8 \times K_{12}$ Bending parallel to the grain $\sigma_{m,g,\|\|}$ = 7.5 N/mm^2 $\sigma_{m,adm,\|\|} = \sigma_{m,g,\|\|} \times K_2 \times K_3 \times K_6 \times K_7 \times K_8$ E_{min} = 7200 N/mm^2	
Clause 2.6.2 Table 16 Clause 2.8	K_2 – wet exposure does apply in this case $K_{2,compression \|\|}$ = 0.6, $K_{2,bending \|\|}$ = 0.8, $K_{2,E}$ = 0.8, K_3 – load duration	
Table 17 (note c)	Medium-term for axial load only K_3 = 1.25 The largest diagonal dimension of the loaded area < 50 m Very short-term – wind load K_3 = 1.75	
Clause 2.10.5 Clause 2.10.6	K_6 – shape factor Rectangular K_6 = 1.0 K_7 – Depth of section 72 mm < h < 300 mm $K_7 = \left(\dfrac{300}{h}\right)^{0.11} = \left(\dfrac{300}{97}\right)^{0.11} = 1.13$	

Contract : Shelter Job Ref. No : Example 7.7	Calcs. by : W.McK.
Part of Structure : Stud Wall	Checked by :
Calc. Sheet No. : 2 of 4	Date :

References	Calculations	Output
Clause 2.9	Load sharing stresses do apply in this case $- K_8 = 1.1$ **Note:** This factor does not apply when evaluating K_{12}	
Clause 2.11.3 Table 21 Clause 2.11.4	**Slenderness:** x–x axis L_{exx} $= (0.85 \times 4000)$ $= 3400$ mm $\dfrac{L_{exx}}{h}$ $= \dfrac{3400}{97}$ $= 35.1$ < 52 (i.e. $L_e/i < 180$)	
Clause 2.11.4	y–y axis L_{eyy} $\geq (1.0 \times 2000)$ $= 2000$ mm *critical or $\geq (0.85 \times 2000)$ $= 1750$ mm $\dfrac{L_{eyy}}{b}$ $= \dfrac{2000}{47}$ $= 42.55$ < 52 (i.e. $L_e/i < 180$) **Critical slenderness = 42.55**	
Table 22	**Medium-term loading** (axial load only): $E/\sigma_{c,\|\|}$ is determined using E_{min} and $\sigma_{c,\|\|} = \sigma_{c,g,\|\|} \times K_2 \times K_3$ E_{min} $= 7200$ N/mm^2 $\sigma_{c,\|\|}$ $= (7.9 \times 0.6 \times 1.25)$ $= 5.93$ N/mm^2 $E/\sigma_{c,\|\|} = \dfrac{7200}{5.93}$ $= 1214.2;$ $\lambda = 42.55$ The value of K_{12} can be determined by interpolation from Table 22 or as above using an approximate value. If necessary a more precise value can be determined.	

Extract from Table 22

L_e/b $E/\sigma_{c,\|\|}$	40.5	46.2
1200	0.288	0.233
1300	0.303	0.247

$K_{12} \approx 0.233$

$\sigma_{c,adm,\|\|} = \sigma_{c,g,\|\|} \times K_2 \times K_3 \times K_8 \times K_{12}$
$\quad = (7.9 \times 0.6 \times 1.25 \times 1.1 \times 0.233)$
$\quad = 1.52$ N/mm^2

Axial stress:
Area $= (47 \times 97)$ $= 4559$ mm^2

$$\sigma_{c,a,||} \quad = \quad \frac{F_c}{Area} \quad = \quad \frac{1.75 \times 10^3}{4559} \quad = \quad 0.38 \text{ N/mm}^2$$

Contract : Shelter Job Ref. No : Example 7.7	Calcs. by : W.McK.
Part of Structure : Stud Wall	Checked by :
Calc. Sheet No. : 3 of 4	Date :

References	Calculations	Output
Table 22	**Very short-term loading** (Axial load combined with bending):	

Very short-term loading (Axial load combined with bending):

$E/\sigma_{c,||}$ is determined using E_{min} and $\sigma_{c,||} = \sigma_{c,g,||} \times K_2 \times K_3$

$E_{min} \quad = \quad 7200 \text{ N/mm}^2$

$\sigma_{c,||} \quad = \quad (7.9 \times 0.6 \times 1.75) \quad = \quad 8.3 \text{ N/mm}^2$

$E/\sigma_{c,||} = \dfrac{7200}{8.3} \quad = \quad 867.5; \qquad \lambda = 42.55$

The value of K_{12} can be determined by interpolation from Table 22 or as above using an approximate value. If necessary a more precise value can be determined.

Extract from Table 22

L_e/b $\diagdown$ $E/\sigma_{c,||}$	40.5	46.2
800	0.217	0.172
900	0.237	0.188

$K_{12} \approx 0.172$

$\sigma_{c,adm,||} = \sigma_{c,g,||} \times K_2 \times K_3 \times K_8 \times K_{12}$
$\qquad\qquad = (7.9 \times 0.6 \times 1.75 \times 1.1 \times 0.172) \quad = 1.57 \text{ N/mm}^2$

$\sigma_{m,adm,||} = \sigma_{m,g,||} \times K_2 \times K_3 \times K_6 \times K_7 \times K_8$
$\qquad\qquad = (7.5 \times 0.8 \times 1.75 \times 1.0 \times 1.13 \times 1.1) = 13.05 \text{ N/mm}^2$

Axial stress:
Area $\quad = \quad (47 \times 97) \quad = \quad 4559 \text{ mm}^2$
$\sigma_{c,a,||} \quad = \quad 0.38 \text{ N/mm}^2$ as before

Bending stress:

Section modulus $\quad = \quad \dfrac{47 \times 97^2}{6} \quad = \quad 73.7 \times 10^3 \text{ mm}^3$

Wind load $\quad = \quad 3.0 \text{ kN/m}^2$

Load carried by one stud $\quad = \quad (0.6 \times 4.4 \times 0.5) \quad = \quad 1.32 \text{ kN}$

Bending moment $\quad = \quad \dfrac{WL}{8} \quad = \quad \dfrac{1.32 \times 4.0}{8} \quad = \quad 0.61 \text{ kNm}$

$\sigma_{m,a,||} \quad = \quad \dfrac{\text{Bending moment}}{\text{Section modulus}} = \dfrac{0.61 \times 10^6}{73.7 \times 10^3} = 8.28 \text{ N/mm}^2$

Clause 2.11.6   Euler crit.ical stress $= \dfrac{\pi^2 E}{\lambda^2}$ where $\lambda = (L_c/i)$

Contract :Shelter Job Ref. No : Example 7.7 Part of Structure : Stud Wall Calc. Sheet No. : 4 of 4	Calcs. by : W.McK. Checked by : Date :

References	Calculations	Output
	$L_c = 2000$ mm; and $i = \sqrt{\dfrac{I_{yy}}{A}}$ $I_{yy} = \dfrac{97 \times 47^3}{12} = 0.839 \times 10^6$ mm^4 $i_{yy} = \sqrt{\dfrac{0.839 \times 10^6}{4559}} = 13.57$ mm; $\lambda = (2000/13.57) = 147.4$ $E = (E_{min} \times K_2) = (7200 \times 0.8) = 5760$ N/mm^2 Euler critical stress $= \dfrac{\pi^2 E}{\lambda^2} = \dfrac{\pi^2 \times 5760}{147.4^2} = 2.62$ N/mm^2 **Combined compression and bending:**	
Clause 2.11.6	$\dfrac{\sigma_{m,a,\|\|}}{\sigma_{m,adm,\|\|}\left(1 - \dfrac{1.5\sigma_{c,a,\|\|}}{\sigma_e} \times K_{12}\right)} + \dfrac{\sigma_{c,a,\|\|}}{\sigma_{c,adm,\|\|}} \leq 1.0$ $\dfrac{1.5\sigma_{c,a,\|\|}}{\sigma_e} \times K_{12} = \left[\dfrac{1.5 \times 0.38}{2.62} \times 0.172\right] = 0.037$ $\dfrac{8.28}{13.05(1 - 0.037)} + \dfrac{0.38}{1.57} = 0.9 \qquad \leq 1.0$	**A 47 mm × 97 mm section of Strength Class C24 is adequate for the studs.**

7.12 Mechanical Fasteners

Traditionally the transfer of forces from one structural timber member to another was achieved by the construction of carpentry joints. In many instances the physical contact or friction between members was relied upon to transfer the forces between them.

The use of mechanical fasteners is now firmly established as an essential part of modern economic design in timber. There are numerous types of fastener available; those considered in BS 5268-2 are:

♦   nailed joints   (Clause 6.4)
♦   screwed joints   (Clause 6.5)
♦   bolted and dowelled joints   (Clause 6.6)
♦   toothed-plate connector joints   (Clause 6.7).

Only nailed connections are considered in this text; additional information for other connection types can be found in ref. 72:

♦   split-ring connector joints   (Clause 6.8)
♦   shear-plate connector joints   (Clause 6.9)
♦   glued joints   (Clause 6.10).

There are many proprietary types of fastener, such as punched metal plate fasteners with or without integral teeth, and splice plates which are also used but are not included in the Code.

Many factors may influence the use of a particular type of fastener, e.g.

♦   method of assembly,
♦   connection details,
♦   purpose of connection,
♦   loading,
♦   permissible stresses,
♦   aesthetics.

Normally there are several methods of connection for any given joint which could provide an efficient structural solution. The choice will generally be dictated by consideration of cost, availability of skills, suitable fabrication equipment, and desired finish required by the client. Glued joints generally require more rigorous conditions of application and control than mechanical fasteners.

A large number of modification factors are associated with the use of mechanical fasteners: these have been summarised in Table 7.8.

7.12.1 Nails (see Figure 7.32)

The most commonly used type of nails are (a) *round plain wire* nails. Two variations of this type are (b) *clout nails* (often referred to as slate, felt or plasterboard nails), which are simply large versions with a larger diameter head, and (c) *lost head nails* in which.

Factors	Application	Clause Number	Value/Location
K_2	service class 3 sections: (wet exposure)	2.6.2	Table 16
K_8	load-sharing: all stresses	2.9	1.1
K_{43}	nails driven into end grain	6.4.4.1	0.7
K_{44}	improved nail lateral loads	6.4.4.4	1.2
K_{45}	threaded part of annular ringed shank and helical shank nails	6.4.4.4	1.5
K_{46}	steel plate to timber joint	6.4.5.2	1.25
K_{48}	duration of loading – *nailed* joint	6.4.9	Given in clause
K_{49}	moisture content – *nailed* joint	6.4.9	Given in clause
K_{50}	number of *nails* in each line	6.4.9	Given in clause
K_{52}	duration of loading – *screwed* joint	6.5.7	Given in clause
K_{53}	moisture content – *screwed* joint	6.5.7	Given in clause
K_{54}	number of *screws* in each line	6.5.7	Given in clause
K_{56}	moisture content – *bolted* joint	6.6.6	Given in clause
K_{57}	number of *bolts* in each line	6.6.6	Given in clause
K_{58}	duration of loading – *toothed-plate connector* joint	6.7.6	Given in clause
K_{59}	moisture content – *toothed-plate connector* joint	6.7.6	Given in clause
K_{60}	end distance / edge distance / spacing of *toothed-plate connectors*	6.7.6	Tables 85,87,88
K_{61}	number of *toothed-plate connectors* in each line	6.7.6	Equation
K_{62}	duration of loading – *split-ring connector* joint	6.8.5	Given in clause
K_{63}	moisture content – *split-ring connector* joint	6.8.5	Given in clause
K_{64}	end distance / edge distance / spacing of *split-ring connectors*	6.8.5	Tables 91,92,93 Tables 93,95,96
K_{65}	number of *split-ring connectors* in each line	6.8.5	Equation
K_{66}	duration of loading – *shear-plate connector* joint	6.9.6	Given in clause
K_{67}	moisture content – *shear-plate connector* joint	6.9.6	Given in clause
K_{68}	end distance / edge distance / spacing of *shear-plate connectors*	6.9.6	Tables 91,92,93 Tables 93,95,96
K_{69}	number of *shear-plate connectors* in each line	6.9.6	Equation
K_{70}	permissible shear stress for *glue-line*	6.10.1.3	0.9
K_C	end distance – connectors	6.7.6	Table 87,88,95
K_S	spacing – connectors	6.7.6	Tables 85,93,87
K_D	edge distance – connectors	6.9.6	Table 96

Table 7.8 Modification Factors

the head is very small. The introduction of *improved nails* such as (d) *square twisted*, and (e) *helically threaded and annular ringed shank nails* with increased lateral and withdrawal resistance has proved very useful; particularly for fixing sheet materials to roofs and floors where 'popping' is often a problem with plain round nails.

Pneumatically driven nails (f), when manufactured from suitably hardened and tempered steel, enable relatively straightforward fixing of timber to materials such as concrete, brickwork and stone.

Pre-drilling of holes may be required to avoid splitting and to enable use with dense hardwoods such as greenheart and keruing timbers. The pre-drilled holes should not be greater than $(0.8 \times$ nail diameter$)$ as indicated in Clause 6.4.1.

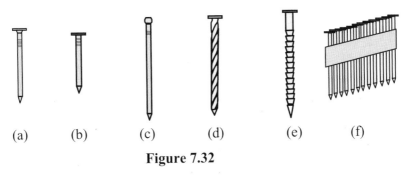

(a)   (b)   (c)   (d)   (e)   (f)

Figure 7.32

Nails are used either for locating timber, e.g. a stud to a wall plate as in Figure 7.33(a), or for transferring forces such as the shear force and bending moment at a knee joint in a portal frame joint, as shown in Figure 7.33(b).

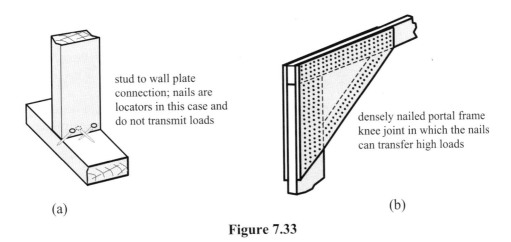

stud to wall plate connection; nails are locators in this case and do not transmit loads

densely nailed portal frame knee joint in which the nails can transfer high loads

(a)   (b)

Figure 7.33

The basic single-shear lateral loads and withdrawal loads for nails in timber-to-timber joints are given in Tables 61 and 62 respectively. The basic single-shear lateral loads for plywood-to-timber and particle board-to-timber joints are given in Tables 63 and 64 respectively.

The design of nailed joints is dependent on a number of factors, two of which are the *headside thickness* and the *pointside thickness* as shown in Figure 7.34:

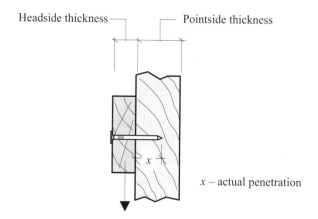

Headside thickness —   ⌐— Pointside thickness

x – actual penetration

Figure 7.34

Standard values for the headside and pointside thicknesses are given in Table 61. In cases where the actual headside or pointside thickness is less than the standard Table 61 values the basic load is multiplied by the smaller of the following two ratios, (a) and (b), as indicated in Clause 6.4.4.1:

(a) $\dfrac{\text{actual headside thickness}}{\text{standard headside thickness}}$,   and

(b) $\dfrac{\text{actual penetration thickness}}{\text{standard pointside thickness}}$

If either (a) or (b) is less than 0.66 for softwoods or 1.0 for hardwoods then *NO load-carrying capacity* should be assumed. The corresponding values when using improved nails are 0.5 for softwoods and 0.75 for hardwoods.

Where the nails are driven into the end grain, the Table 61 values should be multiplied by K_{43} which equals 0.7. It is important to note that *NO withdrawal load* is permitted by a nail driven into the end grain of timber.

In joints where multiple shear planes occur, the total shear load from each nail is equal to the basic value for a single plane multiplied by the number of shear planes. There are additional criteria relating to the timber thicknesses, as indicated in Figure 7.35.

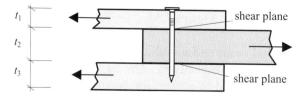

t_1 shear plane

t_2

t_3 shear plane

Figure 7.35

Multiple shear load/nail $=$ $(2 \times$ basic shear load) provided that:

$$t_2 \geq 0.85 \times \text{standard thickness (Table 61)}$$

When t_1 < standard thickness   or
             t_3 < standard thickness   or
             t_2 < $0.85 \times$ standard thickness
Multiple shear load/nail equals the smaller of:
             $(2 \times$ basic shear load) $\times$ ratio (a) and
             $(2 \times$ basic shear load) $\times$ ratio (b)
where ratio (a) and ratio (b) are as before.

The permissible load for a joint is determined by modifying the calculated basic load to allow for duration of loading (K_{48}), moisture content (K_{49}) and the number of nails in each line (K_{50}). Each of the modification factors K_{48}, K_{49} and K_{50} is defined in Clause 6.4.9.

It is necessary to ensure that adequate end and edge distances and spacing between nails is provided to avoid undue splitting. Minimum values are given in Table 53 for all nailed joints. When using Douglas fir the Table 53 values should be multiplied by 0.8.

7.12.2 Example 7.8: Nailed Tension Splice

A tension splice is shown in Figure 7.36. Using the data given design a suitable splice detail considering:

    i)   38 mm × 100 mm softwood splice plates,
    ii)   12 mm Finnish birch plywood,   and
    iii)   pre-drilled 1.5 mm thick mild steel plates.

Timber   Douglas fir
                  Service Class 2
                  Strength Class C16
                  Load duration considered long-term
Nails   3.4 mm diameter round wire × 65 mm long
                  No pre-drilling

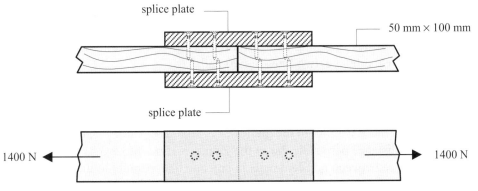

Figure 7.36

(i)   Timber-to-timber joint
        Table 61   For nail diameter 3.4 mm, softwood not pre-drilled
                             Standard penetration   = 41 mm
                             Standard headside thickness   = 41 mm
                             Basic single shear lateral load = 377 kN

                             Actual headside thickness   = 38 mm   < standard

        Clause 6.4.4.1 $\dfrac{\text{actual headside thickness}}{\text{standard headside thickness}}$ $= \dfrac{38}{41}$ = 0.93 > 0.66

                             Actual penetration thickness   = (65 – 38) = 27 mm   < standard

        Clause 6.4.4.1 $\dfrac{\text{actual penetration thickness}}{\text{standard pointside thickness}}$ $= \dfrac{27}{41}$ = 0.66 ≥ 0.66

        Clause 6.4.4.2   Inner member   > 0.85 × standard thickness
        Clause 6.4.4.3   Penetration   = 27 mm   > minimum value of 15 mm

The Table 61 values should be multiplied by the smaller of the ratios in Clause 6.4.4.1, i.e. 0.66.
                             Basic single shear lateral load F = (377 × 0.66) = 248.8 N

        Clause 6.4.9   Permissible load/nail F_{adm} = $(F \times K_{48} \times K_{49} \times K_{50})$
        where :
             F = basic load/nail
             K_{48} = load duration modification factor   long-term load K_{48} = 1.0
             K_{49} = moisture content modification factor   service class 2 K_{49} = 1.0
             K_{50} = number of nails in line modification factor   assume < 10 K_{50} = 1.0
             F_{adm} = $(F \times K_{48} \times K_{49} \times K_{50})$ = (248.8 × 1.0 × 1.0 × 1.0) = 248.8 N

        Number of nails required = $\dfrac{\text{Design load}}{F_{adm}}$ = $\dfrac{1400}{248.8}$ = 5.6 each side of splice
Adopt 6 nails each side i.e. single line of three nails   or alternatively
Adopt 8 nails each side i.e. two lines of two nails

        Clause 6.4.3   Nail spacing
        Table 60   (no pre-drilling)
Minimum end distance parallel to the grain   = 20d = (20 × 3.4)   = 68 mm
Minimum edge distance perpendicular to the grain   = 5d = (5 × 3.4)   = 17 mm
Minimum distance between adjacent nails in any one line, parallel to the grain
                                          = 20d = 68 mm

Minimum distance between lines of nails perpendicular to the grain
$$= 10d = (10 \times 3.4) \quad = 34 \text{ mm}$$

Nailing pattern assuming 6 nails each side

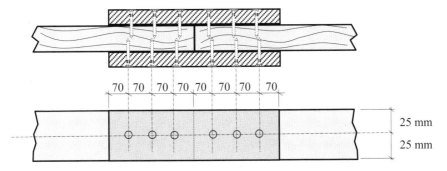

Total length of each splice plate = $[2 \times (4 \times 70)]$ = 560 mm

Alternate pattern using 8 nails each side of the splice:

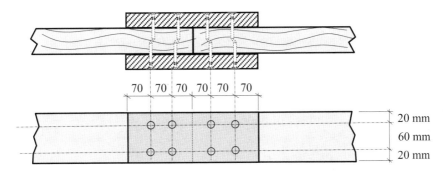

Total length of each splice plate = $[2 \times (3 \times 70)]$ = 420 mm

(ii) 12 mm Finnish birch plywood-to-timber splice

Timber: Douglas fir, Strength Class C16, long-term loading
Nails: 3.4 mm diameter × 65 mm long in softwood, not pre-drilled

Table 63   Plywood is Group II (note 1 in Table 63)
Clause 6.4.6.2   Minimum length of nail = 50 mm < actual length of 65 mm
     Basic single shear lateral load = 355 N
Clause 6.4.9   Permissible load/nail $F_{adm} = (F \times K_{48} \times K_{49} \times K_{50})$
     as before:
     $K_{48} = K_{49} = K_{50} = 1.0$
$$F_{adm} = (355 \times 1.0 \times 1.0 \times 1.0) \quad = 355 \text{ N}$$

$$\text{Number of nails required} \quad = \frac{\text{Design load}}{F_{adm}} \quad = \frac{1400}{355} = 4$$

Adopt 4 nails in line or two lines of two nails similar to the previous splice. The Table 61 values for minimum nail spacings should be satisfied.

(iii)   Steel plate-to-timber splice

Timber:	Douglas fir, Strength Class C16, long-term loading
Nails:	3.4 mm diameter × 65 mm long in softwood, not pre-drilled
Steel plate:	1.5 mm thick mild steel

Clause 6.4.5.1   Minimum thickness of plate $\geq$ 1.2 mm

$$\geq \quad 0.3 \times \text{nail diameter}$$
$$= \quad 0.3 \times 3.4 = 1.02 \text{ mm}$$

Actual thickness of plate   = 1.5 mm is adequate

Clause 6.4.5.2   Basic lateral load   = (Table 61 value × K_{46})

Note: The hole diameter in the steel plate should equal the diameter of the nails, and K_{46} has a value of 1.25.

Table 61   Basic single shear lateral load   = 377 N

Basic lateral load   = (377 × 1.25) =   471.3 N

Clause 6.4.9   Permissible load/nail F_{adm} = ($F \times K_{48} \times K_{49} \times K_{50}$)

as before:

$K_{48} = K_{49} = K_{50} = 1.0$

$$F_{\text{adm}} \quad = \quad (471.3 \times 1.0 \times 1.0 \times 1.0) \quad = \quad 471.3 \text{ N}$$

$$\text{Number of nails required} = \frac{\text{Design load}}{F_{\text{adm}}} = \frac{1400}{471.3} = 3$$

Adopt 4 nails as in the case of Finnish-birch plywood splice plates.

8. Design of Structural Masonry Elements (BS 5628)

Objectives: *To introduce masonry, i.e. brickwork and blockwork as a structural material and to illustrate the process of design for structural masonry elements.*

8.1 Introduction

Despite the use of masonry for construction during many centuries, design techniques based on well-established scientific principles have only been developed during the latter part of the 20th century.

The mechanisation and development of brickmaking occurred in the mid-19th century. Prior to this time the firing of bricks had always been in intermittent kilns. Using this technique, moulded and partially dried bricks were loaded into a kiln and fired. On completion of the firing the fire was put out, the kiln opened and the bricks allowed to cool. This process was then repeated for the next batch.

Modern brickmaking is carried out using a continuous process in which batches of bricks are loaded, fired, cooled and removed in permanent rotation. The shaping of clay to produce bricks is carried out either by extrusion or by moulding/pressing.

The strength of masonry/brickwork is dependent on a number of factors, one of which is the unit strength. (**Note**: the distinction between *brickwork* – an assemblage of bricks and mortar – and *brick* – the individual structural unit. In this text in general, reference to bricks and brickwork also implies blockwork, stonework, etc.)

In civil engineering projects which require high strength characteristics, high density ***engineering bricks*** are frequently used, whilst in general construction ***common bricks*** (commons) are used. Where appearance is a prime consideration ***facing bricks*** are used combining attractive appearance, colour and good resistance to exposure. Bricks which are non-standard size and/or shape are increasingly being used by architects and are known as ***specials*** (see Figure 8.1*)*.

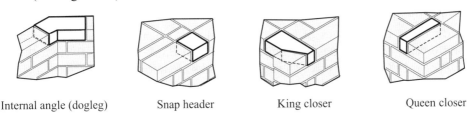

| Internal angle (dogleg) | Snap header | King closer | Queen closer |

Figure 8.1 A selection of *special* bricks

The design of structural masonry/brickwork in the U.K. is governed by the requirements of *BS 5628 – Code of Practice for the Use of Masonry Parts 1, 2 and 3.*

The references to Clause numbers in this text relate to *BS 5628:Part 1:1992* unless stated otherwise (including *Amendment 1 – July 1993* and *Amendment 2 – August 2002*).

8.2   Materials

Masonry can be regarded as an assemblage of structural units, which are bonded together in a particular pattern by mortar or grout.

8.2.1   *Structural Units (Clause 7 and Section 2 Clause 5 of BS 5628 : Part 3)*

There are seven types of structural unit referred to in BS 5628:

♦ calcium silicate (sandlime and flintlime) bricks   (BS 187),
♦ clay bricks   (BS 3921),
♦ dimensions of bricks of special shapes and sizes   (BS 4729),
♦ stone masonry   (BS 5340),
♦ precast concrete masonry units   (BS 6073:Part 1),
♦ reconstructed stone masonry units   (BS 6457),
♦ clay and calcium silicate modular bricks   (BS 6649).

The specification for each of these unit types is given in the appropriate British Standard as indicated. The selection of a particular type of unit for any given structure is dependent on a number of criteria, e.g. strength, durability, adhesion, fire resistance, thermal properties, acoustic properties and aesthetics.

The structural units may be solid, solid with frogs, perforated, hollow or cellular, as indicated in Figure 8.2.

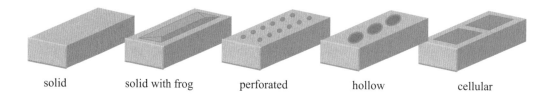

solid   solid with frog   perforated   hollow   cellular

Figure 8.2   Types of structural unit

8.2.1.1 Dimensions and Sizes

The specifications for the sizes of clay bricks, calcium silicate (sandlime and flintlime) bricks and precast concrete masonry units are given in *BS 3921:1985, BS 187:1978* and *BS 6073:Part 1:1981* respectively. The sizes of bricks are normally referred to in terms of work sizes and co-ordinating sizes, as shown in Figure 8.3. When using clay or calcium bricks the standard work sizes for individual units are 215 mm length × 102.5 mm width × 65 mm height. In most cases the recommended joint width is 10 mm, resulting in co-ordination sizes of bricks equal to 225 mm × 112.5 mm × 75 mm.

When designing it is more efficient and economic to specify dimensions of masonry minimising the cutting of brickwork. Wherever possible the dimensions of openings, panels returns, piers etc. should be a multiple of the co-ordinating size, plus or minus the joint thickness where appropriate. Brickwork dimension tables are available from the Brick Development Association (87).

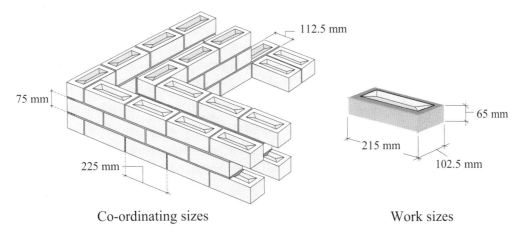

112.5 mm

75 mm

225 mm

65 mm

215 mm

102.5 mm

Co-ordinating sizes

Work sizes

Figure 8.3

8.2.2   Mortar (Clause 15)

Mortar is the medium which binds together the individual structural units to create a continuous structural form, e.g. brickwork, stonework, etc. Mortar serves a number of functions in masonry construction, i.e. to:

♦ bind together the individual units,
♦ distribute the pressures evenly throughout the individual units,
♦ infill the joints between the units and hence increase the resistance to moisture penetration,
♦ maintain the sound characteristics of a wall,
♦ maintain the thermal characteristics of a wall.

Present-day mortars are specifically manufactured to suit the type of construction involved. In most cases they are mixtures of sand, cement and water. The workability is often improved by the inclusion of lime or a mortar plasticiser. Lime is used in mortar for several reasons:

♦ to create a consistency which enables the mortar to *cling and spread*,
♦ to help retain the moisture and prevent the mortar from setting too quickly,
♦ to improve the ability of the mortar to accommodate local movement.

Plasticisers can be used with mortars which have a low cement:sand ratio to improve the workability. Their use introduces air bubbles into the mixture which fill the voids in the sand and increase the volume of the binder paste. The introduction of plasticisers into a mix must be carefully controlled, since the short-term gain in improved workability can be offset in the longer term by creating an excessively porous mortar resulting in reduced durability, strength and bond. This is emphasised in *BS 5628:Part 1:1992*, Clause 17 in which it is stated: *'Plasticisers can only be used with the written permission of the designer.'*

The requirements for mortars in relation to strength, resistance to frost attack during construction, and improvement in bond and consequent resistance to rain penetration, are given in Table 1 of *BS 5628:Part 1*.

Four mortar designation types, (i), (ii), (iii) and (iv) are specified in terms of their cement, lime, sand and plasticiser content, and appropriate 28-day strengths are given. The mortar type is subsequently used in design calculations to determine characteristic masonry strengths (f_k).

8.2.3 Bonds (Appendix B of BS 5628:Part 3:2001)

Walling made from regular-shaped units is constructed by laying the units in definite, specific patterns called **bonds**, according to the orientation of the long sides (**stretchers**) or the short sides (**headers**).

The method of laying structural units is specified in Section 8 of *BS 5628:Part 1* and detailed in Section 32 of Part 3 of the Code. Normally all bricks, solid and cellular blocks are laid on a full bed of mortar with all cross joints and collar joints filled. (*A cross joint is a joint other than a bed joint, at right angles to the face of a wall. A collar joint is a continuous vertical joint parallel to the face of a wall.*)

In situations where units are laid on either their stretcher face or end face, the strength of the units used in design should be based on tests carried out in this orientation. In bricks with frogs, the unit should be laid with the frog or larger frog uppermost. The position and filling of frogs is important since both can affect the resulting strength and sound insulation of a wall. Cellular bricks should be laid with their cavities downwards and unfilled.

It is essential when constructing brickwork walls to ensure that the individual units are bonding together in a manner which will distribute the applied loading throughout the brickwork. This is normally achieved by laying units such that they overlap others in the course below. The resulting pattern of brickwork enables applied loads to be distributed both in the horizontal and vertical directions as shown in Figure 8.4(a) and (b).

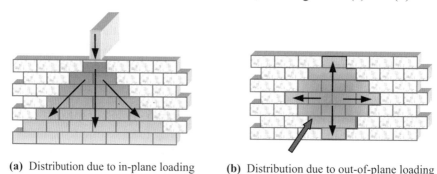

(a) Distribution due to in-plane loading   **(b)** Distribution due to out-of-plane loading

Figure 8.4

A number of bonds have been established which provide brickwork walls with the required characteristics, i.e.:

- vertical and horizontal load distribution for in-plane forces,
- lateral stability to resist out-of-plane forces,
- aesthetically acceptable finishes.

In *BS 5628:Part 3:2001* masonry bonds are defined for both brickwork and blockwork. Examples of several frequently used bonds for **brickwork** i.e. English bond, Flemish bond and Header bond are indicated in Figures 8.5 to 8.7.

♦   *English bond:*   Shows on both faces alternate courses of headers and stretchers.

Figure 8.5

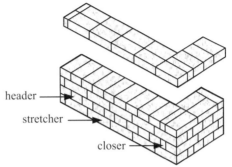

♦   *Flemish bond:*   Shows on the face alternate headers and stretchers in each course. It may be built as a 'single Flemish bond', which shows Flemish bond on both faces of the wall.

Figure 8.6

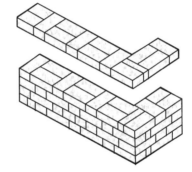

♦   *Header bond:*   Consists of bricks with their ends showing on the face of the wall, laid with a half lap of the brick width.

Figure 8.7

8.2.4   *Joint Finishes (Appendix B of BS 5628:Part 3:2001)*

The final appearance of brickwork is dependent on the finish of the joints between individual units (perpend/vertical-cross joints), and the bed joints between the courses. Various joint finishes can be created depending upon the desired aesthetic effect.

If the finishes are created during construction then the process is called *jointing*; if this is done after completion of the brickwork it is called *pointing*. As indicated in Clause 27.7 of *BS 5628:Part 3*, jointing is preferable since it leaves the bedding mortar undisturbed.

Two of the most commonly used joint finishes are shown in Figures 8.8 and 8.9; others are illustrated in Appendix B of *BS 5628:Part 3*.

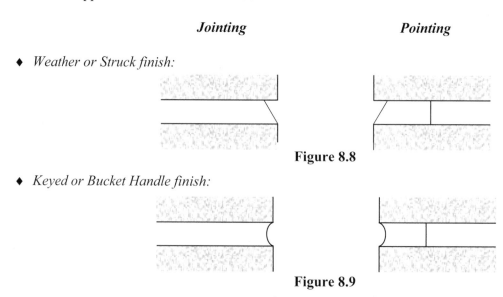

Figure 8.8

Figure 8.9

The type of joint finish selected will be influenced by a number of factors such as exposure conditions and aesthetics. The most effective types to resist rain penetration are weather and keyed finishes. In situations where wind-driven rain is likely, finishes which produce a water-retaining edge, e.g. recessed, should be avoided.

If raking out of a joint is carried out as a finishing feature, the reduction in wall thickness should be considered when calculating design stresses as indicated in a footnote to Clauses 8.2, 8.3 and 8.4 of *BS 5628:Part 1:1992*.

8.2.5   *Damp-Proof Courses (Clause 12 and Clause 10 of BS 5628:Part 3:2001)*

The purpose of a damp-proof course (d.p.c.) is to provide an impermeable barrier to the movement of moisture into a building. The passage of water may be horizontal, upward or downward. The material properties required of d.p.cs are set out in *BS 5628:Part 3*, Clause 21.4.3, they are:

- an expected life at least equal to that of the building,
- resistance to compression without extrusion,
- resistance to sliding where necessary,
- adhesion to units and mortar where necessary,
- resistance to accidental damage during installation and subsequent building operations,
- workability at temperatures normally encountered during building operations, with particular regard to ease of forming and sealing joints, fabricating junctions, steps and stop ends, and ability to retain shape.

There is a wide range of materials which are currently used as d.p.c.s, which fall into three categories:

- flexible   e.g. lead, copper, polyethylene, bitumen, bitumen polymers,
- semi-rigid e.g. mastic asphalt,
- rigid   e.g. epoxy resin/sand, three courses of engineering brick, and bonded slate.

The physical properties and performance of each type are given in *BS 5628:Part 3*, Table 12. The relevant British Standards which apply to each type are:

- bitumen   (BS 6398),
- brick   (BS 3921),
- polyethylene   (BS 6515),
- all others   (BS 743).

The correct positioning of d.p.cs is important to ensure continuity of the impervious barrier throughout a building. In Clause 21.5 of *BS 5628:Part 3*, detailed advice is given relating to the use of d.p.c.s in most common situations, i.e. below ground level, immediately above ground level, under cills, at jambs of openings, over openings, at balcony thresholds, in parapets and in chimneys. A typical detail for a steel lintel over an opening in a cavity wall is shown in Figure 8.10.

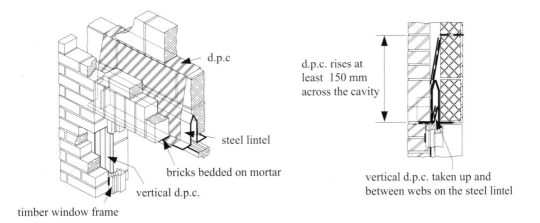

d.p.c

d.p.c. rises at least 150 mm across the cavity

steel lintel

bricks bedded on mortar

vertical d.p.c.

vertical d.p.c. taken up and between webs on the steel lintel

timber window frame

Figure 8.10

8.2.6   *Rendering (Clause 21.3.2 of BS 5628:Part 3:2001)*

The rendering of masonry involves applying an additional surface to the external walls to improve the weather resistance and in some instances to provide a decorative finish. There are five commonly used types of render:

- pebbledash (dry dash)   this produces a rough finish with exposed stones which are thrown onto a freshly applied coat of mortar,

- ♦ roughcast (wet dash)   this produces a rough finish by throwing a wet mix containing a proportion of small stones,
- ♦ scraped or textured   the final coat of mortar is treated using one of a variety of tools to produce a desired finish,
- ♦ plain coat   the final coat is a smooth/level coat finished with wood-cork or felt-faced pad,
- ♦ machine applied finish   the final coat is thrown on the wall by a machine producing the desired texture.

It is important when applying a rendered finish to give consideration to the possibility of cracking. In circumstances where strong, dense mixes are used for the render to protect against severe exposure conditions, there is the risk of considerable shrinkage cracking. The tendency is to use weaker renders based on cement:lime mixes which accommodate higher levels of shrinkage, are more absorbent and reduce the flow of water over the surface cracks.

8.2.7   *Wall Ties (Clause 13 and Clause 19.5 of BS 5628:Part 3:2001)*

Wall ties (see Figure 8.11) are required to tie together a masonry wall and another structural component/element such that they behave compositely to resist/transfer the applied loading. The design of wall ties should comply with the requirements of *BS 1243 Specification for metal ties for cavity wall construction*. In situations where corrosion is likely, the ties should be manufactured from either stainless steel or a non-ferrous material.

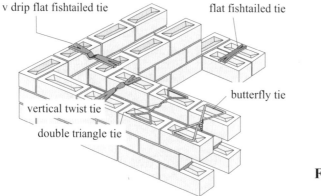

Figure 8.11

Some of these ties must have a 'drip' to prevent the movement of water through a cavity (as shown in Figure 8.12) and be designed such that they minimise the retention of mortar droppings during construction.

Figure 8.12

The most commonly used wall ties are those used in cavity-wall construction, i.e. vertical twist, double triangle and butterfly ties as shown in Figure 8.11. The strength and spacing of ties in cavity walls must be sufficient to develop the combined stiffness of both leaves if they are to be effective. The requirements for the provision of ties are given in Clause 19.5 and Table 9 (see Figure 8.13) of *BS 5628:Part 3:2001*.

Extract from BS 5628:Part 3:2001

Table 9. Wall ties

(A) Spacing of ties

Least leaf thickness (one or both)	Type of tie	Cavity width	Equivalent no. of ties per square metre	Spacing of ties	
				Horizontally	Vertically
mm 65 to 90 90 or more	All See table 9 (b)	mm 50 to 75 50 to 150	4.9 2.5	mm 450 900	mm 450 450

(B) Selection of ties

		Type of tie in BS 1243	Cavity width
Increasing strength	Increasing flexibility and sound insulation	Vertical twist	mm 150 or less
		Double triangle	75 or less
		Butterfly	75 or less

Figure 8.13

8.3   Material Properties

Masonry is a non-homogeneous, non-isotropic composite material which exists in many forms comprising units of varying shape, size and physical characteristics. The parameters which are most significant when considering structural design relate to strength and elastic properties, e.g. compressive, flexural and shear strengths, modulus of elasticity, coefficient of friction, creep, moisture movement and thermal expansion; these are discussed individually in Sections 8.3.1 to 8.3.7. Tensile strength is generally ignored in masonry design.

The workmanship involved in constructing masonry is more variable than is normally found when using most other structural materials, and consideration must be given to this at the design stage.

8.3.1   *Compressive Strength*

The compressive strength of masonry is dependent on numerous factors such as:

♦   the mortar strength,
♦   the unit strength,
♦   the relative values of unit and mortar strength,
♦   the aspect ratio of the units (ratio of height to least horizontal dimension),
♦   the orientation of the units in relation to the direction of the applied load,
♦   the bed-joint thickness.

This list gives an indication of the complexity of making an accurate assessment of masonry strength. Unit strengths and masonry strengths are given in *BS 5628:Part 1:1992* in Figures 1(a) to 1(d) and Tables 2(a) to 2(d). These values are derived from research data carried out on individual units, small wall units (wallettes) and full-scale testing of storey height walls. The tabulated values are intended for use with masonry in which the structural units are laid on their normal bed faces in the attitude in which their compressive strengths are determined and in which they are normally loaded. Variations to this can be accommodated and testing procedures are specified in Appendix A of the code for the experimental determination of the characteristic compressive strength of masonry. Full-scale testing of storey height panels is considered to give the most accurate estimate of potential strength of masonry walls.

8.3.2   *Flexural Strength*

The non-isotropic nature of masonry results in two principal modes of flexural failure which must be considered:

◆   failure parallel to the bed-joints, and
◆   failure perpendicular to the bed-joints,

as shown in Figures 8.14 (a) and (b).

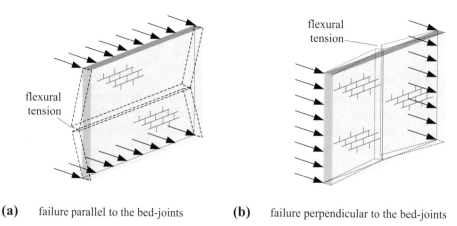

(a)   failure parallel to the bed-joints   **(b)**   failure perpendicular to the bed-joints

Figure 8.14

The ratio of (flexural strength parallel to the bed-joints) to (flexural strength perpendicular to the bed-joints) is known as the orthogonal ratio (μ) and has a typical value of 0.33 for clay, calcium silicate and concrete bricks, and 0.6 for concrete blocks. Research indications are that the flexural strengths of clay bricks are significantly influenced by the water absorption characteristics of the units.

In the case of concrete blocks the flexural strength *perpendicular* to the bed-joints is significantly influenced by the compressive strength of the units. There does not appear to be any meaningful correlation between the strength of calcium silicate bricks, concrete bricks or concrete blocks parallel to the bed-joints with any standard physical property.

In all cases the flexural strength in both directions is dependent on the strength of the mortar used and in particular the adhesion between the units and the mortar. The adhesion is very variable and research has shown it to be dependent on properties such as the pore structure of the units and mortar, the grading of the mortar sand and the moisture content of the mortar at the time of laying.

These consideration have been included during the development of the values given in *BS 5628:Part 1*: Table 3, *'Characteristic flexural strength of masonry'*.

8.3.3   Tensile Strength

As mentioned previously, the tensile strength of masonry is generally ignored in design. However, in Clause 24.1 the code indicates that a designer is permitted to assume 50% of the flexural strength values given in Table 3 when considering direct tension induced by suction forces arising from wind loads on roof structures, or by the probable effects of misuse or accidental damage.

Note: *In no circumstances may the combined flexural and direct tensile stresses exceed the values given in Table 3.*

8.3.4   Shear Strength

The shear strength of masonry is important when considering wall panels subject to lateral forces and structural forms such as diaphragm and fin walls where there is the possibility of vertical shear failure between the transverse ribs and flanges during bending.

Shear failure is most likely to be due to in-plane horizontal shear forces, particularly at the level of damp-proof courses.

The characteristic shear strength is dependent on the mortar strength and any precompression which exists. In *BS 5628:Part 1:1992*, Clause 25 linear relationships are given for the characteristic shear strength and the precompression as follows:

(a)   *Shear in a horizontal direction in a horizontal plane*

Mortar designations (i) and (ii)
$$f_v = 0.35 + 0.6g_A$$
$$\leq 1.75 \text{ N/mm}^2$$
Mortar designations (iii) and (iv)
$$f_v = 0.15 + 0.6g_A$$
$$\leq 1.4 \text{ N/mm}^2$$

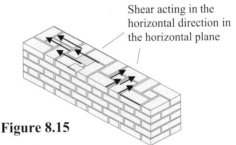

Shear acting in the horizontal direction in the horizontal plane

Figure 8.15

where:
f_v is the characteristic shear strength in the horizontal direction in the horizontal plane (see Figure 2 of the code),
g_A is the design vertical load per unit area of wall cross-section due to the vertical loads calculated for the appropriate loading conditions specified in Clause 22.

(b)   *Shear in bonded masonry in the vertical direction in the vertical plane*
For brick:
Mortar designations (i) and (ii)
$$f_v = 0.7 \text{ N/mm}^2$$

Mortar designations (iii) and (iv)
$f_v = 0.5 \text{ N/mm}^2$

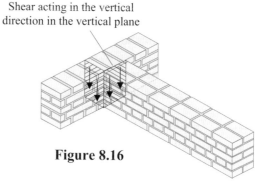

Shear acting in the vertical
direction in the vertical plane

For dense aggregate solid concrete block with a minimum strength of 7 N/mm²:
Mortar designations (i) and (ii)
$f_v = 0.7 \text{ N/mm}^2$
Mortar designations (iii) and (iv)
$f_v = 0.5 \text{ N/mm}^2$

Figure 8.16

The design shear strength is obtained by applying the partial safety factor γ_{mv} given in Clause 27.4 of the code and hence:

$$\text{Design shear strength of masonry} = \frac{f_v}{\gamma_{mv}}$$

where γ_{mv} = 2.5   when mortar not weaker than designation (iv) is used, and
     = 1.25   when considering the probable effects of misuse or accidental damage.

In the case of wall panels which are subject to lateral loads and restrained by concrete supports, the shear forces can be transmitted by metal wall ties. The characteristic shear strength of various types of tie which are engaged in dovetail slots in structural concrete are given in Table 8 of the code. As with masonry, the design values can be obtained by applying a partial safety factor γ_m as given in Clause 27.5, i.e.
γ_m = 3.0 under normal conditions and
  = 1.5 when considering the probable effects of misuse or accidental damage.

8.3.5   Modulus of Elasticity

Since masonry is an anisotropic, composite material, the value of its elastic modulus, E_m, is variable depending on several factors such as materials used, the direction and type of loading etc. A typical stress–strain curve is shown in Figure 8.17.

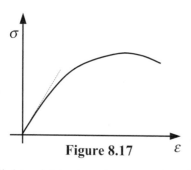

Figure 8.17

The actual value of E_m varies considerably and is approximated in *BS 5628:Part 2:2000*, Clause 7.4.1.7 as being equal to $0.9 f_k \text{ kN/mm}^2$. In practice this may range from $0.5 f_k$ to $2.0 f_k$; the value given, however, is sufficiently accurate for design purposes.

In the long-term the value of elastic modulus allowing for creep and shrinkage may be taken as: $E_m = 0.45 f_k \text{ kN/mm}^2$ for clay and dense aggregate concrete masonry and $E_m = 0.3 f_k \text{ kN/mm}^2$ for calcium silicate, autoclaved aerated concrete (a.a.c.) and lightweight concrete masonry as indicated in Appendix C of Part 2 of the code. In service masonry stresses are normally very low when compared with the ultimate value, and consequently the use of linear elastic analysis techniques to determine structural deformations is acceptable. (**Note:** f_k is defined in section 8.4.1)

8.3.6 Coefficient of Friction

The value of the coefficient of friction between clean concrete and masonry faces is given in Clause 26 as 0.6.

8.3.7 Creep, Moisture Movement and Thermal Expansion

The effects of creep, shrinkage, moisture and thermal movement are all significant, particularly when considering the design of prestressed masonry. In each case the loss of prestressing force can occur and at low levels of strain in the tendon the effects of prestress can be eliminated. They can all induce fine cracking and opening up of joints and may require the provision of movement joints.

The creep characteristics can be estimated as indicated in *BS 5628:Part 2:2000*, Clause 9.4.2.5, i.e. creep is numerically equal to 1.5 × the elastic deformation of the masonry for fired-clay or calcium silicate bricks and 3.0 × the elastic deformation of the masonry for dense aggregate concrete blocks.

The moisture movement of masonry, [expansion (+) or contraction (−)], can be estimated using the values of shrinkage strain according to Clause 9.4.2.4 of *BS 5628:Part 2:2000*, where shrinkage/expansion strain is equal to:

$$\varepsilon = -500 \times 10^{-6} \ (-0.5 \text{ mm/m}) \qquad \text{for calcium silicate and concrete masonry.}$$

8.4 Axially Loaded Walls

8.4.1 Introduction

Load-bearing walls resisting primarily vertical, in-plane loading, are often referred to as *axially loaded walls* whilst wall panels resisting wind loading are known as *laterally loaded wall panels*. The most commonly used types of axially loaded elements are:

- single-leaf (solid) walls,
- cavity walls,
- walls stiffened with piers, and
- columns,

as indicated in Figure 8.18 in this text and in Figure 3 of *BS 5628:Part 1:1992*.

Each of these types of element may be subject to *concentric* axial loading inducing compression only, or *eccentric* axial loading resulting in combined compressive and bending forces.

The magnitude of loading which can be sustained by a load-bearing wall or column is dependent on a number of factors, such as the:

- characteristic compressive strength of masonry, i.e. combined units and mortar (f_k),
- partial safety factor for the material strength (γ_m),
- plan area,
- thickness of the wall (t),
- slenderness of the element (h_{eff}/t_{eff}),

- ◆ eccentricity of the applied load (e_x),
- ◆ combined slenderness and eccentricity,
- ◆ eccentricities about both axes of a column,
- ◆ type of structural unit, and
- ◆ laying of structural units.

These factors are discussed separately in Sections 8.4.2 to 8.4.11.

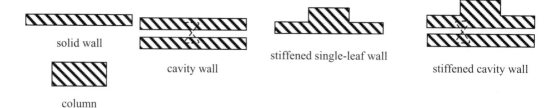

solid wall

cavity wall

stiffened single-leaf wall

stiffened cavity wall

column

Figure 8.18

8.4.2   *Characteristic Compressive Strength of Masonry (f_k)*

The characteristic compressive strength of masonry is given in Figures 1(a) to 1(d) and in Tables 2(a) to 2(d) of the code for a variety of different types of structural unit including standard format bricks, solid and hollow concrete blocks. Alternatively, as indicated in Clause 2.3.1, the value of f_k of any masonry '*... may be determined by tests on wall specimens, following the procedure laid down in A.2*', where A.2 refers to Appendix A.2 in *BS 5628:Part 1:1992*. Figure 1(a) and Table 2(a) from the code are reproduced in this text in Figure 8.19 and Figure 8.20 respectively.

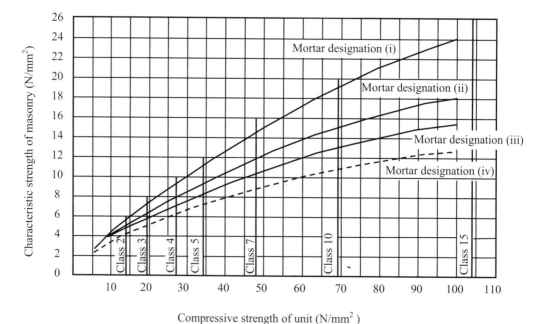

Figure 8.19

The value of f_k for any known combination of mortar designation and unit compressive strength can be determined from the tables. Linear interpolation within the tables is permitted, facilitated by the graphs given in Figures 1(a) to 1(d) of the code. The desired compressive strength of units and the mortar designation are normally specified by the designer.

Table 2(a) applies to masonry built with standard format bricks complying with the requirements of :

- BS 187:1987   *Specification for Calcium Silicate Bricks,*
- BS 6073-1:1981   *Specification for Precast Concrete Masonry Units*, or
- BS 3921:1985   *Specification for Clay Bricks*

Table 2. Characteristic compressive strength of masonry, f_k in N/mm^2

(a) Constructed with standard format bricks

Mortar desig- nation	Compressive strength of unit (N/mm^2)								
	5	10	15	20	27.5	35	50	70	100
(i)	2.5	4.4	6.0	7.4	9.2	11.4	15.0	19.2	24.0
(ii)	2.5	4.2	5.3	6.4	7.9	9.4	12.2	15.1	18.2
(iii)	2.5	4.1	5.0	5.8	7.1	8.5	10.6	13.1	15.5
(iv)	2.2	3.5	4.4	5.2	6.2	7.3	9.0	10.8	12.7

Figure 8.20

8.4.2.1 *Compressive Strength of Unit*

The compressive strength of masonry units may be given in N/mm^2, determined from standard quality assurance testing during production, or be assigned to a specific strength class by the manufacturer.

In BS 187, units are classified in terms of a strength class as shown in Table 8.1.

Designation	Class	Compressive strength (N/mm^2)
Facing brick or Load-bearing brick	3	20.5
	4	27.5
	5	34.5
	6	41.5
	7	48.5

Table 8.1

Manufacturers are required to identify their products on relevant documentation, providing the following information:

♦ the name, trade mark or other means of identification of the manufacturer,
♦ the strength class of brick as designated in Table 2 of the code (Table 8.1 in this text),
♦ the work size, length, width and height, and whether with or without a frog,
♦ the number of the British Standard used e.g. BS 187.

In *BS 6073-1:1981*, unit strengths are specified as shown in Table 8.2. The graphs in Figures 1(a) to 1(d) in *BS 5628:Part 1:1992* can be used for interpolation between these compressive strengths to determine the characteristic strength of the masonry.

Unit strengths as given in *BS 6073:Part 2:1981*							
Bricks (N/mm^2)	7.0	10.0	15.0	20.0	30.0	40.0	
Blocks (N/mm^2)	2.8	3.5	5.0	7.0	10.0	15.0	20.0	35.0

Table 8.2

In Table 4 of *BS 3921:1985* the classification of bricks by compressive strength and water absorption is given as shown in Table 8.3.
Note: It is important to note that there is no direct relationship between compressive strength and water absorption, as given in this table, and durability.

Table 4. Classification of bricks by compressive strength and water absorption		
Class	Compressive strength (see 2.1)	Water absorption (see 2.2)
	N/mm^2	% by mass
Engineering A	≥ 70	≤ 4.5
Engineering B	≥ 50	≤ 7.0
Damp-proof course 1	≥ 5	≤ 4.5
Damp-proof course 2	≥ 5	≤ 7.0
All others	≥ 5	No limits
NOTE 1. There is no direct relationship between compressive strength and water absorption as given in this table and durability. NOTE 2. Damp-proof course 1 bricks are recommended for use in buildings whilst damp-proof course 2 bricks are recommended for use in external works (see table 13 of *BS 5628:Part 3:1985*).		

Table 8.3

8.4.2.2 Compressive Strength of Mortar

The most appropriate mortar for any particular application is dependent on a number of factors such as strength, durability and resistance to frost attack. In Table 13 of *BS 5628:Part 3:2001*, guidance is given on the choice of masonry units and mortar designations most appropriate for particular situations regarding durability.

The mortar designations given in Table 1 of *BS 5628:Part 1:1992* (see Table 8.4), have been selected to provide the most suitable mortar which will be readily workable to enable the production of satisfactory work at an economic rate and to provide adequate durability.

The mortars are designated (i) to (iv), (i) being the strongest and most durable. Careful consideration is required when selecting a combination of structural unit and mortar designation; whilst the strongest mix may be suitable for clay brickwork in exposed situations, the inherent shrinkage in calcium silicate bricks caused by moisture movement can lead to cracking if strong, inflexible joints are present.

Table 1. Requirements for mortar

	Mortar designa-tion	Type of mortar (proportion by volume)			Mean compressive strength at 28 days	
		Cement:lime:sand	Masonry cement:sand	Cement: sand with plasticizer	Prelim-inary (laboratory) tests	Site tests
					N/mm^2	N/mm^2
Increasing strength Increasing ability to accommodate movement, e.g. due to settlement, temperature and moisture changes	(i) (ii) (iii) (iv)	1 : 0 to ¼ : 3 1 : ½ : 4 to 4¼ 1 : 1 : 5 to 6 1 : 2 : 8 to 9	– 1 : 2 ½ to 3 ½ 1 : 4 to 5 1 : 5 ½ : 6 ½	– 1 : 3 to 4 1 : 5 to 6 1 : 7 to 8	16.0 6.5 3.6 1.5	11.0 4.5 2.5 1.0
Direction of change in properties is shown by the arrows		Increasing resistance to frost attack during construction ⟶ Improvement in bond and consequent resistance to rain penetration ⟵				

Table 8.4

The required 28 days' mean compressive strength of the designated mortars, based on site tests, ranges from 11.0 N/mm^2 for designation (i) to 1.0 N/mm^2 for designation (iv). The site tests should be carried out according to the requirements of BS 4551 as indicated in Clause A.1.3 of *BS 5628:Part 1:1992*.

8.4.3 Partial Safety Factor for Material Strength (γ_m)

The design compressive strength of masonry is determined by dividing the characteristic strength (f_k), by a partial safety factor (γ_m), which is given in *BS 5628:Part 1:1992* in Table 4(a) for compression and in Table 4(b) for flexure, and is shown in Tables 8.5(a) and (b).

γ_m for Compression		Category of Construction control	
		Special	Normal
Category of manufacturing control of structural units	Special	2.5	3.1
	Normal	2.8	3.5

Table 8.5(a)

γ_m for Flexure	Category of Construction control	
	Special	Normal
	2.5	3.0

Table 8.5(b)

The γ_m factor makes allowance for the inherent differences between the estimated strength characteristics as determined using laboratory tested masonry specimens and the actual strength of masonry constructed under site conditions, and in addition allows for variations in the quality of materials produced during the manufacturing process.

The value of γ_m adopted is dependent on the degree of quality control practised by manufacturers and the standard of site supervision, testing and workmanship achieved during construction. There are two categories of control adopted in the code:

♦ normal control, and
♦ special control.

8.4.3.1 Normal Control   (Clause 27.2.1.1 and Clause 27.2.2.1)

In **manufacturing**, normal control *'... should be assumed when the supplier is able to meet the requirements for compressive strength in the appropriate British Standard, but does not meet the requirements for the special category...'*.

In **construction**, normal control *'... should be assumed whenever the work is carried out following the recommendations for workmanship in section four of BS 5628:Part 3: 2001, or BS 5390, including appropriate supervision and inspection.'*

8.4.3.2 Special Control (Clause 27.2.1.2 and Clause 27.2.2.2)

In **manufacturing**, special control *'... may be assumed where the manufacturer:*
(a) agrees to supply consignments of structural units to a specified strength limit, referred to as the acceptance limit, for compressive strength, such that the average compressive strength of a sample of structural units, taken from any consignment and tested in accordance with the appropriate British Standard specification, has a probability of not more than 2.5% of being below the acceptance limit, and
*(b) operates a quality control scheme, the results of which can be made available to demonstrate to the satisfaction of the purchaser that the acceptance limit is consistently being met in practice, with the probability of failing to meet the limit being never greater than that stated in **27.2.1.2 (a)**.'*

In **construction**, special control *'... may be assumed where the requirements of the normal category control are complied with and in addition:*
(a) the specification, supervision and control ensure that the construction is

compatible with the use of the appropriate partial safety factors given in Table 4(a);

(b) *preliminary compressive strength tests carried out on the mortar to be used, in accordance with A.1, indicate compliance with the strength requirements given in table 1 and regular testing of the mortar used on site, in accordance with A.1, shows that compliance with the strength requirements given in table 1 is being maintained.'*

The values of γ_m in Tables 4(a) and (b) apply to **compressive** and **flexural failure**. When considering the probable effects of misuse or accident, these values may be halved except where a member is deemed to be a '*protected member*' as defined in Clause 37.1.1 of the code. As indicated in Clause 27.3, in circumstances where wall tests have been carried out in accordance with Appendix A of the code to determine the characteristic strengths, the γ_m values in Tables 4(a) and (b) can be multiplied by 0.9.

When considering **shear failure**, the partial safety factor for masonry strength (γ_{mv}), should be taken as 2.5 as indicated in Clause 27.4.

The value of γ_m to be applied to the **strength of wall ties** is given in Clause 27.5 as 3.0; as with compressive and flexural failure, when considering the effects of misuse or accidental damage this value of γ_m can be halved.

8.4.4 Plan Area (Clause 23.1.7)

In Clause 23.1.7 allowance is made for the increased possibility of low-strength units having an adverse effect on the strength of a wall or column with a small plan area. For example, consider the two walls A and B shown in Figure 8.21.

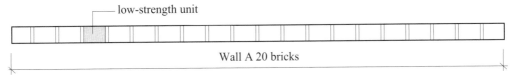

The proportion of the cross-sectional area affected by the low-strength unit in wall A is 5%

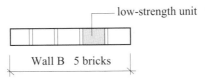

The proportion of the cross-sectional area affected by the low-strength unit in wall B is 20%

Figure 8.21

Although this effect is statistical it is essentially a geometrical effect on the compressive strength. The allowance for this is made by multiplying the characteristic compressive strength, (f_k), by the factor **(0.70 + 1.5A)**
where:
A is the horizontal loaded cross-sectional area of the wall or column.

This factor applies to all walls and columns where the cross-sectional area is less than 0.2 m². Clearly when $A = 0.2$ m² the factor is equal to 1.0.

8.4.5   *Thickness of Wall t (Clause 23.1.2)*

The compressive failure of walls occurs predominantly by the development of vertical cracks induced by the Poisson's ratio effect. The existence of vertical joints (as shown in Figure 8.22), reduces the resistance of brickwork to vertical cracking.

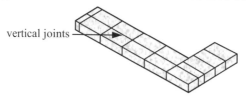

vertical joints

Figure 8.22

The indication from experimental evidence is that greater resistance to crack development is afforded by continuity in the cross-section, such as occurs in narrow (half-brick) walls, than is the case when vertical joints are present. This increased resistance to compressive failure is reflected in a modification factor equal to 1.15 which is applied to the characteristic compressive strength (f_k), obtained from Table 2(a) for standard format bricks. The factor applies to single walls and loaded inner leaves of cavity walls, as indicated in Clause 23.1.2 of the code. (**Note:** this does not apply when both leaves are loaded.)

8.4.5.1 *Effective Thickness t_{ef} (Clause 28.4)*

The concept of effective thickness was introduced to determine the *slenderness ratio* when considering buckling due to compression. In Figure 3 of *BS 5628:Part 1:1992*, values of effective thickness are given for various plan arrangements, as shown in Figures 8.23 and 8.24. The modifications to the actual thicknesses accounts for the stiffening effects of piers, intersecting and cavity walls, which enhance the wall stability.

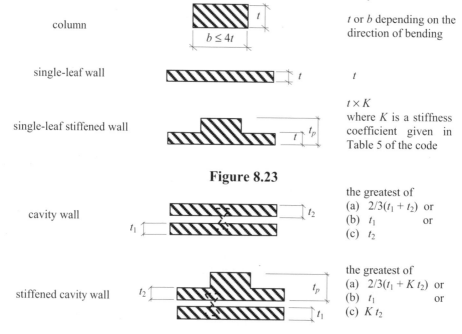

column t or b depending on the direction of bending

$b \leq 4t$

single-leaf wall t

single-leaf stiffened wall $t \times K$
                                 where K is a stiffness coefficient given in Table 5 of the code

Figure 8.23

cavity wall   the greatest of
                                 (a) $2/3(t_1 + t_2)$ or
                                 (b) t_1 or
                                 (c) t_2

stiffened cavity wall   the greatest of
                                 (a) $2/3(t_1 + K t_2)$ or
                                 (b) t_1 or
                                 (c) $K t_2$

Figure 8.24

The stiffness coefficient *K* is given in Table 5 of the code, as indicated in Figure 8.25.

Table 5. Stiffness coefficient for walls stiffened by piers			
Ratio of pier spacing (centre to centre) to pier width	Ratio t_p/t of pier thickness to actual thickness of wall to which it is bonded		
	1	2	3
6	1.0	1.4	2.0
10	1.0	1.2	1.4
20	1.0	1.0	1.0

NOTE. Linear interpolation between the values given in table 5 is permissible, but not extrapolation outside the limits given.

Figure 8.25

As indicated in Clause 28.4.2 where a wall is stiffened by intersecting walls, the value of *K* can be determined assuming that the intersecting walls are equivalent to piers of width equal to the thickness of the intersecting wall and of thickness equal to 3 × the thickness of the stiffened wall, as shown in Figure 8.26.

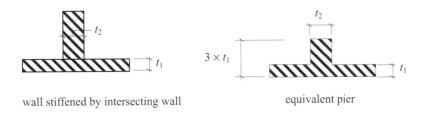

wall stiffened by intersecting wall   equivalent pier

Figure 8.26

8.4.6   *Slenderness (h_{ef}/t_{ef}) (Clause 28)*

The slenderness of a structural element is a mathematical concept used to assess the tendency of that element to fail by buckling when subjected to compressive forces. In many cases this is defined as a ratio of effective buckling length (l_e) to radius of gyration (*r*) about the axis of buckling.

The effective buckling length is related to the type and degree of end fixity of the element, and the radius of gyration is related to the cross-sectional geometry.

In rectangular cross-sections such as frequently encountered in masonry and timber design, the actual thickness is equal to the radius of gyration multiplied by $\sqrt{12}$; consider a rectangular element of width *d* and thickness *t*:

$$r_{yy} = \sqrt{\frac{I_{yy}}{\text{Area}}} = \sqrt{\frac{dt^3}{12} \times \frac{1}{dt}} = \frac{t}{\sqrt{12}}$$

$$t = \sqrt{12}\, r_{yy} \qquad \therefore\ t \propto r_{yy}$$

Since t is directly proportional to r then slenderness can equally well be expressed in terms of thickness instead of radius of gyration.

In masonry design, in Clause 3.19 of *BS 5628:Part 1:1992* the slenderness ratio is defined as '*The ratio of effective height or length to the effective thickness.*' The effective thickness as described in section 2.1.4.1 is a modification of the actual thickness to account for different plan layouts. The slenderness ratio should *not exceed 27*, except in the case of walls less than 90 mm thick, in buildings more than two storeys, where it should *not exceed 20* (see Clause 28.1).

8.4.6.1 Effective Height (Clause 28.3.1)

Walls (Clause 28.3.1.1)
'*The effective height of a wall may be taken as:*

(a) *0.75 times the clear distance between lateral supports which provide enhanced resistance* (see sections 8.4.6.4 and 8.4.6.6) *to lateral movement, or*

(b) *the clear distance between lateral supports which provide simple resistance to lateral movement.*'

Columns (Clause 28.3.1.2)
'*The effective height of a column should be taken as the distance between lateral supports or twice the height of the column in respect of a direction in which lateral support is not provided.*'

Columns formed by adjacent openings in walls (Clause 28.3.1.3)
'*Where openings occur in a wall such that the masonry between any two openings is, by definition[1], a column, the effective height of the column should be taken as follows.*

(a) *Where an enhanced resistance to lateral movement of the wall containing the column is provided, the effective height should be taken as 0.75 times the distance between the supports plus 0.25 times the height of the taller of the two openings.*

(b) *Where a simple resistance to lateral movement of the wall containing the column is provided, the effective height should be taken as the distance between the supports*'

These conditions are illustrated in Figure 8.27.
Note: It is important to ensure that the **column** and not only the wall is provided with the assumed restraint condition, particularly when it extends to the level of the support.

---

[1] Clause 3.7 An isolated vertical member whose width is not more than four times its thickness.

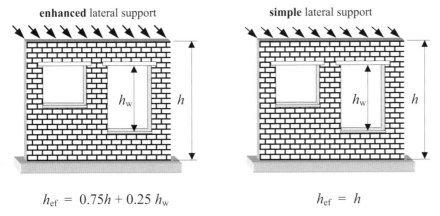

enhanced lateral support   simple lateral support

$h_{ef} = 0.75h + 0.25\,h_w$ $h_{ef} = h$

Figure 8.27

Piers (Clause 28.3.1.4)

'Where the thickness of a pier is not greater than 1.5 times the thickness of the wall of which it forms a part, it may be treated as a wall for effective height considerations; otherwise the pier should be treated as a column in the plane at right angles to the wall.

NOTE. The thickness of a pier, t_p, is the overall thickness of the wall or, when bonded into one leaf of a cavity wall, the thickness obtained by treating this leaf as an independent wall.'

8.4.6.2 Effective Length (Clause 28.3.2)

'The effective length of a wall may be taken as:

    *(a) 0.75 times the clear distance between vertical lateral supports or twice the distance between a support and a free edge, where lateral supports provide **enhanced resistance** to lateral movement* [see Figure 8.28].

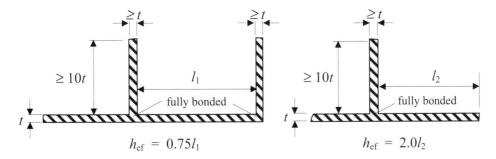

$h_{ef} = 0.75l_1$ $h_{ef} = 2.0l_2$

Figure 8.28

    *(b) the clear distance between lateral supports or 2.5 times the distance between a support and a free edge where lateral supports provide **simple resistance** to lateral movement'* [see Figure 8.29].

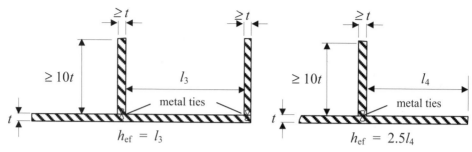

Figure 8.29

These definitions require an assessment of support conditions at each end of the element being considered, i.e. horizontal lateral support in the case of effective height, and vertical lateral support in the case of effective length. As indicated in Clause 28.2.1 the supports should be capable of transmitting the sum of the following design forces to the principal elements providing lateral stability to the structure:

(a) *'the simple static reactions to the total applied design horizontal forces at the line of lateral support, and*

(b) *2.5% of the total design vertical load that the wall or column is designed to carry at the line of the lateral support; the elements of construction that provide lateral stability to the structure as a whole need not be designed to support this force.'*

Two types of horizontal and vertical lateral supports are defined in the code:

♦   simple supports, and
♦   enhanced supports.

8.4.6.3 Horizontal Simple Supports (Clause 28.2.2.1)

All types of floors and roofs which abutt walls provide simple support if they are detailed as indicated in Appendix C of the code; a few typical examples are given in Figures 8.30 and 8.31.

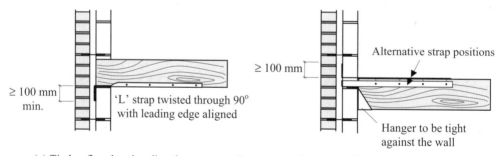

(a) Timber floor bearing directly on to a wall   (b) Timber floor using typical floor hanger

Figure 8.30

Note: In case (a) in houses up to three storeys no straps are required, provided that the joist spacing is not greater than 1.2 m and the joist bearing is 90 mm minimum.

In case (b) in houses up to three storeys no straps are required, provided that the joist is effectively tied to the hanger.

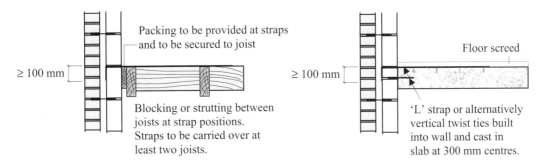

(a) Timber floor abutting external cavity wall

(b) In-situ and precast concrete floor abutting external cavity wall

Figure 8.31

8.4.6.4 Horizontal Enhanced Supports (Clause 28.2.2.2)

Enhanced lateral support can be assumed where:

'(a) floors and roofs of any form of construction span on to the wall or column from both sides at the same level;

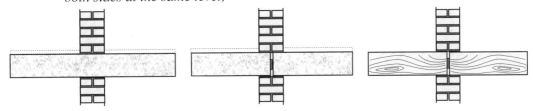

Figure 8.32

(b) an in-situ concrete floor or roof, or a precast concrete floor or roof giving equivalent restraint, irrespective of the direction of span, has a bearing of at least one-half the thickness of the wall or inner leaf of a cavity wall or column on to which it spans but in no case less than 90 mm;

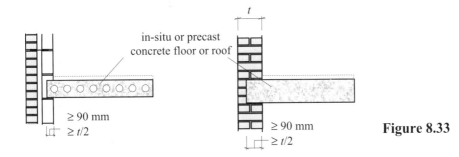

Figure 8.33

(c) *in the case of houses of not more than three storeys, a timber floor spans on to a wall from one side and has a bearing of not less than 90 mm.*

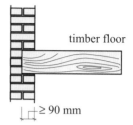

timber floor

≥ 90 mm

Figure 8.34

Preferably, columns should be provided with lateral support in both horizontal directions.'

8.4.6.5 Vertical Simple Supports (Clause 28.2.3.1)

Simple lateral support may be assumed where '... *an intersecting or return wall not less than the thickness of the supported wall or loadbearing leaf of a cavity wall extends from the intersection at least ten times the thickness of the supported wall or loadbearing leaf and is connected to it by metal anchors calculated in accordance with 28.2.1,* [see 8.4.6.2], *and evenly distributed throughout the height at not more than 300 mm centres.'*

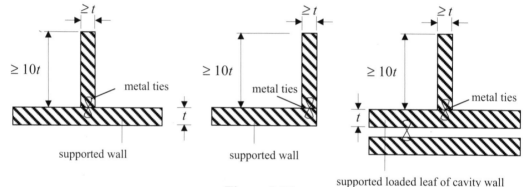

Figure 8.35

8.4.6.6 Vertical Enhanced Support (Clause 28.3.2)

Vertical enhanced support may be assumed as the cases indicated in 8.4.6.5 where the supporting walls are fully bonded providing restraint against rotation, as shown in Figure 8.36.

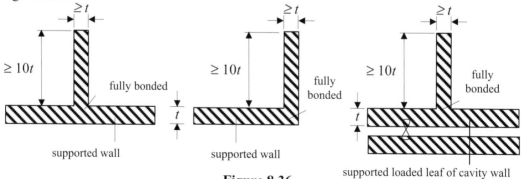

Figure 8.36

8.4.7   *Eccentricity of Applied Loading (Clauses 30, 31 and 32.1)*

In most cases the applied loading on a wall is not concentric. This is due to factors such as construction details, tolerances and non-uniformity of materials. The resultant eccentricity may be in the plane of the wall, as in the case of masonry shear-walls resisting combined lateral wind loading and vertical floor/roof loading, or perpendicular to the plane of the wall, as in the case of walls supporting floor/roof slabs and/or beams spanning on to them.

8.4.7.1 *Eccentricity in the Plane of a Wall*

This type of eccentricity is considered in Clauses 30 and 32.1 and Figure 4 of the code. The load distribution along a length of wall subject to both lateral and vertical loading can be determined using statics by combining both axial and bending effects, as shown in Figure 8.37.

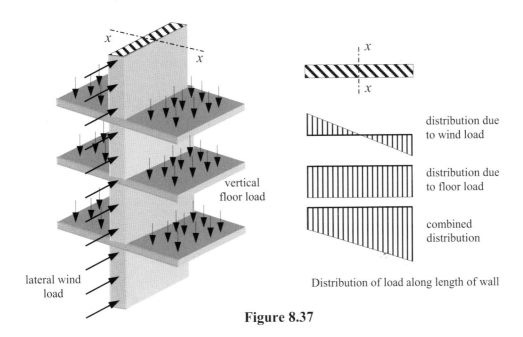

Figure 8.37

Standard methods of elastic analysis, assuming rigid-floor plate deformation and consequently load distribution in proportion to the relative stiffness of walls, can be used to determine the lateral loading on any particular shear-wall within a building. If the plan layout of the walls in a building is asymmetric, it may be necessary to consider the effects of torsion in the analysis procedure to calculate the lateral load distribution.

8.4.7.2 *Eccentricity Perpendicular to the Plane of a Wall*

In Clause 31 of the code it is stated that '*Preferably, eccentricity of loading on walls and columns should be calculated but, at the discretion of the designer it may be assumed that...*'. Whilst this calculation is possible (ref. 41), most engineers adopt the values suggested in the remainder of this clause, i.e.

◆ Where the load is transmitted to a
   wall by a single floor or roof it is
   assumed to act at one-third of the
   depth of the bearing area from the
   loaded face of the wall or
   loadbearing leaf.

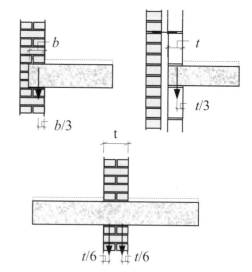

◆ Where a uniform floor is continuous
   over a wall, each side of the floor
   may be taken as being supported
   individually on half the total bearing
   area.

◆ Where joist hangers are used, the load should be assumed to be applied at the face
   of the wall.

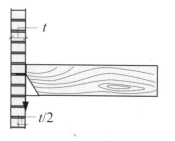
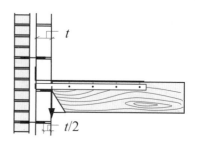

8.4.8   *Combined Slenderness and Eccentricity*

The effect of slenderness (Section 8.4.6), and eccentricity (Section 8.4.7), is to reduce the
loadbearing capacity of a loaded wall or column. The combined effects of both these
characteristics is allowed for in the code by evaluating a capacity reduction factor β as
shown in Figure 8.38 (Table 7 of the code).

The derivation of the capacity reduction factor β is given in Appendix B of the code.
The application of β is based on several assumptions:

◆ only braced vertical walls/columns are considered,
◆ additional moments caused by buckling effects are included,
◆ at failure, the stress distribution at the critical section can be represented by a
   rectangular stress block.

Consideration of the assumptions given above result in the location of the maximum
eccentricity of load (and hence critical section) occurring at a distance of $0.4h$ below the
top of the wall.

Slender- ness ratio $h_{\text{ef}}/t_{\text{ef}}$	Eccentricity at top of wall, e_x			
	Up to 0.05t (see note 1)	0.1t	0.2t	0.3t
0	1.00	0.88	0.66	0.44
6	1.00	0.88	0.66	0.44
8	1.00	0.88	0.66	0.44
10	0.97	0.88	0.66	0.44
12	0.93	0.87	0.66	0.44
14	0.89	0.83	0.66	0.44
16	0.83	0.77	0.64	0.44
18	0.77	0.70	0.57	0.44
20	0.70	0.64	0.51	0.37
22	0.62	0.56	0.43	0.30
24	0.53	0.47	0.34	
26	0.45	0.38		
27	0.40	0.33		

Table 7. Capacity reduction factor, β

NOTE 1. It is not necessary to consider the effects of eccentricities up to and including 0.05t.

NOTE 2. Linear interpolation between eccentricities and slenderness ratios is permitted.

NOTE 3. The derivation of β is given in Appendix B.

Figure 8.38

8.4.9   Eccentricities in Columns   (Clause 32.2.2)

Generally, the eccentricity of loading on a wall is relative to the axis parallel to the centre line, as indicated in Figure 8.39(a). In columns, it is common for the applied loading to be eccentric relative to both the major and minor axes, as indicated in Figure 8.39(b).

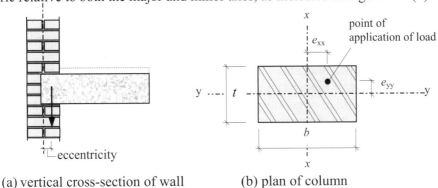

(a) vertical cross-section of wall   (b) plan of column

Figure 8.39

In Clause 32.2.2 the design vertical load resistance of rectangular columns is given by:

$$\frac{\beta \, btf_k}{\gamma_m}$$

where:
β is the capacity reduction factor,
b is the width of the column,
t is the thickness of the column,
f_k is the characteristic compressive strength of masonry, and
γ_m is the material partial safety factor.

The value of the capacity reduction factor β is dependent on the magnitudes of the eccentricities e_{xx} and e_{yy} of the applied load about the two principal axes.

8.4.10 *Type of Structural Unit (Clause 23.1, 23.1.3 to 23.1.9)*

The standard format structural units are 215 mm wide × 102.5 mm thick × 65 mm high, as shown in Figure 8.40.

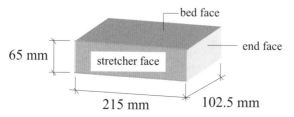

Figure 8.40

The characteristic strengths f_k of various types of masonry unit are specified in Tables 2(a) to 2(d) and Clauses 23.1.3 to 23.1.9 of the code; these are summarised in Figure 8.41.

8.4.11 *Laying of Structural Units (Clause 8)*

Structural units are normally laid on their bed face and the strengths given in Tables 2(a) to 2(d) relate to this. Where units are laid on a different face, i.e. the stretcher or end face, their strength in this orientation should be determined by testing according to the requirements of BS 187, BS 6073-1 and BS 3921 as appropriate.

To develop the full potential strength of units with frogs it is necessary to ensure that a full and correct bedding of mortar is made and the frogs are filled. In circumstances where a lower frog cannot be filled or is only partially filled, tests should be carried according to BS 187 and BS 3921 to determine the masonry strength.

Note: When bed joints are raked out for pointing it is important to make allowance for the resulting loss of strength, as indicated in Clause 8 of the code.

The design parameters discussed in section 8.4 are illustrated in Examples 8.1 to 8.8.

Type of Structural Unit	Reference	Dimensions H–height L–least horizontal dimension	Characteristic Strength (f_k)
standard format bricks	Clause 23.1	215 mm × 102.5 mm × 65 mm	Table 2(a)
90 mm wide × 90 mm high modular bricks	Clause 23.1.3	thickness equal to width	Table 2(a) × 1.25
		other thicknesses	Table 2(a) × 1.10
wide bricks	Clause 23.1.4	$H : L < 0.6$	Testing according to A.2
blocks	Clause 23.1	$H : L = 0.6$	Table 2(b)
hollow blocks	Clause 23.1.5	$0.6 < H : L < 2.0$	Interpolation between Tables 2(b) & 2(c)
solid concrete blocks (i.e. no cavities)	Clause 23.1.6	$0.6 < H : L < 2.0$	Interpolation between Tables 2(b) & 2(d)
hollow blocks	Clause 23.1	$2.0 < H : L < 4.0$	Table 2(c)
hollow concrete blocks filled with concrete	Clause 23.1.7	see Note 1	Interpolation between Tables 2(b) & 2(d)
natural stone	Clause 23.1.8	Design on the basis of solid concrete blocks of equivalent compressive strength; see Clause 23.1.6	
random rubble	Clause 23.1.9	0.75 × natural stone masonry of similar materials; if built with lime-mortar use 0.5 × masonry in mortar designation (iv)	

Note 1: The compressive strength of the blocks should be based on net area, and the 28-day cube strength of the concrete infill should not be less than this value.

Figure 8.41

8.5   Example 8.1:   Single-Leaf Masonry Wall – Concentric Axial Load

A series of internal walls in a braced building are indicated in Figure 8.42. Using the design data given and assuming all concrete slabs to be fully loaded, determine a suitable structural unit / mortar combination for the walls using:

i)   standard format bricks,
ii)   hollow blocks 100 mm × 440 mm long × 215 mm deep.

Design data:

Characteristic dead load including self-weight of slabs and finishes	7.0 kN/m²
Characteristic imposed on slabs	4.0 kN/m²
Assume the characteristic self-weight of standard format brick	2.2 kN/m²
Assume the characteristic self-weight of hollow blocks	1.5 kN/m²
Category of manufacturing/construction control	normal/normal

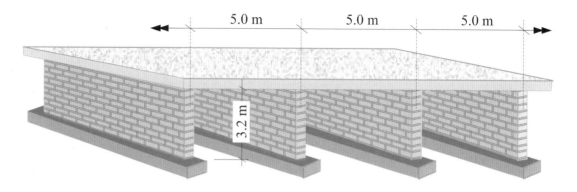

Figure 8.42

8.5.1   Solution to Example 8.1

Contract: Masonry Wall Job Ref. No. : Example 8.1 Part of Structure : Concentric Axial Load Calc. Sheet No. : 1 of 5		Calcs. by : W.McK. Checked by : Date :
References	**Calculations**	**Output**

	Consider a 1.0 metre length of wall
	
	(i) Standard format bricks
	Characteristic self-weight of wall = $(0.4 \times 3.2 \times 2.2)$*
	                                              = **2.82 kN/m**
	*Note: the critical section occurs at a distance of $0.4h$ down the wall, i.e. maximum eccentricity.
	
	Characteristic dead load due to the slab:
	                              = $(7.0 \times 5.0 \times 1.0)$
	                              = **35.0 kN/m per wall**
	Characteristic imposed load on the slab:
	                              = $(4.0 \times 5.0 \times 1.0)$
	                              = **20.0 kN/m per wall**	

Clause 22(a)	Partial Safety Factor for Loads (γ_f) Dead loads γ_f = 1.4 Imposed loads γ_f = 1.6 Design load/metre length of wall: = $[(1.4 (2.82 + 35.0) + (1.6 \times 20.0)]$ = **85.0 kN**	
Clause 32.2.1	Design Vertical Load Resistance of Walls Design vertical load resistance/unit length = $\dfrac{\beta\, t\, f_k}{\gamma_m}$ this value must be $\geq$ 85.0 kN The required unknown is the characteristic compressive strength of the masonry f_k : $$\therefore \quad f_k \geq \frac{85.0 \times \gamma_m}{\beta\, t}$$	

Contract: Masonry Wall Job Ref. No. : Example 8.1 Part of Structure : Concentric Axial Load Calc. Sheet No. : 2 of 5	Calcs. by : W.McK. Checked by : Date :

References	Calculations	Output
Clause 27.3	**Partial Safety Factor for Material Strength (γ_m)**	
	Category for manufacturing control is normal Category for construction control is normal	
Table 4(a)	**Table 4(a) Partial safety factors for material strength, γ_m**	

		Category of construction control	
		Special	Normal
Category of manufacturing control of structural units	**Special**	2.5	3.1
	Normal	2.8	3.5

References	Calculations	Output
	Partial safety factor $\gamma_m = 3.5$	
	The capacity reduction factor β is given in Table 7 and requires values for both the slenderness ratio and the eccentricity of the load.	
Clause 28	Consideration of slenderness of walls and columns slenderness ratio $(SR) = h_{ef}/t_{ef} \leq 27$	
Clause 28.2.2	Horizontal Lateral Support Since the walls have a concrete roof with a bearing length of at least one-half the thickness of the wall, enhanced resistance to lateral movement can be assumed.	
Clause 28.3.1	Effective Height $h_{ef} = 0.75 \times$ clear distance between lateral supports $= (0.75 \times 3200) = 2400$ mm	
Clause 28.4.1	Effective Thickness For single-leaf walls the effective thickness is equal to the actual thickness, as indicated in Figure 3 of the code. $t_{ef} = 102.5$ mm $SR = \dfrac{2400}{102.5} \approx 23.4 < 27$	
Clause 31	Eccentricity Perpendicular to the Wall The load may be assumed to act at an eccentricity equal to one-third of the depth of the bearing area from the loaded face of the wall. Since the slabs are supported from both sides with equal loads, the resultant eccentricity is equal to zero; concentric load.	

Contract: Masonry Wall Job Ref. No. : Example 8.1 Part of Structure : Concentric Axial Load Calc. Sheet No. : 3 of 5	Calcs. by : W.McK. Checked by : Date :

References	Calculations	Output
Table 7	**Capacity Reduction Factor** Linear interpolation between slenderness and eccentricity values is permitted when using Table 7. S.R. = 23.4, and $e \leq 0.05t$	

Table 7. Capacity reduction factor, β	
Slender- ness ratio h_{ef}/t_{ef}	Eccentricity at top of wall, e_x Up to 0.05t (see note 1)
18	0.77
20	0.70
22	0.62
24	0.53
26	0.45
27	0.40

$\beta = [0.62 - (0.09 \times 1.4)/2.0)] = 0.56$

References	Calculations	Output
Clause 32.2.1	**Design Vertical Load Resistance** $f_k \geq \dfrac{85.0 \times \gamma_m}{\beta t} = \dfrac{85.0 \times 3.5}{0.56 \times 102.5} = 5.2 \text{ N/mm}^2$	
Clause 23.1.2	**Narrow Brick Walls** When using standard format bricks to construct a wall one brick (i.e. 102.5 mm) wide, the values of f_k obtained from Table 2(a) can be multiplied by 1.15. f_k required $\geq \dfrac{5.2}{1.15} = \mathbf{4.52 \text{ N/mm}^2}$	
Table 2	Units of compressive strength 15 N/mm^2 (or greater) combined with mortar designation (i), (ii) or (iii) will be adequate. Although this thickness of wall will satisfy the ultimate limit state requirement, consideration should also be given to other limit states e.g. resistance to rain penetration, frost attack and/or fire, ability to accommodate movement. Advice regarding these criteria can be found in: *BS 5628:Part 3:2001.*	**Adopt standard format bricks with compressive strength $\geq$ 15 N/mm^2 and mortar designation (i), (ii) or (iiii).**

Contract: Masonry Wall Job Ref. No. : Example 8.1 Part of Structure : Concentric Axial Load Calc. Sheet No. : 4 of 5	Calcs. by : W.McK. Checked by : Date :

References	Calculations	Output

Consider a 1.0 metre length of wall

(ii) Hollow blocks:
100 mm wide × 440 mm long × 215 mm deep

$$\text{Aspect ratio} \quad = \quad \frac{height}{least\ horizontal\ dimension} \quad = \quad \frac{215}{100}$$
$$= \quad 2.15$$

Characteristic self-weight of wall $\quad = \quad (0.4 \times 3.2 \times 1.5)$
$$= \quad \textbf{1.92 kN/m}$$

Characteristic dead load due to the slab:
as before $\quad = \quad$ **35.0 kN/m per wall**
Characteristic imposed load on the slab:
as before $\quad = \quad$ **20.0 kN/m per wall**

Clause 22(a)

Partial Safety Factor for Loads (γ_f)
Dead loads $\quad \gamma_f \ = \ 1.4$
Imposed loads $\quad \gamma_f \ = \ 1.6$
Design load/metre length of wall:
$= \quad [(1.4\,(1.92 + 35.0) + (1.6 \times 20.0)] \quad = \quad \textbf{83.7 kN}$

Clause 32.2.1

Design Vertical Load Resistance of Walls

Design vertical load resistance/unit length $= \dfrac{\beta\,t\,f_k}{\gamma_m}$ this value

must be $\quad \geq \quad$ 83.7 kN

$$\therefore \quad f_k \ \geq \ \frac{83.7 \times \gamma_m}{\beta\,t}$$

Clause 27.3

Partial Safety Factor for Material Strength (γ_m)
Category for manufacturing control is normal
Category for construction control is normal

Table 4(a)

Table 4(a) Partial safety factors for material strength, γ_m		Category of construction control	
		Special	Normal
Category of manufacturing control of structural units	Special	2.5	3.1
	Normal	2.8	3.5

Partial safety factor $\gamma_m = \quad 3.5$

Contract: Masonry Wall Job Ref. No. : Example 8.1 Part of Structure: Concentric Axial Load Calc. Sheet No. : 5 of 5	Calcs. by : W.McK. Checked by : Date :

References	Calculations	Output
	The capacity reduction factor β is given in Table 7 and requires values for both the slenderness ratio and the eccentricity of the load.	
Clause 28	Consideration of slenderness of walls and columns slenderness ratio (SR) $=$ h_{ef}/t_{ef} $\leq$ 27	
Clause 28.2.2	Horizontal Lateral Support As before enhanced resistance to lateral movement can be assumed.	
Clause 28.3.1	Effective Height $h_{ef} = (0.75 \times 3200) = 2400$ mm	
Clause 28.4.1	Effective Thickness For single-leaf walls the effective thickness is equal to the actual thickness, as indicated in Figure 3 of the code. $t_{ef} = 100$ mm $SR = \dfrac{2400}{100} = 24.0 < 27$	
Clause 31	Eccentricity Perpendicular to the Wall As before, since the slabs are supported from both sides with equal loads the resultant eccentricity is equal to zero; concentric load.	
Table 7	Capacity Reduction Factor Linear interpolation between slenderness and eccentricity values when using Table 7 is not required in this case. S.R. = 24.0, and $e \leq 0.05t$ $\beta = 0.53$	
Clause 32.2.1	Design Vertical Load Resistance $f_k \geq \dfrac{83.7 \times \gamma_m}{\beta t} = \dfrac{83.7 \times 3.5}{0.53 \times 100} = 5.53$ N/mm^2	
Clause 23.1.2	Narrow Brick Wall: **Only applies to walls made from standard format bricks which are 102.5 mm wide, i.e. not in this case.** Aspect ratio = 2.15 $\therefore$ Use Table 2(c)	
Table 2(c)	Units of compressive strength 10 N/mm^2 combined with mortar designation (i) or (ii) will be adequate.	**Adopt hollow blocks with compressive strength ≥ 10 N/mm^2 and mortar designation (i) or (ii)**

8.6 Example 8.2: Single-Leaf Masonry Wall – Eccentric Axial Load

A 215 mm thick brickwork wall supports two precast floor slabs as indicated in Figure 8.43. Using the design data given, determine a suitable structural unit/mortar combination for the walls using:

Design data:
W_1:

Characteristic dead load (per metre length of wall)	10.0 kN
Characteristic imposed	25.0 kN

W_2:

Characteristic dead load	8.0 kN
Characteristic imposed	15.0 kN
Assume the characteristic self-weight of standard format brick	4.7 kN/m^2
Category of manufacturing/construction control	special/normal
Clear height between lateral supports providing enhanced resistance to lateral movement	2.8 m

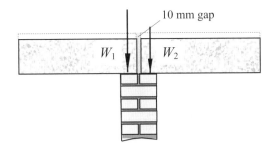

Figure 8.43

8.6.1 Solution to Example 8.2

Contract: Masonry Wall Job Ref. No. : Example 8.2 Part of Structure: Eccentric Axial Load Calc. Sheet No. : 1 of 4	Calcs. by : W.McK. Checked by : Date :

References	Calculations	Output
Clause 22(a)	Consider a 1.0 metre length of wall Partial Safety Factor for Loads (γ_f) Dead loads γ_f = 1.4 Imposed loads γ_f = 1.6 W_1: Design load/metre length of wall: = $[(1.4 \times 10) + (1.6 \times 25.0)]$ = **54.0 kN** W_2: Design load/metre length of wall: = $[(1.4 \times 8) + (1.6 \times 15.0)]$ = **35.2 kN** Self-weight of 0.4 × height of wall = $(0.4 \times 2.8 \times 1.0 \times 4.7)$ = **5.3 kN** **Total design load = W_{total} = (54.0 + 35.2 + 5.3) = 94.5 kN**	

Contract: Masonry Wall Job Ref. No. : Example 8.2 **Part of Structure: Eccentric Axial Load** **Calc. Sheet No. : 2 of 4**		**Calcs. by : W.McK.** **Checked by :** **Date :**

References	Calculations	Output
Clause 32.2.1	Design Vertical Load Resistance of Walls Design vertical load resistance/unit length $= \dfrac{\beta\, t\, f_k}{\gamma_m}$ this value $\qquad\qquad$ must be $\geq$ 94.5 kN The required unknown is the characteristic compressive strength of the masonry f_k : $\qquad\qquad \therefore \quad f_k \geq \dfrac{94.5 \times \gamma_m}{\beta\, t}$	
Clause 27.3	Partial Safety Factor for Material Strength (γ_m) Category for manufacturing control is special Category for construction control is normal	
Table 4(a)	**Table 4(a) Partial safety factors for material strength, γ_m** |		Category of construction control	| |---|---|---|---|
		Special	Normal
Category of manufacturing control of structural units	Special	2.5	3.1
	Normal	2.8	3.5 | Partial safety factor $\gamma_m =$ 3.1 The capacity reduction factor β is given in Table 7 and requires values for both the slenderness ratio and the eccentricity of the load.	
Clause 28	Consideration of slenderness of walls and columns: slenderness ratio (SR) $= h_{ef}/t_{ef} \leq 27$	
Clause 28.2.2	Horizontal Lateral Support Since the walls have enhanced resistance to lateral movement:	
Clause 28.3.1	Effective Height $h_{ef} = 0.75 \times$ clear distance between lateral supports $\quad = (0.75 \times 2800) = 2100$ mm	
Clause 28.4.1	Effective Thickness For single-leaf walls the effective thickness is equal to the actual thickness, as indicated in Figure 3 of the code. $t_{ef} = 215$ mm	

Contract: Masonry Wall Job Ref. No. : Example 8.2 Part of Structure: Eccentric Axial Load Calc. Sheet No. : 3 of 4	Calcs. by : W.McK. Checked by : Date :

References	Calculations	Output

$$SR = \frac{2100}{215} = 9.8 \quad < 27$$

Clause 31

Eccentricity Perpendicular to the Wall
Each of the loads may be assumed to act at an eccentricity equal to one-third of the depth of the bearing area from the loaded face of the wall.

Eccentricity of each load from the centre-line:
$$e_1 = e_2 = [(2/3) \times (0.5 \times 205)] + 5 = 73.68 \text{ mm}$$

Resultant moment of the loads about the centre-line:
$$= [(54 - 35.2) \times 73.68] = 1385 \text{ kNmm}$$

The equivalent eccentricity at the top of the wall
$$= \text{Moment/Total load} = (1385/89.2) = 15.6 \text{ mm}$$

The resultant eccentricity e_x at the top of the wall is given in Table 7 in terms of the thickness t of the wall.

$$e_x = (15.6/215)t = 0.073t$$

Table 7

Capacity Reduction Factor
Use linear interpolation between eccentricities of $0.05t$ and $0.1t$ for S.R. = 8 and S.R = 10:

$$SR = 8 \quad \beta_{ex = 0.07} = [1.00 - (0.12 \times 0.02/0.05)] = 0.95$$
$$SR = 10 \quad \beta_{ex = 0.07} = [0.97 - (0.09 \times 0.02/0.05)] = 0.93$$

SR	$\beta_{ex = 0.073}$
8	0.95
10	0.93

References	Calculations	Output
	Contract: Masonry Wall Job Ref. No. : Example 8.2 Part of Structure: Eccentric Axial Load Calc. Sheet No. : 4 of 4	Calcs. by : W.McK. Checked by : Date :

References	Calculations	Output
Clause 32.2.1	$SR = 9.8$ $\beta = [0.95 - (0.01 \times 1.8/2.0)] = 0.93$ Design Vertical Load Resistance $f_k \geq \dfrac{94.5 \times \gamma_m}{\beta t} = \dfrac{94.5 \times 3.1}{0.93 \times 215} = 1.47 \text{ N/mm}^2$	**Adopt standard format bricks with compressive strength ≥ 5 N/mm^2 and mortar designation (i), (ii), (iiii) or (iv).**
Table 2(a)	Standard format bricks: Units of compressive strength 5 N/mm^2 combined with mortar designation (i), (ii), (iii) or (iv) will be adequate.	

In Table 7 no values of β corresponding with slenderness ratios > 26 and eccentricities > 0.1t have been given, i.e. in the bottom right-hand corner. In circumstances where high eccentricities and high slenderness ratios exist, the capacity reduction factor can be evaluated using the equations in Appendix B. If possible, it is advisable to avoid this situation by using higher effective thicknesses and/or specifying details which will reduce the eccentricity.

8.7   Columns (Clause 32.2.2)

An isolated, axially loaded masonry element in which the width is not greater than four times the thickness is defined as a column as indicated in Figure 3 of the code. In addition the masonry between any two openings in a wall is, by definition, a column, as illustrated in Section 8.4.6, Figure 8.27. In most cases columns are solid but they may have a cavity, as shown in Figure 8.44.

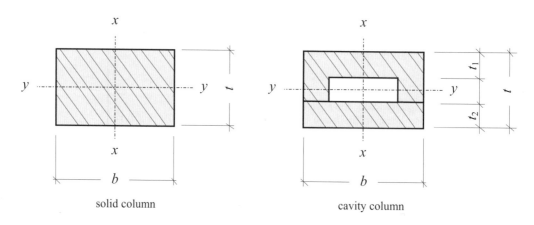

solid column   cavity column

Figure 8.44

In cavity columns:

(i) $b \leq (4 \times$ the overall thickness) when both leaves are loaded,   and

(ii) $b \leq (4 \times$ thickness of the loaded leaf) when only one leaf is loaded .

The design of axially loaded columns is very similar to that for walls, with additional consideration being given to the possibility of buckling about both the x–x axis and the y–y axis. (In walls support is provided in the longitudinal direction by adjacent material and only the y–y axis need be considered.) In Clause 32.2.2 the design vertical load resistance of a rectangular column is given by:

$$\frac{\beta \, b t \, f_k}{\gamma_m}$$

where:

b is the width of the column,

t is the thickness of the column,

β, f_k and γ_m are as before.

The value of the capacity reduction factor β is determined by considering four possible cases of eccentricity and slenderness, as indicated in Section 8.4.9. The effective height of columns is defined in Clause 28.3.1.2 and Clause 28.3.1.3 (see Section 8.4.6).

Piers in which the thickness t_p, is greater than $1.5 \times$ the thickness of the wall of which they form a part should be treated as columns for effective height considerations.

The effective height of columns should be taken as the distance between lateral supports or twice the height of the column in respect of a direction in which lateral support is not provided, as indicated in Clause 28.3.1.2. (**Note:** *Enhanced* resistance to lateral movement applies to walls, not columns.)

As indicated in Clause 28.2.2.2 resistance to lateral movement may be assumed where:

(a)   '*floors or roofs of any form of construction span on to the wall or column, from both sides at the same level;*'

(b)   *an in-situ concrete floor or roof, or a precast concrete floor or roof giving equivalent restraint, irrespective of the direction of the span, has a bearing of at least one-half the thickness of the wall or inner leaf of a cavity wall or column on to which it spans but in no case less than 90 mm;*'

(c)   *in the case of houses of not more than three storeys, a timber floor spans on to a wall from one side and has a bearing of not less than 90 mm.*'

'*Preferably, columns should be provided with lateral support in both horizontal directions.*'

8.8   Example 8.3: Masonry Column

A solid masonry column is required to support an eccentric axial load of 200 kN, as shown in Figure 8.45. Assume that lateral support is provided at the top of the column about both axes. Using the design data given, check the suitability of the proposed brick/mortar combination.

Design data:

Assume the category of manufacturing control	special
Assume the category of construction control	special
Characteristic strength of unit (standard format bricks)	20.0 N/mm^2
Mortar designation	Type (i)
Clear height about both axis	3000 mm

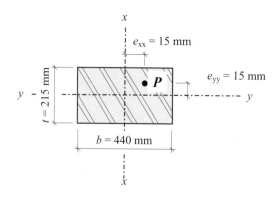

Figure 8.45

8.8.1   Solution to Example 8.3

Contract : Column Job Ref. No. : Example 8.3 Part of Structure : Eccentrically Loaded Column Calc. Sheet No. : 1 of 2	Calcs. by : W.McK. Checked by : Date :

References	Calculations	Output
Clause 32.2.2	Design vertical load resistance $\quad = \dfrac{\beta\, b\, t\, f_k}{\gamma_m}$	
Table 2(a)	Characteristic compressive strength $\quad f_k = 7.4$ N/mm^2	
Clause 23.1.1	Small plan area: Column area $\quad = \quad (0.215 \times 0.44) = \quad 0.095$ m$^2 \qquad < 0.2$ m^2 Use the multiplying factor $= \quad (0.7 + 1.5A)$ $\qquad\qquad\qquad\quad = \quad [0.7 + (1.5 \times 0.095)] \quad = 0.84$ Modified compressive strength $\quad = \quad (0.84 \times 7.4)$ $\qquad\qquad\qquad\qquad\qquad\quad = \quad 6.2$ N/mm^2	
Table 4(a)	Control; special/special $\quad \gamma_m = \quad 2.5$ $b \; = \; 440$ mm $\qquad\qquad\quad t \; = \; 215$ mm The capacity reduction factor β, is determined according to the actual eccentricities e_{xx} and e_{yy} with respect to $0.05b$ and $0.05t$ as indicated in Clause 32.2.2 (a), (b), (c) and (d). $0.05b = \quad (0.05 \times 440) \quad = \quad 22.0$ mm $0.05t = \quad (0.05 \times 215) \quad = \quad 10.8$ mm	

References	Calculations	Output
	$e_{xx} = 15.0$ mm $< 0.05b$ $e_{yy} = 15.0$ mm $\geq 0.05t$	
Clause 32.2.2 (b)	Use Table 7	
	Slenderness ratio $= \dfrac{h_{ef} \text{ (relative to the minor axis)}}{t_{ef} \text{ (based on column thickness } t)}$	
	$SR = \dfrac{3000}{215} = 14$	
	Use eccentricity appropriate to the minor axis: $e_x = e_{yy} = \dfrac{15.0}{215}t = 0.07t$	
Table 7	Using interpolation $\beta = [0.89 - (0.06 \times 0.02)/0.05] = 0.87$	**Proposed column section and brick/mortar combination are adequate**
	$P \leq \dfrac{\beta\,b\,t\,f_k}{\gamma_m} = \dfrac{0.87 \times 440 \times 215 \times 6.2}{2.5 \times 10^3} = 204$ kN ≥ 200 kN	

8.9 Stiffened Walls

The introduction of piers to stiffen walls reduces their slenderness and increases the loadbearing capacity. The reduction in slenderness results from an effective thickness which is greater than the unstiffened wall and is evaluated using a stiffness coefficient as indicated in Figure 3, Table 5 and Clause 28.4.2 in the code, i.e.

$$t_{ef} = t \times K \quad \text{where } K \geq 1.0$$

The effective thickness is subsequently used to determine the slenderness ratio and hence the capacity reduction factor β. Clearly since K is ≥ 1.0 the slenderness will be less, and consequently the value of β will be greater, than that for the unstiffened wall. In addition to this an equivalent wall thickness can be estimated since the pier will support some of the applied load. This can be carried out in most cases as follows, using a 'rule of thumb' as indicated in Figure 8.46.

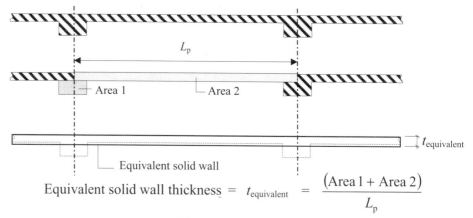

$$\text{Equivalent solid wall thickness} = t_{\text{equivalent}} = \frac{(\text{Area 1} + \text{Area 2})}{L_p}$$

Figure 8.46

The value of $t_{\text{equivalent}}$ can be used in the equation given in Clause 32.2.1, i.e. $\dfrac{\beta\, t\, f_k}{\gamma_m}$.

8.10   Example 8.4: Stiffened Single-Leaf Masonry Wall

An alternative structural form to the walls in Example 8.1 is to stiffen them by the inclusion of piers, as indicated in Figure 8.47. Using the same design data as given in Example 8.1 determine a new brick/mortar combination which can be used.

Design data:

Design load/metre length of wall	85.0 kN
Category of manufacturing/construction control	normal/normal
Effective height h_{ef}	2400 mm

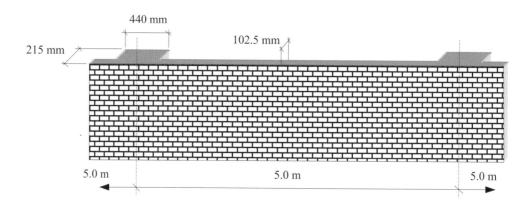

Figure 8.47

8.10.1 Solution to Example 8.4

References	Calculations	Output
	Contract : Column Job Ref. No. : Example 8.4 **Part of Structure : Stiffened Single-Leaf Wall** **Calc. Sheet No. : 1 of 2**	**Calcs. by : W.McK.** **Checked by :** **Date :**

References	Calculations	Output
Clause 32.2.1	Design vertical load resistance = $\dfrac{\beta\, t\, f_k}{\gamma_m}$	
Table 4(a)	Control; special/special γ_m = 2.5	102.5 mm 215 mm
	440 mm 5000 mm	
Clause 28.4.2 Figure 3	Effective Thickness t_{ef} = $t \times K$	
Table 5	Stiffness Coefficient for Walls Stiffened by Piers	

Table 5. Stiffness coefficient for walls stiffened by piers

Ratio of pier spacing (centre to centre) to pier width	Ratio t_p/t of pier thickness to actual thickness of wall to which it is bonded		
	1	2	3
6	1.0	1.4	2.0
10	1.0	1.2	1.4
20	1.0	1.0	1.0

NOTE. Linear interpolation between the values given in table 5 is permissible, but not extrapolation outside the limits given.

References	Calculations	Output
	$\dfrac{\text{pier spacing}}{\text{pier width}} = \dfrac{5000}{440} \approx 11.4;$ $\dfrac{\text{pier thickness}}{\text{actual thickness}} = \dfrac{215}{102.5} = 2.1$	
	Using linear interpolation in Table 5 K = 1.19 t_{ef} = (1.19×102.5) = 122 mm	
Clause 28	Slenderness: SR = $\dfrac{2400}{122}$ = 19.7 < 27 Eccentricity e_x = zero $\leq$ 0.05t	

References	Calculations	Output
	Contract : Column Job Ref. No. : Example 8.4 **Part of Structure : Stiffened Single-Leaf Wall** **Calc. Sheet No. : 2 of 2**	**Calcs. by : W.McK.** **Checked by :** **Date :**

References	Calculations	Output
Table 7	Capacity Reduction Factor (β) Using linear interpolation in Table 7 $\quad \beta = 0.71$ The equivalent thickness to allow for the additional load carried by the piers can be estimated: Equivalent solid wall thickness = $t_{equivalent} = \dfrac{(\text{Area } 1 + \text{Area } 2)}{L_p}$ Area 1 $\quad = \quad [(215 - 102.5) \times 440] \quad = \quad 49{,}500 \text{ mm}^2$ Area 2 $\quad = \quad (102.5 \times 5000) \quad = \quad 512{,}500 \text{ mm}^2$ $t_{equivalent} \quad = \quad \dfrac{(49500 + 512500)}{5000} \quad = \quad 112.4 \text{ mm}$	
Clause 32.2.1	**Design Vertical Load Resistance** $f_k \geq \dfrac{85.0 \times \gamma_m}{\beta\, t_{equivalent}} \quad = \quad \dfrac{85.0 \times 3.5}{0.71 \times 112.4} \quad = \quad 3.73 \text{ N/mm}^2$ **Note:** Ignoring the contribution of the piers to the strength i.e. using t instead of $t_{equivalent}$ is conservative and will produce a slightly higher required value of f_k.	**Adopt standard format bricks with compressive strength ≥ 10 N/mm^2 and mortar designation (i), (ii) or (iiii).**

8.11 Cavity Walls

The fundamental requirements of external walls in buildings include the provision of adequate strength, stability, thermal and sound insulation, fire resistance and resistance to rain penetration. Whilst a single skin-wall can often satisfy the strength and stability requirements, in many cases the thickness of wall necessary to satisfy some of the other requirements (e.g. resistance to rain penetration, cf. *BS 5628:Part 3*) can be uneconomic and inefficient.

In most cases of external wall design, cavity wall construction comprising two leaves with a gap between them is used. Normally the outer leaf is a half-brick (102.5 mm thick) common or facing brick and the inner leaf is either the same or light-weight, thermally efficient, concrete block.

The minimum thickness of each leaf is specified in Clause 29.1.2 of *BS 5628:Part 1* as 75 mm. The width of the cavity between the leaves may vary between 50 mm and 150 mm but should not be greater than 75 mm where either of the leaves is less than 90 mm in thickness, as indicated in Clause 29.1.3 of the code. These criteria are illustrated in Figure 8.48.

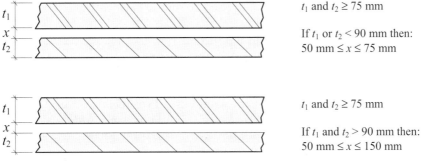

t_1 and $t_2 \geq 75$ mm

If t_1 or $t_2 < 90$ mm then:
50 mm $\leq x \leq 75$ mm

t_1 and $t_2 \geq 75$ mm

If t_1 and $t_2 > 90$ mm then:
50 mm $\leq x \leq 150$ mm

Figure 8.48

The minimum width requirement is to reduce the possibility of bridging across the cavity, e.g. by mortar droppings, which could cause the transmission of moisture. The maximum width requirement is to limit the length of the wall ties connecting the two leaves and hence reduce their tendency to buckle if subjected to compressive forces.

The effect of ties is to stiffen each of the leaves in a wall and consequently to reduce the slenderness ratio.

Where the applied vertical load is supported by one leaf only, the influence of the unloaded leaf is considered when evaluating the effective thickness of the wall, as indicated in Figure 3 of the code. The cross-sectional area used to determine the load resistance is that of the loaded leaf only.

Where the applied vertical load acts between the centroids of the two leaves, it should be replaced by a statically equivalent axial load in each leaf, as indicated in Clause 32.2.3 of the code and illustrated in Figure 8.49.

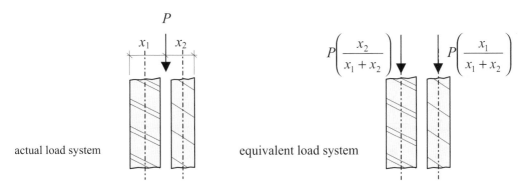

Figure 8.49

Each leaf is designed seperately to resist the equivalent axial load, using the stiffening effect of the other leaf to determine the effective thickness and hence the slenderness ratio.

The provision of ties is governed by the requirements of Clause 29.1.5 and suitable minimum values are given Table 6 of *BS 5628:Part 1* and also shown in Figure 8.50.

The ties should be staggered and evenly distributed over a wall area. Additional ties should be provided at a rate of one tie per 300 mm height or equivalent, located not more than 225 mm from the vertical edges of openings and at vertical unreturned or unbonded edges such as at movement joints or the sloping verge of gable walls. As indicated in Clause 29.1.6, ties should be embedded at least 50 mm in each leaf.

Selection of wall ties: types and lengths				
Least leaf thickness (one or both)	Nominal cavity width	Permissible type of tie		Tie length
mm	mm	Shape name* in accordance with BS 1243	Type member** in accordance with DD 140 : Part 2	mm
75	75 or less	(a), (b) or (c)	1, 2, 3, 4	175
90	75 or less	(a), (b) or (c)	1, 2, 3, 4	200
90	76 to 90	(b) or (c)	1 or 2	225
90	91 to 100	(b) or (c)	1 or 2	225
90	101 to 125	(c)	1 or 2	250
90	126 to 150	(c)	1 or 2	275
* (a) – butterfly (b) – double triangle (c) – vertical twist				
** type 1 is the stiffest and type 4 the least stiff				

Figure 8.50

Since the two leaves of a cavity wall may be of different materials with differing physical properties and/or subject to different thermal effects, differential movement is inevitable. To prevent potentially damaging loosening of the embedded ties, the uninterrupted height of the outer leaf of external cavity walls can be limited or suitable detailing of the construction incorporated to accommodate the movement.

8.11.1 Limitation on Uninterrupted Height (Clause 29.2.2)

The code specifies that:'… *the outer leaf should be supported at intervals of not more than every third storey or every 9 m, whichever is less. … for buildings not exceeding four storeys or 12 m in height, whichever is less, the outer leaf may be uninterrupted for its full height.*' This is illustrated in Figure 8.51.

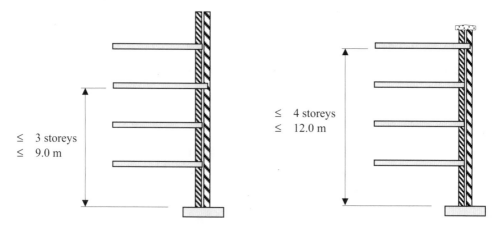

$\leq$ 3 storeys
$\leq$ 9.0 m

$\leq$ 4 storeys
$\leq$ 12.0 m

Figure 8.51

8.11.2 Accommodation of Differential Vertical Movement (Clause 29.2.3)

Where differential movement is accommodated by calculation this should include all factors such as elastic movement, moisture movement, thermal movement, creep etc., and the value should not exceed 30 mm. Detailing which includes separate lintels for both leaves, appropriate ties which can accommodate movement and the fixing of window frames cills etc. must be considered. In most cases engineers adopt the method of limiting the uninterrupted height.

8.11.3 Accommodation of Differential Horizontal Movement (Clause 29.2.4)

Advice is given in BS 5628:Part 3 on the provision of vertical joints to accommodate horizontal movements.

8.12 Example 8.5: Cavity Wall

The external cavity brick wall of a braced building is shown in Figure 8.52. The wall supports a reinforced concrete slab at the first level which imposes characteristic loads as indicated.

  Using the design data given, check the suitability of the proposed brick/mortar combination for the inner leaf.

Design data:
W_1:

Characteristic dead load (per metre length of wall)	50.0 kN
Characteristic imposed	80.0 kN

W_3:

Characteristic dead load	15.0 kN
Characteristic imposed	25.0 kN
Characteristic self-weight of wall	2.2 kN/m^2
Category of manufacturing control	special
Category of construction control	normal
Mortar designation	Type (ii)
Compressive strength of unit	15 N/mm^2

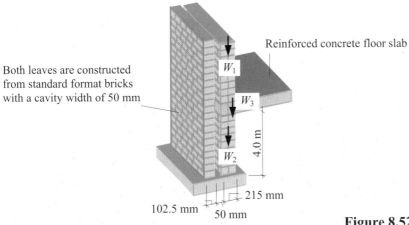

Both leaves are constructed from standard format bricks with a cavity width of 50 mm

Reinforced concrete floor slab

W_1

W_3

W_2

4.0 m

215 mm

102.5 mm   50 mm

Figure 8.52

8.12.1 Solution to Example 8.5

Contract : Column Job Ref. No. : Example 8.5 Part of Structure : Cavity Wall Calc. Sheet No. : 1 of 2	Calcs. by : W.McK. Checked by : Date :

References	Calculations	Output
Clause 22(a)	Partial Safety Factor for Loads (γ_f) Dead loads $\quad \gamma_f = 1.4$ Imposed loads $\gamma_f = 1.6$ Consider a 1.0 metre length of wall $W_1:$ concentric load Design load/metre length of wall from above: $= [(1.4 \times 50.0) + (1.6 \times 80.0)] = $ **198.0 kN** $W_2:$ concentric load Characteristic self-weight of $(0.4 \times$ height of wall): $= (0.4 \times 4.0 \times 2.2) \qquad\qquad = $ **3.52 kN**	
Clause 31	$W_3:$ eccentric load Design load/metre length of wall from reinforced concrete slab: $= [(1.4 \times 15.0) + (1.6 \times 25.0)] = $ **61.0 kN** **Total design load** $= (W_1 + W_2 + W_3)$ $= (198.0 + 61.0 + 3.52) = $ **262.5 kN**	
Clause 32.2.1	Design vertical load resistance/unit length $= \dfrac{\beta\, t\, f_k}{\gamma_m}$	
Clause 27.3	Partial Safety Factor for Material Strength (γ_m) Category for manufacturing control is special Category for construction control is normal	
Table 4(a)	$\gamma_m = 3.1$	
Clause 28	Consideration of Slenderness of Walls and Columns slenderness ratio (SR) $= h_{ef}/t_{ef} \quad \le \quad 27$	
Clause 28.2.2	Horizontal Lateral Support Since this structure has a concrete roof with a bearing length of at least one-half the thickness of the wall, enhanced resistance to lateral movement can be assumed.	
Clause 28.3.1	Effective Height $h_{ef} = 0.75 \times$ clear distance between lateral supports $= (0.75 \times 4000) = 3000$ mm	
Clause 28.4.1	Effective Thickness For cavity walls the effective thickness is as indicated in Figure 3 of the code and equal to the greatest of: (a) $2/3(t_1 + t_2) = 2/3(102.5 + 215) = 211.7$ mm, (b) $\quad t_1 \quad = 102.5$ mm $\quad$ or (c) $\quad t_2 \quad = 215$ mm $\qquad\qquad\qquad \therefore t_{ef} = 215$ mm	

Contract : Column Job Ref. No. : Example 8.5 Part of Structure : Cavity Wall Calc. Sheet No. : 2 of 2	Calcs. by : W.McK. Checked by : Date :

References	Calculations	Output
Clause 28	Slenderness ratio $= SR = \dfrac{3000}{215} \approx 14 < 27$	
Clause 31	Eccentricity Perpendicular to the wall The load from the slab may be assumed to act at an eccentricity equal to one-third of the depth of the bearing area from the loaded face of the wall, i.e. $t/6$ from the centre-line. The resultant eccentricity e_x required for the capacity reduction factor β can be determined as follows: $(W_1 + W_3) \times e_x = (W_3 \times t/6)$ $(198.0 + 61.0)\, e_x = (61.0 \times t)/6$ $e_x = (10.2t\,/\,259)$ $= 0.04t$ $\leq 0.05t$	
Table 7	Capacity Reduction Factor: $\quad \beta = 0.89$	
Table 2(a)	Compressive strength of unit $= 15 \text{ N/mm}^2$ Mortar designation $= \text{Type (ii)}$ Compressive strength of masonry $\quad f_k = 5.3 \text{ N/mm}^2$	
Clause 32.2.1	Design Vertical Load Resistance $= \dfrac{\beta\, t\, f_k}{\gamma_m}$ $= \dfrac{0.89 \times 215 \times 5.3}{3.1}$ $= \mathbf{327.1 \text{ kN/m}}$ $\geq 262.5 \text{ kN}$	**The proposed brick/mortar combination for the inner leaf is suitable**

8.13   Laterally Loaded Walls

8.13.1   Introduction

In many instances, for example cladding panels, masonry must resist forces induced by lateral loading such as wind pressure. The geometric dimensions and support conditions of such panels frequently result in two-way bending as shown in Figure 8.53.

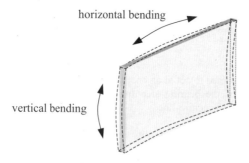

horizontal bending

vertical bending

Figure 8.53

Masonry is a non-isotropic material resulting in flexural strengths and modes of failure which are different in the vertical and horizontal directions. The failure mode due to simple vertical bending occurs with cracking developing along the bed joints, as shown in Figure 8.54(a), and in simple horizontal bending with cracking developing through the vertical joints, as shown in Figure 8.54(b).

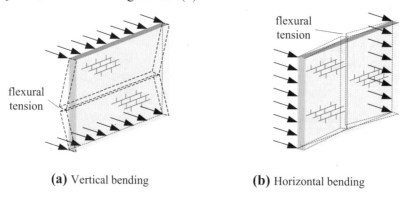

flexural tension

flexural tension

(a) Vertical bending   **(b)** Horizontal bending

Figure 8.54

In *BS 5628:Part 1:1992* Table 3, the characteristic flexural strength of masonry (f_{kx}) is given relative to both individual failure modes. As indicated in Clause 24.2 of the code, linear interpolation between the values of f_{kx} is permitted for:

(a) concrete block walls of thickness between 100 mm and 250 mm,
(b) concrete blocks of compressive strength between 2.8 N/mm^2 and 7.0 N/mm^2
      in a wall of given thickness.

Since the development of flexural tension is clearly an important factor in the flexural strength of masonry, any loading which tends to reduce this will enhance the strength of a panel. In loadbearing walls such as the lower levels of multi-storey buildings, the pre-compression caused by the dead load from above will increase the resistance to failure parallel to the bed joints since the flexural tension is reduced by the axial compression.

Note: (It is important to recognise that the increased compressive stresses due to combined flexure and axial load should not exceed their specified limits; and in addition, that the pre-compression will be significantly lower at the upper levels of the building.)

The ratio of the flexural strength due to vertical bending to that due to horizontal bending is known as the **orthogonal ratio**, i.e.:

$$\mu = \frac{f_{kx\,par}}{f_{kx\,perp}}$$

where:
$f_{kx\,par}$ is the characteristic flexural strength parallel to the bed joints

$f_{kx\,perp}$ is the characteristic strength perpendicular to the bed joints

As indicated in Clause 36.4.2 of the code, the value of the orthogonal ratio can be modified to reflect the enhanced strength due to pre-compression by increasing the value of $f_{kx\,par}$ accordingly, i.e.:

$$\text{Modified orthogonal ratio} = \frac{\left(\dfrac{f_{kx\,par}}{\gamma_m} + g_d\right)}{\dfrac{f_{kx\,perp}}{\gamma_m}} = \frac{\left(f_{kx\,par} + \gamma_m g_d\right)}{f_{kx\,perp}}$$

where:
$f_{kx\,par}$, $f_{kx\,perp}$ and γ_m are as before
g_d is the design vertical load/unit area.

This should be evaluated assuming $\gamma_f = 0.9$ as indicated in Clause 22 of the code.

Wall panels with a high *height to length* ratio and with one vertical edge unsupported are particularly sensitive to failure parallel to the bed joints, and it is often advantageous to utilise any existing pre-compression to enhance the flexural strength.

8.13.2 Design Strength of Panels (Clause 36.4 and Table 9)

The design of panels as given in Clause 36.4 of the code requires that the design moment of resistance is equal to or greater than the calculated design moment due to the applied loads, as determined using the bending moment coefficients from Table 9, i.e.

(i)   Consider failure perpendicular to the plane of the bed joints:

Moment of resistance of the panel $= \dfrac{f_{kx\ perp}}{\gamma_m} Z$

Moment due to applied loads $= \alpha W_k \gamma_f L^2$ per unit height

$$\frac{f_{kx\ perp}}{\gamma_m} Z \geq \alpha W_k \gamma_f L^2 \qquad\qquad \text{Equation (1)}$$

(ii)   Consider failure parallel to the plane of the bed joints:

Moment of resistance of the panel $= \dfrac{f_{kx\ par}}{\gamma_m} Z$

Moment due to applied loads $= \mu\alpha W_k \gamma_f L^2$ per unit height

$$\frac{f_{kx\ par}}{\gamma_m} Z \geq \mu\alpha W_k \gamma_f L^2 \qquad\qquad \text{Equation (2)}$$

where:
α is the bending moment coefficient taken from Table 9
γ_f is the partial safety factor for loads (Clause 22)
μ is the orthogonal ratio
L is the length of the panel
W_k is the characteristic wind load/unit area
Z is the elastic section modulus (**note:** based on the **net** area for hollow blocks)
$f_{kx\ perp}$, $f_{kx\ par}$, and γ_m are as before.

The calculation to determine the required flexural strength can be carried out using either equation (1) or equation (2); it is not necessary to evaluate both since the effect of the orthogonal ratio is included in the bending moment coefficients given in Table 9, i.e. in Equation (2):

$$\frac{f_{kx\ par}}{\gamma_m} = \frac{\mu\, f_{kx\ perp}}{\gamma_m} = \mu \times (\alpha W_k \gamma_f L^2 \text{ per unit height}) = \mu\alpha W_k \gamma_f L^2 \text{ per unit height}$$

The values of the bending moment coefficients given in Table 9 are dependent on:

i)   the edge restraint conditions of panels,
ii) the orthogonal ratio (μ) and
iii) the aspect ratio h/L, where h is the height of the panel and L is the length of the panel.

The edge restraint conditions are summarised in Figure 8.55.

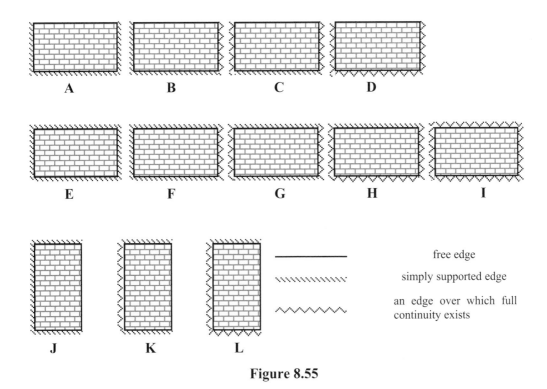

Figure 8.55

The values of the orthogonal ratio vary from 0.3 to 1.0, and the values of the aspect ratio vary from 0.3 to 1.75. Linear interpolation of μ and h/L is permitted. When the dimensions of a wall are outside the range of h/L given in Table 9, it will usually be sufficient to calculate the moments on the basis of a simple span e.g. a panel of type A.

A wall having h/L less than 0.3 will tend to act as a free-standing wall, whilst the same panel having h/L greater than 1.75 will tend to span horizontally.

8.13.3 *Edge Support Conditions and Continuity (Clause 36.2)*

The lateral resistance of masonry panels is dependent on the degree of rotational and lateral restraint at the edges of the panel and/or the continuity past a support such as a pier or a column. In most cases, with the exception of a free edge, unless masonry is fully bonded into return walls or is in intimate and permanent contact with a roof or floor, a simple support should be assumed.

In *BS 5628:Part 3* guidance is given to assess fixed support conditions for both single-leaf solid walls and cavity walls. In all cases a wall should be adequately connected to its support and all supports should be sufficiently strong and rigid to carry the transmitted load.

In addition to the Part 3 guidelines, *BS 5628:Part 1:1992*, Figures 7 and 8 provide examples of typical support conditions and continuity over supports, as shown in Figures 8.56 and 8.57. The connection to a support may be in the form of ties or by shear resistance of the masonry taking into account the damp-proof course, if it exists.

Values of characteristic tensile and shear strength which should be used for various types of wall ties used in panel supports are given in Part 1, Table 8 of the code. Where it is necessary to transmit compression, provided that any gap between the wall and the supporting structure is not greater than 75 mm, the values for tension given in Table 8 may be used for ties other than butterfly or double triangle types.

Metal ties to columns.
Simple support – direct force restraint restricted to values given in Table 8.

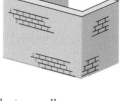

Bonded return walls.
Restrained support – direct force and moment restraint limited by tensile strength of masonry as given in Clause 24.1.

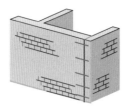

Metal ties to columns or unbonded return walls. Shear and possibly moment restraint. Shear limited to values given in Table 8.

Bonded to piers.
Intermediate pier – Direct force and moment restraint limited by tensile strength of masonry as given in Clause 24.1.
End pier – Simple support: direct force restraint restricted to values given in Table 8.

Figure 8.56   Vertical support conditions

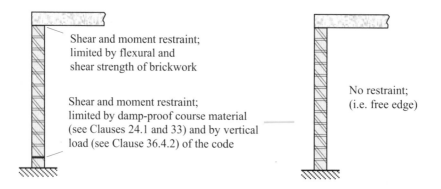

Shear and moment restraint; limited by flexural and shear strength of brickwork

Shear and moment restraint; limited by damp-proof course material (see Clauses 24.1 and 33) and by vertical load (see Clause 36.4.2) of the code

No restraint; (i.e. free edge)

(a)   In-situ floor slab cast on to wall span parallel to wall

(c)   Wall built up to but **not** pinned to the slab above

Figure 8.57   Vertical support conditions

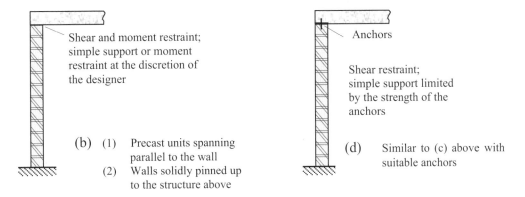

Shear and moment restraint; simple support or moment restraint at the discretion of the designer

Anchors

Shear restraint; simple support limited by the strength of the anchors

(b) (1) Precast units spanning parallel to the wall
(2) Walls solidly pinned up to the structure above

(d) Similar to (c) above with suitable anchors

Figure 8.57 (continued) Horizontal support conditions

8.13.4 *Limiting Dimensions (Clause 36.3)*

The limiting dimensions of laterally loaded walls and free-standing walls, as set out in Clause 36.3, are indicated in Figure 8.58.

(a) *Panel supported on three edges*

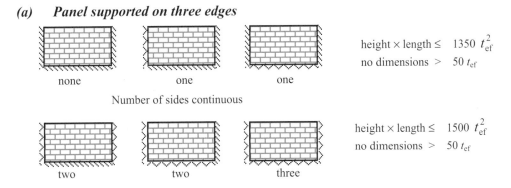

none — one — one

Number of sides continuous

$\text{height} \times \text{length} \leq 1350\ t_{ef}^2$
$\text{no dimensions} > 50\ t_{ef}$

two — two — three

Number of sides continuous

$\text{height} \times \text{length} \leq 1500\ t_{ef}^2$
$\text{no dimensions} > 50\ t_{ef}$

(b) *Panel supported on four edges*

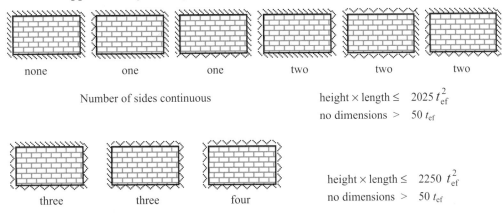

none — one — one — two — two — two

Number of sides continuous

$\text{height} \times \text{length} \leq 2025\ t_{ef}^2$
$\text{no dimensions} > 50\ t_{ef}$

three — three — four

$\text{height} \times \text{length} \leq 2250\ t_{ef}^2$
$\text{no dimensions} > 50\ t_{ef}$

Number of sides continuous

Figure 8.58

(c)   Panel supported top and bottom

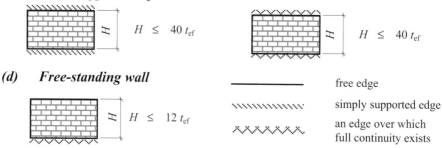

$$H \leq 40\, t_{ef} \qquad\qquad H \leq 40\, t_{ef}$$

(d)   Free-standing wall

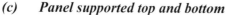

$$H \leq 12\, t_{ef}$$

—————————   free edge

〜〜〜〜〜〜〜〜〜   simply supported edge

ʌʌʌʌʌʌʌʌ   an edge over which
full continuity exists

Figure 8.58 (continued)

8.13.5   Design Lateral Strength of Cavity Walls (Clause 36.4.5)

The design lateral strength of cavity walls is normally assumed to be equal to the sum of that for each of the two leaves. This is based on the assumption that vertical twist ties, ties with equivalent strength or double-triangle/wire butterfly ties as indicated in Clause 36.2, are used. Since both leaves are assumed to deflect together, assuming the same orthogonal ratio, the total applied load will be distributed between the walls in proportion to their stiffnesses (**note:** the relative stiffness I is adequate for this purpose), e.g.

Assume the total applied lateral load $= W$

Applied load on $leaf_1 = \left[\dfrac{I_1}{I_{total}}\right]W$

Applied load on $leaf_2 = \left[\dfrac{I_2}{I_{total}}\right]W$

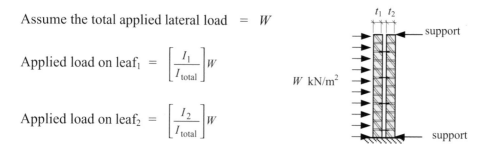

$W \ \text{kN/m}^2$

where:

I_1 is the second moment of area of $leaf_1$ [i.e. $(t_1^3/12)$/unit length]

I_2 is the second moment of area of $leaf_2$ [i.e. $(t_2^3/12)$/unit length]

$I_{total} = (I_1 + I_2)$

These proportions can be expressed in terms of the leaf thicknesses:

$$\text{applied load on } leaf_1 = \left[\frac{t_1^3}{t_1^3 + t_2^3}\right]W \quad ; \quad \text{applied load on } leaf_2 = \left[\frac{t_2^3}{t_1^3 + t_2^3}\right]W$$

Where the orthogonal ratio of the two leaves differ, the proportions can be expressed in terms of the bending moment resistance of each leaf.

8.14 Example 8.6: Laterally Loaded Single-Leaf Wall

A two-storey steel-framed building is to be clad in a single-leaf 102.5 mm thick brickwork wall, as shown in Figure 8.59. Panel A is simply supported at the bottom, with continuity existing over the vertical sides. Using the design data given, check the suitability of the brick/mortar combination for Panel A.

Assume:
Panel A is free at roof level and is a non-loadbearing wall.
Neglect any precompression due to the self-weight of the brickwork.

Design data:

Characteristic wind load (W_k)	0.7 kN/m^2
Category of construction control	special
Mortar designation	(ii)
Use clay bricks with water absorption	$\geq$ 7%
	$\leq$ 12%

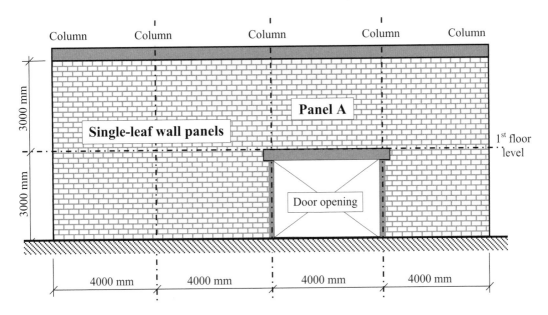

Figure 8.59

8.14.1 Solution to Example 8.6

Contract : Framed Building Job Ref. No. : Example 8.6	Calcs. by : W.McK.
Part of Structure : Laterally Loaded Single-Leaf Wall Panel	Checked by :
Calc. Sheet No. : 1 of 2	Date :

References	Calculations	Output
Clause 36.2 Figures 7 and 8	Assume simple supports at the bottom of the wall. The continuity of the wall over the columns provides fixity to the vertical edges of each panel. Panel considered for design: 	
Clause 36.3 (Figure 8.58)	Limiting dimensions: The panel is supported on three edges, with two sides continuous. height × length ≤ $1500\,t_{ef}^2$ and no dimension > $50t_{ef}$ t_{ef} = 102.5 mm, ∴ $1500\,t_{ef}^2$ = 15.76×10^6 mm^2 $50t_{ef}$ = 5125 mm height × length = (3000 × 4000) = 12.0×10^6 mm^2 < $1500\,t_{ef}^2$ height = 3000 mm < $50t_{ef}$ length = 4000 mm < $50t_{ef}$	**Limiting dimensions are satisfied**
Table 9 (Figure 8.55)	The panel considered corresponds to Type C. The orthogonal ratio (μ) is required to obtain a value for the bending moment coefficient: this is dependent on the type of brick and mortar used. Assume a value equal to 0.35 and check at the end of the calculation.	
Table 3	For clay bricks with water absorption ≥ 7% and ≤ 12% and using mortar designation (ii): $f_{kx\,par}$ = 0.4 N/mm^2 $f_{kx\,perp}$ = 1.1 N/mm^2 $\mu = \dfrac{0.4}{1.1}$ = 0.36; $h/L = (3000/4000)$ = 0.75	

Contract : Framed Building Job Ref. No. : Example 8.6	Calcs. by : W.McK.
Part of Structure : Laterally Loaded Single-Leaf Wall Panel	Checked by :
Calc. Sheet No. : 2 of 2	Date :

References	Calculations	Output

References: Table 9

Calculations:

(Extract relating to panel type C)

	h/L				
μ	0.3	0.5	**0.75**	1.00	1.25
1.00	-	-	0.037	-	-
0.90	-	-	0.038	-	-
0.80	-	-	0.039	-	-
0.70	-	-	0.040	-	-
0.60	-	-	0.041	-	-
0.50	-	-	0.043	-	-
0.40	-	-	0.044	-	-
0.35	-	-	0.045	-	-
0.30	-	-	0.046	-	-

Interpolation (for $\mu = 0.36$) gives $\alpha \approx 0.045$

References: Clause 36.4.2

Consider bending **perpendicular** to the bed joints.

Design bending moment $= (\alpha\, W_k \gamma_f L^2)/$ metre height.

References: Clause 22

$\gamma_f = 1.4$

Note:
'*In the particular case of freestanding walls and laterally loaded wall panels, whose removal would in no way affect the stability of the remaining structure, γ_f applied on the wind load may be taken as 1.2*'

$$\text{Design bending moment} = (0.045 \times 0.7 \times 1.2 \times 4.0^2)$$
$$= 0.6 \text{ kNm /metre height}$$

References: Clause 36.4.3

Design moment of resistance $= \left(\dfrac{f_{kx\,perp}}{\gamma_m} Z\right) \geq 0.6 \times 10^6$

Category for construction control is normal

References: Table 4(b)

$\gamma_m = 2.5$

$$Z = \frac{1000 \times 102.5^2}{6} = (1.751 \times 10^6) \text{ mm}^3 / \text{metre height}$$

$$\left(\frac{f_{kx\,perp}}{\gamma_m} Z\right) = \frac{1.1 \times 1.751 \times 10^6}{2.5}$$

$$= 0.77 \times 10^6 \text{ Nmm / metre height}$$

$$\geq 0.6 \times 10^6 \text{ Nmm / metre height}$$

Output:

Proposed brick/mortar combination is adequate

8.15   Example 8.7: Laterally Loaded Cavity Wall

A two-storey steel-framed building is to be clad in a cavity wall comprising 102.5 mm thick brickwork inner and outer leaves, as shown in Figure 8.60. The brickwork leaf of Panel A is free on the edge adjacent to the door opening, simply supported at the base, and continuous along the other two sides. Using the design data given, determine a suitable brick/mortar combination for Panel A.

Assume:
The cavity wall is loadbearing.
Neglect any precompression due to the self-weight of the brickwork.
The two leaves are tied together using cavity wall ties in accordance with the relevant British Standards.

Design data:
Characteristic wind load (W_k)   1.0 kN/m^2
Category of construction control   normal

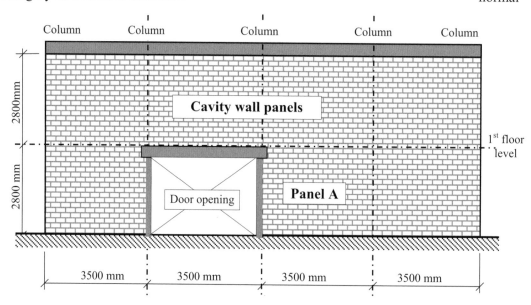

Figure 8.60

8.15.1 Solution to Example 8.7

Contract : Steel Frame Job Ref. No. : Example 8.7	Calcs. by : W.McK.
Part of Structure : Cavity Wall Panel	Checked by :
Calc. Sheet No. : 1 of 2	Date :

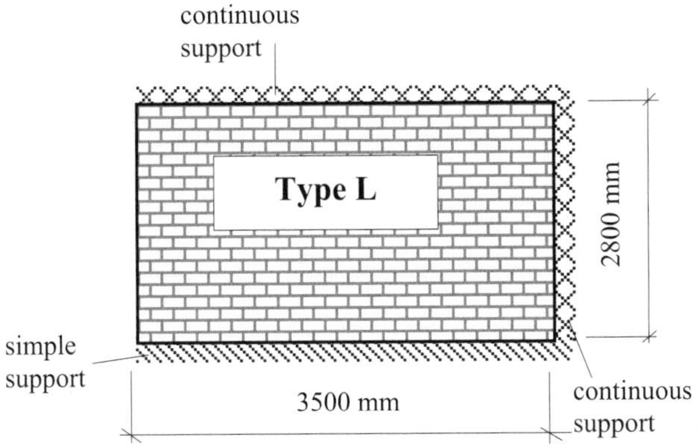

References	Calculations	Output
Clause 36.2	In the case of cavity walls, only one leaf need be continuous (the thicker of the two), provided wall ties in accordance with Table 6 are used between the two leaves and between each section of the discontinuous leaf and the support. Panel considered for design: 	
Clause 36.3 (Figure 8.58)	Limiting dimensions: The panel is supported on three edges with two sides continuous and one simply supported height × length ≤ $1500\,t_{ef}^2$ and no dimension > $50t_{ef}$	
Clause 28.4.1	Effective thickness For cavity walls the effective thickness is as indicated in Figure 3 of the code and equal to the greatest of: (a) $2/3(t_1 + t_2)$ = $2/3(102.5 + 102.5)$ = 136.7 mm or (b) t_1 = 102.5 mm or (c) t_2 = 75 mm ∴ t_{ef} = 136.7 mm ∴ $1500\,t_{ef}^2$ = 28.03×10^6 mm^2 $50t_{ef}$ = 6835 mm height × length = (2800×3500) = 9.8×10^6 mm^2 < $1500\,t_{ef}^2$ height = 2800 mm < $50t_{ef}$ length = 3500 mm < $50t_{ef}$	Limiting dimensions are satisfied

Contract : Steel Portal Frame Job Ref. No. : Example 8.7	Calcs. by : W.McK.
Part of Structure : Cavity Wall Panel	Checked by :
Calc. Sheet No. : 2 of 2	Date :

References	Calculations	Output
Table 9 (Figure 8.55)	The panel considered corresponds to Type L. The orthogonal ratio (μ) is required to obtain a value for the bending moment coefficient, since this is dependent on the type of brick and mortar used. A assume a value equal to 0.35 and check at the end of the calculation. $\mu = 0.35$; $h/L = (2800/3500) = 0.8$	
Table 9	(Extract relating to panel type L)	

	h /L				
μ	0.3	0.5	**0.75**	**1.00**	1.25
0.40	-	-	0.055	0.078	-
0.35	-	-	0.060	0.084	-
0.30	-	-	0.066	0.092	-

References	Calculations	Output
	Interpolation (for $h/L = 0.8$) gives $\alpha = 0.065$	
Clause 36.4.2	Consider bending perpendicular to the bed joints. Design bending moment $= (\alpha W_k \gamma_f L^2)/$ metre height.	
Clause 22	$\gamma_f = 1.4$ Design bending moment $= (0.065 \times 1.0 \times 1.4 \times 3.5^2)$ $= 1.11$ kNm /metre height	
Clause 36.4.5	Since both leaves are the same, the bending moment carried by one leaf $= (0.5 \times 1.11) = 0.555$ kNm /metre height	
Clause 36.4.3	Design moment of resistance $= \left(\dfrac{f_{kx\,perp}}{\gamma_m} Z \right) \geq 0.555 \times 10^6$	
Table 4(b)	$\gamma_m = 3.0$ $Z = \dfrac{1000 \times 102.5^2}{6} = (1.751 \times 10^6)$ mm^3 / metre height $\dfrac{f_{kx\,perp} \times 1.751 \times 10^6}{3.0} \geq 0.555 \times 10^6$ Nmm / metre height $f_{kx\,perp} \geq 0.95$ N/mm^2	
Table 3	Use clay bricks having a water absorption between 7% and 12% with a mortar designation Type (ii) or (iii) ($f_{kx\,perp} = 1.1$ N/mm^2) Check the assumed value of the orthogonal ratio μ (i.e. = 0.35) $\mu = \dfrac{f_{kx\,par}}{f_{kx\,perp}} = \dfrac{1.1}{0.4} = 0.36 \geq$ assumed value	**Adopt clay bricks: with water absorption between 7% and 12%. Mortar designation Type (ii) or (iii)**

8.16 Concentrated Loads on Walls

When considering relatively flexible members bearing on a wall, the stress distribution due to the reaction is assumed to be triangular as shown in Figure 8.61.

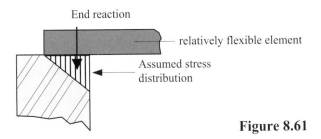

Figure 8.61

This is evident in Clause 31 when assessing the eccentricity, i.e. assuming '*that the load transmitted to a wall by a single floor or roof acts at one-third of the depth of the bearing area from the loaded face of the wall or loadbearing leaf.*'

In the case of loads (e.g. beam end reactions, column loads) which are applied to a wall through relatively stiff elements such as deep reinforced concrete beams, **pad-stones** or **spreader beams** as shown in Figure 8.62, provision is made in Clause 34 of the code for enhanced local bearing stresses.

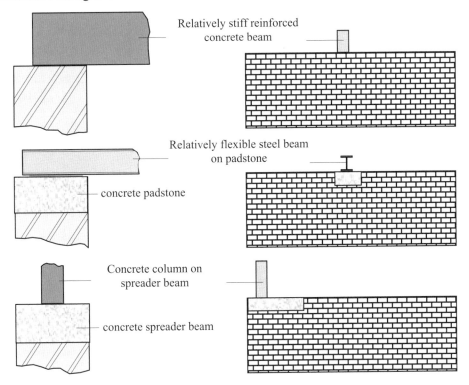

Figure 8.62

Three types of bearing are considered in Figure 5 of the code and are illustrated in Table 8.6 and Table 8.7 in this text.

Bearing Type 1	x	y	z	Local Design Strength
	$\geq \frac{1}{2} t$	$\leq 3 t$		$\dfrac{1.25 f_k}{\gamma_m}$
	$\geq \frac{1}{2} t$	$\leq 2 t$		$\dfrac{1.25 f_k}{\gamma_m}$
	≥ 50 mm $\leq \frac{1}{2} t$	No restriction	Edge distance may be zero	$\dfrac{1.25 f_k}{\gamma_m}$
	$> \frac{1}{2} t$ $\leq t$	$\leq 6 x$	Edge distance $\geq x$	$\dfrac{1.25 f_k}{\gamma_m}$

Table 8.6

Bearing Type 2	x	y	z	**Local Design Strength**
	≥ 50 mm $\leq \frac{1}{2}t$	$\leq 8x$	Edge distance $\geq x$	$\dfrac{1.5f_k}{\gamma_m}$
	$\geq \frac{1}{2}t$	$\leq 2t$		$\dfrac{1.5f_k}{\gamma_m}$
	$> \frac{1}{2}t$ $\leq t$	$\leq 4t$	Edge distance $\geq x$	$\dfrac{1.5f_k}{\gamma_m}$
Bearing Type 3				**Local Design Strength**
  Spreader beam	The distribution of stress under the spreader beam should be derived from an acceptable elastic theory, e.g. (i) using simple elastic theory assuming a triangular stress block with zero tension, or (ii) using Timoshenko's analysis for elastic foundations (46)			$\dfrac{2.0f_k}{\gamma_m}$

Table 8.7

In ***Bearing Type 1*** an increase of 25% in the characteristic compressive strength (f_k) is permitted immediately beneath the bearing surface, whilst in ***Bearing Type 2*** and ***Bearing Type 3*** 50% and 100% increases respectively are permitted.

This local increase in strength at the bearing of a concentrated load is primarily due to the development of a triaxial state of stress in the masonry. Factors relating to the precise bearing details (e.g. relative dimensions of wall and bearing area, or proximity of the bearing to the wall surface or wall end) also influence the ultimate failure load. The stress concentrations which occur in these circumstances rapidly disperse throughout the remainder of the masonry. It is normally assumed that this dispersal occurs at an angle of 45°, as shown in Figure 8.63.

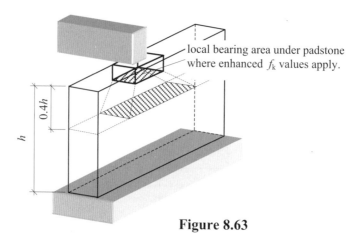

Figure 8.63

In Clause 34 the code requires that the strength of masonry be checked both locally under a concentrated load assuming the enhanced strength and at a distance of $0.4h$ below the level of the bearing where the strength should be checked in accordance with Clause 32. The effects of other loads, which are applied above the level of the bearing, should also be included in the calculated design stress. The assumed stress distributions for Bearing Types 1, 2 and 3 are illustrated in Figure 6 of the code and are shown in Figure 8.64.

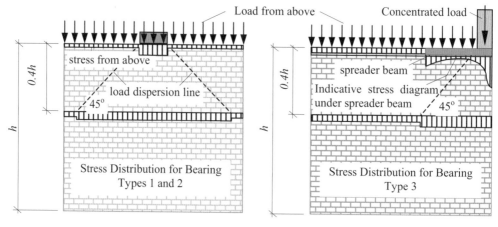

Figure 8.64

In Bearings Types 1 and 2 it is reasonable to assume a uniform stress distribution beneath the concentrated loads. Since the enhancement of f_k is partly due to the development of a triaxial state of stress beneath the load, in most cases it is assumed that restraint is provided at each side of the bearing. **Note:** *It is important to recognise that strength enhancement cannot be justified in situations in which a minimum edge distance occurs on BOTH sides of a bearing.*

8.16.1 Example 8.8: Concentrated Load Due to Reinforced Concrete Beam

A 255 mm thick brickwork cavity wall in a masonry office block supports a reinforced concrete beam on its inner leaf, as shown in Figure 8.65. Using the data provided, determine a suitable brick/mortar combination.

Design data:

Category of manufacturing control	special
Category of construction control	normal
Ultimate load from above the level of the beam	21.0 kN/m
Ultimate load from the floor slab	8.0 kN/m
Ultimate end reaction from the beam	75 kN
Characteristic unit-weight of plaster	21.0 kN/m^3
Characteristic unit-weight of brickwork	18.0 kN/m^3

Assume that the walls are part of a braced structure

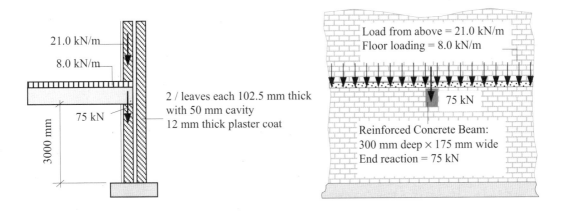

Figure 8.65

8.16.2 Solution to Example 8.8

Contract : Office Block Job Ref. No. : Example 8.8 Part of Structure : Masonry Wall Under Beam Calc. Sheet No. : 1 of 5	Calcs. by : W.McK. Checked by : Date :

References	Calculations	Output
	Self-weight of 12 mm thick plaster layer $\;=\;$ (0.012×21) $\qquad\qquad\qquad\qquad\qquad\qquad\quad\; =\;$ 0.25 kN/m^2 Self-weight of 102.5 mm thick cavity wall $=$ (0.1025×18) $\qquad\qquad\qquad\qquad\qquad\qquad\quad\; =\;$ 1.85 kN/m^2 Ultimate weight of wall and plaster $\quad=\;$ $1.4 \times (0.25+1.85)$ $\qquad\qquad\qquad\qquad\qquad\qquad\quad\; =\;$ 2.94 kN/m^2 	
Clause 34	The bearing stress should be considered at two locations; section x–x for local stresses and section y–y for distributed stresses.	
Clause 27.3 Table 4(a)	Category of manufacturing control – special Category of construction control – normal $\gamma_m \;=\; 3.1$	
Figure 5 (Table 8.7 of this text)	**Consider section x–x** width of bearing $x \;=\;$ 102.5 mm $\geq \tfrac{1}{2}t,\; \leq t$ length of bearing $y \;=\;$ 175 mm $\leq 4t$ edge distance $z \;\geq\; x$ Assume Bearing Type 2 at location of the beam Local strength $\;=\; \dfrac{1.5 f_k}{\gamma_m}$	

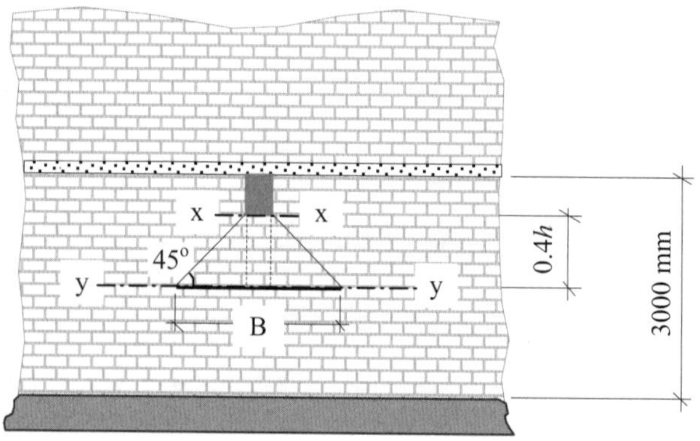

References	Calculations	Output
	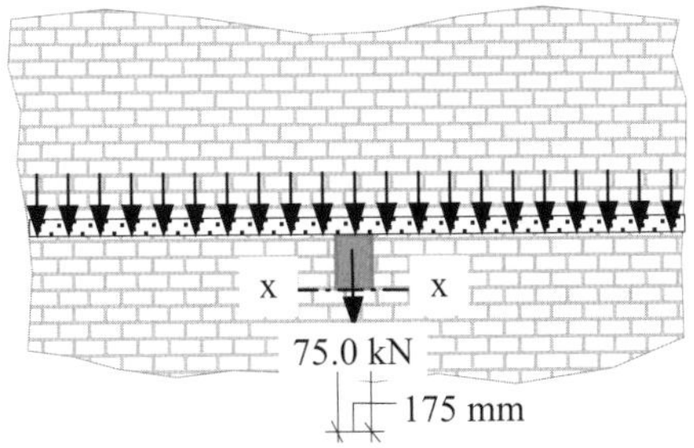	

Design load from above $\quad = (29.0 \times 0.175) = \quad 5.1$ kN

Design load from beam $\quad = 75.0$ kN

Total design load at x–x $\quad = (5.1 + 75.0) \quad = \quad 80.1$ kN

Bearing Type 2:

$$\text{Local resistance} \quad = \left[\frac{1.5 f_k}{\gamma_m} \times A_b\right] \geq \quad 80.1 \text{ kN}$$

$$A_b = (102.5 \times 175) \quad = \quad 179.38 \times 10^2 \text{ mm}^2$$

$$f_k \geq \left[\frac{80.1 \times 10^3 \times 3.1}{1.5 \times 179.38 \times 10^2}\right] \quad = \quad 9.2 \text{ N/mm}^2$$

Clause 23.1.2   The narrow brick wall factor applies $\therefore f_k$ required $= (9.2 / 1.15)$

$$= 8.0 \text{ N/mm}^2$$

Consider section y–y

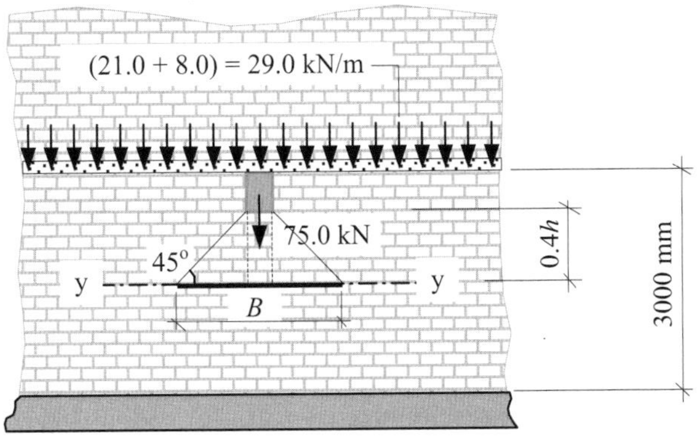

$0.4h \quad = \quad (0.4 \times 3000) \quad = \quad 1200$ mm

$B \quad = \quad [175 + (2 \times 0.4\,h)] = \quad [175 + (2 \times 1200)]$

$\quad = \quad 2575$ mm

Load due to self-weight of $0.4h$ of wall $\quad = \quad (1.2 \times 2.9)$

$\quad = 3.5$ kN/m length

Contract : Office Block Job Ref. No. : Example 8.8 Part of Structure : Masonry Wall Under Beam Calc. Sheet No. : 3 of 5	Calcs. by : W.McK. Checked by : Date :

References	Calculations	Output
	Design load from above = 29.0 kN/m length Design load from beam = 75.0 kN This load is distributed over the length $B =$ 2575 mm $\qquad = \dfrac{75.0}{2.575} = $ 29.1 kN/m length Design load at section y–y = (3.5 + 29.0 +29.1) $\qquad\qquad\qquad = $ 61.6 kN/m length	
Clause 32.2.1 Table 4(a)	Design vertical load resistance/unit length $= \dfrac{\beta\, t\, f_k}{\gamma_m}$ $\gamma_m = $ 3.1	
Clause 28	Consideration of Slenderness of Walls and Columns slenderness ratio (SR) $=$ h_{ef}/t_{ef} $\leq$ 27	
Clause 28.2.2	Horizontal Lateral Support Since this structure has a concrete floor with a bearing length of at least one-half the thickness of the wall, enhanced resistance to lateral movement can be assumed.	
Clause 28.3.1	Effective Height $h_{ef} = $ $0.75 \times$ clear distance between lateral supports $\qquad = $ $(0.75 \times 3000) = $ 2250 mm	
Clause 28.4.1	Effective Thickness For cavity walls the effective thickness is as indicated in Figure 3 of the code and equal to the greatest of: (a) $2/3(t_1 + t_2) = $ $2/3(102.5 + 102.5) = $ 136.7 mm or (b) t_1 $= $ 102.5 mm or (c) t_2 $= $ 102.5 mm $\qquad \therefore$ $t_{ef} = $ 136.7 mm Slenderness ratio $= $ SR $= $ $\dfrac{2250}{136.7} = $ 16.5 < 27	
Clause 31	Eccentricity Perpendicular to the Wall 29.0 kN 29.1 kN $e_x \approx \dfrac{(29.1 \times t/6)}{61.6} = $ 0.08t 3.5 kN	

Contract : Office Block Job Ref. No. : Example 8.8	Calcs. by : W.McK.
Part of Structure : Masonry Wall Under Beam	Checked by :
Calc. Sheet No. : 4 of 5	Date :

References	Calculations	Output

Table 7 — Capacity Reduction Factor

Table 7. Capacity reduction factor, β

Slenderness ratio h_{ef}/t_{ef}	Eccentricity at top of wall, e_x			
	Up to 0.05t (see note 1)	0.1t	0.2t	0.3t
0	1.00	0.88	0.66	0.44
16	0.83	0.77	0.64	0.44
18	0.77	0.70	0.57	0.44

$\beta_{(SR=16.5;\ ex=0.08t)} \approx 0.8$

Clause 23.1.2 — Narrow Brick Walls

When using standard format bricks to construct a wall one brick (i.e. 102.5 mm) wide, the values of f_k obtained from Table 2(a) can be multiplied by 1.15.

Clause 32.2.1

Design vertical load resistance $= \dfrac{\beta t (1.15 \times f_k)}{\gamma_m} \geq 61.6$ kN/m

$$f_k = \frac{(61.6 \times 3.1)}{(0.8 \times 102.5 \times 1.15)}$$
$$= 2.0 \text{ N/mm}^2$$

From the two values obtained, i.e. 8.0 N/mm² and 2.0 N/mm², it is evident that the selection of unit strength is based on the local bearing strength at section x–x.

Since this is a localised problem the introduction of a bearing pad under the beam may result in a more economic solution.

Assume units with compressive strength of 10 N/mm² and Type (ii) mortar are to be used.

Table 2

$f_k = 4.2$ N/mm²

Local design resistance $= \left(\dfrac{1.5 f_k}{\gamma_m}\right) \times A_b \geq 61.6$ kN/m

$$A_b \geq \left(\frac{61.6 \times 10^3 \times 3.1}{1.5 \times 4.2}\right) = 30.3 \times 10^3 \text{ mm}^2$$

References	Calculations	Output
	Contract : Office Block Job Ref. No. : Example 8.8	**Calcs. by : W.McK.**

Contract : Office Block Job Ref. No. : Example 8.8
Part of Structure :   Masonry Wall Under Beam
Calc. Sheet No. : 5 of 5

Calcs. by : W.McK.
Checked by :
Date :

References	Calculations	Output
	Min. length of bearing required $= \dfrac{30.3 \times 10^3}{102.5} = $ 296 mm For Type 2 bearing with $\frac{1}{2}t \le x \le t$ the length $y \le 4t$ $\therefore y \le (4 \times 102.5) = $ 410 mm > required length of 296 mm 320 mm 102.5 mm Adopt a bearing pad (assume 45° dispersion of load) (320 mm long × 102.5 mm wide × 150 mm thick)	**Adopt:** **Units with compressive strength ≥ 10.0 N/mm²** **in mortar Type (ii)** **Use concrete padstone** **320 mm long ×** **102.5 mm wide ×** **150 mm thick**

8.17 Glossary of Commonly Used Terms

Bat

A portion of brick manufactured or formed on site by cutting a whole brick across its length, e.g. a snapheader.

Bed face

The face of a structural unit which is normally laid on the mortar bed.

Bed joint

A mortar layer between the bed faces of masonry units.

Block

A masonry unit exceeding in length, width or height the dimensions specified for a brick. To avoid confusion with slabs and panels, the height of a block should not exceed either its length or six times its width.

Bond	An arrangement of structural units in an element (e.g. a wall) designed to ensure that vertical, horizontal and transverse distribution of load occurs throughout the element.
Brick	A masonry unit, including joint material, which does not exceed 337.5 mm in length, 225 mm in width and 112.5 mm in height. In addition, to avoid confusion with tile work, the height should not be less than 38 mm.
Brickwork/Blockwork	An assemblage of bricks or blocks bonded together to create a structural element.
Cavities	Holes which are closed at one end.
Cavity wall	Two parallel single-leaf walls, usually at least 50 mm apart, and effectively tied together with wall ties, the space between being left as a continuous cavity or filled with non-loadbearing material (usually thermal insulation).
Cellular bricks	Bricks having holes closed at one end which exceed 20% of the volume of the brick.
Chase	Channel formed in the face of masonry.
Closer:	A portion of brick manufactured or formed on site by cutting a whole brick across its length, and used to maintain bond, e.g. *kingcloser* and *queencloser.*
Collar-jointed wall	Two parallel single-leaf walls spaced at least 25 mm apart, with the space between them filled with mortar and so tied together as to result in composite action under load.
Common bricks	Masonry unit suitable for general construction but with no particular surface finish or attractive appearance.
Coordinating size	The size of a coordinating space allocated to a masonry unit, including allowances for joints and tolerances.
Corbel	A unit cantilevered from the face of a wall to form a bearing.
Cornice	A continuous projection from the facade of a building, part of a building or a wall.
Course:	A layer of masonry which includes a layer of mortar and masonry units.

Damp-proof course **(dpc)**	A layer or layers of material laid or inserted in a structure to prevent the passage of water.
Double-leaf wall	See collar-jointed wall.
Dowel	A device such as a flat strip or round bar, of uniform cross-section, embedded in the mortar of some of the horizontal joints (beds) at the ends of a panel, and fixed rigidly to an adjacent structure to provide lateral restraint, thus preventing in-plane horizontal movement of the panel.
Efflorescence	The resulting white bloom left on the surface of a wall after soluble salts, which are present in the bricks/mortars, are washed out by excess water. These salts will subsequently re-dissolve in rain and be washed away in a relatively short period of time.
Engineering bricks	Dense, semi-vitreous fired-clay bricks having minimum compressive strength and maximum absorption characteristics conforming to the requirements of *BS 3921:1985*.
Faced wall	A wall in which the facing and backing are bonded such that they behave compositely under load.
Facing bricks	Masonry units which are specially manufactured to provide an aesthetically attractive appearance.
Fair faced	Work built with particular care, with respect to line and with even joints, where it is visible when finished.
Flashing	A sheet of impervious material (e.g. lead, bituminous felt) applied to a structure and dressed to cover an intersection or joint where water would otherwise penetrate.
Frogged bricks	Bricks having depressions formed in one or more bed faces, the volume of which does not exceed 20% of the gross volume of the brick.
Grip hole	A formed void in a masonry unit to enable it to be more readily grasped and lifted with one or both hands or by machine.
Grouted cavity wall	Two parallel single-leaf walls, spaced at least 50 mm apart, effectively tied together with wall ties and with the intervening cavity filled with fine aggregate concrete (grout), which may be reinforced, so as to result in composite action under load.
Header	A structural unit with its end showing on the face of the wall.

Hollow bricks	Bricks having holes in excess of 25% and larger than perforated bricks.
Indenting	The omission of structural units to form recesses into which future work can be bonded.
Jamb (Reveal)	The visible part of each side of a recess or opening in a wall.
Joint: **Bed joint**	The mortar layer upon which the structural units are set.
Cross joint	A joint, other than a bed joint, normal to the face of the wall.
Wall joint	A joint parallel to the face of a wall.
Jointing:	The filling and finishing of raked-out joints during construction.
Lime Bleeding	The resulting staining left on the surface of a wall after soluble lime, which is produced during the hydration of the mortar, is deposited by the movement of rain water through freshly set and hardened mortar. These disfiguring stains will not weather off but will require removal with dilute acid to restore the appearance of the masonry.
Loadbearing masonry	Masonry which is suitable for supporting significant vertical/lateral loads in addition to its own self-weight.
Longitudinal joint	A vertical mortar joint within the thickness of a wall, parallel to the face of the wall.
Movement joint	A joint specifically designed and provided to permit relative movement of a wall and its adjacent structure to occur without impairing the functional integrity of the structure as a whole.
Padstone	A strong block, usually concrete, bedded on a wall to distribute a concentrated load.
Panel	An area of masonry with defined boundaries which may or may not contain openings.
Partition	An internal wall intended for visual sub-division of space, i.e. non-loadbearing.
Perforated bricks	Bricks having holes in excess of 25% of the brick's volume, provided the holes are less than 20 mm wide or 500 mm^2 in area with up to three handholds within the 25% total.
Perpend joint	A mortar joint perpendicular to the bed joint and to the face of the wall (a vertical cross-joint).

Pigment	Inert mineral additives used to extend the colour range of mortar beyond that which can be achieved using various natural sands, cement and lime.
Pointing	The filling and finishing of raked-out joints after construction.
Quoin block	An external corner block.
Shear wall	A wall to resist lateral forces in its plane.
Shell bedded wall	A wall in which the masonry units are bedded on two general purpose mortar strips at the outside edges of the bed face of the units (general purpose mortar as defined in EC6).
Single-leaf wall	A wall of structural units laid to overlap (see bond) in one or more directions and set solidly in mortar.
Sleeper wall	A dwarf wall, usually honeycombed, to carry a plate supporting a floor.
Slip	A masonry unit either manufactured or cut, of the same height and length as a header or stretcher, and normally with a thickness of between 20 mm or 50 mm.
Solid brick	A brick which has no holes, cavities or depressions.
Squint	A special brick manufactured for an oblique quoin (i.e. on an external corner).
Stiffening wall	A wall set perpendicular to another wall to give it support against lateral forces or to resist buckling and thus to provide stability to the building.
Stock bricks	Bricks originally hand-made in the south-east of England, so called from the timber 'stock' fixed to the bench that forms the *frog*. Sometimes used to describe bricks held in stock by brick-makers or merchants.
Strap	A device for connecting masonry members to other adjacent components, such as floors and roofs.
Stretcher	A structural unit with its length in the direction of the wall.
String course	A distinctive course of brickwork in a wall, usually projecting from the wall and used as an architectural feature.

Structural units	Bricks or blocks used in combination with mortar to construct masonry.
Toothing	Masonry units left projecting to bond with future work.
Veneered wall	A wall having a face that is attached to the backing, but not so bonded as to result in composite action under load.
Wall ties	Metal strips used to connect the two separate leaves of a cavity wall, increasing the stiffness of each one.
Weathering	(a) The cover applied to, or the geometrical form of, a part of a structure to enable it to shed rainwater. (b) The effect of climatic and atmospheric conditions on the external surface of materials.
Wire-cut bricks	Bricks shaped by extruding a column of clay through a die, the column being subsequently cut to the size of a brick.
Work size	The size of a building component specified for its manufacture, to which its actual size should conform within specified permissible deviations.

Appendix 1

Greek Alphabet

SI Prefixes

Approximate Values of Coefficients of Static Friction

Approximate Values of Material Properties

Greek Alphabet

alpha	α	A	a		nu	ν	N	n
beta	β	B	b		xi	ξ	Ξ	x
gamma	γ	Γ	g		omicron	o	O	o
delta	δ	Δ	d		pi	π	Π	p
epsilon	ε	E	e		rho	ρ	P	r
zeta	ζ	Z	z		sigma	σ	χ	s
eta	η	H	e		tau	τ	T	t
theta	θ	Θ	t		upsilon	υ	Y	u
iota	ι	I	i		phi	ϕ	Φ	p
kappa	κ	K	k		chi	χ	X	k
lamda	λ	Λ	l		psi	ψ	Ψ	p
mu	μ	M	m		omega	ω	Ω	o

SI Prefixes

T	**tera**	10^{12}		d	**deci**	10^{-1}
G	**giga**	10^{9}		c	**centi**	10^{-2}
M	**mega**	10^{6}		m	**milli**	10^{-3}
k	**kilo**	10^{3}		μ	**micro**	10^{-6}
h	**hecto**	10^{2}		n	**nano**	10^{-9}
da	**deda**	10^{1}		p	**pico**	10^{-12}

Approximate Coefficients of Static Friction

Surfaces in Contact	(μ)
Timber on Timber (fibres parallel to the motion)	0.4
Timber on Timber (fibres perpendicular to the motion)	0.5
Metal on Concrete	0.3
Metal on Timber	0.2
Metal on Metal	$0.15 - 0.2$
Timber on Stone	0.4
Metal on Masonry	$0.3 - 0.5$
Masonry on Masonry (hard)	$0.2 - 0.3$
Masonry on Masonry (soft)	$0.4 - 0.6$
Well Lubricated Hard Smooth Surfaces	0.05

Approximate Values of Material Properties

Material	Elastic Modulus E (N/mm^2)	Poisson's Ratio ν	Coefficient of Thermal Expansion ($\alpha/^{\circ}$C)
Concrete	$18.0 - 38.0 \times 10^{6}$ (see Figure 5.10)	0.2	$10.0 - 14.0 \times 10^{-6}$
Steel	205×10^{6}	0.3	12.0×10^{-6}
Timber	$4,600 - 18,000$	0.3	$3.6 - 5.4 \times 10^{-6}$
Masonry	$0.5 f_k - 2.0 f_k$	–	Clay: $5.0 - 7.0 \times 10^{-6}$ Calcium Silicate: $14 - 15 \times 10^{-6}$

Appendix 2

Properties of Geometric Figures

A = Cross-sectional area

y_1 or y_2 = Distance to centre of gravity

z_{xx} = Elastic Section Modulus about the $x-x$ axis

r_{xx} = Radius of Gyration about the $x-x$ axis

I_{xx} = Second Moment of Area about the $x-x$ axis

Square:

$A = d^2$

$y = d/2$

$I_{xx} = \dfrac{d^4}{12}$

$Z_{xx} = \dfrac{d^3}{6}$

$r_{xx} = \dfrac{d}{\sqrt{12}}$

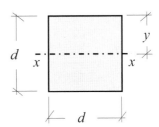

Square:

$A = d^2$

$y = d/2$

$I_{xx} = \dfrac{d^4}{3}$

$Z_{xx} = \dfrac{d^3}{3}$

$r_{xx} = \dfrac{d}{\sqrt{3}}$

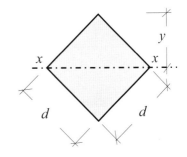

Square:

$A = d^2$

$y = \dfrac{d}{\sqrt{2}}$

$I_{xx} = \dfrac{d^4}{12}$

$Z_{xx} = \dfrac{d^3}{6\sqrt{2}}$

$r_{xx} = \dfrac{d}{\sqrt{12}}$

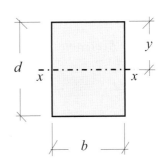

Rectangle :

$A = bd$

$y = d/2$

$I_{xx} = \dfrac{bd^3}{12}$

$Z_{xx} = \dfrac{bd^2}{6}$

$r_{xx} = \dfrac{d}{\sqrt{12}}$

Rectangle:

$A = bd$

$I_{xx} = \dfrac{bd^3}{3}$

$r_{xx} = \dfrac{d}{\sqrt{3}}$

$y = d/2$

$Z_{xx} = \dfrac{bd^2}{3}$

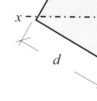

Rectangle:

$A = bd$

$I_{xx} = \dfrac{b^3 d^3}{6(b^2 + d^2)}$

$r_{xx} = \dfrac{bd}{\sqrt{6(b^2 + d^2)}}$

$y = \dfrac{bd}{\sqrt{b^2 + d^2}}$

$Z_{xx} = \dfrac{b^2 d^2}{6\sqrt{b^2 + d^2}}$

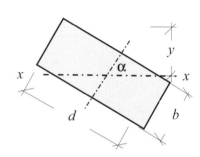

Rectangle:

$A = bd$

$I_{xx} = \dfrac{bd(b^2 \sin^2\alpha + d^2 \cos^2\alpha)}{12}$

$Z_{xx} = \dfrac{bd(b^2 \sin^2\alpha + d^2 \cos^2\alpha)}{6(b \sin\alpha + d \cos\alpha)}$

$r_{xx} = \sqrt{\dfrac{b^2 \sin^2\alpha + d^2 \cos^2\alpha}{12}}$

$y = \dfrac{b \sin\alpha + d \cos\alpha}{2}$

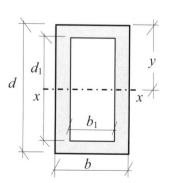

Hollow Rectangle

$A = (bd - b_1 d_1)$

$I_{xx} = \dfrac{(bd^3 - b_1 d_1^3)}{12}$

$r_{xx} = \sqrt{\dfrac{bd^3 - b_1 d_1^3}{12A}}$

$y = d/2$

$Z_{xx} = \dfrac{(bd^3 - b_1 d_1^3)}{6d}$

Trapezoid:

$$A = \frac{d(b+b_1)}{2}$$

$$y = \frac{d(2b+b_1)}{3(b+b_1)}$$

$$I_{xx} = \frac{d^3(b^2+4bb_1+b_1^2)}{36(b+b_1)}$$

$$Z_{xx} = \frac{d^2(b^2+4bb_1+b_1^2)}{12(2b+b_1)}$$

$$r_{xx} = \frac{d}{6(b+b_1)}\sqrt{2(b^2+4bb_1+b_1^2)}$$

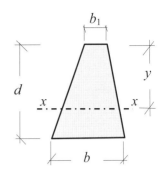

Circle:

$$A = \pi R^2$$

$$y = R = \frac{d}{2}$$

$$I_{xx} = \frac{\pi d^4}{64} = \frac{\pi R^4}{4}$$

$$Z_{xx} = \frac{\pi d^3}{32} = \frac{\pi R^3}{4}$$

$$r_{xx} = \frac{d}{4} = \frac{R}{2}$$

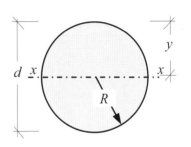

Hollow Circle:

$$A = \frac{\pi(d^2-d_1^2)}{4}$$

$$y = R = \frac{d}{2}$$

$$I_{xx} = \frac{\pi(d^4-d_1^4)}{64}$$

$$Z_{xx} = \frac{\pi(d^4-d_1^4)}{32d}$$

$$r_{xx} = \frac{\sqrt{d^2+d_1^2}}{4}$$

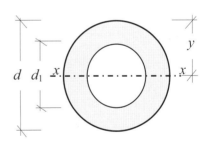

Semi-Circle:

$$A = \frac{\pi R^2}{2}$$

$$y = R\left(1-\frac{4}{3\pi}\right)$$

$$I_{xx} = R^4\left(\frac{\pi}{8}-\frac{8}{9\pi}\right)$$

$$Z_{xx} = \frac{R^3(9\pi^2-64)}{24(3\pi-4)}$$

$$r_{xx} = R\frac{\sqrt{9\pi^2-64}}{6\pi}$$

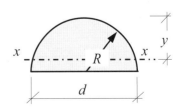

Equal Rectangles:

$A = b(d - d_1)$ $y = d/2$

$$I_{xx} = \frac{b(d^3 - d_1^3)}{12}$$ $$Z_{xx} = \frac{b(d^3 - d_1^3)}{6d}$$

$$r_{xx} = \sqrt{\frac{d^3 - d_1^3}{12(d - d_1)}}$$

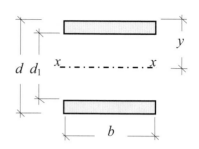

Unequal Rectangles:

$A = bt + b_1t_1$

$$y = \frac{0.5bt^2 + b_1t_1(d - 0.5t_1)}{A}$$

$$I_{xx} = \left\{ \left(\frac{bt^3}{12} + btc^2 \right) + \left(\frac{b_1t_1^3}{12} + b_1t_1c_1^2 \right) \right\}$$

$$Z_{xx} = \frac{I}{y} \qquad Z_{xx1} = \frac{I}{y_1}$$

$$r_{xx} = \sqrt{\frac{I_{xx}}{A}}$$

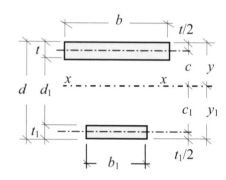

Triangle:

$$A = \frac{bd}{2}$$ $$y = \frac{2d}{3}$$

$$I_{xx} = \frac{bd^3}{36}$$ $$Z_{xx} = \frac{bd^2}{24}$$

$$r_{xx} = \frac{d}{\sqrt{18}}$$

Triangle:

$$A = \frac{bd}{2}$$ $$y = \frac{2d}{3}$$

$$I_{xx} = \frac{bd^3}{12}$$ $$Z_{xx} = \frac{bd^2}{12}$$

$$r_{xx} = \frac{d}{\sqrt{6}}$$

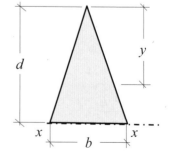

Appendix 3

Beam Reactions, Bending Moments and Deflections

Simply Supported Beams

Cantilever Beams

Propped Cantilevers

Fixed-End Beams

w = **Distributed load (kN/m)** and W = **Total load (kN)**

Simply Supported Beams:

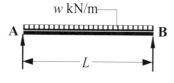

$V_A = wL/2$ $\qquad\qquad$ $V_B = wL/2$
Maximum bending moment at centre = $wL^2/8$
Maximum deflection = $(5wL^4/384EI)$

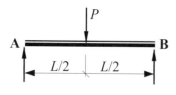

$V_A = P/2$ $\qquad\qquad$ $V_B = P/2$
Maximum bending moment at centre = $PL/4$
Maximum deflection = $(PL^3/48EI)$

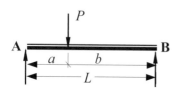

$V_A = Pb/L$ $\qquad\qquad$ $V_B = Pa/L$
Maximum bending moment at centre = Pab/L
Mid-span deflection = $PL^3[(3a/L) - (4a^3/L^3)]/48EI$
(This value will be within 2.5% of the maximum)

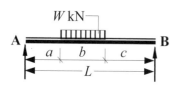

$V_A = W(0.5b + c)/L$ $\quad$ $V_B = W(0.5b + a)/L$
Maximum bending moment at $x = W(x^2 - a^2)/2b$
where $\quad x = [a + (V_A b/W)]$ from **A**
Maximum deflection $\approx W(8L^3 - 4Lb^2 + b^3)/384EI$
(This is the value at the centre when $a = c$)

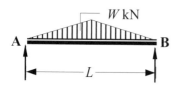

$V_A = W/2$ $\qquad\qquad$ $V_B = W/2$
Maximum bending moment at centre = $WL/6$
Maximum deflection = $WL^3/60EI$

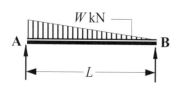

$V_A = 2W/3$ $\qquad\qquad$ $V_B = W/3$
Maximum bending moment at $x = 0.128WL$
where $\quad x = 0.4226L$ from **A**
Maximum deflection $\approx 0.01304WL^3/384EI$
where $\quad x = 0.4807L$ from **A**

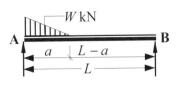

$$V_A = W(3L - a)/3L \qquad V_B = Wa/3L$$

Maximum bending moment at x:

$$= \frac{Wa}{3}\left(1 - \frac{a}{L} + \sqrt{\frac{4a^3}{27L^3}}\right)$$

where $\quad x = a\left(1 - \sqrt{\frac{a}{3L}}\right)$ from **A**

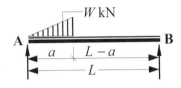

$$V_A = W(3L - 2a)/3L \qquad V_B = 2Wa/3L$$

Maximum bending moment at x:

$$= \frac{2Wa}{3}\left(1 - \frac{2a}{3L}\right)^{\frac{3}{2}}$$

where $\quad x = a\sqrt{1 - \frac{2a}{3L}}$ from **A**

Cantilever Beams:
Anti-clockwise support moments considered negative.

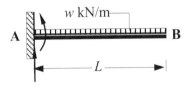

$V_A = wL$

Maximum (–ve) bending moment $M_A = -wL^2/2$

Maximum deflection $= wL^4/8EI$

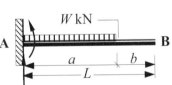

$V_A = W$

Maximum (–ve) bending moment $M_A = -Wa/2$

Maximum deflection at B $= Wa^3\left(1 + \dfrac{4b}{3a}\right)\bigg/8EI$

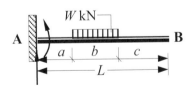

$V_A = W$

Maximum (–ve) bending moment $M_A = -W(a + b/2)$

Maximum deflection at B $= (W/24EI) \times k$

where $k =$

$(8a^3 + 18a^2b + 12ab^2 + 3b^3 + 12a^2c + 12abc + 4b^2c)$

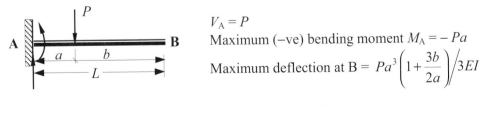

$V_A = P$

Maximum (–ve) bending moment $M_A = -Pa$

Maximum deflection at B $= Pa^3\left(1 + \dfrac{3b}{2a}\right)\bigg/3EI$

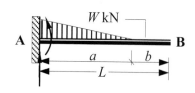

$V_A = W$

Maximum (−ve) bending moment $M_A = -Wa/3$

Maximum deflection at B $= Wa^3\left(1+\dfrac{5b}{4a}\right)\Big/15EI$

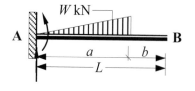

$V_A = W$

Maximum (−ve) bending moment $M_A = -2Wa/3$

Maximum deflection at B $= 11Wa^3\left(1+\dfrac{15b}{11a}\right)\Big/60EI$

Propped Cantilevers:

Where the support moment (M_A) is included in an expression for reactions, its value should be assumed positive.

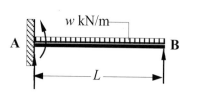

$V_A = 5wL/8 \qquad\qquad V_B = 3wL/8$

Maximum (−ve) bending moment $M_A = -wL^2/8$

Maximum (+ve) bending moment at $x = +9wL^2/128$

where $x = 0.625L$ from **A**

Maximum deflection at $y = wL^4/185EI$

where $y = 0.5785L$ from **A**

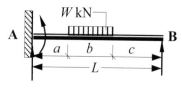

$V_A = W(0.5b + c)/L + M_A/L$

$V_B = W(0.5b + a)/L - M_A/L$

Maximum (−ve) bending moment M_A:

$\quad = -Wb(b + 2c)\,[2(L^2 - c^2 - bc) - b^2)]/8L^2b$

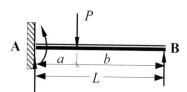

$V_A = (P - V_B)$

$V_B = Pa^2[(b + 2L)]/2L^3$

Maximum (−ve) bending moment M_A:

$\quad = -Pb[(L^2 - b^2)]/2L^2$

Maximum (+ve) bending moment at point load:

$\quad = -\dfrac{Pb}{2}\left(2 - \dfrac{3b}{L} + \dfrac{b^3}{L^3}\right)$

Maximum deflection at point load position:

$\quad = \dfrac{Pa^3b^2}{12EIL^3}(4L - a)$

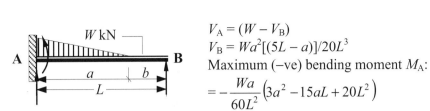

$V_A = (W - V_B)$
$V_B = Wa^2[(5L - a)]/20L^3$
Maximum (–ve) bending moment M_A:
$$= -\frac{Wa}{60L^2}\left(3a^2 - 15aL + 20L^2\right)$$
$= $ Maximum (+ve) bending moment at x:
$= [V_B x - W(x - b)^3/3a^2]$
where $\quad x = b + \dfrac{a^2}{2L}\sqrt{1 - \dfrac{a}{5L}}$ from **B**

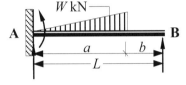

$V_A = (W - V_B)$
$V_B = Wa^2[(15L - 4a)]/20L^3$
Maximum (–ve) bending moment M_A:
$$= -Wa\left(\frac{a^2}{5L^2} - \frac{3a}{4L} + \frac{2}{3}\right)$$

Fixed-End Beams:

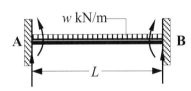

$V_A = wL/2 \qquad\qquad V_B = wL/2$
Maximum (–ve) bending moment M_A:
$\quad = - wL^2/12$
Maximum (+ve) bending moment at mid-span:
$\quad = + wL^2/24$
Maximum deflection at point load:
$\quad = \quad wL^4/384EI$

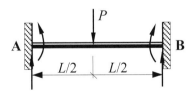

$V_A = P/2 \qquad\qquad V_B = P/2$
Support bending moments:
$M_A = - PL/8 \qquad$ and $\qquad M_B = + PL/8$
Maximum (+ve) bending moment at mid-span:
$\quad = + PL/8$
Maximum deflection at mid-span $\quad = \quad PL^3/192EI$

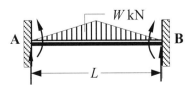

$V_A = W/2 \qquad\qquad V_B = W/2$
Support bending moments:
$M_A = - 5WL/48 \qquad$ and $\qquad M_B = + 5WL/48$
Maximum (+ve) bending moment at mid-span:
$\quad = + WL/16$
Maximum deflection at mid-span $\quad = \quad 1.4WL^3/384EI$

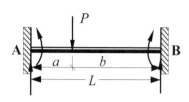

$$V_A = Pb^2 (1 + 2a/L)/L^2$$
$$V_B = Pa^2 (1 + 2b/L)/L^2$$
Support bending moments:
$$M_A = - Pab^2/L^2 \quad \text{and} \quad M_B = + Pa^2b/L^2$$
Maximum (+ve) bending moment at point load:
$$= + 2Pa^2b^2/L^3$$

Maximum deflection at point $x = \dfrac{2Pa^2b^3}{3EI(3L-2a)^2}$

where $x = \dfrac{L^2}{(3L-2a)}$ from **A**

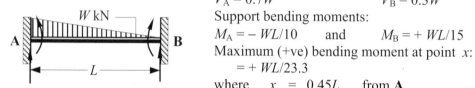

$$V_A = 0.7W \qquad\qquad V_B = 0.3W$$
Support bending moments:
$$M_A = - WL/10 \quad \text{and} \quad M_B = + WL/15$$
Maximum (+ve) bending moment at point x:
$$= + WL/23.3$$
where $x = 0.45L$ from **A**
Maximum deflection at point $y = WL^3/382EI$
where $y = 0.475L$ from **A**

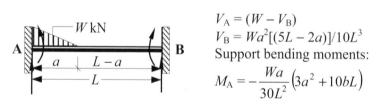

$$V_A = (W - V_B)$$
$$V_B = Wa^2[(5L - 2a)]/10L^3$$
Support bending moments:
$$M_A = -\frac{Wa}{30L^2}\left(3a^2 + 10bL\right) \quad \text{and}$$

$$M_B = +\frac{Wa^2}{30L^2}(5L - 3a)$$

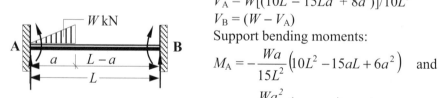

$$V_A = W[(10L^3 - 15La^2 + 8a^3)]/10L^3$$
$$V_B = (W - V_A)$$
Support bending moments:
$$M_A = -\frac{Wa}{15L^2}\left(10L^2 - 15aL + 6a^2\right) \quad \text{and}$$

$$M_B = +\frac{Wa^2}{10L^2}(5L - 4a)$$

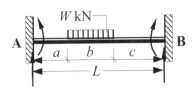

$V_A = W(0.5b + c)/L + (M_A - M_B)/L$

$V_B = W(0.5b + a)/L - (M_A - M_B)/L$

Support bending moments:

$$M_A = -\frac{W}{12L^2b}\left\{\left[(L-a)^3 \times (L+3a)\right] - c^3(4L-3c)\right\}$$

$$M_B = +\frac{W}{12L^2b}\left\{\left[(L-c)^3 \times (L+3c)\right] - a^3(4L-3a)\right\}$$

Maximum deflection at mid-span when $a = c$

$$= \frac{W}{384EI}\left(L^3 + 2L^2a + 4La^2 - 8a^3\right)$$

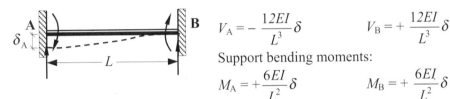

$V_A = -\dfrac{12EI}{L^3}\delta \qquad\qquad V_B = +\dfrac{12EI}{L^3}\delta$

Support bending moments:

$M_A = +\dfrac{6EI}{L^2}\delta \qquad\qquad M_B = +\dfrac{6EI}{L^2}\delta$

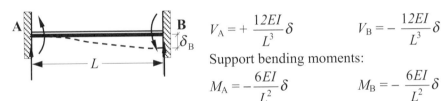

$V_A = +\dfrac{12EI}{L^3}\delta \qquad\qquad V_B = -\dfrac{12EI}{L^3}\delta$

Support bending moments:

$M_A = -\dfrac{6EI}{L^2}\delta \qquad\qquad M_B = -\dfrac{6EI}{L^2}\delta$

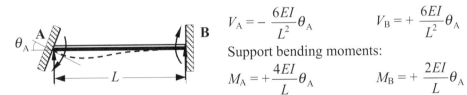

$V_A = -\dfrac{6EI}{L^2}\theta_A \qquad\qquad V_B = +\dfrac{6EI}{L^2}\theta_A$

Support bending moments:

$M_A = +\dfrac{4EI}{L}\theta_A \qquad\qquad M_B = +\dfrac{2EI}{L}\theta_A$

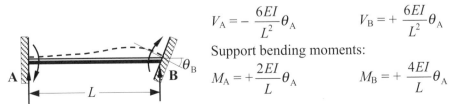

$V_A = -\dfrac{6EI}{L^2}\theta_A \qquad\qquad V_B = +\dfrac{6EI}{L^2}\theta_A$

Support bending moments:

$M_A = +\dfrac{2EI}{L}\theta_A \qquad\qquad M_B = +\dfrac{4EI}{L}\theta_A$

Appendix 4

Continuous Beam Coefficients

Uniformly Distributed Loads

It is assumed that all spans are equal and W is the **total uniformly distributed load/span**. Positive reactions are upwards and positive bending moments induce tension on the underside of the beams.

Two-span beams

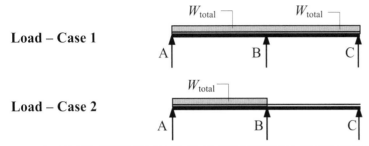

Load	Support	Support	Support	Span AB	Support B	Span BC
Case	V_A	V_B	V_C	M_{AB}	M_B	M_{BC}
1	0.375	1.250	0.375	0.070	− 0.125	0.070
2	0.438	0.625	− 0.063	0.096	− 0.063	

Coefficients for Reactions and Bending Moments

Vertical Reaction = **coefficient** $\times$ W
Bending Moment = **coefficient** $\times$ $W \times$ **span (L)**

Example: Determine the reaction at A and the bending moment at support B for the beam shown in Figure A4.1.

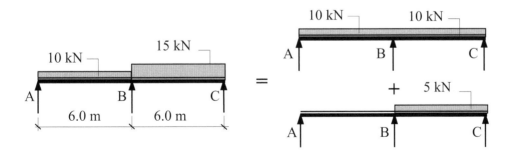

Figure A4.1

V_A = $(0.375 \times 10) - (0.063 \times 5)$ = + 3.345 kN
M_B = $- (0.125 \times 10 \times 6.0) - (0.063 \times 5 \times 6.0)$ = − 9.39 kNm

Three-span beams

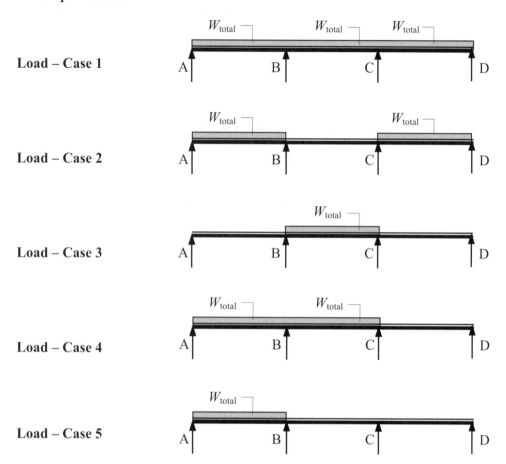

Load – Case 1

Load – Case 2

Load – Case 3

Load – Case 4

Load – Case 5

Load Case	Support V_A	Support V_B	Support V_C	Support V_D	Span AB M_{AB}	Support B M_B	Span BC M_{BC}	Support C M_C	Span CD M_{CD}
1	0.4	1.1	1.1	0.4	0.080	− 0.100	0.025	− 0.100	0.080
2	0.45	0.55	0.55	0.45	0.101	− 0.050	-	− 0.050	0.101
3	− 0.05	0.55	0.55	− 0.05	-	− 0.05	0.075	− 0.05	-
4	0.383	1.2	0.45	− 0.033	0.073	− 0.117	0.054	−0.033	-
5	0.433	0.65	−0.10	0.017	0.094	− 0.067	-	−0.017	-

Coefficients for Reactions and Bending Moments

Four-span beams

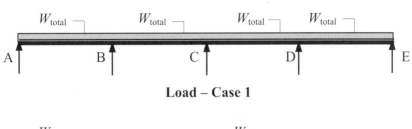

Load – Case 1

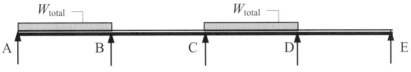

Load – Case 2

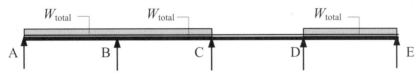

Load – Case 3

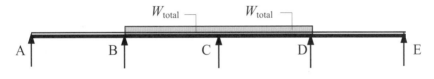

Load – Case 4

Load – Case 5

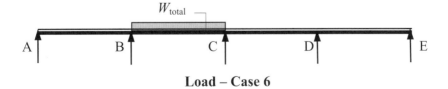

Load – Case 6

(Four-span beams cont.)

Load Case	Support V_A	Support V_B	Support V_C	Support V_D	Support V_D
1	0.393	1.143	0.929	1.143	0.393
2	0.446	0.572	0.464	0.572	− 0.054
3	0.38	1.223	0.357	0.598	0.442
4	− 0.036	0.464	1.143	0.464	− 0.036
5	0.433	0.652	0.107	0.027	− 0.005
6	− 0.049	0.545	0.571	− 0.080	0.013

Coefficients for Reactions

Load Case	Span AB M_{AB}	Support B M_B	Span BC M_{BC}	Support C M_C	Span CD M_{CD}	Support D M_D	Span DE M_{DE}
1	0.077	− 0.107	0.036	− 0.071	0.036	− 0.107	0.077
2	0.1	− 0.054	-	− 0.036	0.081	− 0.054	-
3	0.072	− 0.121	0.061	− 0.018	-	− 0.058	0.098
4	-	− 0.036	0.056	− 0.107	0.056	− 0.036	-
5	0.094	-	-	-	-	-	-
6	-	− 0.049	0.074	− 0.054	-	+ 0.014	-

Coefficients for Bending Moments

Point Loads

It is assumed that all spans are equal and P is the **total central point-load/span**. Positive reactions are upwards and positive bending moments induce tension on the underside of the beams.

Two-span beams

Load – Case 1

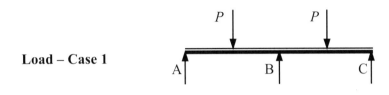

Load – Case 2

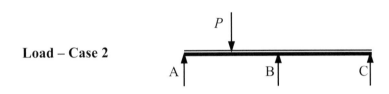

Load Case	Support V_A	Support V_B	Support V_C	Span AB M_{AB}	Support B M_B	Span BC M_{BC}
1	0.313	1.375	0.313	0.156	− 0.188	0.156
2	0.406	0.688	− 0.094	0.203	− 0.094	-

Coefficients for Reactions and Bending Moments

Vertical Reaction = **coefficient** $\times$ ***P***
Bending Moment = **coefficient** $\times$ ***P*** $\times$ **span (*L*)**

Three-span beams

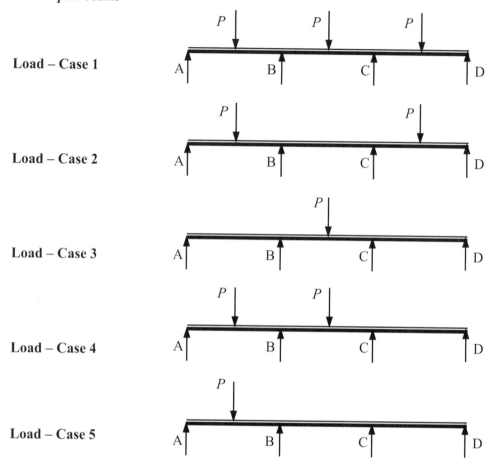

Load – Case 1

Load – Case 2

Load – Case 3

Load – Case 4

Load – Case 5

Load Case	Support V_A	Support V_B	Support V_C	Support V_D	Span AB M_{AB}	Support B M_B	Span BC M_{BC}	Support C M_C	Span CD M_{CD}
1	0.35	1.15	1.15	0.35	0.175	− 0.150	0.100	− 0.150	0.175
2	0.425	0.575	0.575	0.425	0.213	− 0.075	-	− 0.075	0.213
3	− 0.075	0.575	0.575	− 0.075	-	− 0.075	0.175	− 0.075	-
4	0.325	1.3	0.425	− 0.05	0.163	− 0.175	0.138	− 0.050	-
5	0.4	0.725	− 0.15	0.025	0.2	− 0.100	-	+ 0.025	0.025

Coefficients for Reactions and Bending Moments

Four-span beams

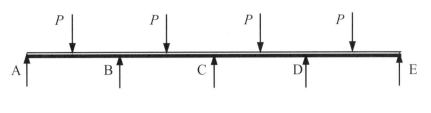

Load – Case 1

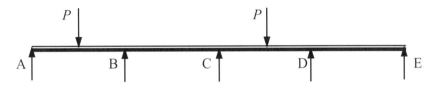

Load – Case 2

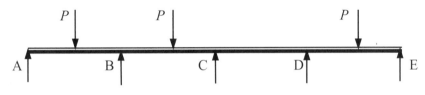

Load – Case 3

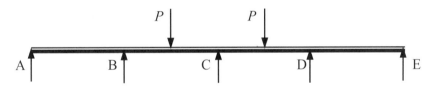

Load – Case 4

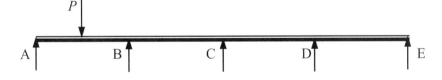

Load – Case 5

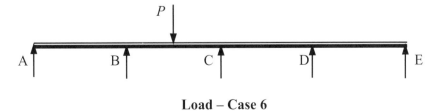

Load – Case 6

(Four-span beams cont.)

Load Case	Support V_A	Support V_B	Support V_C	Support V_D	Support V_D
1	0.339	1.214	0.893	1.214	0.339
2	0.420	0.607	0.446	0.607	− 0.08
3	0.319	1.335	0.286	0.647	0.413
4	− 0.054	0.446	1.214	0.446	− 0.054
5	0.4	0.728	− 0.161	0.04	− 0.007
6	− 0.074	0.567	0.607	− 0.121	0.02

Coefficients for Reactions

Load Case	Span AB M_{AB}	Support B M_B	Span BC M_{BC}	Support C M_C	Span CD M_{CD}	Support D M_D	Span DE M_{DE}
1	0.170	− 0.161	0.116	− 0.107	0.116	− 0.161	0.170
2	0.210	− 0.080	-	− 0.054	0.183	− 0.080	-
3	0.160	− 0.181	0.146	− 0.027	-	− 0.087	0.207
4	-	− 0.054	0.143	− 0.161	0.143	− 0.054	-
5	0.200	− 0.100	-	+ 0.027	-	− 0.007	-
6	-	− 0.074	0.173	− 0.080	-	+ 0.020	-

Coefficients for Bending Moments

Appendix 5

Self-weights of Construction Materials

Approximate Self-weights of Materials (ref. BS 648)	
Material	**Self-weight (kN/m^3)**
Reinforced Concrete	24.0
Steel	78.5
Timber	7.9 (ave.)
Granite	25.1 – 28.7
Sandstone	22.0 – 24.0
Slate	28.3
	Self-weight (kN/m^2)
Asphalt 2 layers (19 mm thick)	0.41
Mineral-surfaced bitumen roofing felt	0.034
Blockwork walling /(25 mm thick)	
Hollow clay units	0.259
Medium-density clay units	0.283
High-density clay units	0.327
Solid concrete units	0.547
Hollow concrete units	0.347
Cellular concrete units	0.405
Brickwork /(25 mm thick)	
Solid — Low-density	0.505
Solid — Medium-density	0.547
Solid — High-density	0.591
Perforated — Low-density	0.508
Perforated — Medium-density	0.547
Perforated — High-density	0.591
Gypsum Plaster and Partitions	
Building panels (75 mm thick)	0.439
Dry partition (64 mm thick)	0.259
Solid-core plaster board (13 mm thick)	0.112
Two coats of 13 mm thick plaster	0.220
Glass	
Sheet — 4.0 mm thick	0.073
Sheet — 2.8 mm thick	0.098
Cast, clear plate and armoured plate — 3.2 mm thick	0.088
Cast, clear plate and armoured plate — 25.0 mm thick	0.649
Wired cast — 6.4 mm	0.171

Appendix 6

Areas of Reinforcing Steel

Areas for individual groups of bars

Bar diameter (mm)		6	8	10	12	16	20	25	32	40
Bar perimeter (mm)		18.9	25.1	31.4	37.7	50.3	62.8	78.5	101	126
Weight (kg/m)		0.222	0.395	0.616	0.888	1.58	2.47	3.85	6.31	9.86
	No. of Bars									
	1	28.3	50.3	78.5	113	201	314	491	804	1260
	2	56.6	101	157	226	402	628	982	1610	2510
	3	84.9	151	236	339	603	943	1470	2410	3770
	4	113	201	314	452	804	1260	1960	3220	5030
	5	142	252	393	566	1010	1570	2450	4020	6280
	6	170	302	471	679	1210	1890	2950	4830	7540
	7	198	352	550	792	1410	2200	3440	5630	8800
	8	226	402	628	905	1610	2510	3930	6430	10100
	9	255	453	707	1020	1810	2830	4420	7240	11300
	10	283	503	785	1130	2010	3140	4910	8040	12600

(Cross-sectional area for group (mm²))

Areas/metre width for various pitches of bars

Bar diameter (mm)		6	8	10	12	16	20	25	32	40
	Pitch (mm)									
	50	566	1010	1570	2260	4020	6280	9820	16100	-
	75	377	671	1050	1510	2680	4190	6550	10700	16800
	100	283	503	785	1130	2010	3140	4910	8040	12600
	125	226	402	628	905	1610	2510	3930	6430	10100
	150	189	335	523	754	1340	2090	3270	5360	8380
	175	162	287	449	646	1150	1800	2810	4600	7180
	200	142	252	393	566	1010	1570	2450	4020	6280
	250	113	201	314	452	804	1260	1960	3220	5030
	300	94.3	168	262	377	670	1050	1640	2680	4190
	400	71	126	196	283	503	786	1230	2010	3140
	500	56.6	101	157	226	402	628	982	1610	2510

(Cross-sectional area / metre width for various pitches of bars (mm²))

Appendix 7

Bolt and Weld Capacities

Non-Preloaded Ordinary Bolts – Grade 4.6 in S 275					
Diameter of Bolt	Tensile Stress Area	Tension Capacity		Shear Capacity	
		Nominal $0.8A_tp_t$	Exact A_tp_t	Single Shear	Double Shear
D (mm)	A_t (mm^2)	P_{nom} (kN)	P_t (kN)	P_s (kN)	$2P_s$ (kN)
12	84.3	16.2	20.2	13.5	27.0
16	157	30.1	37.7	25.1	50.2
20	245	47.0	58.8	39.2	78.4
22	303	58.2	72.7	48.5	97.0
24	353	67.8	84.7	56.5	113
27	459	88.1	110	73.4	147
30	561	108	135	89.8	180

Non-Preloaded Ordinary Bolts – Grade 8.8 in S 275					
Diameter of Bolt	Tensile Stress Area	Tension Capacity		Shear Capacity	
		Nominal $0.8A_tp_t$	Exact A_tp_t	Single Shear	Double Shear
D (mm)	A_t (mm^2)	P_{nom} (kN)	P_t (kN)	P_s (kN)	$2P_s$ (kN)
12	84.3	37.8	47.2	31.6	63.2
16	157	70.3	87.9	58.9	118
20	245	110	137	91.9	184
22	303	136	170	114	227
24	353	158	198	132	265
27	459	206	257	172	344
30	561	251	314	210	421

Fillet Weld Capacities with E35 Electrode – S 275			
Leg Length s (mm)	Throat Thickness a (mm)	Longitudinal Capacity P_L (kN/mm)	Transverse Capacity P_T (kN/mm)
3.0	2.1	0.462	0.577
4.0	2.8	0.616	0.770
5.0	3.5	0.770	0.963
6.0	4.2	0.924	1.155
8.0	5.6	1.232	1.540
10.0	7.0	1.540	1.925
12.0	8.4	1.848	2.310
15.0	10.5	2.310	2.888
18.0	12.6	2.772	3.465
20.0	14.0	3.080	3.850
22.0	15.4	3.388	4.235
25.0	17.5	3.850	4.813

Non-Preloaded H.S.F.G. Bolts – General Grade in S 275					
Diameter of Bolt	Tensile Stress Area	Tension Capacity		Shear Capacity	
		Nominal $0.8A_t p_t$	Exact $A_t p_t$	Single Shear	Double Shear
D (mm)	A_t (mm²)	P_{nom} (kN)	P_t (kN)	P_s (kN)	$2P_s$ (kN)
12	84.3	39.8	49.7	33.7	67.4
16	157	74.1	92.6	62.8	126
20	245	116	145	98.0	196
22	303	143	179	121	242
24	353	167	208	141	282
27	459	189	236	161	321
30	561	231	289	196	393

Preloaded H.S.F.G. Bolts:Non-Slip in Service – General Grade in S 275								
Diameter of Bolt	Tensile Stress Area	Bolt Proof Load	Tension Capacity		Shear Capacity		Slip Resistance for $\mu = 0.5$	
					Single Shear	Double Shear	Single Shear	Double Shear
D (mm)	A_t (mm²)	P_o (kN)	$1.1P_o$ (kN)	$A_t p_t$ (kN)	(kN)	(kN)	(kN)	(kN)
12	84.3	49.4	54.3	49.7	33.7	67.4	27.2	54.3
16	157	92.1	101	92.6	62.8	126	50.7	101
20	245	144	158	145	98.0	196	79.2	158
22	303	177	195	179	121	242	97.4	195
24	353	207	228	208	141	282	114	228
27	459	234	257	236	161	321	129	257
30	561	286	315	289	196	393	157	315

Preloaded H.S.F.G. Bolts:Non-Slip under Factored Loads – General Grade in S 275										
Diameter of Bolt	Bolt Proof Load	Bolt Tension Capacity	Slip Resistance P_o							
			$\mu = 0.2$		$\mu = 0.3$		$\mu = 0.4$		$\mu = 0.5$	
			Single Shear	Double Shear	Single Shear	Double Shear	Single Shear	Double Shear	Single Shear	Double Shear
D (mm)	P_o (kN)	$0.9P_o$ (kN)	(kN)	(kN)	(kN)	(kN)	(kN)	(kN)	(kN)	(kN)
12	49.4	44.5	8.89	17.8	13.3	26.7	17.8	35.6	22.2	44.5
16	92.1	82.9	16.6	33.2	24.9	49.7	33.2	66.3	41.4	82.9
20	144	130	25.9	51.8	38.9	77.8	51.8	104	64.8	130
22	177	159	31.9	63.7	47.8	95.6	63.7	127	79.7	159
24	207	186	37.3	74.5	55.9	112	74.5	149	93.2	186
27	234	211	42.1	84.2	63.2	126	84.2	168	105	211
30	286	257	51.5	103	77.2	154	103	206	129	257

Bibliography

British Standards

1. **BS 187:** *Specification for calcium silicate (sandlime and flintlime) bricks*
    BSI, 1978

2. **BS 648:** *Schedule of weights of building materials*
    BSI, 1964

3. **BS 743:** *Specification for materials for damp-proof courses*
    BSI, 1970

4. **BS 1217:** *Specification for cast stone*
    BSI, 1997

5. **BS 1243:** *Specification for metal ties for cavity wall construction*
    BSI, 1978

6. **BS 3921:** *Specification for clay bricks and blocks*
    BSI, 1985

7. **BS 4449:** *Specification for carbon steel bars for the reinforcement of concrete*
    BSI, 1997

8. **BS 4729:** *Specification for dimensions of bricks of special shapes and sizes*
    BSI, 1990

9. **BS 4887:** *Mortar plasticizers*
    BSI, Part 1:1986,   Part 2:1987

10. **BS 4978:** *Specification for softwood grades for structural use*
    BSI, 1996

11. **BS 5224:** *Specification for masonry cement*
    BSI, 1995

12. **BS 5262:** *Code of practice for external renderings*
    BSI, 1991

13. **BS 5268-2:2002:** *Structural use of timber: Code of practice for permissible stress design, materials and workmanship*
    BSI, 2002

14. **BS 5268-3:1998:** *Code of practice for trussed rafter roofs*
   BSI, 1998

15. **BS 5390:** *Code of practice for stone masonry*
   BSI, 1976

16. **BS 5628-1:1992:** *Code of practice for use of masonry: Structural use of unreinforced masonry*
   BSI, 1992

17. **BS 5628-3:2001:** *Code of practice for use of masonry: Materials and components, design and workmanship*
   BSI, 2001

18. **BS 5950-1: 2000** *Code of practice for design – Rolled and welded sections,*
   BSI, 2000

19. **BS 5950-2: 2001** *Specification for materials, fabrication and erection – Rolled and welded sections*
   BSI, 2001

20. **BS 6073:** *Precast concrete masonry units – Part 1: Specification for precast masonry units*
   BSI, 1981

21. **BS 6399-1:1996:** *Loading for buildings: Code of practice for dead and imposed loads*
   BSI, 1996

22. **BS 6399-2:1997:** *Loading for buildings: Code of practice for wind loads*
   BSI, 1997

23. **BS 6399-3:1996:** *Loading for buildings: Code of practice for imposed roof loads*
   BSI, 1996

24. **BS 6446:1997:** *Specification for manufacture of glued structural components of timber and wood-based panel products*
   BSI, 1997

25. **BS 6457:1984:** *Specification for reconstructed stone masonry units*
   BSI, 1984

26. **BS 6649:1985:** *Specification for clay and calcium silicate modular bricks*
   BSI, 1985

27.   ***BS 8110-1:1997:*** *Structural use of concrete: Code of practice for design and construction*
      BSI, 1997

28.   ***BS 8110-2:1985:*** *Structural use of concrete: Code of practice for special circumstances*
      BSI, 1985

29.   ***BS 8110-3:1985:*** *Structural use of concrete: Design charts for singly reinforced beams, doubly reinforced beams and column*
      BSI, 1985

30.   ***DD 86-3:*** *Damp-proof courses – Guide to characteristic strengths of damp-proof course material used in masonry*
      BSI, 1990

31.   ***DD 140-2: 1987:*** *Wall ties – Part 2: Recommendations for design of wall ties*
      BSI, 1987

32.   ***prEN 1990, Eurocode:*** *Basis of structural design*
      BSI, 2001

33.   ***prEN 1991, Eurocode 1:*** *Actions on structures*
      BSI, 2001

34.   ***prEN 1992, Eurocode 2:*** *Design of concrete structures*
      BSI, 2002

35.   ***prEN 1993, Eurocode 3:*** *Design of steel structures*
      BSI, 2001

36.   ***prEN 1994, Eurocode 4:*** *Design of composite steel and concrete structures*
      BSI, 2002

37.   ***prEN 1995, Eurocode 5:*** *Design of timber structures*
      BSI, 2002

38.   ***prEN 1996, Eurocode 6:*** *Design of masonry structures*
      BSI, 2001

39.   ***Extracts from British Standards for students of structural design***
      5th *Edition* (PP 7312:2002)
      BSI, 2002

General

40.   **Freudenthal, A.M.**
      The Safety of Structures
      Proceedings of the American Society of Civil Engineers, October 1945

41.   **Hinks, John, and Cook, Geoff**
      The Technology of Building Defects
      E & FN SPON, 1997

42.   **Johansen, K.W.**
      Yield Line Formulae for Slabs,
      Cement and Concrete Association

43.   **Kaminetzky, Dov**
      Design and Construction Failures: lessons from forensic investigations
      McGraw-Hill, 1991

44.   **Levi, Matthys, and Salvadori, Mario**
      Why Buildings Fall Down
      W. W. Norton & Company, 1994

45.   **Roark, R.J.**
      Formulas for Stress and Strain, 4th Edition
      McGraw-Hill, New York, 1956

46.   **Timoshenko**
      Strength of Materials: Part 2
      Van Nostrand Reinhold, New York

47.   ***Appraisal of Existing Structures***
      Institution of Structural Engineers, 1996

48.   ***Report of the inquiry into the collapse of flats at Ronan Point, Canning Town, London***
      HMSO, 1968

49.   ***Stability of Buildings***
      Institution of Structural Engineers, 1998

50.   ***Surveys and Inspections of Buildings and Similar Structures***
      Institution of Structural Engineers, 1991

51.   ***The collapse of a precast concrete building under construction***
      Technical statement by the Building Research Station, London, HMSO, 1963

Reinforced Concrete

52. Neville, A. M.
Properties of Concrete, 4th Edition
Longman, Scientific and Technical, Harlow 1998

53. Concise Eurocode for the Design of Concrete Buildings
British Cement Association, 1993

54. Reinforced Concrete Designer's Handbook, 10th Edition
E & F N Spon, 1988

55. Standard Method of Detailing Structural Concrete
The Institution of Structural Engineers / The Concrete Society, 1989

56. Worked Examples for the Design of Concrete Buildings
British Cement Association, 1994

Structural Steelwork

57. Baddoo, N.R., Morrow, A.W., and Taylor, J.C.
C-EC3 – Concise Eurocode 3 for the design of steel buildings in the United Kingdom
The Steel Construction Institute, Publication Number 116, 1993

58. Hayward, Alan, and Weare, Frank
Steel Detailers' Manual
Blackwell Science (UK), 1992

59. McKenzie, W.M.C.
Design of Structural Steelwork
Palgrave, 1998

60. Narayanan, R., Lawless, V., Naji, F.J., and Taylor, J.C.
Introduction to Concise Eurocode 3 (C - EC3) – with Worked Examples
The Steel Construction Institute, Publication Number 115, 1993

61. Nethercot, D.A., Salter, P.R., and Malik, A.S.
Design of members subject to combined bending and torsion (SCI-P057)
The Steel Construction Institute, 1989

62. **Taylor, J.C., Baddoo, N.R., Morrow, A.W., and Gibbons, C.**
Steelwork design guide to Eurocode 3: Part 1.1 – Introducing Eurocode 3.
A comparison of EC3: Part 1.1 with BS 5950: Part 1
The Steel Construction Institute, Publication Number 114, 1993

63. ***Brick Cladding to Steel Framed Buildings***
Brick Development Association and British Steel Corporation, 1986

64. ***Design for Manufacture Guidelines***
The Steel Construction Institute, Publication Number 150, 1995

65. ***Handbook of Structural Steelwork, 3rd Edition***
The Steel Construction Institute, Publication Number P201, 2002

66. ***Introduction to Steelwork Design to BS 5950:Part 1.***
The Steel Construction Institute, Publication Number P069, 1994

67. ***Joints in Steel Construction: Moment connections***
The Steel Construction Institute, 1995

68. ***Steelwork Design Guide to BS 5950-1:2000–Volume 1 Section Properties, Member Capacities, 6th Edition***
The Steel Construction Institute, Publication Number 202, 2001

69. ***Steelwork Design Guide to BS 5950: Part 1:1990–Volume 2, Worked Examples (Revised Edition)***
The Steel Construction Institute, Publication Number P002, 1993

70. ***Steel Designers' Manual, 6th Edition***
Blackwell Science (UK), 2002

Structural Timber

71. **Baird, J.A., and Ozelton, E.C.**
Timber Designers' Manual, 3nd Edition
Blackwell Science (UK), 2002

72. **McKenzie, W.M.C.**
Design of Structural Timber
Palgrave, 2000

73. ***Fir Plywood Web Beam Design***
Publication by COFI: Council of Forest Industries of British Columbia

74.   ***Step 1: Structural Timber Education Programme***
        Volume 1 First Edition, Centrum Hout, The Netherlands, 1995

75.   ***Step 2: Structural Timber Education Programme***
        Volume 2 First Edition, Centrum Hout, The Netherlands, 1995

Structural Masonry

76.   **Curtin, W.G., Shaw, G., Beck J.K., and Bray W.A.**
        Loadbearing Brickwork Crosswall Construction
        Brick Development Association, 1983

77.   **Curtin, W.G., Shaw, G., Beck, J.K., and Bray W.A.**
        Structural Masonry Designers' Manual,
        Second Edition, BSP Professional Books, 1987

78.   **Curtin, W.G., Shaw, G., Beck, J.K., and Parkinson G.I.**
        Structural Masonry Detailing
        Granad Publishing Ltd.,1984

79.   **Hammett, Michael**
        A Basic Guide To Brickwork Mortars
        Brick Development Association, 1988

80.   **Hammett, Michael**
        Resisting Rain Penetration With Facing Brickwork
        Brick Development Association, 1997

81.   **Hammett, Michael**
        Bricks – Notes on their properties
        Brick Development Association, 1999

82.   **Haseltine, B.A., and Moore, J.F.A.**
        Handbook to BS 5628: Structural Use of Masonry:Part 1: Unreinforced Masonry
        Brick Development Association Design Guide 10, 1981

83.   **Haseltine, B.A., and Tutt, J.N.**
        Handbook to BS 5628: Part 2: Section 1: Background and Materials
        Brick Development Association, 1991.

84.   **Haseltine, B.A., and Tutt, J.N.**
        External Walls: Design for Wind Loads
        Brick Development Association, 1984

85. **McKenzie, W.M.C**
Design of Structural Masonry
Palgrave, 2001

86. **Parkinson, G., Shaw, G., Beck, J.K., and Knowles, D.**
Appraisal & Repair of Masonry
Thomas Telford Ltd, 1996

87. **BDA Design Note 3**
Brickwork Dimension Tables
Brick Development Association

Index